LINEAR PROGRAMMING

A Series of Books in the Mathematical Sciences
Victor Klee, Editor

LINEAR PROGRAMMING

Vašek Chvátal
McGill University

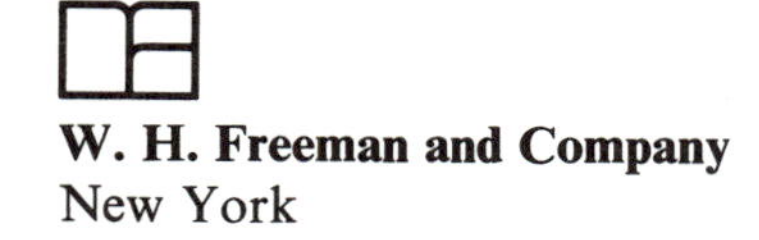

W. H. Freeman and Company
New York

Paperback Cover Art: Painting © François de Lucy

AMS Classification Scheme (1980)
Primary Classification: 90C05, Linear Programming
Secondary Classification: 9001

Library of Congress Cataloging in Publication Data

Chvátal, Vašek, 1946–
Linear programming.

(A Series of books in the mathematical sciences)
Bibliography: p.
Includes index.
1. Linear programming. I. Title. II. Series.
T57.74.C54 1983 519.7′2 82-21132
ISBN 0-7167-1195-8
ISBN 0-7167-1587-2 (pbk.)

Printed in the United States of America

12 13 14 15 16 VB 9 9 8 7 6 5

To my mother

'I don't want to bore you,' Harvey said, 'but you should understand that these heaps of wire can practically think—linear programming—which means that instead of going through all the alternatives they have a hunch which is the right one.'

From Billion-Dollar Brain *by Len Deighton.*

Contents

Preface

Linear Programming was written primarily for upper-division and graduate courses in operations research/management science, mathematics, and computer science. It may serve not only as an introduction to the subject but also as a reference and guide to applications. When I taught in the Operations Research Department at Stanford University, I found that none of the available texts quite fit my presentation, so I wrote lecture notes and distributed them in class. The students' enthusiasm encouraged me to rework the notes into a manuscript. Victor Klee gave me further encouragement and suggested additional topics to be included in the text. In the following several years, the manuscript grew to nearly five times its original length and was extensively revised as suggested by my further use in courses at McGill University and in a series of lectures at Université de Paris–Sud. In addition, I benefited greatly from the comments of instructors who used the text at the University of California at Berkeley, Carnegie-Mellon University, the University of Denver, Georgia Institute of Technology, the University of Houston, the University of Kentucky, Université de Montréal, the University of Rochester, Simon Fraser University, Stanford University, Tohoku University, Vanderbilt University, and the University of Waterloo.

For me, writing this book was a long and enjoyable exercise in exposition. I would return again and again to Paul Halmos' essay* on writing mathematics whenever I needed a bit of extra stimulation to carry on. Speak to someone, says Halmos; and I tried to follow his advice by always keeping a few of my students in mind. At the same time, however, I could not help thinking of some of my friends and colleagues, excellent mathematicians who never got around to learning linear programming. For them, I tried to make the text as appetizing as I could by not making simple ideas sound esoteric. Furthermore, the exposition is accessible to readers with only a minimal mathematical background: a reasonably bright and motivated high school student could follow and understand. No prerequisites are needed; even elementary

* P. Halmos, How to write mathematics, *L'Enseignment Mathématique*, vol. 16, 1970, pp. 123–152.

facts concerning matrices are developed from scratch. Finally, I tried to avoid fussy formalism by using the reader's ability to generalize from examples. As R. P. Boas put it,

> Suppose that you want to teach the "cat" concept to a very young child. Do you explain that a cat is a relatively small, primarily carnivorous mammal with retractile claws, a distinctive sonic output, etc.? I'll bet not. You probably show the kid a lot of different cats, saying "kitty" each time, until it gets the idea. To put it more generally, generalizations are best made by abstraction from experience.*

In this book, most of the concepts and techniques are introduced by illustrative examples; cumbersome notation with its plethora of subscripts and superscripts is often avoided altogether. This improves clarity without sacrificing rigor.

Further features of my presentation include the following:

(i) Other approaches frequently use a geometric setting to introduce and motivate linear programming; the algebraic mechanism of the simplex method is explained only after the student has learned to draw polygons in the plane and to solve two-variable problems by the graphic method. I had several reasons for abandoning this strategy. First, the graphic method is a dead end: the student's satisfaction in having learned something is, or at least should be, instantly spoiled by the realization that this bit of learning is useless in practice—unless, of course, one is making two kinds of sausages from three kinds of meat. Second, the student with a less solid mathematical background may become confused by the constant switching between geometric and algebraic descriptions. The insight, if any, provided by the geometric point of view may not be worth the mental effort of keeping both descriptions simultaneously in mind. Finally, I like to teach my students something nontrivial as soon as possible, before they lose interest or at least begin to suspect that they are not going to learn much in the course. Trivial preliminaries, such as the graphic method, put off the substantial material.

For these reasons, I have refrained from even mentioning the geometric point of view in the first part of the book and postponed its presentation until Chapter 17. At this point, the student will have digested the fundamental algebraic techniques; geometric illustrations will now come as a pleasant afterthought. These illustrations constitute only the first half of the chapter; the second half consists of applications of the algebraic results to the geometry of convex sets. Juxtaposing these two different subjects may help the student appreciate the two-way correspondence between geometry and algebra.

Those instructors who do not share my inclinations in this regard have the easy option of using material from the self-contained first half of Chapter 17 with the

---

* R. P. Boas, Can we make mathematics intelligible? *American Mathematical Monthly*, vol. 88, 1981, pp. 727–731.

first five chapters. The sections entitled "Geometric Interpretation of LP Problems" and "The Graphic Method" might be read with Chapter 1, "Geometric Interpretation of the Simplex Method" with Chapter 2, "Geometric Interpretation of the Perturbation Method" with Chapter 3, "Klee–Minty Examples Reviewed" with Chapter 4, and "Geometric Interpretation of the Duality Theorem" with Chapter 5.

(ii) Simplex tableaus, which once seemed to be a compulsory feature of linear programming books, are not used in this text at all. In the first five chapters, where the simplex method is motivated and explained, I prefer to write out in full the system of linear equations expressing the basic variables and the objective function in terms of nonbasic variables. These "dictionaries" reveal all the information they carry at first glance, and updating is a natural procedure impossible to forget. Tableaus have neither of these two virtues. To understand a tableau, one has to interpret it as the array of coefficients in the corresponding dictionary, a mental process complicated by the fact that the sign conventions in extracting these coefficients are not unified; to update a tableau, one has to follow mechanical rules learned by rote.

In subsequent chapters, where efficient implementations of the simplex method are studied, dictionaries are abandoned in favor of the revised simplex format. On typical large problems solved in practice, whose arrays are very sparse and have a special structure, this format requires much less work per simplex iteration than the dictionary/tableau format. This is why practitioners oppose the use of tableaus so vehemently and why I considered using the revised simplex format from the very start. What swayed my decision was classroom experience: my students took to dictionaries like fish to water, but initially many of them felt less at ease with the matrix manipulations of the revised simplex method. I suspect that even those who found the revised simplex method quite natural at once might have felt differently had they not first been eased into the simplex method through the intuitive concept of a dictionary.

(iii) The inverse of the basis is never used as a computational device in this text: each iteration of the revised simplex method revolves around solving two systems of equations rather than premultiplying and postmultiplying the inverse by a row and a column vector, respectively. In particular, the conventional "product form of the inverse" has been replaced by the essentially equivalent "eta factorization of the basis." This treatment simplifies the exposition and leads naturally to efficient implementations. The inverse of the basis is an anachronism that has no place in modern versions of the revised simplex method.

(iv) The use of transportation tableaus is avoided in the discussion of transshipment problems. This frequently used device is actually quite harmful in several ways. First, it can be used only in the special case of transportation problems, even though more general transshipment problems can be handled just as easily. Second, it obscures the intuitive combinatorial notions of paths, cycles, and trees. Finally, it distracts attention from the fact that large networks found in applications tend to be very sparse. In this text, transshipment problems are treated in their natural setting of networks.

Linear Programming provides flexibility in its use. A one-semester course might be based on Chapters 1–12 and 19–20. In fact, these chapters probably contain more than can be covered at a comfortable pace in one semester. Note that throughout the book, passages in small print may be skipped on the first reading without loss of continuity. However, small print does not signify inferiority; quite the contrary: a sophisticated reader may enjoy these passages more than the rest of the text.

Chapter 6 consists mainly of facts about systems of linear equations and matrices that should be, but often are not, known to the intended readers. Most of my students would enroll in the course unaware of the advantages of a triangular factorization (an LU-decomposition in a product form) over the inverse of a matrix as a device for solving systems of linear equations. Similarly, students are often unaware of the distinction between Gaussian and Gauss–Jordan elimination methods, not to mention the superiority of the former. (One does not have to look far to find a source of this confusion: at least one linear programming textbook describes Gauss–Jordan elimination under the name of Gaussian elimination; at least three others present the inferior Gauss–Jordan elimination without so much as mentioning Gaussian elimination.) For these reasons, I decided to keep this material in its present place rather than relegate it to an appendix.

Chapters 11–18 cover selected applications loosely arranged in order of decreasing practical interest. Chapter 13 on the cutting-stock problem can usefully be included in a management science/operations research course if time permits; Chapter 14 on data approximations is, perhaps, a little drier and more mathematical. Practical applicability of Chapter 15 aside, game theory is fun and students enjoy it. Chapter 16 on linear inequalities and the second half of Chapter 17 on geometry would fit nicely into a mathematically oriented course; Chapter 18 on finding vertices of polyhedra might appeal to students in computer science.

Chapters 19–23 on network-flow problems form a self-contained unit that can be read independently of the rest of the book. I have used these five chapters repeatedly in a graduate course in combinatorial optimization. At a leisurely pace, this material took up most of the semester.

The last three chapters cover computational techniques usually reserved for advanced courses. Nevertheless, it might be a good idea to discuss the triangular factorizations presented in Chapter 24 just after the presentation of the revised simplex method in Chapter 7: this material is not part of the basic curriculum, though perhaps it should be. The details of the recent ellipsoid method are left for the appendix: the preliminary discussion of the role of this method in Chapter 4 should be adequate for management science/operations research courses.

Each chapter concludes with a set of problems. These are of two kinds: some simply test the reader's understanding of the chapter and others extend the material beyond the text proper. Those marked by small triangles are answered in the back; detailed solutions to all the problems are collected in the Solutions Manual.

I was fortunate to have many friends help me in writing this book; now I would like to thank them all. The warm encouragement and the valuable suggestions from Victor Klee meant much to me; Vic also put me in touch with W. H. Freeman and Company. There, I had the advantage of having an editor who is also a mathematician, Peter Renz. His advice was very helpful in improving the manuscript. I thank the following readers, who gave me many helpful comments; all of them read various parts or versions of the manuscript and some of them used the manuscript in their own courses: David Avis, Adrian Bondy, David Clark, Scot Drysdale, Franci Fiala, Peter Gács, Geňa Hahn, Dorit Hochbaum, Alan Hoffman, Petr Kubát, Harold Kuhn, Hang Tong Lau, Leonid Levin, Mike Martin, Takao Nishizeki, Chris Paige, André Perold, Kevin Phelps, Maurice Queyranne, J. K. Reid, Vojta Rödl, Ivo Rosenberg, Mike Saunders, Bill Steiger, Hunter Swanson, Bob Tarjan, John Tomlin, Klaus Truemper, Sue Whitesides, Gene Woolsey, and Boris Yamnitsky. In particular, Dorit Hochbaum spent grueling hours with Peter Renz and myself as we restructured the first ten chapters. I would especially like to thank André Perold and Mike Saunders for the limitless patience with which they kept explaining implementations of the simplex method to me. Judith Wilson, Project Editor at Freeman, gave me competent and cheerful support as the manuscript was being transformed into a book. Paul Higham, Minoru Ishii, and Nick Tsikopoulos helped me with correcting the galley proofs and, even at that late stage, suggested a number of improvements.

Finally, I want to thank my friend François de Lucy for letting us reproduce one of his paintings on the cover of the paperback edition. Even though the painting does not represent a decaying polyhedron as some might believe, it does provide a fitting illustration: it was inspired by a crumpled sheet of paper, an object frequently sighted on this author's floor during years of writing.

January 1983 *Vašek Chvátal*

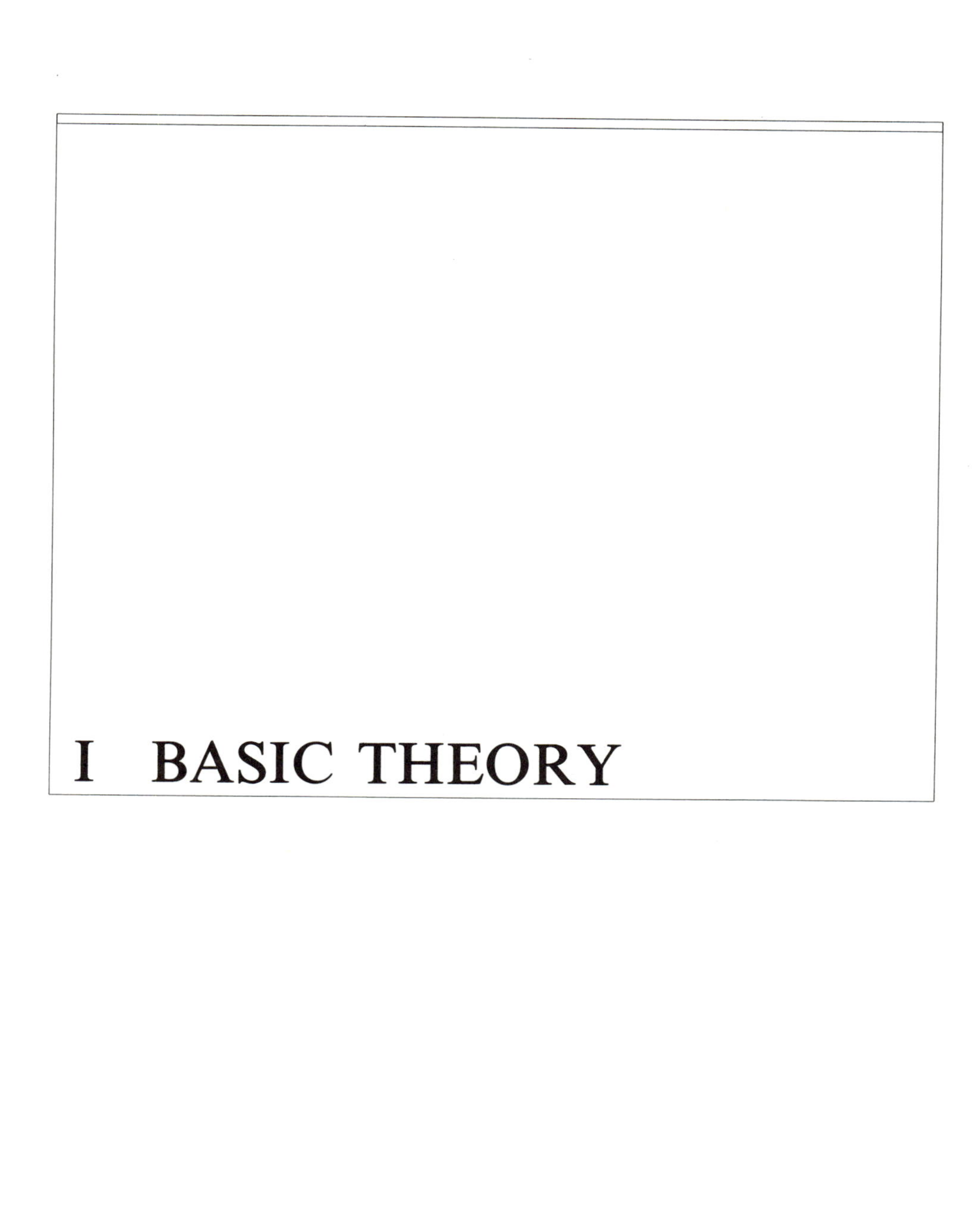

I BASIC THEORY

1

Introduction

In this short chapter, we shall explain what is meant by linear programming and sketch a history of this subject.

A DIET PROBLEM

Polly wonders how much money she must spend on food in order to get all the energy (2,000 kcal), protein (55 g), and calcium (800 mg) that she needs every day. (For iron and vitamins, she will depend on pills. Nutritionists would disapprove, but the introductory example ought to be simple.) She chooses six foods that seem to be cheap sources of the nutrients; her data are collected in Table 1.1.

Table 1.1 Nutritive Value per Serving

Food	Serving size	Energy (kcal)	Protein (g)	Calcium (mg)	Price per serving (cents)
Oatmeal	28 g	110	4	2	3
Chicken	100 g	205	32	12	24
Eggs	2 large	160	13	54	13
Whole milk	237 cc	160	8	285	9
Cherry pie	170 g	420	4	22	20
Pork with beans	260 g	260	14	80	19

Then she begins to think about her menu. For example, 10 servings of pork with beans would take care of all her needs for only (?) $1.90 per day. On the other hand, 10 servings of pork with beans is a lot of pork with beans—she would not be able to stomach more than 2 servings a day. She decides to impose servings-per-day limits on all six foods:

Oatmeal	at most 4 servings per day
Chicken	at most 3 servings per day
Eggs	at most 2 servings per day
Milk	at most 8 servings per day
Cherry pie	at most 2 servings per day
Pork with beans	at most 2 servings per day.

Now, another look at the data shows Polly that 8 servings of milk and 2 servings of cherry pie every day will satisfy the requirements nicely and at a cost of only $1.12. In fact, she could cut down a little on the pie or the milk or perhaps try a different combination. But so many combinations seem promising that one could go on and on, looking for the best one. Trial and error is not particularly helpful here. To be systematic, we may speculate about some as yet unspecified menu consisting of x_1 servings of oatmeal, x_2 servings of chicken, x_3 servings of eggs, and so on. In order to stay below the upper limits, that menu must satisfy

$$\begin{aligned} 0 &\le x_1 \le 4 \\ 0 &\le x_2 \le 3 \\ 0 &\le x_3 \le 2 \\ 0 &\le x_4 \le 8 \\ 0 &\le x_5 \le 2 \\ 0 &\le x_6 \le 2. \end{aligned} \tag{1.1}$$

And, of course, there are the requirements for energy, protein, and calcium; they lead to the inequalities

$$\begin{aligned} 110x_1 + 205x_2 + 160x_3 + 160x_4 + 420x_5 + 260x_6 &\ge 2{,}000 \\ 4x_1 + 32x_2 + 13x_3 + 8x_4 + 4x_5 + 14x_6 &\ge 55 \\ 2x_1 + 12x_2 + 54x_3 + 285x_4 + 22x_5 + 80x_6 &\ge 800. \end{aligned} \tag{1.2}$$

If some numbers $x_1, x_2, \ldots, x_6$ satisfy inequalities (1.1) and (1.2), then they describe a satisfactory menu; such a menu will cost, in cents per day,

$$3x_1 + 24x_2 + 13x_3 + 9x_4 + 20x_5 + 19x_6. \tag{1.3}$$

In designing the most economical menu, Polly wants to find numbers $x_1, x_2, \ldots, x_6$ that satisfy (1.1) and (1.2), and make (1.3) as small as possible. As a mathematician would put it, she wants to

$$
\begin{array}{ll}
\text{minimize} & 3x_1 + 24x_2 + 13x_3 + 9x_4 + 20x_5 + 19x_6 \\
\text{subject to} & 0 \le x_1 \le 4 \\
& 0 \le x_2 \le 3 \\
& 0 \le x_3 \le 2 \\
& 0 \le x_4 \le 8 \\
& 0 \le x_5 \le 2 \\
& 0 \le x_6 \le 2 \\
\end{array}
$$
$$
\begin{aligned}
110x_1 + 205x_2 + 160x_3 + 160x_4 + 420x_5 + 260x_6 &\ge 2000 \\
4x_1 + 32x_2 + 13x_3 + 8x_4 + 4x_5 + 14x_6 &\ge 55 \\
2x_1 + 12x_2 + 54x_3 + 285x_4 + 22x_5 + 80x_6 &\ge 800.
\end{aligned} \tag{1.4}
$$

Her problem is known as a *diet problem.*

LINEAR PROGRAMMING

Problems of this kind are called "linear programming problems," or "LP problems" for short; linear programming is the branch of applied mathematics concerned with these problems. Here are other examples:

$$
\begin{array}{ll}
\text{maximize} & 5x_1 + 4x_2 + 3x_3 \\
\text{subject to} & 2x_1 + 3x_2 + x_3 \le 5 \\
& 4x_1 + x_2 + 2x_3 \le 11 \\
& 3x_1 + 4x_2 + 2x_3 \le 8 \\
& x_1, x_2, x_3 \ge 0
\end{array} \tag{1.5}
$$

(with "$x_1, x_2, x_3 \ge 0$" used as shorthand for "$x_1 \ge 0, x_2 \ge 0, x_3 \ge 0$") or

$$
\begin{array}{ll}
\text{minimize} & 3x_1 - x_2 \\
\text{subject to} & -x_1 + 6x_2 - x_3 + x_4 \ge -3 \\
& 7x_2 + 2x_4 = 5 \\
& x_1 + x_2 + x_3 = 1 \\
& x_3 + x_4 \le 2 \\
& x_2, x_3 \ge 0.
\end{array} \tag{1.6}
$$

In general, if $c_1, c_2, \ldots, c_n$ are real numbers, then the function f of real variables $x_1, x_2, \ldots, x_n$ defined by

$$f(x_1, x_2, \ldots, x_n) = c_1x_1 + c_2x_2 + \cdots + c_nx_n = \sum_{j=1}^{n} c_jx_j$$

is called a *linear function.* If f is a linear function and if b is a real number, then the equation

$$f(x_1, x_2, \ldots, x_n) = b$$

is called a *linear equation* and the inequalities

$$f(x_1, x_2, \ldots, x_n) \leq b$$
$$f(x_1, x_2, \ldots, x_n) \geq b$$

are called *linear inequalities.* Linear equations and linear inequalities are both referred to as *linear constraints.* Finally, a *linear programming problem* is the problem of maximizing (or minimizing) a linear function subject to a finite number of linear constraints. We shall usually attach different subscripts i to different constraints and different subscripts j to different variables. For simplicity of exposition, we shall restrict ourselves in Chapters 1–7 to LP problems of the following form:

$$\begin{aligned} \text{maximize} \quad & \sum_{j=1}^{n} c_j x_j \\ \text{subject to} \quad & \sum_{j=1}^{n} a_{ij} x_j \leq b_i \qquad (i = 1, 2, \ldots, m) \\ & x_j \geq 0 \qquad (j = 1, 2, \ldots, n). \end{aligned} \tag{1.7}$$

These problems will be referred to as LP problems in the *standard form.* (The reader should be warned that the terminology is far from unified; several authors prefer the terms *canonical* or *symmetric* form, and others reserve these adjectives for altogether different problems.) For example, (1.5) is a problem in the standard form (with $n = 3, m = 3, a_{11} = 2, a_{12} = 3$, and so on). What distinguishes the problems in the standard form from the rest? First, all of their constraints are linear *inequalities.* Secondly, the last n of the $m + n$ constraints in (1.7) are very special: they simply stipulate that none of the n variables may assume negative values. Such constraints are called *nonnegativity constraints.* (Note that problem (1.6) differs from the standard form on both counts: two of its constraints are linear equations and the variables x_1, x_4 may assume negative values.)

The linear function that is to be maximized or minimized in an LP problem is called the *objective function* of that problem. For example, the function z of variables $x_1, x_2, x_3, x_4, x_5, x_6$ defined by

$$z(x_1, x_2, \ldots, x_6) = 3x_1 + 24x_2 + 13x_3 + 9x_4 + 20x_5 + 19x_6$$

is the objective function of Polly's diet problem (1.4). Numbers $x_1, x_2, \ldots, x_n$ that satisfy all the constraints of an LP problem are said to constitute a *feasible solution* of that problem. For instance, we have observed that

$$x_1 = 0, \quad x_2 = 0, \quad x_3 = 0, \quad x_4 = 8, \quad x_5 = 2, \quad x_6 = 0$$

is a feasible solution of (1.4). Finally, a feasible solution that maximizes the objective function (or minimizes it, depending on the form of the problem) is called an *optimal solution*; the corresponding value of the objective function is called the *optimal value* of the problem. As it turns out, the unique optimal solution of (1.4) is

$$x_1 = 4, \quad x_2 = 0, \quad x_3 = 0, \quad x_4 = 4.5, \quad x_5 = 2, \quad x_6 = 0$$

or simply (4, 0, 0, 4.5, 2, 0). Accordingly, the optimal value of (1.4) is 92.5. Not every LP problem has a unique optimal solution; some problems have many different optimal solutions and others have no optimal solutions at all. The latter may occur for one of two radically different reasons: either there are no feasible solutions at all or there are, in a sense, too many of them. The first case may be illustrated on the problem

$$\begin{array}{lrcl} \text{maximize} & 3x_1 - x_2 & & \\ \text{subject to} & x_1 + x_2 & \leq & 2 \\ & -2x_1 - 2x_2 & \leq & -10 \\ & x_1, x_2 & \geq & 0 \end{array} \qquad (1.8)$$

which has no feasible solutions at all. Such problems are called *infeasible*. On the other hand, even though the problem

$$\begin{array}{lrcl} \text{maximize} & x_1 - x_2 & & \\ \text{subject to} & -2x_1 + x_2 & \leq & -1 \\ & -x_1 - 2x_2 & \leq & -2 \\ & x_1, x_2 & \geq & 0 \end{array} \qquad (1.9)$$

does have feasible solutions, none of them is optimal: for every number M there is a feasible solution x_1, x_2 such that $x_1 - x_2 > M$. In a sense, (1.9) has such an abundance of feasible solutions that none of them can aspire to be the best. Problems with this property are called *unbounded*. As we shall prove later (Theorem 3.4), every linear programming problem belongs to one of the three categories noted here: it has an optimal solution, is infeasible, or is unbounded.

HISTORY OF LINEAR PROGRAMMING

As mathematical disciplines go, linear programming is quite young. It started in 1947 when G. B. Dantzig designed the "simplex method" for solving linear programming formulations of U.S. Air Force planning problems. What followed was an exciting period of rapid development in this new field. It soon became clear that a surprisingly wide range of apparently unrelated problems in production management could be stated in linear programming terms and, most importantly, solved by the simplex method. Such problems, if noticed at all, had traditionally been tackled by a hit-or-miss approach guided only by experience and intuition. The use of linear programming often brought about a considerable increase in the efficiency of the whole operation. (Until then, expansion of the efficiency frontier usually came from technological innovations. This new way to increase efficiency—*under existing technological conditions*—by improvements in organization and planning, made many managers appreciate the practical importance of mathematics. At least, it made them aware of the advantage of stating their decision problems in clear-cut and well-defined terms.) As the popularity of linear programming theory increased, applications in new areas occurred, many of them far from obvious. In turn, these applications stimulated further theoretical research by pointing out the need for solving problems that would have otherwise seemed uninteresting. In this fascinating interplay between theory and applications, a new branch of applied mathematics established itself.

As calculus developed from the seventeenth century's need to solve problems of mechanics, linear programming developed from the twentieth century's need to solve problems of management. Yet other profound influences stimulated the evolution of the new field from its very inception. Economics was one of them: as early as 1947, T. C. Koopmans began pointing out that linear programming provided an excellent framework for the analysis of classical economic theories, such as the renowned system proposed in 1874 by L. Walras. On the other hand, linear programming brought together previously known theorems of pure mathematics concerning such diverse topics as the geometry of convex sets, extremal problems of combinatorial nature, and the theory of two-person games. Finally, it was fortunate and perhaps even inevitable that linear programming developed concurrently with modern computer technology: without electronic computers, present-day large-scale linear programming would be unthinkable.

Scientific fields are rarely born overnight; with the advantage of hindsight, one can often track down the sources that paved the way for the decisive breakthrough. The field of linear programming is no exception. At the core of its mathematical theory is the study of systems of linear inequalities; such systems were investigated by Fourier as far back as 1826. Since then, quite a few other mathematicians have considered the subject, although none of them has devised an algorithm whose efficiency has come close to that of the simplex method. Nevertheless, some of them proved various special cases of a fundamental theorem that is now called the *duality theorem* of linear programming. On the applied side, L. V. Kantorovich pointed out the practical significance of a restricted class of LP problems, and proposed a rudimentary algorithm for their solution as early as 1939. Regrettably, this effort remained neglected in the U.S.S.R. and unknown elsewhere until long after linear programming became an elegant theory through the independent work of Dantzig and others.

In the 1970s, linear programming came twice to public attention. On October 14, 1975, the Royal Sweden Academy of Sciences awarded the Nobel Prize in economic science to L. V. Kantorovich and T. C. Koopmans "for their contributions to the theory of optimum allocation of resources." (As the reader may know, there is no Nobel Prize in mathematics. Apparently the Academy regarded the work of G. B. Dantzig, who is universally recognized as the father of linear programming, as being too mathematical.) The second event was even more dramatic. Ever since the invention of the simplex method, mathematicians had been looking for a *theoretically* satisfactory algorithm to solve LP problems. (A word of explanation is in order: theoretical criteria for judging the efficiency of algorithms are quite different from practical ones. Thus, an algorithm like the simplex method, which is eminently satisfactory in practical applications, may be found theoretically unsatisfactory. The converse is also true: theoretically satisfactory algorithms may be thoroughly useless in practice. We shall return to this distinction in

Chapter 4.) The breakthrough came in 1979 when L. G. Khachian published a description of such an algorithm (based on earlier works by Shor, and by Judin and Nemirovskii). Newspapers around the world published reports of this result, some of them full of hilarious misinterpretations. We shall present the algorithm in the appendix.

For a thorough survey of the history of linear programming, the reader is referred to Chapter 2 of Dantzig's monograph (1963). References to many applications of linear programming may be found in Riley and Gass (1958). Some of the more recent applications are referenced in Gass (1975). □

PROBLEMS

Answers to problems marked with the symbol △ are found at the back of the book.

1.1 Which of the problems below are in the standard form?

a. Maximize $3x_1 - 5x_2$

subject to
$$\begin{aligned} 4x_1 + 5x_2 &\geq 3 \\ 6x_1 - 6x_2 &= 7 \\ x_1 + 8x_2 &\leq 20 \\ x_1, x_2 &\geq 0. \end{aligned}$$

b. Minimize $3x_1 + x_2 + 4x_3 + x_4 + 5x_5$

subject to
$$\begin{aligned} 9x_1 + 2x_2 + 6x_3 + 5x_4 + 3x_5 &\leq 5 \\ 8x_1 + 9x_2 + 7x_3 + 9x_4 + 3x_5 &\leq 2 \\ x_1, x_2, x_3, x_4 &\geq 0. \end{aligned}$$

c. Maximize $8x_1 - 4x_2$

subject to
$$\begin{aligned} 3x_1 + x_2 &\leq 7 \\ 9x_1 + 5x_2 &\leq -2 \\ x_1, x_2 &\geq 0. \end{aligned}$$

1.2 State in the standard form:

minimize $-8x_1 + 9x_2 + 2x_3 - 6x_4 - 5x_5$

subject to
$$\begin{aligned} 6x_1 + 6x_2 - 10x_3 + 2x_4 - 8x_5 &\geq 3 \\ x_1, x_2, x_3, x_4, x_5 &\geq 0. \end{aligned}$$

1.3 Prove that (1.8) is infeasible and (1.9) is unbounded.

△**1.4** Find necessary and sufficient conditions for the numbers s and t to make the LP problem

maximize $x_1 + x_2$

subject to
$$\begin{aligned} sx_1 + tx_2 &\leq 1 \\ x_1, x_2 &\geq 0 \end{aligned}$$

a. have an optimal solution,

b. be infeasible,

c. be unbounded.

△**1.5** Prove or disprove: If problem (1.7) is unbounded, then there is a subscript k such that the problem

$$\begin{array}{lll} \text{maximize} & x_k & \\ \text{subject to} & \sum_{j=1}^{n} a_{ij}x_j \leq b_i & (i = 1, 2, \ldots, m) \\ & x_j \geq 0 & (j = 1, 2, \ldots, n) \end{array}$$

is unbounded.

△**1.6** [Adapted from Greene et al. (1959).] A meat packing plant produces 480 hams, 400 pork bellies, and 230 picnic hams every day; each of these products can be sold either fresh or smoked. The total number of hams, bellies, and picnics that can be smoked during a normal working day is 420; in addition, up to 250 products can be smoked on overtime at a higher cost. The *net* profits are as follows:

	Fresh	Smoked on regular time	Smoked on overtime
Hams	\$8	\$14	\$11
Bellies	\$4	\$12	\$7
Picnics	\$4	\$13	\$9

For example, the following schedule yields a total net profit of \$9,965:

	Fresh	Smoked	Smoked (overtime)
Hams	165	280	35
Bellies	295	70	35
Picnics	55	70	105

The objective is to find the schedule that maximizes the total net profit. Formulate as an LP problem in the standard form.

1.7 [Adapted from Charnes et al. (1952).] An oil refinery produces four types of raw gasoline: alkylate, catalytic-cracked, straight-run, and isopentane. Two important characteristics of each gasoline are its performance number PN (indicating antiknock properties) and its vapor pressure RVP (indicating volatility). These two characteristics, together with the production levels in barrels per day, are as follows:

	PN	RVP	Barrels produced
Alkylate	107	5	3,814
Catalytic-cracked	93	8	2,666
Straight-run	87	4	4,016
Isopentane	108	21	1,300

These gasolines can be sold either raw, at \$4.83 per barrel, or blended into aviation gasolines (Avgas A and/or Avgas B). Quality standards impose certain requirements on the aviation gasolines; these requirements, together with the selling prices, are as follows:

	PN	RVP	Price per barrel
Avgas A	at least 100	at most 7	\$6.45
Avgas B	at least 91	at most 7	\$5.91

The PN and RVP of each mixture are simply weighted averages of the PNs and RVPs of its constituents. For example, the refinery could adopt the following strategy:

- Blend 2,666 barrels of alkylate and 2,666 barrels of catalytic into 5,332 barrels of Avgas A with

$$\text{PN} = \frac{(2{,}666 \times 107) + (2{,}666 \times 93)}{5{,}332} = 100$$

$$\text{RVP} = \frac{(2{,}666 \times 5) + (2{,}666 \times 8)}{5{,}332} = 6.5.$$

- Blend 1,148 barrels of alkylate, 4,016 barrels of straight-run, and 1,024 barrels of isopentane into 6,188 barrels of Avgas B with

$$\text{PN} = \frac{(1{,}148 \times 107) + (4{,}016 \times 87) + (1{,}024 \times 108)}{6{,}188} \doteq 94.2$$

$$\text{RVP} = \frac{(1{,}148 \times 5) + (4{,}016 \times 4) + (1{,}024 \times 21)}{6{,}188} \doteq 7.$$

Sell 276 barrels of isopentane raw.

This sample plan yields a total profit of

$$(5{,}332 \times 6.45) + (6{,}188 \times 5.91) + (276 \times 4.83) \doteq \$72{,}296.$$

The refinery aims for the plan that yields the largest possible profit. Formulate as an LP problem in the standard form.

1.8 An electronics company has a contract to deliver 20,000 radios within the next four weeks. The client is willing to pay \$20 for each radio delivered by the end of the first week, \$18 for those delivered by the end of the second week, \$16 by the end of the third week, and \$14 by the end of the fourth week. Since each worker can assemble only 50 radios per week, the company cannot meet the order with its present labor force of 40; hence it must hire and train temporary help. Any of the experienced workers can be taken off the assembly line to instruct a class of three trainees; after one week of instruction, each of the trainees can either proceed to the assembly line or instruct additional new classes.

At present, the company has no other contracts; hence some workers may become idle once the delivery is completed. All of them, whether permanent or temporary, must be kept on the payroll till the end of the fourth week. The weekly wages of a worker, whether assembling, instructing, or being idle, are \$200; the weekly wages of a trainee are \$100. The production costs, excluding the worker's wages, are \$5 per radio.

For example, the company could adopt the following program.

First week: 10 assemblers, 30 instructors, 90 trainees
Workers' wages: \$8,000

Trainees' wages: \$9,000
Profit from 500 radios: \$7,500
Net loss: \$9,500

Second week: 120 assemblers, 10 instructors, 30 trainees
Workers' wages: \$26,000
Trainees' wages: \$3,000
Profit from 6,000 radios: \$78,000
Net profit: \$49,000

Third week: 160 assemblers
Workers' wages: \$32,000
Profit from 8,000 radios: \$88,000
Net profit: \$56,000

Fourth week: 110 assemblers, 50 idle
Workers' wages: \$32,000
Profit from 5,500 radios: \$49,500
Net profit: \$17,500

This program, leading to a total net profit of \$113,000, is one of many possible programs. The company's aim is to maximize the total net profit. Formulate as an LP problem (not necessarily in the standard form).

△ **1.9** [S. Masuda (1970); see also V. Chvátal (1983).] The *bicycle problem* involves n people who have to travel a distance of ten miles, and have one single-seat bicycle at their disposal. The data are specified by the walking speed w_j and the bicycling speed b_j of each person j ($j = 1, 2, \ldots, n$); the task is to minimize the arrival time of the last person. (Can you solve the case of $n = 3$ and $w_1 = 4$, $w_2 = w_3 = 2$, $b_1 = 16$, $b_2 = b_3 = 12$?) Show that the optimal value of the LP problem

$$\begin{aligned} \text{minimize} \quad & t \\ \text{subject to} \quad & t - x_j - x'_j - y_j - y'_j \geq 0 \quad (j = 1, 2, \ldots, n) \\ & t - \sum_{j=1}^{n} y_j - \sum_{j=1}^{n} y'_j \geq 0 \\ & w_j x_j - w_j x'_j + b_j y_j - b_j y'_j = 10 \quad (j = 1, 2, \ldots, n) \\ & \sum_{j=1}^{n} b_j y_j - \sum_{j=1}^{n} b_j y'_j \leq 10 \\ & x_j, x'_j, y_j, y'_j \geq 0 \quad (j = 1, 2, \ldots, n) \end{aligned}$$

provides a lower bound on the optimal value of the bicycle problem.

2

How the Simplex Method Works

In this chapter, we shall learn to solve LP problems in the standard form by the simplex method. A rigorous analysis of the details will be deferred to Chapter 3.

FIRST EXAMPLE

We shall illustrate the simplex method on the following example:

$$\begin{array}{ll} \text{maximize} & 5x_1 + 4x_2 + 3x_3 \\ \text{subject to} & 2x_1 + 3x_2 + x_3 \le 5 \\ & 4x_1 + x_2 + 2x_3 \le 11 \\ & 3x_1 + 4x_2 + 2x_3 \le 8 \\ & x_1, x_2, x_3 \ge 0. \end{array} \tag{2.1}$$

A preliminary step of the method consists of introducing so-called slack variables.

In order to motivate this concept, let us consider the first of our constraints,

$$2x_1 + 3x_2 + x_3 \leq 5. \tag{2.2}$$

For every feasible solution x_1, x_2, x_3, the value of the left-hand side of (2.1) is at most the value of the right-hand side; often, there may be a slack between the two values. We shall denote the slack by x_4. That is, we shall *define* $x_4 = 5 - 2x_1 - 3x_2 - x_3$; with this notation, inequality (2.2) may now be written as $x_4 \geq 0$. In an analogous way, the next two constraints give rise to variables x_5 and x_6. Finally, following a time-honored convention, we shall denote the objective function $5x_1 + 4x_2 + 3x_3$ by z. To summarize: for every *choice* of numbers x_1, x_2, and x_3, we shall *define* numbers x_4, x_5, x_6, and z by the formulas

$$\begin{aligned} x_4 &= 5 - 2x_1 - 3x_2 - x_3 \\ x_5 &= 11 - 4x_1 - x_2 - 2x_3 \\ x_6 &= 8 - 3x_1 - 4x_2 - 2x_3 \\ z &= 5x_1 + 4x_2 + 3x_3. \end{aligned} \tag{2.3}$$

With this notation, our problem may be restated as

$$\text{maximize} \quad z \quad \text{subject to} \quad x_1, x_2, x_3, x_4, x_5, x_6 \geq 0. \tag{2.4}$$

The new variables x_4, x_5, x_6 defined by (2.3) are called *slack variables*; the old variables x_1, x_2, x_3 are usually referred to as the *decision variables.* It is crucial to note that the equations in (2.3) spell out an equivalence between (2.1) and (2.4). More precisely:

- Every feasible solution x_1, x_2, x_3 of (2.1) can be extended, in the unique way determined by (2.3), into a feasible solution $x_1, x_2, \ldots, x_6$ of (2.4).
- Every feasible solution $x_1, x_2, \ldots, x_6$ of (2.4) can be restricted, simply by deleting the slack variables, into a feasible solution x_1, x_2, x_3 of (2.1).
- This correspondence between feasible solutions of (2.1) and feasible solutions of (2.4) carries optimal solutions of (2.1) onto optimal solutions of (2.4), and vice versa.

The grand strategy of the simplex method is that of *successive improvements*: having found some feasible solution $x_1, x_2, \ldots, x_6$ of (2.4), we shall try to proceed to another feasible solution $\bar{x}_1, \bar{x}_2, \ldots, \bar{x}_6$, which is better in the sense that

$$5\bar{x}_1 + 4\bar{x}_2 + 3\bar{x}_3 > 5x_1 + 4x_2 + 3x_3.$$

Repeating this process a finite number of times, we shall eventually arrive at an optimal solution.

To begin with, we need some feasible solution $x_1, x_2, \ldots, x_6$. Finding one in our example presents no difficulty: setting the decision variables x_1, x_2, x_3 at zero, we evaluate the slack variables x_4, x_5, x_6 from (2.3). Hence our initial solution,

$$x_1 = 0, \quad x_2 = 0, \quad x_3 = 0, \quad x_4 = 5, \quad x_5 = 11, \quad x_6 = 8 \tag{2.5}$$

yields $z = 0$.

In the spirit of the grand strategy sketched above, we should now look for a feasible solution that yields a higher value of z. Finding such a solution is not difficult. For example, if we keep $x_2 = x_3 = 0$ and increase the value of x_1, we obtain $z = 5x_1 > 0$. Thus, if we keep $x_2 = x_3 = 0$ and set $x_1 = 1$, we obtain $z = 5$ (and $x_4 = 3, x_5 = 7, x_6 = 5$). Better yet, if we keep $x_2 = x_3 = 0$ and set $x_1 = 2$, we obtain $z = 10$ (and $x_4 = 1$, $x_5 = 3$, $x_6 = 2$). However, if we keep $x_2 = x_3 = 0$ and set $x_1 = 3$, we obtain $z = 15$ and $x_4 = x_5 = x_6 = -1$; this won't do, since feasibility requires $x_i \geq 0$ for every i. The moral is that we cannot increase x_1 too much. The question is: *Just how much can we increase* x_1 *(keeping* $x_2 = x_3 = 0$ *at the same time) and still maintain feasibility* ($x_4, x_5, x_6 \geq 0$)?

The condition $x_4 = 5 - 2x_1 - 3x_2 - x_3 \geq 0$ implies $x_1 \leq \frac{5}{2}$; similarly, $x_5 \geq 0$ implies $x_1 \leq \frac{11}{4}$ and $x_6 \geq 0$ implies $x_1 \leq \frac{8}{3}$. Of these three bounds, the first is the most stringent. Increasing x_1 up to that bound we obtain our next solution,

$$x_1 = \frac{5}{2}, \quad x_2 = 0, \quad x_3 = 0, \quad x_4 = 0, \quad x_5 = 1, \quad x_6 = \frac{1}{2}. \tag{2.6}$$

Note that this solution yields $z = \frac{25}{2}$, which is indeed an improvement over $z = 0$.

Next, we should look for a feasible solution that is even better than (2.6). However, this task seems a little more difficult. What made the first iteration so easy? We had at our disposal not only the feasible solution (2.5), but also the system of linear equations (2.3), which guided us in our quest for an improved feasible solution. If we wish to continue in a similar way, we should manufacture a new system of linear equations that relates to (2.6) much as system (2.3) relates to (2.5).

What properties should the new system have? Note that (2.3) expresses the variables that assume positive values in (2.5) in terms of the variables that assume zero values in (2.5). Similarly, the new system should express those variables that assume positive values in (2.6) in terms of the variables that assume zero values in (2.6): in short, it should express x_1, x_5, x_6 (as well as z) in terms of x_2, x_3, and x_4. In particular, the variable x_1, which just changed its value from zero to positive should change its position from the right-hand side to the left-hand side of the system of equations. Similarly, the variable x_4, which just changed its value from positive to zero, should move from the left-hand side to the right-hand side.

To construct the new system, we shall begin with the newcomer to the left-hand side, namely, the variable x_1. The desired formula for x_1 in terms of x_2, x_3, x_4 is obtained easily from the first equation in (2.3):

$$x_1 = \frac{5}{2} - \frac{3}{2}x_2 - \frac{1}{2}x_3 - \frac{1}{2}x_4. \tag{2.7}$$

Next, in order to express x_5, x_6, and z in terms of x_2, x_3, x_4, we simply substitute from (2.7) into the corresponding rows of (2.3):

$$\begin{aligned}
x_5 &= 11 - 4\left(\frac{5}{2} - \frac{3}{2}x_2 - \frac{1}{2}x_3 - \frac{1}{2}x_4\right) - x_2 - 2x_3 \\
&= 1 + 5x_2 + 2x_4, \\
x_6 &= 8 - 3\left(\frac{5}{2} - \frac{3}{2}x_2 - \frac{1}{2}x_3 - \frac{1}{2}x_4\right) - 4x_2 - 2x_3 \\
&= \frac{1}{2} + \frac{1}{2}x_2 - \frac{1}{2}x_3 + \frac{3}{2}x_4, \\
z &= 5\left(\frac{5}{2} - \frac{3}{2}x_2 - \frac{1}{2}x_3 - \frac{1}{2}x_4\right) + 4x_2 + 3x_3 \\
&= \frac{25}{2} - \frac{7}{2}x_2 + \frac{1}{2}x_3 - \frac{5}{2}x_4.
\end{aligned}$$

Hence our new system reads

$$\begin{aligned}
x_1 &= \frac{5}{2} - \frac{3}{2}x_2 - \frac{1}{2}x_3 - \frac{1}{2}x_4 \\
x_5 &= 1 + 5x_2 + 2x_4 \\
x_6 &= \frac{1}{2} + \frac{1}{2}x_2 - \frac{1}{2}x_3 + \frac{3}{2}x_4 \\
z &= \frac{25}{2} - \frac{7}{2}x_2 + \frac{1}{2}x_3 - \frac{5}{2}x_4.
\end{aligned} \tag{2.8}$$

As we did in the first iteration, we shall now try to increase the value of z by increasing the value of a suitably chosen right-hand side variable, while at the same time keeping the remaining right-hand side variables fixed at zero. Note that increases in the values of x_2 or x_4 would bring about *decreases* in the value of z, which is very much against our intentions. Thus, we have no choice: the right-hand side variable to increase its value is necessarily x_3. How much can we increase x_3? The answer can be read directly from system (2.8): with $x_2 = x_4 = 0$, the constraint $x_1 \geq 0$ implies $x_3 \leq 5$, the constraint $x_5 \geq 0$ imposes no restriction at all, and the constraint $x_6 \geq 0$ implies $x_3 \leq 1$. Hence, $x_3 = 1$ is the best we can do; our new solution is

$$x_1 = 2, \quad x_2 = 0, \quad x_3 = 1, \quad x_4 = 0, \quad x_5 = 1, \quad x_6 = 0. \tag{2.9}$$

(Note that the value of z just increased from 12.5 to 13.)

As we have learned, getting just the improved solution isn't good enough; we also want a system of linear equations to go with (2.9). In this system, the positive-valued variables x_1, x_3, x_5 will appear on the left, whereas the zero-valued variables x_2, x_4, x_6

will appear on the right. To construct the system, we begin again with the newcomer to the left-hand side, namely, the variable x_3. From the third equation in (2.8), we have $x_3 = 1 + x_2 + 3x_4 - 2x_6$; substituting for x_3 into the remaining equations in (2.8), we obtain

$$\begin{aligned} x_3 &= 1 + x_2 + 3x_4 - 2x_6 \\ x_1 &= 2 - 2x_2 - 2x_4 + x_6 \\ x_5 &= 1 + 5x_2 + 2x_4 \\ z &= 13 - 3x_2 - x_4 - x_6. \end{aligned} \tag{2.10}$$

Now it's time for the third iteration. First of all, from the right-hand side of (2.10) we have to choose a variable whose increase brings about an increase of the objective function. However, there is no such variable: indeed, if we increase any of the right-hand side variables x_2, x_4, x_6, we will make the value of z *decrease*. Thus, it seems that we have come to a standstill. In fact, the very presence of this standstill indicates that we are done; we have solved our problem; the solution described by the last table is optimal. Why? The answer lies hidden in the last row of (2.10):

$$z = 13 - 3x_2 - x_4 - x_6. \tag{2.11}$$

Our last solution (2.9) yields $z = 13$; proving that this solution is optimal amounts to proving that every feasible solution satisfies the inequality $z \leq 13$. Since every feasible solution $x_1, x_2, \ldots, x_6$ satisfies, among other relations, the inequalities $x_2 \geq 0$, $x_4 \geq 0$, and $x_6 \geq 0$, the desired inequality $z \leq 13$ follows directly from (2.11).

DICTIONARIES

In general, given a problem

$$\begin{aligned} &\text{maximize} && \sum_{j=1}^{n} c_j x_j \\ &\text{subject to} && \sum_{j=1}^{n} a_{ij} x_j \leq b_i \qquad (i = 1, 2, \ldots, m) \\ & && x_j \geq 0 \qquad (j = 1, 2, \ldots, n) \end{aligned} \tag{2.12}$$

we first introduce the slack variables $x_{n+1}, x_{n+2}, \ldots, x_{n+m}$ and denote the objective function by z. That is, we define

$$\begin{aligned} x_{n+i} &= b_i - \sum_{j=1}^{n} a_{ij} x_j \qquad (i = 1, 2, \ldots, m) \\ z &= \sum_{j=1}^{n} c_j x_j. \end{aligned} \tag{2.13}$$

In the framework of the simplex method, each feasible solution $x_1, x_2, \ldots, x_n$ of (2.12) is represented by $n + m$ *nonnegative* numbers $x_1, x_2, \ldots, x_{n+m}$, with $x_{n+1}, x_{n+2}, \ldots, x_{n+m}$ defined by (2.13). In each iteration, the simplex method moves from some feasible solution $x_1, x_2, \ldots, x_{n+m}$ to another feasible solution $\bar{x}_1, \bar{x}_2, \ldots, \bar{x}_{n+m}$, which is better than the previous one in the sense that

$$\sum_{j=1}^{n} c_j\bar{x}_j > \sum_{j=1}^{n} c_j x_j.$$

(Actually, the last statement is not quite correct: the inequality is not always strict. This point and other subtleties will be discussed in Chapter 3.)

As we have seen, it is convenient to associate a system of linear equations with each of the feasible solutions: such systems make it easier to find the improved feasible solutions. They do so by translating any choice of values of the right-hand side variables into the corresponding values of the left-hand side variables and of the objective function. Following J. E. Strum (1972), we shall refer to these systems as *dictionaries.* Thus, every dictionary associated with (2.12) will be a system of linear equations in the variables $x_1, x_2, \ldots, x_{n+m}$ and z. However, not every system of linear equations in these variables constitutes a dictionary. To begin with, we have defined $x_{n+1}, x_{n+2}, \ldots, x_{n+m}$ and z in terms of $x_1, x_2, \ldots, x_n$, and so the $n + m + 1$ variables are heavily interdependent. This interdependence must be captured by every dictionary associated with (2.12): the translations must be correct. More precisely, we shall insist that:

Every solution of the set of equations comprising a dictionary must be also a solution of (2.13), and vice versa. (2.14)

For example, for every choice of numbers $x_1, x_2, \ldots, x_6$ and z, the following three statements are equivalent:

- $x_1, x_2, \ldots, x_6, z$ constitute a solution of (2.3),
- $x_1, x_2, \ldots, x_6, z$ constitute a solution of (2.8),
- $x_1, x_2, \ldots, x_6, z$ constitute a solution of (2.10).

In that sense, the three dictionaries (2.3), (2.8), and (2.10) contain the same information concerning the interdependence among the seven variables. Nevertheless, each of the three dictionaries presents this information in its very own way. The form of (2.3) suggests that we are free to choose the numerical values of x_1, x_2, and x_3 at will, whereupon the values of x_4, x_5, x_6, and z are determined: in this dictionary, the decision variables x_1, x_2, x_3 act as independent variables, while z and the slack variables x_4, x_5, x_6 are dependent on them. Dictionary (2.8) presents x_2, x_3, x_4 as independent and x_1, x_5, x_6, z as dependent. In dictionary (2.10), the independent variables are x_2, x_4, x_6 and the dependent ones are x_3, x_1, x_5, z. In general:

The equations of every dictionary must express m of the variables x_1,

$x_2, \ldots, x_{n+m}$ and the objective function z in terms of the remaining n variables. (2.15)

The properties (2.14) and (2.15) are the defining properties of dictionaries.

In addition to these two properties, dictionaries (2.3), (2.8), and (2.10) have the following property:

Setting the right-hand side variables at zero and evaluating the left-hand side variables, we arrive at a *feasible* solution.

Dictionaries with this additional property will be called *feasible dictionaries*. Hence, every feasible dictionary describes a feasible solution. However, not every feasible solution is described by a feasible dictionary; for instance, no dictionary describes the feasible solution $x_1 = 1, x_2 = 0, x_3 = 1, x_4 = 2, x_5 = 5, x_6 = 3$ of (2.1). Feasible solutions that can be described by dictionaries are called *basic*. The characteristic feature of the simplex method is the fact that it works exclusively with basic feasible solutions and ignores all other feasible solutions.

SECOND EXAMPLE

We shall complete our preview of the simplex method by applying it to another LP problem:

$$
\begin{array}{lrcrcrl}
\text{maximize} & 5x_1 & + & 5x_2 & + & 3x_3 & \\
\text{subject to} & x_1 & + & 3x_2 & + & x_3 & \leq 3 \\
& -x_1 & & & + & 3x_3 & \leq 2 \\
& 2x_1 & - & x_2 & + & 2x_3 & \leq 4 \\
& 2x_1 & + & 3x_2 & - & x_3 & \leq 2 \\
& & & & & x_1, x_2, x_3 & \geq 0.
\end{array}
$$

In this case, the initial feasible dictionary reads

$$
\begin{array}{rcrcrcrcr}
x_4 & = & 3 & - & x_1 & - & 3x_2 & - & x_3 \\
x_5 & = & 2 & + & x_1 & & & - & 3x_3 \\
x_6 & = & 4 & - & 2x_1 & + & x_2 & - & 2x_3 \\
x_7 & = & 2 & - & 2x_1 & - & 3x_2 & + & x_3 \\
\hline
z & = & & & 5x_1 & + & 5x_2 & + & 3x_3.
\end{array}
\tag{2.16}
$$

(Even though the order of the equations in a dictionary is quite irrelevant, we shall make a habit of writing the formula for z last and separating it from the rest of the table by a solid line. Of course, that does *not* mean that the last equation is the sum of the previous ones.) This feasible dictionary describes the feasible solution

$$x_1 = 0, \quad x_2 = 0, \quad x_3 = 0, \quad x_4 = 3, \quad x_5 = 2, \quad x_6 = 4, \quad x_7 = 2.$$

However, there is no need to write this solution down, as we just did: the solution is implicit in the dictionary.

In the first iteration, we shall attempt to increase the value of z by making one of the right-hand side variables positive. At this moment, any of the three variables x_1, x_2, x_3 would do. In small examples, it is common practice to choose the variable that, in the formula for z, has the largest coefficient: the increase in that variable will make z increase at the fastest rate (but not necessarily to the highest level). In our case, this rule leaves us a choice between x_1 and x_2; choosing arbitrarily, we decide to make x_1 positive. As the value of x_1 increases, so does the value of x_5. However, the values of x_4, x_6, and x_7 decrease, and none of them is allowed to become negative. Of the three constraints $x_4 \geq 0$, $x_6 \geq 0$, $x_7 \geq 0$ that impose upper bounds on the increment of x_1, the last constraint $x_7 \geq 0$ is the most stringent: it implies $x_1 \leq 1$. In the improved feasible solution, we shall have $x_1 = 1$ and $x_7 = 0$. Without writing the new solution down, we shall now construct the new dictionary. All we need to know is that x_1 just made its way from the right-hand side to the left, whereas x_7 went in the opposite direction. From the fourth equation in (2.16), we have

$$x_1 = 1 - \frac{3}{2}x_2 + \frac{1}{2}x_3 - \frac{1}{2}x_7. \tag{2.17}$$

Substituting from (2.17) into the remaining equations of (2.16), we arrive at the desired dictionary

$$\begin{aligned}
x_1 &= 1 - \frac{3}{2}x_2 + \frac{1}{2}x_3 - \frac{1}{2}x_7 \\
x_4 &= 2 - \frac{3}{2}x_2 - \frac{3}{2}x_3 + \frac{1}{2}x_7 \\
x_5 &= 3 - \frac{3}{2}x_2 - \frac{5}{2}x_3 - \frac{1}{2}x_7 \\
x_6 &= 2 + 4x_2 - 3x_3 + x_7 \\
\hline
z &= 5 - \frac{5}{2}x_2 + \frac{11}{2}x_3 - \frac{5}{2}x_7.
\end{aligned} \tag{2.18}$$

The construction of (2.18) completes the first iteration of the simplex method.

Digression on Terminology

The variables x_j that appear on the left-hand side of a dictionary are called *basic*; the variables x_j that appear on the right-hand side are *nonbasic*. The basic variables are said to constitute a *basis*. Of course, the basis changes with each iteration: for example, in the first iteration, x_1 entered the basis whereas x_7 left it. In each iteration, we first choose the nonbasic variable that is to enter the basis and then we find out which basic variable must leave the basis. The choice of the *entering* variable is motivated by our desire to increase the value of z; the determination of the *leaving* variable is based on the requirement that all variables must assume nonnegative values. The leaving variable is that basic variable whose nonnegativity imposes the most stringent upper bound on the increment of the entering variable. The formula for the leaving variable appears in the *pivot row* of the dictionary; the computational process of constructing the new dictionary is referred to as *pivoting*.

Back to the Second Example

In our example, the variable to enter the basis during the second iteration is quite unequivocally x_3. This is the only nonbasic variable in (2.18) whose coefficient in the last row is positive. Of the four basic variables, x_6 imposes the most stringent upper bound on the increase of x_3, and, therefore, has to leave the basis. Pivoting, we arrive at our third dictionary,

$$\begin{aligned}
x_3 &= \frac{2}{3} + \frac{4}{3}x_2 + \frac{1}{3}x_7 - \frac{1}{3}x_6 \\
x_1 &= \frac{4}{3} - \frac{5}{6}x_2 - \frac{1}{3}x_7 - \frac{1}{6}x_6 \\
x_4 &= 1 - \frac{7}{2}x_2 \qquad\quad + \frac{1}{2}x_6 \\
x_5 &= \frac{4}{3} - \frac{29}{6}x_2 - \frac{4}{3}x_7 + \frac{5}{6}x_6 \\
\hline
z &= \frac{26}{3} + \frac{29}{6}x_2 - \frac{2}{3}x_7 - \frac{11}{6}x_6.
\end{aligned} \tag{2.19}$$

In the third iteration, the entering variable is x_2 and the leaving variable is x_5. Pivoting yields the dictionary

$$\begin{aligned}
x_2 &= \frac{8}{29} - \frac{8}{29}x_7 + \frac{5}{29}x_6 - \frac{6}{29}x_5 \\
x_3 &= \frac{30}{29} - \frac{1}{29}x_7 - \frac{3}{29}x_6 - \frac{8}{29}x_5 \\
x_1 &= \frac{32}{29} - \frac{3}{29}x_7 - \frac{9}{29}x_6 + \frac{5}{29}x_5 \\
x_4 &= \frac{1}{29} + \frac{28}{29}x_7 - \frac{3}{29}x_6 + \frac{21}{29}x_5 \\
\hline
z &= 10 - 2x_7 - x_6 - x_5.
\end{aligned} \tag{2.20}$$

At this point, no nonbasic variable can enter the basis without making the value of z decrease. Hence, the last dictionary describes an optimal solution of our example. That solution is

$$x_1 = \frac{32}{29}, \quad x_2 = \frac{8}{29}, \quad x_3 = \frac{30}{29}$$

and it yields $z = 10$.

FURTHER REMARKS

The reader may have noticed that, having first carefully laid down the definition of a dictionary, we then proceeded to refer to (2.18), (2.19), and (2.20) as dictionaries, without bothering to verify that they do indeed have property (2.14). Such carelessness can be easily justified. Take, for example, system (2.18). Since (2.18) arises from (2.16) by arithmetical operations (namely, pivoting with x_1 entering and x_7 leaving), every solution of (2.16) must be also a solution of (2.18). The converse is also true, since (2.16) can be obtained from (2.18) by pivoting with x_7 entering and x_1 leaving. Hence, *every solution of* (2.18) *is a solution of* (2.16), *and vice versa.* Similar arguments show that *every solution of* (2.19) *is a solution of* (2.18), *and vice versa*; and that *every solution of* (2.20) *is a solution of* (2.19), *and vice versa.*

Another point of concern is the question of the *uniqueness*, as opposed to the *existence*, of optimal solutions. This question will be of no great interest to us; nevertheless, it is easy to deal with and so we will get it out of the way now. Note that in each of our two examples, we not only found an optimal solution, but we also collected the evidence to prove that there is only one optimal solution. For instance, the final dictionary for our first problem reads

$$\begin{array}{l} x_3 = 1 + x_2 + 3x_4 - 2x_6 \\ x_1 = 2 - 2x_2 - 2x_4 + x_6 \\ x_5 = 1 + 5x_2 + 2x_4 \\ \hline z = 13 - 3x_2 - x_4 - x_6. \end{array}$$

The last row shows that every feasible solution with $z = 13$ satisfies $x_2 = x_4 = x_6 = 0$; the rest of the dictionary shows that every such solution satisfies $x_3 = 1$, $x_1 = 2$, $x_5 = 1$; therefore, there is just one optimal solution. A similar argument applies to the second problem.

Of course, there are LP problems with more than just one optimal solution; having solved such problems by the simplex method, we can effectively describe all the optimal solutions. For example, consider the following dictionary:

$$\begin{aligned}
x_4 &= 3 + x_2 - 2x_5 + 7x_3 \\
x_1 &= 1 - 5x_2 + 6x_5 - 8x_3 \\
x_6 &= 4 + 9x_2 + 2x_5 - x_3 \\
\hline
z &= 8 \qquad\qquad\qquad - x_3.
\end{aligned}$$

The last row shows that every optimal solution satisfies $x_3 = 0$ (but not necessarily $x_2 = 0$ or $x_5 = 0$). For such solutions, the rest of the dictionary implies

$$\begin{aligned}
x_4 &= 3 + x_2 - 2x_5 \\
x_1 &= 1 - 5x_2 + 6x_5 \\
x_6 &= 4 + 9x_2 + 2x_5.
\end{aligned} \tag{2.21}$$

We conclude that every optimal solution arises by the substitution formulas (2.21) from some x_2 and x_5 such that

$$\begin{aligned}
-x_2 + 2x_5 &\leq 3 \\
5x_2 - 6x_5 &\leq 1 \\
-9x_2 - 2x_5 &\leq 4 \\
x_2, x_5 &\geq 0.
\end{aligned}$$

(In fact, the inequality $-9x_2 - 2x_5 \leq 4$ is clearly redundant; its validity is forced by $x_2 \geq 0$ and $x_5 \geq 0$.)

There are a few other rough spots we deliberately failed to point out in our overview of the simplex method. We shall discuss them in Chapter 3.

TABLEAU FORMAT

The simplex method is often introduced in a format differing from ours. To outline the more popular *tableau format*, we shall return to the first example of this chapter. To begin, let us write down the equations of the first dictionary in a slightly modified form:

$$\begin{array}{rcl}
2x_1 + 3x_2 + x_3 + x_4 \qquad\qquad & = & 5 \\
4x_1 + x_2 + 2x_3 \qquad + x_5 \qquad & = & 11 \\
3x_1 + 4x_2 + 2x_3 \qquad\qquad + x_6 & = & 8 \\
\hline
-z + 5x_1 + 4x_2 + 3x_3 \qquad\qquad\qquad & = & 0.
\end{array}$$

Recording just the coefficients at the x_i's, together with the right-hand sides, we obtain our first *tableau*:

$$\begin{array}{cccccc|c}
2 & 3 & 1 & 1 & 0 & 0 & 5 \\
4 & 1 & 2 & 0 & 1 & 0 & 11 \\
3 & 4 & 2 & 0 & 0 & 1 & 8 \\
\hline
5 & 4 & 3 & 0 & 0 & 0 & 0.
\end{array}$$

In a similar way, the equations of the second dictionary,

$$\begin{array}{rcl}
x_1 + \frac{3}{2}x_2 + \frac{1}{2}x_3 + \frac{1}{2}x_4 & = & \frac{5}{2} \\
- 5x_2 \qquad - 2x_4 + x_5 & = & 1 \\
- \frac{1}{2}x_2 + \frac{1}{2}x_3 - \frac{3}{2}x_4 \qquad + x_6 & = & \frac{1}{2} \\
\hline
-z \qquad - \frac{7}{2}x_2 + \frac{1}{2}x_3 - \frac{5}{2}x_4 & = & -\frac{25}{2}
\end{array}$$

give rise to a second tableau:

$$\begin{array}{cccccccc}
1 & \frac{3}{2} & \frac{1}{2} & \frac{1}{2} & 0 & 0 & & \frac{5}{2} \\
0 & -5 & 0 & -2 & 1 & 0 & & 1 \\
0 & -\frac{1}{2} & \frac{1}{2} & -\frac{3}{2} & 0 & 1 & & \frac{1}{2} \\
\hline
0 & -\frac{7}{2} & \frac{1}{2} & -\frac{5}{2} & 0 & 0 & & -\frac{25}{2}.
\end{array}$$

It is a routine matter to translate the pivoting rules, previously derived in terms of dictionaries, into the language of tableaus. The following steps describe the procedure; the reader should have no trouble verifying its correctness. (At any rate, the procedure is not important for our exposition since we do not use the tableau format.)

Step 1. Examine all numbers in the last row (except the one farthest right, which equals the current value of $-z$). If all of them are negative or zero, stop: the tableau describes an optimal solution. Otherwise find the largest of these numbers; the column in which it appears is called the *pivot column* and corresponds to the entering variable.

For example, the pivot column in our first tableau is the first one:

$$\begin{array}{|c|cccccc}
2 & 3 & 1 & 1 & 0 & 0 & 5 \\
4 & 1 & 2 & 0 & 1 & 0 & 11 \\
3 & 4 & 2 & 0 & 0 & 1 & 8 \\
\hline
5 & 4 & 3 & 0 & 0 & 0 & 0
\end{array}$$

Step 2. For each row whose entry r in the pivot column is positive, look up the entry s in the rightmost column. The row with the smallest ratio $\frac{s}{r}$ is called the *pivot row* and corresponds to the leaving variable. (If all the entries in the pivot column are negative or zero, then the problem is unbounded; more on that in Chapter 3.)

In our example, the pivot row is the first row $\left(\text{with } \frac{s}{r} = \frac{5}{2}\right)$:

2	3	1	1	0	0	5
4	1	2	0	1	0	11
3	4	2	0	0	1	8
5	4	3	0	0	0	0.

Step 3. Divide every entry in the pivot row by the *pivot number*, found in the intersection of the pivot row with the pivot column:

1	$\frac{3}{2}$	$\frac{1}{2}$	$\frac{1}{2}$	0	0	$\frac{5}{2}$
4	1	2	0	1	0	11
3	4	2	0	0	1	8
5	4	3	0	0	0	0.

Step 4. From every remaining row, subtract a suitable multiple of the new pivot row. This operation is designed to make every entry in the pivot column (except for the pivot number) become zero; hence, the "suitable multiple" results when the new pivot row is multiplied by the entry appearing in the pivot column and in the row in question. (In our example, step 4 results in the second tableau.)

A tableau is nothing but a cryptic recording of a dictionary with all the variables collected on the left-hand side and the symbols for these variables omitted. We shall continue to use dictionaries instead, since they are more explicit. (Of course, nothing prevents the reader tired of writing the same symbols $x_1, x_2, \ldots$ over and over again from using the tableau shorthand.) □

A WARNING

There is often more than one way of describing a particular algorithm; descriptions aimed at clarifying underlying concepts are often quite different from those that suggest efficient computer implementations. The simplex method is no exception. Dictionaries may provide a convenient tool for explaining its basic principles. However, in implementing the method for computer solutions of large problems, considerations of computational efficiency and numerical accuracy overshadow such didactic niceties. We shall begin to study efficient implementations of the simplex method in Chapters 7 and 8.

PROBLEMS

△ **2.1** Solve the following problems by the simplex method:

a. maximize $3x_1 + 2x_2 + 4x_3$

subject to
$$\begin{aligned} x_1 + x_2 + 2x_3 &\le 4 \\ 2x_1 + 3x_3 &\le 5 \\ 2x_1 + x_2 + 3x_3 &\le 7 \\ x_1, x_2, x_3 &\ge 0 \end{aligned}$$

b. maximize $5x_1 + 6x_2 + 9x_3 + 8x_4$

subject to
$$\begin{aligned} x_1 + 2x_2 + 3x_3 + x_4 &\le 5 \\ x_1 + x_2 + 2x_3 + 3x_4 &\le 3 \\ x_1, x_2, x_3, x_4 &\ge 0 \end{aligned}$$

c. maximize $2x_1 + x_2$

subject to
$$\begin{aligned} 2x_1 + 3x_2 &\le 3 \\ x_1 + 5x_2 &\le 1 \\ 2x_1 + x_2 &\le 4 \\ 4x_1 + x_2 &\le 5 \\ x_1, x_2 &\ge 0. \end{aligned}$$

2.2 Use the simplex method to describe *all* the optimal solutions of the following problem:

maximize $2x_1 + 3x_2 + 5x_3 + 4x_4$

subject to
$$\begin{aligned} x_1 + 2x_2 + 3x_3 + x_4 &\le 5 \\ x_1 + x_2 + 2x_3 + 3x_4 &\le 3 \\ x_1, x_2, x_3, x_4 &\ge 0. \end{aligned}$$

3

Pitfalls and How to Avoid Them

The examples illustrating the simplex method in the preceding chapter were purposely smooth. They did not point out the dangers that can occur. The purpose of the present chapter, therefore, is to rigorously analyze the method by scrutinizing its every step.

THREE KINDS OF PITFALLS

Three kinds of pitfalls can occur in the simplex method.

(i) INITIALIZATION. We might not be able to start: How do we get hold of a feasible dictionary?

(ii) ITERATION. We might get stuck in some iteration: Can we always choose an entering variable, find the leaving variable, and construct the next feasible dictionary by pivoting?

(iii) TERMINATION. We might not be able to finish: Can the simplex method construct an endless sequence of dictionaries without ever reaching an optimal solution?

In the preceding chapter, INITIALIZATION never came up. Given a problem

$$\begin{aligned} \text{maximize} \quad & \sum_{j=1}^{n} c_j x_j \\ \text{subject to} \quad & \sum_{j=1}^{n} a_{ij} x_j \leq b_i \qquad (i = 1, 2, \ldots, m) \\ & x_j \geq 0 \qquad (j = 1, 2, \ldots, n) \end{aligned} \tag{3.1}$$

we constructed the initial feasible dictionary by simply writing down the formulas defining the slack variables and the objective function,

$$\begin{array}{lcl} x_{n+i} & = & b_i - \sum_{j=1}^{n} a_{ij} x_j \qquad (i = 1, 2, \ldots, m) \\ \hline z & = & \sum_{j=1}^{n} c_j x_j \, . \end{array}$$

In general, this dictionary is feasible if and only if each right-hand side, b_i, in (3.1) is nonnegative. This is the case if and only if

$$x_1 = 0, \quad x_2 = 0, \ldots, x_n = 0$$

is a feasible solution of (3.1). Since the set of zero values is sometimes called the "origin," problems (3.1) with each right-hand side b_i nonnegative are referred to as problems with a *feasible origin.* For the moment, we shall avoid the pitfalls of INITIALIZATION by default: we shall restrict ourselves to problems with a feasible origin. Problems with an infeasible origin are discussed on pages 39–42.

Iteration

Given some feasible dictionary, we have to select an entering variable, to find a leaving variable, and to construct the next feasible dictionary by pivoting.

Choosing an entering variable. The entering variable is *a nonbasic variable x_j with a positive coefficient $\bar{c}_j$ in the last row of the current dictionary.* This rule is ambiguous in the sense that it may provide more than one candidate for entering the basis, or no candidate at all. The latter alternative implies that the current dictionary describes an optimal solution, at which point the method may terminate. More precisely, consider the last row of our current dictionary,

$$z = z^* + \sum_{j \in N} \bar{c}_j x_j$$

with N standing for the set of subscripts j of nonbasic variables x_j. Our current solution, with $x_j = 0$ whenever $j \in N$, gives the objective function the numerical value of z^*. If $\bar{c}_j \leq 0$ whenever $j \in N$, then every feasible solution, with $x_j \geq 0$

whenever $j \in N$, gives the objective function a numerical value of at most z^*; hence the current solution is optimal. On the other hand, if there is more than one candidate for entering the basis, then any of these candidates may serve. (In hand calculations involving small problems, it is customary to choose the candidate x_j that has the largest coefficient $\bar{c}_j$. In most computer implementations of the simplex method, however, this practice is abandoned. More on this subject in Chapter 7.)

Finding the leaving variable. The leaving variable is *that basic variable whose nonnegativity imposes the most stringent upper bound on the increase of the entering variable.* Again, this rule is ambiguous in the sense that it may provide more than one candidate for leaving the basis, or no candidate at all. The latter alternative is illustrated on the dictionary

$$\begin{array}{rcl} x_2 &=& 5 + 2x_3 - x_4 - 3x_1 \\ x_5 &=& 7 \qquad\quad - 3x_4 - 4x_1 \\ \hline z &=& 5 + x_3 - x_4 - x_1. \end{array}$$

The entering variable is x_3, but neither of the two basic variables x_2, x_5 imposes an upper bound on its increase. Therefore, we can make x_3 as large as we wish (maintaining $x_1 = x_4 = 0$) and still retain feasibility: setting $x_3 = t$ for any positive t, we obtain a feasible solution with $x_1 = 0$, $x_2 = 5 + 2t$, $x_4 = 0$, $x_5 = 7$, and $z = 5 + t$. Since t can be made arbitrarily large, z can be made arbitrarily large. We conclude that the problem is *unbounded*: for every number M, there is a feasible solution $x_1, x_2, \ldots, x_5$ such that $x_3 - x_4 - x_1 > M$. The same conclusion can be reached in general: if there is no candidate for leaving the basis, then we can make the value of the entering variable, and therefore also the value of the objective function, as large as we wish. In that case, the problem is unbounded. On the other hand, if there is more than one candidate for leaving the basis, then any of these candidates may serve. Once the entering and leaving variables have been selected, pivoting is a straightforward matter.

Degeneracy. The presence of more than one candidate for leaving the basis has interesting consequences. For illustration, consider the dictionary

$$\begin{array}{rcl} x_4 &=& 1 \qquad\qquad\qquad\; - 2x_3 \\ x_5 &=& 3 - 2x_1 + 4x_2 - 6x_3 \\ x_6 &=& 2 + x_1 - 3x_2 - 4x_3 \\ \hline z &=& \qquad 2x_1 - x_2 + 8x_3. \end{array}$$

Having chosen x_3 to enter the basis, we find that each of the three basic variables x_4, x_5, x_6 limits the increase of x_3 to $\frac{1}{2}$. Hence each of these three variables is a candidate for leaving the basis. We arbitrarily choose x_4. Pivoting as usual, we obtain the dictionary

$$
\begin{array}{rcl}
x_3 &=& 0.5 \qquad\qquad\qquad\quad - 0.5x_4 \\
x_5 &=& \quad - 2x_1 + 4x_2 + 3x_4 \\
x_6 &=& \qquad\;\; x_1 - 3x_2 + 2x_4 \\
\hline
z &=& 4 + 2x_1 - x_2 - 4x_4.
\end{array}
$$

This dictionary differs from all the dictionaries we have encountered so far in one important respect: along with the nonbasic variables, the basic variables x_5 and x_6 have value zero in the associated solution. Basic solutions with one or more basic variables at zero are called *degenerate.*

Although harmless in its own right, degeneracy may have annoying side effects. These are illustrated on the next iteration in our example. There, x_1 enters the basis and x_5 leaves; because of degeneracy, the constraint $x_5 \geq 0$ limits the increment of x_1 to *zero.* Hence the value of x_1 will remain unchanged, and so will the values of the remaining variables and the value of the objective function z. This is annoying, for the motivation behind the simplex method is a desire to increase the value of z in each iteration. In this particular iteration, that desire remains unfulfilled: pivoting changes the dictionary into

$$
\begin{array}{rcl}
x_1 &=& \qquad\quad 2x_2 + 1.5x_4 - 0.5x_5 \\
x_3 &=& 0.5 \qquad\qquad - 0.5x_4 \\
x_6 &=& \quad - x_2 + 3.5x_4 - 0.5x_5 \\
\hline
z &=& 4 + 3x_2 - x_4 - x_5
\end{array}
$$

but it does not affect the associated solution at all. Simplex iterations that do not change the basic solution are called *degenerate.* (As the reader may verify, the next iteration is degenerate again, but the one after that turns out to be nondegenerate and brings us to the optimal solution.)

In a sense, degeneracy is something of an accident: a basic variable may vanish only if the results of successive pivot operations just happen to cancel each other out. And yet degeneracy abounds in LP problems arising from practical applications. It has been said that nearly all such problems yield degenerate basic feasible solutions at some stage of the simplex method. Whenever that happens, the simplex method may stall by going through a few (and sometimes quite a few) degenerate iterations in a row. Typically, such a block of degenerate iterations ends with a breakthrough represented by a nondegenerate iteration; an example of the atypical case is presented next.

Termination: Cycling

Can the simplex method go through an endless sequence of iterations without ever finding an optimal solution? Yes, it can. To justify this claim, let us consider the initial dictionary

$$
\begin{array}{rcl}
x_5 & = & \quad - 0.5x_1 + 5.5x_2 + 2.5x_3 - 9x_4 \\
x_6 & = & \quad - 0.5x_1 + 1.5x_2 + 0.5x_3 - x_4 \\
x_7 & = & 1 - x_1 \\
\hline
z & = & 10x_1 - 57x_2 - 9x_3 - 24x_4
\end{array}
$$

and let us agree on the following:

(i) The entering variable will always be the nonbasic variable that has the largest coefficient in the z-row of the dictionary.

(ii) If two or more basic variables compete for leaving the basis, then the candidate with the smallest subscript will be made to leave.

Now the sequence of dictionaries constructed in the first six iterations goes as follows.

After the first iteration:

$$
\begin{array}{rcl}
x_1 & = & \quad 11x_2 + 5x_3 - 18x_4 - 2x_5 \\
x_6 & = & \quad - 4x_2 - 2x_3 + 8x_4 + x_5 \\
x_7 & = & 1 - 11x_2 - 5x_3 + 18x_4 + 2x_5 \\
\hline
z & = & 53x_2 + 41x_3 - 204x_4 - 20x_5.
\end{array}
$$

After the second iteration:

$$
\begin{array}{rcl}
x_2 & = & \quad - 0.5x_3 + 2x_4 + 0.25x_5 - 0.25x_6 \\
x_1 & = & \quad - 0.5x_3 + 4x_4 + 0.75x_5 - 2.75x_6 \\
x_7 & = & 1 + 0.5x_3 - 4x_4 - 0.75x_5 - 13.25x_6 \\
\hline
z & = & 14.5x_3 - 98x_4 - 6.75x_5 - 13.25x_6.
\end{array}
$$

After the third iteration:

$$
\begin{array}{rcl}
x_3 & = & \quad 8x_4 + 1.5x_5 - 5.5x_6 - 2x_1 \\
x_2 & = & \quad - 2x_4 - 0.5x_5 + 2.5x_6 + x_1 \\
x_7 & = & 1 \qquad - x_1 \\
\hline
z & = & 18x_4 + 15x_5 - 93x_6 - 29x_1.
\end{array}
$$

After the fourth iteration:

$$
\begin{array}{rcl}
x_4 & = & \quad - 0.25x_5 + 1.25x_6 + 0.5x_1 - 0.5x_2 \\
x_3 & = & \quad - 0.5x_5 + 4.5x_6 + 2x_1 - 4x_2 \\
x_7 & = & 1 \qquad - x_1 \\
\hline
z & = & 10.5x_5 - 70.5x_6 - 20x_1 - 9x_2.
\end{array}
$$

After the fifth iteration:

$$
\begin{array}{rcl}
x_5 &=& 9x_6 + 4x_1 - 8x_2 - 2x_3 \\
x_4 &=& -\ x_6 - 0.5x_1 + 1.5x_2 + 0.5x_3 \\
x_7 &=& 1 - x_1 \\
\hline
z &=& 24x_6 + 22x_1 - 93x_2 - 21x_3.
\end{array}
$$

After the sixth iteration:

$$
\begin{array}{rcl}
x_6 &=& -\ 0.5x_1 + 1.5x_2 + 0.5x_3 - x_4 \\
x_5 &=& -\ 0.5x_1 + 5.5x_2 + 2.5x_3 - 9x_4 \\
x_7 &=& 1 - x_1 \\
\hline
z &=& 10x_1 - 57x_2 - 9x_3 - 24x_4.
\end{array}
$$

Since the dictionary constructed after the sixth iteration is identical with the initial dictionary, the method will go through the same six iterations again and again without ever finding the optimal solution (which, as we shall see later, has $z = 1$). This phenomenon is known as cycling. More precisely, we say that the simplex method *cycles* if one dictionary appears in two different iterations (and so the sequence of iterations leading from the dictionary to itself can be repeated over and over without end). Note that cycling can occur only in the presence of degeneracy: since the value of the objective function increases with each nondegenerate iteration and remains unchanged after each degenerate one, all the iterations in the sequence leading from a dictionary to itself must be degenerate. Cycling is one reason why the simplex method may fail to terminate; the following theorem shows that it is the only reason.

THEOREM 3.1. If the simplex method fails to terminate, then it must cycle.

PROOF. To begin, note that there are only finitely many ways of choosing m basic variables from all the $n + m$ variables. Thus, if the simplex method fails to terminate, then some basis must appear in two different iterations. Now it only remains to be proved that any two dictionaries with the same basis must be identical. (This fact becomes trivial as soon as one describes dictionaries in terms of matrices, as we shall do in Chapter 7. Nevertheless, we can and shall present an easy proof from scratch right now.) Consider two dictionaries

$$\begin{array}{ll} x_i = b_i - \sum_{j \notin B} a_{ij} x_j & (i \in B) \\ \hline z = v + \sum_{j \notin B} c_j x_j & \end{array} \tag{3.2}$$

and

$$\begin{array}{ll} x_i = b_i^* - \sum_{j \notin B} a_{ij}^* x_j & (i \in B) \\ \hline z = v^* + \sum_{j \notin B} c_j^* x_j & \end{array} \tag{3.3}$$

with the same set of basic variables x_i $(i \in B)$. It is a defining property of dictionaries that every solution $x_1, x_2, \ldots, x_{n+m}, z$ of (3.2) is a solution of (3.3) and vice versa. In particular, if x_k is a nonbasic variable and if t is a number, then the numbers

$$x_k = t, \quad x_j = 0 \, (j \notin B \quad \text{and} \quad j \neq k), \quad x_i = b_i - a_{ik} t \, (i \in B), \quad z = v + c_k t,$$

constituting a solution of (3.2), must satisfy (3.3). Hence,

$$b_i - a_{ik} t = b_i^* - a_{ik}^* t \quad \text{for all} \quad i \in B, \quad \text{and} \quad v + c_k t = v^* + c_k^* t.$$

Since these identities must hold for all numbers t, we have

$$b_i = b_i^*, a_{ik} = a_{ik}^* \quad \text{for all} \quad i \in B, \quad \text{and} \quad v = v^*, c_k = c_k^*.$$

Since x_k was an arbitrary nonbasic variable, the two dictionaries are identical.

Cycling is a rare phenomenon. In fact, constructing an LP problem on which the simplex method may cycle is difficult. [Our example is adapted from K. T. Marshall and J. W. Suurballe (1969). The first example of this size was constructed by E. M. L. Beale (1955) and the first example ever was constructed by A. J. Hoffman (1953). Incidentally, Marshall and Suurballe (1969) proved that if the simplex method cycles off-optimum on a problem that has an optimal solution, then the dictionaries must involve at least six variables and at least three equations.] P. Wolfe (1963) and T. C. T. Kotiah and D. I. Steinberg (1978) reported having come across practical problems that cycled (in 25 and 18 iterations, respectively) but such reports are scarce. For this reason, the remote possibility of cycling is disregarded in most computer implementations of the simplex method.

There are ways of preventing the occurrence of cycling altogether. The classic *perturbation method and lexicographic method* avoid cycling by a judicious choice of the leaving variable in each simplex iteration; the more recent *smallest-subscript rule* does so by an easy choice of *both* the entering and the leaving variables. The former alternative maintains the freedom of choice among different candidates for entering the basis, but it requires extra computations to choose the leaving variable; the latter alternative requires no extra work at all, but it gives up the multitude of choices for the entering variable. We shall explain the details of both.

The perturbation method and the lexicographic method. The perturbation and the lexicographic methods are closely related. The perturbation method, suggested first by A. Orden and developed independently by A. Charnes (1952), provides an intuitive motivation for the lexicographic method of G. B. Dantzig, A. Orden, and P. Wolfe (1955). The lexicographic method can be seen as an implementation of the perturbation method.

The starting point relies on the observations that cycling can be stamped out by stamping out degeneracy and that degeneracy itself is something of an accident. To elaborate on the second observation, consider a degenerate dictionary. The basic variables currently at zero would most likely assume small nonzero values if the initial right-hand sides, b_i, were changed slightly; at the same time, if these changes were truly microscopic, then the problem could be considered unchanged for all practical purposes. One way of exploiting these observations is to add a small positive ε to each b_i, and then to apply the simplex method to the resulting problem. This trick (with $\varepsilon = 10^{-6}$ or so) is actually used in some computer implementations of the simplex method; it helps to reduce the number of degenerate iterations. Nevertheless, it does not constitute a reliable safeguard against cycling: for instance, if the simplex method is applied to the problem

$$\begin{array}{lrcl}
\text{maximize} & 10x_1 - 57x_2 - 9x_3 - 24x_4 + 100x_5 & & \\
\text{subject to} & x_5 & \le & 1 + \varepsilon \\
& 0.5x_1 - 5.5x_2 - 2.5x_3 + 9x_4 + x_5 & \le & 1 + \varepsilon \\
& 0.5x_1 - 1.5x_2 - 0.5x_3 + x_4 + x_5 & \le & 1 + \varepsilon \\
& x_1 + x_5 & \le & 2 + \varepsilon \\
& x_1, x_2, \ldots, x_5 & \ge & 0
\end{array}$$

then the degenerate dictionary

$$\begin{array}{rcl}
x_5 & = & 1 + \varepsilon - x_6 \\
x_7 & = & - 0.5x_1 + 5.5x_2 + 2.5x_3 - 9x_4 + x_6 \\
x_8 & = & - 0.5x_1 + 1.5x_2 + 0.5x_3 - x_4 + x_6 \\
x_9 & = & 1 - x_1 + x_6 \\
\hline
z & = & 100 + 100\varepsilon + 10x_1 - 57x_2 - 9x_3 - 24x_4 - 100x_6
\end{array}$$

is obtained after the first iteration and, as the reader may verify, the simplex method cycles in the next six iterations. (The cycle is essentially the same as that of the preceding example.)

What went wrong here was that the small amounts ε added to the right-hand sides cancelled each other out in the first iteration. To guarantee that such cancellations will never take place (and therefore all the dictionaries will remain nondegenerate), we shall perturb the different right-hand sides $b_1, b_2, \ldots, b_m$ by radically different amounts $\varepsilon_1, \varepsilon_2, \ldots, \varepsilon_m$. More precisely, we shall choose a very small ε_1 and then make each ε_{i+1} much smaller than the preceding ε_i: in symbols,

$$0 < \varepsilon_m \ll \varepsilon_{m-1} \ll \cdots \ll \varepsilon_2 \ll \varepsilon_1 \ll 1. \tag{3.4}$$

Then we shall apply the simplex method to the perturbed problem

$$\text{maximize} \quad \sum_{j=1}^{n} c_j x_j$$

$$\text{subject to} \quad \sum_{j=1}^{n} a_{ij}x_j \le b_i + \varepsilon_i \qquad (i = 1, 2, \ldots, m)$$
$$x_j \ge 0 \qquad (j = 1, 2, \ldots, n).$$

This is the *perturbation method.* (The perturbation method is usually presented with $\varepsilon_1, \varepsilon_2, \ldots, \varepsilon_m$ equal to the powers $\varepsilon, \varepsilon^2, \ldots, \varepsilon^m$ of the same small ε. Our version makes the subsequent analysis a little more transparent.) For illustration, let us return to our first example on which the simplex method cycled. There, the initial dictionary reads

$$\begin{array}{rcl}
x_5 &=& \varepsilon_1 - 0.5x_1 + 5.5x_2 + 2.5x_3 - 9x_4 \\
x_6 &=& \varepsilon_2 - 0.5x_1 + 1.5x_2 + 0.5x_3 - x_4 \\
x_7 &=& 1 + \varepsilon_3 - x_1 \\
\hline
z &=& 10x_1 - 57x_2 - 9x_3 - 24x_4.
\end{array}$$

Again, the entering variable is x_1. The constraints $x_5 \ge 0$, $x_6 \ge 0$, and $x_7 \ge 0$ limit the increase of x_1 to $2\varepsilon_1$, $2\varepsilon_2$, and $1 + \varepsilon_3$, respectively. Since $2\varepsilon_2 < 2\varepsilon_1 < 1 + \varepsilon_3$, the leaving variable is x_6, and the next dictionary reads

$$\begin{array}{rcl}
x_1 &=& 2\varepsilon_2 + 3x_2 + x_3 - 2x_4 - 2x_6 \\
x_5 &=& \varepsilon_1 - \varepsilon_2 + 4x_2 + 2x_3 - 8x_4 + x_6 \\
x_7 &=& 1 - 2\varepsilon_2 + \varepsilon_3 - 3x_2 - x_3 + 2x_4 + 2x_6 \\
\hline
z &=& 20\varepsilon_2 - 27x_2 + x_3 - 44x_4 - 20x_6.
\end{array}$$

Now the only candidate for the entering variable is x_3 and the only candidate for the leaving variable is x_7. The resulting dictionary,

$$\begin{array}{rcl}
x_3 &=& 1 - 2\varepsilon_2 + \varepsilon_3 - 3x_2 + 2x_4 + 2x_6 - x_7 \\
x_1 &=& 1 + \varepsilon_3 - x_7 \\
x_5 &=& 2 + \varepsilon_1 - 5\varepsilon_2 + 2\varepsilon_3 - 2x_2 - 4x_4 + 5x_6 - 2x_7 \\
\hline
z &=& 1 + 18\varepsilon_2 + \varepsilon_3 - 30x_2 - 42x_4 - 18x_6 - x_7
\end{array}$$

is the optimal dictionary for the perturbed problem. It may be converted into the optimal dictionary for the original problem by simply disregarding all the terms involving $\varepsilon_1, \varepsilon_2, \varepsilon_3$.

How should we choose the numerical values of $\varepsilon_1, \varepsilon_2, \ldots, \varepsilon_m$? The simplest answer is that we do not have to do that at all: rather than committing ourselves to definite values of $\varepsilon_1, \varepsilon_2, \ldots, \varepsilon_m$, we may just think of these symbols as representing indefinite quantities, which satisfy (3.4). After several iterations of the simplex method, these symbols spread throughout the various rows of the dictionary, but they remain confined to the absolute terms in each of the $m + 1$ rows; the coefficients at the nonbasic variables in the dictionary are unaffected by the perturbation. Now when it comes to finding the leaving variable, each of the constraints $x_i \ge 0$ for a nonbasic x_i limits the increase of the entering x_j to a quantity such as $2\varepsilon_1$, $2\varepsilon_2$, $1 + \varepsilon_3$, or, more generally,

$$r = r_0 + r_1\varepsilon_1 + \cdots + r_m\varepsilon_m, \quad s = s_0 + s_1\varepsilon_1 + \cdots + s_m\varepsilon_m \tag{3.5}$$

and so on. As we are about to explain, assumption (3.4) allows us to compare the numerical values of such quantities without referring to the precise values of $\varepsilon_1, \varepsilon_2, \ldots, \varepsilon_m$. If r and s in (3.5) are distinct, then there is the smallest subscript k such that $r_k \ne s_k$. It is customary to say that r is *lexicographically smaller* than s if $r_k < s_k$. (The choice of the term *lexicographically* is explained by observing that, for instance, $2 + 21\varepsilon_1 + 19\varepsilon_2 + 20\varepsilon_2$ is lexicographically smaller than $2 + 21\varepsilon_1 + 20\varepsilon_2 + 20\varepsilon_3 + 15\varepsilon_4 + 14\varepsilon_5$ for the same reason that "bust" comes before "button" in a dictionary.) It is easy to prove that r is lexicographically smaller than s if and only if r is numerically smaller than s for all values of $\varepsilon_1, \varepsilon_2, \ldots, \varepsilon_m$ that satisfy (3.4). This statement has to be made precise by specifying just what is meant by the symbol $\ll$ in (3.4); we leave the details for problem 3.7.

The *lexicographic method* is that implementation of the perturbation method in which $\varepsilon_1, \varepsilon_2, \ldots, \varepsilon_m$ are treated as symbols, and quantities such as r and s in (3.5) are compared by the lexicographic rule. Note that it is always possible to choose the leaving variable by the lexicographic rule: in every finite set of expressions such as r and s in (3.5), there is always one that is lexicographically smaller than or equal to all the others. Even though this fact may be taken for granted intuitively, rigor requires that it be proved; we leave the details for problem 3.6. Another fine point concerns the behavior of the objective function z. The value of z, equal to some expression $v_0 + v_1\varepsilon_1 + \cdots + v_m\varepsilon_m$, remains unchanged in each degenerate iteration and increases, in the lexicographic sense, with each nondegenerate one. (In our example, the increase from 0 to $20\varepsilon_2$ in the first iteration was followed by the increase from $20\varepsilon_2$ to $1 + 18\varepsilon_2 + \varepsilon_3$ in the second iteration.) It is intuitively obvious that the total of two or more lexicographic increases is a lexicographic increase; a rigorous proof of this fact follows from the result of problem 3.5. Now it follows that, even in the generalized context of the lexicographic method, cycling is possible only in the presence of degeneracy. Finally, note that the only function of the terms involving $\varepsilon_1, \varepsilon_2, \ldots, \varepsilon_m$ is to guide us toward the appropriate choice of a leaving variable whenever two or more candidates present themselves in the original problem. If, at any moment, these terms are deleted, then the dictionary for the perturbed problem reduces to a dictionary for the original problem.

THEOREM 3.2. The simplex method terminates as long as the leaving variable is selected by the lexicographic rule in each iteration.

PROOF. In view of the preceding remarks, we need merely prove that no degenerate dictionary will be constructed. (If all dictionaries are nondegenerate, then all iterations are nondegenerate. In that case, cycling cannot occur and the desired conclusion follows from Theorem 3.1.) Thus, we need only consider an arbitrary row

$$x_k = (r_0 + r_1\varepsilon_1 + \cdots + r_m\varepsilon_m) - \sum_{j \notin B} d_j x_j \tag{3.6}$$

of an arbitrary dictionary and to prove that at least one of the $m + 1$ numbers $r_0, r_1, \ldots, r_m$ is distinct from zero. (Actually, we shall prove that at least one of the m numbers $r_1, r_2, \ldots, r_m$ is distinct from zero.) Writing $d_k = 1$ and $d_i = 0$ for all basic variables x_i distinct from x_k, we record (3.6) as

$$\sum_{j=1}^{n+m} d_j x_j = r_0 + \sum_{i=1}^{m} r_i \varepsilon_i. \tag{3.7}$$

Since this equation has been obtained by algebraic manipulations from the definitions of the slack variables,

$$x_{n+i} = b_i + \varepsilon_i - \sum_{j=1}^{n} a_{ij} x_j \qquad (i = 1, 2, \ldots, m) \tag{3.8}$$

it must hold for all choices of numbers $x_1, x_2, \ldots, x_{n+m}$ and $\varepsilon_1, \varepsilon_2, \ldots, \varepsilon_m$ that satisfy (3.8). Hence, the equation

$$\sum_{j=1}^{n} d_j x_j + \sum_{i=1}^{m} d_{n+i}\left(b_i + \varepsilon_i - \sum_{j=1}^{n} a_{ij}x_j\right) = r_0 + \sum_{i=1}^{m} r_i\varepsilon_i$$

which is obtained by substituting from (3.8) into (3.7), must hold for all choices of numbers x_1, $x_2, \ldots, x_n$ and $\varepsilon_1, \varepsilon_2, \ldots, \varepsilon_m$. Writing this identity as

$$\sum_{j=1}^{n} \left(d_j - \sum_{i=1}^{m} d_{n+i}a_{ij}\right)x_j + \sum_{i=1}^{m} (d_{n+i} - r_i)\varepsilon_i = r_0 - \sum_{i=1}^{m} d_{n+i}b_i$$

we observe that the coefficient at each x_j, the coefficient at each ε_i, and the right-hand side must equal zero. Thus

$$\begin{aligned} d_{n+i} &= r_i && \text{for all} \quad i = 1, 2, \ldots, m \\ d_j &= \sum_{i=1}^{m} d_{n+i}a_{ij} && \text{for all} \quad j = 1, 2, \ldots, n. \end{aligned} \tag{3.9}$$

If all the numbers $r_1, r_2, \ldots, r_m$ were equal to zero, then (3.9) would imply $d_{n+i} = 0$ for all $i = 1, 2, \ldots, m$ and $d_j = 0$ for all $j = 1, 2, \ldots, n$, contradicting the fact that $d_k = 1$. ■

With the hindsight provided by Theorem 3.2, it becomes easy to prove that every LP problem in the standard form can be perturbed by adding suitable small *numbers* $\varepsilon_1, \varepsilon_2, \ldots, \varepsilon_m$ to the right-hand sides $b_1, b_2, \ldots, b_m$ in such a way that the simplex method applied to the perturbed problem will terminate. In fact, the numbers $\varepsilon_1, \varepsilon_2, \ldots, \varepsilon_m$ may be chosen as the powers $\varepsilon, \varepsilon^2, \ldots, \varepsilon^m$ of any sufficiently small positive ε. We leave the details for problem 3.8.

As we have observed, the terms involving $\varepsilon_1, \varepsilon_2, \ldots, \varepsilon_m$ are needed only when a tie has to be broken between two or more candidates for leaving the basis. Thus we might just as well wait until such a need arises, and only then introduce an ad hoc perturbation. This idea was developed by P. Wolfe (1963); its lexicographic counterpart comes from G. B. Dantzig (1960).

Smallest-subscript rule. This term will refer to breaking ties in the choice of the entering and leaving variables by always choosing the candidate x_k that has the smallest subscript k. The motivation for this elegant concept is provided by the following result.

THEOREM 3.3. [R. G. Bland (1977).] The simplex method terminates as long as the entering and leaving variables are selected by the smallest-subscript rule in each iteration.

PROOF. By virtue of Theorem 3.1, we need only show that cycling is impossible when the smallest-subscript rule is used. We shall do this by deriving a contradiction from the assumption that the smallest-subscript rule leads from some dictionary D_0 to itself in a sequence of degenerate iterations. For definiteness, let us say that this sequence of iterations produces dictionaries D_1, $D_2, \ldots, D_k$ such that $D_k = D_0$. A variable will be called *fickle* if it is nonbasic in some of these dictionaries and basic in others. Among all the fickle variables, let x_t have the largest subscript. In the sequence $D_0, D_1, \ldots, D_k$, there is a dictionary D with x_t leaving (basic in D but nonbasic in the next dictionary), and some other fickle variable x_s entering (nonbasic in D but basic in the next dictionary). Further along in the sequence $D_0, D_1, \ldots, D_k, D_1, D_2, \ldots, D_k$, there must be a dictionary D^* with x_t entering. Let us record D as

$$\begin{array}{ll} x_i = b_i - \sum_{j \notin B} a_{ij} x_j & (i \in B) \\ \hline z = v + \sum_{j \notin B} c_j x_j \,. & \end{array}$$

Since all the iterations leading from D to D^* are degenerate, the objective function z must have the same value v in both dictionaries. Thus, the last row of D^* may be recorded as

$$z = v + \sum_{j=1}^{n+m} c_j^* x_j$$

with $c_j^* = 0$ whenever x_j is basic in D^*. Since this equation has been obtained from D by algebraic manipulations, it must be satisfied by every solution of D. In particular, it must be satisfied by $x_s = y$, $x_j = 0$ ($j \notin B$ but $j \neq s$), $x_i = b_i - a_{is}y$ ($i \in B$) and $z = v + c_s y$ for every choice of y. Thus we have

$$v + c_s y = v + c_s^* y + \sum_{i \in B} c_i^* (b_i - a_{is} y)$$

and, after simplification,

$$\left(c_s - c_s^* + \sum_{i \in B} c_i^* a_{is}\right) y = \sum_{i \in B} c_i^* b_i$$

for every choice of y. Since the right-hand side of the last equation is a constant independent of y, we conclude that

$$c_s - c_s^* + \sum_{i \in B} c_i^* a_{is} = 0. \tag{3.10}$$

The rest is easy. Since x_s is entering in D, we have $c_s > 0$. Since x_s is not entering in D^* and yet $s < t$, we have $c_s^* \leq 0$. Hence (3.10) implies that

$$c_r^* a_{rs} < 0 \qquad \text{for some} \quad r \in B. \tag{3.11}$$

Since $r \in B$, the variable x_r is basic in D; since $c_r^* \neq 0$, the same variable is nonbasic in D^*. Hence, x_r is fickle and we have $r \leq t$. Actually, x_r is different from x_t: since x_t is leaving in D, we have $a_{ts} > 0$ and so $c_t^* a_{ts} > 0$. Now $r < t$ and yet x_r is not entering in D^*. Thus, we cannot have $c_r^* > 0$. From (3.11), we conclude that

$$a_{rs} > 0.$$

Since all the iterations leading from D to D^* are degenerate, the two dictionaries describe the same solution. In particular, the value of x_r is zero in both dictionaries (x_r is nonbasic in D^*) and so $b_r = 0$. Hence x_r was a candidate for leaving the basis of D—yet we picked x_t, even though $r < t$. This contradiction completes the proof. ■

One further point: termination of the simplex method can be guaranteed even without abiding by the smallest-subscript rule in every single iteration. We might resort to the smallest-subscript rule, for instance, only when the last fifty or so iterations were degenerate, and abandon it after the next nondegenerate iteration in favor of any other way of choosing the entering and leaving variables. Although cycling might conceivably take place in this case, each block of consecutive degenerate iterations would be followed by a nondegenerate iteration, and so each dictionary would be constructed only a finite number of times. □

Initialization

The only remaining point that needs to be explained is getting hold of the initial feasible dictionary in a problem

$$\begin{aligned} \text{maximize} \quad & \sum_{j=1}^{n} c_j x_j \\ \text{subject to} \quad & \sum_{j=1}^{n} a_{ij} x_j \le b_i \qquad (i = 1, 2, \ldots, m) \\ & x_j \ge 0 \qquad (j = 1, 2, \ldots, n) \end{aligned}$$

with an infeasible origin. The trouble with an infeasible origin is twofold. First, it may not be clear that our problem has any feasible solutions at all. Second, even if a feasible solution is apparent, a feasible dictionary may not be. One way of getting around both obstacles uses a so-called *auxiliary problem*,

$$\begin{aligned} \text{minimize} \quad & x_0 \\ \text{subject to} \quad & \sum_{j=1}^{n} a_{ij} x_j - x_0 \le b_i \qquad (i = 1, 2, \ldots, m) \\ & x_j \ge 0 \qquad (j = 0, 1, \ldots, n). \end{aligned}$$

A feasible solution of the auxiliary problem is readily available: it suffices to set the value of each x_j with $1 \le j \le n$ at zero and make the value of x_0 sufficiently large. Furthermore, it is easy to see that the original problem has a feasible solution *if and only if* the auxiliary problem has a feasible solution with $x_0 = 0$. To put it differently, the original problem has a feasible solution if and only if the optimum value of the auxiliary problem is zero. Hence our plan is to solve the auxiliary problem first; the technical details are illustrated on the problem

$$\begin{aligned} \text{maximize} \quad & x_1 - x_2 + x_3 \\ \text{subject to} \quad & 2x_1 - x_2 + 2x_3 \le 4 \\ & 2x_1 - 3x_2 + x_3 \le -5 \\ & -x_1 + x_2 - 2x_3 \le -1 \\ & x_1, x_2, x_3 \ge 0. \end{aligned}$$

To avoid unnecessary confusion, we write the auxiliary problem in its maximization form:

$$\begin{aligned} \text{maximize} \quad & -x_0 \\ \text{subject to} \quad & 2x_1 - x_2 + 2x_3 - x_0 \le 4 \\ & 2x_1 - 3x_2 + x_3 - x_0 \le -5 \\ & -x_1 + x_2 - 2x_3 - x_0 \le -1 \\ & x_0, x_1, x_2, x_3 \ge 0. \end{aligned}$$

Writing down the formulas defining the slack variables x_4, x_5, x_6 and the objective function w, we obtain the dictionary

$$\begin{array}{rcl} x_4 &=& 4 - 2x_1 + x_2 - 2x_3 + x_0 \\ x_5 &=& -5 - 2x_1 + 3x_2 - x_3 + x_0 \\ x_6 &=& -1 + x_1 - x_2 + 2x_3 + x_0 \\ \hline w &=& - x_0 \end{array}$$

which is infeasible. Nevertheless, this infeasible dictionary can be transformed into a feasible one by a single pivot, with x_0 entering and x_5 leaving the basis:

$$\begin{array}{rcl} x_0 &=& 5 + 2x_1 - 3x_2 + x_3 + x_5 \\ x_4 &=& 9 - 2x_2 - x_3 + x_5 \\ x_6 &=& 4 + 3x_1 - 4x_2 + 3x_3 + x_5 \\ \hline w &=& -5 - 2x_1 + 3x_2 - x_3 - x_5. \end{array}$$

In general, the auxiliary problem may be written as

$$\begin{array}{lll} \text{maximize} & -x_0 & \\ \text{subject to} & \sum_{j=1}^{n} a_{ij}x_j - x_0 \leq b_i & (i = 1, 2, \ldots, m) \\ & x_j \geq 0 & (j = 0, 1, \ldots, n). \end{array}$$

Writing down the formulas defining the slack variables $x_{n+1}, x_{n+2}, \ldots, x_{n+m}$ and the objective function w gives us the dictionary

$$\begin{array}{rcll} x_{n+i} &=& b_i - \sum_{j=1}^{n} a_{ij}x_j + x_0 & (i = 1, 2, \ldots, m) \\ \hline w &=& - x_0 & \end{array}$$

which is infeasible. Nevertheless, this infeasible dictionary can be transformed into a feasible one by a single pivot, with x_0 entering and the "most infeasible" x_{n+i} leaving the basis. More precisely, the leaving variable is that x_{n+k} whose negative value, b_k, has the largest magnitude among all the negative numbers b_i. After pivoting, the variable x_0 assumes the positive value of $-b_k$, whereas each basic x_{n+i} assumes the nonnegative value of $b_i - b_k$. Now we are set to solve the auxiliary problem by the simplex method. In our illustrative example, the computations go as follows.

After the first iteration, with x_2 entering and x_6 leaving:

$$\begin{array}{rcl} x_2 &=& 1 + 0.75x_1 + 0.75x_3 + 0.25x_5 - 0.25x_6 \\ x_0 &=& 2 - 0.25x_1 - 1.25x_3 + 0.25x_5 + 0.75x_6 \\ x_4 &=& 7 - 1.5x_1 - 2.5x_3 + 0.5x_5 + 0.5x_6 \\ \hline w &=& -2 + 0.25x_1 + 1.25x_3 - 0.25x_5 - 0.75x_6. \end{array}$$

After the second iteration, with x_3 entering and x_0 leaving:

$$\begin{array}{rcl} x_3 &=& 1.6 - 0.2x_1 + 0.2x_5 + 0.6x_6 - 0.8x_0 \\ x_2 &=& 2.2 + 0.6x_1 + 0.4x_5 + 0.2x_6 - 0.6x_0 \\ x_4 &=& 3 - x_1 \qquad\qquad - x_6 + 2x_0 \\ \hline w &=& \qquad\qquad\qquad\qquad\qquad - x_0. \end{array} \tag{3.12}$$

The last dictionary (3.12) is optimal. Since the optimal value of the auxiliary problem is zero, dictionary (3.12) points out a feasible solution of the original problem: $x_1 = 0, x_2 = 2.2, x_3 = 1.6$. Furthermore, (3.12) can be easily converted into the desired feasible dictionary of the original problem. To obtain the first three rows of the desired dictionary, we simply copy down the first three rows of (3.12), omitting all the terms involving x_0:

$$\begin{array}{rcl} x_3 &=& 1.6 - 0.2x_1 + 0.2x_5 + 0.6x_6 \\ x_2 &=& 2.2 + 0.6x_1 + 0.4x_5 + 0.2x_6 \\ x_4 &=& 3 - x_1 \qquad\qquad - x_6. \end{array} \tag{3.13}$$

To obtain the last row, we have to express the original objective function

$$z = x_1 - x_2 + x_3 \tag{3.14}$$

in terms of the nonbasic variables x_1, x_5, x_6. For this purpose, we simply substitute from (3.13) into (3.14), obtaining

$$\begin{aligned} z &= x_1 - (2.2 + 0.6x_1 + 0.4x_5 + 0.2x_6) + (1.6 - 0.2x_1 + 0.2x_5 + 0.6x_6) \\ &= -0.6 + 0.2x_1 - 0.2x_5 + 0.4x_6. \end{aligned}$$

In short, the desired dictionary reads

$$\begin{array}{rcl} x_3 &=& 1.6 - 0.2x_1 + 0.2x_5 + 0.6x_6 \\ x_2 &=& 2.2 + 0.6x_1 + 0.4x_5 + 0.2x_6 \\ x_4 &=& 3 - x_1 \qquad\qquad - x_6 \\ \hline z &=& -0.6 + 0.2x_1 - 0.2x_5 + 0.4x_6. \end{array}$$

Clearly, the same procedure will transform an optimal dictionary of the auxiliary problem into a feasible dictionary of the original problem whenever x_0 is nonbasic in the former.

Now, let us review the general situation. We have learned how to construct the auxiliary problem and its first feasible dictionary. In the process of solving the auxiliary problem, we may encounter a dictionary where x_0 competes with other variables for leaving the basis. If and when that happens, it is only natural to choose x_0 as the actual leaving variable; immediately after pivoting, we obtain a dictionary where

$$x_0 \text{ is nonbasic, and so the value of } w \text{ is zero.} \tag{3.15}$$

Clearly, a feasible dictionary with this property is optimal. However, we may also reach the optimum of the auxiliary problem while x_0 is still basic. Thus, we may obtain an optimal dictionary where

x_0 is basic and the value of w is nonzero (3.16)

or, conceivably, an optimal dictionary where

x_0 is basic and the value of w is zero. (3.17)

Let us examine case (3.17). Since the next-to-last dictionary was not yet optimal, the value of $w = -x_0$ must have changed from some negative level to zero in the last iteration. To put it differently, the value of the basic variable x_0 must have dropped from some positive level to zero in the last iteration. But then x_0 was a candidate for leaving the basis; yet, contrary to our policy, we did not pick it. This contradiction shows that (3.17) cannot occur. Hence the optimal dictionary of the auxiliary problem has either property (3.15) or property (3.16). In the former case, we construct a feasible dictionary of the original problem as illustrated previously and proceed to solve the original problem by the simplex method; in the latter case, we simply conclude that the original problem is infeasible.

This strategy is known as the *two-phase simplex method.* In the *first phase*, we set up and solve the auxiliary problem; if the optimal dictionary turns out to have property (3.15) then we proceed to the *second phase*, solving the original problem itself. We shall return to the two-phase simplex method in Chapter 8.

THE FUNDAMENTAL THEOREM OF LINEAR PROGRAMMING

This name is given to the following result.

THEOREM 3.4. Every LP problem in the standard form has the following three properties:

(i) If it has no optimal solution, then it is either infeasible or unbounded.

(ii) If it has a feasible solution, then it has a basic feasible solution.

(iii) If it has an optimal solution, then it has a basic optimal solution.

PROOF. The first phase of the two-phase simplex method either discovers that the problem is infeasible or else it delivers a basic feasible solution. The second phase of the two-phase simplex method either discovers that the problem is unbounded or else it delivers a basic optimal solution. ■

Note that the first property is not shared by problems whose constraints may include *strict* linear inequalities $\sum a_j x_j < b$. To take a trivial example, the problem

maximize x subject to $x < 0$

is neither infeasible nor unbounded and yet it has no optimal solution. The remaining two properties (ii) and (iii) tell us that, when looking for feasible or optimal solutions of an LP problem in the standard form, we may confine our search to a *finite* set. These two properties, easy to establish from scratch, are often used to motivate the simplex method. Our exposition has followed the reverse pattern, with an emphasis placed on actually solving the problem—and the fundamental theorem of linear programming obtained as an effortless afterthought. □

PROBLEMS

△ **3.1** Maximize $x_1 + 3x_2 - x_3$

subject to

$$\begin{aligned} 2x_1 + 2x_2 - x_3 &\le 10 \\ 3x_1 - 2x_2 + x_3 &\le 10 \\ x_1 - 3x_2 + x_3 &\le 10 \\ x_1, x_2, x_3 &\ge 0. \end{aligned}$$

3.2 In the tableau format, a natural tie-breaking rule for the choice of the pivot row favors the rows that appear higher up in the tableau. Show that in the following example (constructed by H. W. Kuhn), this tie-breaking rule leads to cycling:

maximize $2x_1 + 3x_2 - x_3 - 12x_4$

subject to

$$\begin{aligned} -2x_1 - 9x_2 + x_3 + 9x_4 &\le 0 \\ \frac{1}{3}x_1 + x_2 - \frac{1}{3}x_3 - 2x_4 &\le 0 \\ x_1, x_2, x_3, x_4 &\ge 0. \end{aligned}$$

3.3 Solve problem 3.2 by the perturbation technique.

3.4 Arrange the following expressions in a sequence from lexicographically smallest to lexicographically largest:

$3 - \varepsilon_1$
3
$2 + 10\varepsilon_1$
$3 - 4\varepsilon_1 + \varepsilon_2$
$\varepsilon_2 + 3\varepsilon_3$
$3 + 4\varepsilon_1 + \varepsilon_3$
$3 - 4\varepsilon_1 + \varepsilon_2 + \varepsilon_3.$

3.5 Prove: If $r = r_0 + r_1\varepsilon_1 + \cdots + r_m\varepsilon_m$ is lexicographically smaller than $s = s_0 + s_1\varepsilon_1 + \cdots + s_m\varepsilon_m$ and if s is lexicographically smaller than $t = t_0 + t_1\varepsilon_1 + \cdots + t_m\varepsilon_m$, then r is lexicographically smaller than t.

3.6 Use the result of problem 3.5 to prove that, in every finite set of distinct expressions, such as r and s in (3.5), there is an expression that is lexicographically smaller than all the others.

3.7 Prove that for every pair of expressions in (3.5) there is a positive number δ such that the following two statements are equivalent: (i) r is lexicographically smaller than s; (ii) for every choice of numbers $\varepsilon_1, \varepsilon_2, \ldots, \varepsilon_m$ such that

$$0 < \varepsilon_1 < \delta \quad \text{and} \quad 0 < \varepsilon_i < \delta\varepsilon_{i-1} \qquad \text{for all } i = 2, 3, \ldots, m$$

r is numerically smaller than s.

3.8 Use Theorem 3.2 and the result of problem 3.7 to prove the following. For every LP problem

$$\begin{aligned} \text{maximize} \quad & \sum_{j=1}^{n} c_j x_j \\ \text{subject to} \quad & \sum_{j=1}^{n} a_{ij} x_j \le b_i \qquad (i = 1, 2, \ldots, m) \\ & x_j \ge 0 \qquad (j = 1, 2, \ldots, n) \end{aligned}$$

there is a positive number δ such that the simplex method used to

$$\begin{aligned} \text{maximize} \quad & \sum_{j=1}^{n} c_j x_j \\ \text{subject to} \quad & \sum_{j=1}^{n} a_{ij} x_j \le b_i + \varepsilon^i \qquad (i = 1, 2, \ldots, m) \\ & x_j \ge 0 \qquad (j = 1, 2, \ldots, n) \end{aligned}$$

terminates whenever $0 < \varepsilon < \delta$.

△ **3.9** Solve the following problems by the two-phase simplex method:

a. $$\begin{aligned} \text{maximize} \quad & 3x_1 + x_2 \\ \text{subject to} \quad & x_1 - x_2 \le -1 \\ & -x_1 - x_2 \le -3 \\ & 2x_1 + x_2 \le 4 \\ & x_1, x_2 \ge 0 \end{aligned}$$

b. $$\begin{aligned} \text{maximize} \quad & 3x_1 + x_2 \\ \text{subject to} \quad & x_1 - x_2 \le -1 \\ & -x_1 - x_2 \le -3 \\ & 2x_1 + x_2 \le 2 \\ & x_1, x_2 \ge 0 \end{aligned}$$

c. $$\begin{aligned} \text{maximize} \quad & 3x_1 + x_2 \\ \text{subject to} \quad & x_1 - x_2 \le -1 \\ & -x_1 - x_2 \le -3 \\ & 2x_1 - x_2 \le 2 \\ & x_1, x_2 \ge 0. \end{aligned}$$

3.10 Prove or disprove: A feasible dictionary whose last row reads $z = z^* + \sum \bar{c}_j x_j$ describes an optimal solution if and only if $\bar{c}_j \le 0$ for all j.

4

How Fast Is the Simplex Method?

The subject of this chapter is the number of iterations in the simplex method. We shall also comment on the distinction between theoretically satisfactory and practically satisfactory algorithms, with a particular regard to linear programming.

TYPICAL NUMBER OF ITERATIONS

For *practical* problems of the form

$$\begin{aligned} \text{maximize} \quad & \sum_{j=1}^{n} c_j x_j \\ \text{subject to} \quad & \sum_{j=1}^{n} a_{ij} x_j \leq b_i \qquad (i = 1, 2, \ldots, m) \\ & x_j \geq 0 \qquad (j = 1, 2, \ldots, n) \end{aligned} \tag{4.1}$$

with $m < 50$ and $m + n < 200$, Dantzig (1963, p. 160) reported the number of iterations as being usually less than $3m/2$ and only rarely going to $3m$. This observation agrees with empirical findings obtained more recently for much larger problems: the typical number of iterations increases proportionally to m (with the proportionality constant in the range suggested by Dantzig) and only very slowly with n. (It is sometimes said that, for a fixed m, the typical number of iterations is proportional to the logarithm of n.) Theoretical explanations of this phenomenon were proposed by G. B. Dantzig (1980), K.-H. Borgwardt (1982) and S. Smale (1982). It is this remarkable efficiency of the simplex method that accounts for its staggering success. At the current level of computer technology, typical problems with about 100 constraints and variables are solved in a few seconds; even problems with several thousands of constraints can be handled successfully. (To attain this level of efficiency, the simplex method has to be implemented properly, so that the time *per iteration* is reduced as much as possible. Consequently, the format of dictionaries has to be abandoned in favor of less time-consuming ways of organizing the necessary computations. We shall begin to study this matter in Chapter 7.) For problems with some particular structure amenable to specialized versions of the simplex method (such as the network simplex method of Chapter 19 or generalized upper bounding of Chapter 25), this limit can be pushed even further.

Monte Carlo simulation studies of the number of iterations were pioneered by H. W. Kuhn and R. E. Quandt (1963), who solved a number of problems (4.1) with $c_j = 1$ for all j, $b_i = 10{,}000$ for all i, and each a_{ij} selected at random from the set of positive integers between 1 and 1,000. A small part of these experiments has been reproduced, on a slightly larger scale, with the results exhibited in Table 4.1. (Each entry in the table represents the average number of iterations over 100 problems.) In each simplex iteration, the entering variable was that nonbasic variable that had the largest coefficient in the z-row of the dictionary. We shall refer to this selection rule as the *largest-coefficient* rule.

TABLE 4.1 Average Number of Iterations Required by the Largest-Coefficient Rule

m \ n	10	20	30	40	50
10	9.40	14.2	17.4	19.4	20.2
20		25.2	30.7	38.0	41.5
30			44.4	52.7	62.9
40				67.6	78.7
50					95.2

Source: D. Avis and V. Chvátal (1978).

The production management problems solved in practice are very much different from such randomly generated examples. Typically, most of their coefficients a_{ij} are zeros, the remaining nonzero coefficients occur in clusters that are very far from random, and the range of distinct numerical values of the coefficients is often very small. In spite of these differences, the Monte Carlo simulation results are in striking agreement with the empirical observations quoted above: for instance, if $n = 50$, then the average number of iterations is about $2m$.

PROBLEMS REQUIRING AN UNUSUALLY LARGE NUMBER OF ITERATIONS

From a purist point of view, it would be even more reassuring to have a proof that, for *every* problem (4.1), the simplex method would require no more than, say, $10mn$ iterations to find an optimal solution. However, there is no such proof. Worse than that, there are examples of LP problems that make the simplex method go through an enormous number of iterations. V. Klee and G. J. Minty (1972) have shown that in the process of solving the problem

$$\begin{aligned} \text{maximize} \quad & \sum_{j=1}^{n} 10^{n-j} x_j \\ \text{subject to} \quad & \left(2 \sum_{j=1}^{i-1} 10^{i-j} x_j\right) + x_i \leq 100^{i-1} \qquad (i = 1, 2, \ldots, n) \\ & x_j \geq 0 \qquad (j = 1, 2, \ldots, n) \end{aligned} \tag{4.2}$$

the simplex method goes through $2^n - 1$ iterations. (A proof is outlined in problems 4.2 and 4.3.) This number is quite frightening. For example, at the rate of 100 iterations per second (a reasonably generous estimate), problem (4.2) with $n = 50$ would take more than 300,000 years to solve! (The empirical and simulation results just quoted do *not* contradict this result. They simply suggest that problems requiring large numbers of iterations must be rare. For this reason, the Klee–Minty examples (4.2) and other similar examples are sometimes referred to as "pathological.")

As our starting point for further discussion, we choose the Klee–Minty problem with $n = 3$,

$$\begin{aligned} \text{maximize} \quad & 100x_1 + 10x_2 + x_3 \\ \text{subject to} \quad & x_1 \leq 1 \\ & 20x_1 + x_2 \leq 100 \\ & 200x_1 + 20x_2 + x_3 \leq 10{,}000 \\ & x_1, x_2, x_3 \geq 0. \end{aligned} \tag{4.3}$$

Using the largest-coefficient rule, we construct the following sequence of dictionaries. The initial dictionary:

$$\begin{array}{rcl} x_4 &=& 1 - x_1 \\ x_5 &=& 100 - 20x_1 - x_2 \\ x_6 &=& 10{,}000 - 200x_1 - 20x_2 - x_3 \\ \hline z &=& 100x_1 + 10x_2 + x_3. \end{array}$$

After the first iteration:

$$\begin{array}{rcl} x_1 &=& 1 - x_4 \\ x_5 &=& 80 + 20x_4 - x_2 \\ x_6 &=& 9{,}800 + 200x_4 - 20x_2 - x_3 \\ \hline z &=& 100 - 100x_4 + 10x_2 + x_3. \end{array}$$

After the second iteration:

$$\begin{array}{rcl} x_1 &=& 1 - x_4 \\ x_2 &=& 80 + 20x_4 - x_5 \\ x_6 &=& 8{,}200 - 200x_4 + 20x_5 - x_3 \\ \hline z &=& 900 + 100x_4 - 10x_5 + x_3. \end{array}$$

After the third iteration:

$$\begin{array}{rcl} x_4 &=& 1 - x_1 \\ x_2 &=& 100 - 20x_1 - x_5 \\ x_6 &=& 8{,}000 + 200x_1 + 20x_5 - x_3 \\ \hline z &=& 1{,}000 - 100x_1 - 10x_5 + x_3. \end{array}$$

After the fourth iteration:

$$\begin{array}{rcl} x_4 &=& 1 - x_1 \\ x_2 &=& 100 - 20x_1 - x_5 \\ x_3 &=& 8{,}000 + 200x_1 + 20x_5 - x_6 \\ \hline z &=& 9{,}000 + 100x_1 + 10x_5 - x_6. \end{array}$$

After the fifth iteration:

$$\begin{array}{rcl} x_1 &=& 1 - x_4 \\ x_2 &=& 80 + 20x_4 - x_5 \\ x_3 &=& 8{,}200 - 200x_4 + 20x_5 - x_6 \\ \hline z &=& 9{,}100 - 100x_4 + 10x_5 - x_6. \end{array}$$

After the sixth iteration:

$$\begin{aligned}
x_1 &= 1 - x_4 \\
x_5 &= 80 + 20x_4 - x_2 \\
x_3 &= 9{,}800 + 200x_4 - 20x_2 - x_6 \\
\hline
z &= 9{,}900 + 100x_4 - 10x_2 - x_6.
\end{aligned}$$

After the seventh iteration:

$$\begin{aligned}
x_4 &= 1 - x_1 \\
x_5 &= 100 - 20x_1 - x_2 \\
x_3 &= 10{,}000 - 200x_1 - 20x_2 - x_6 \\
\hline
z &= 10{,}000 - 100x_1 - 10x_2 - x_6.
\end{aligned}$$

In the first iteration, we were led to an unfortunate choice of the entering variable: had we made x_3 rather than x_1 enter the basis, we would have pivoted directly to the final dictionary. In view of this blunder, it is natural to question the expediency of the largest-coefficient rule: perhaps the simplex method would *always* go through only a small number of iterations if it were directed by some other rule. In fact, the largest-coefficient rule is not quite natural. More specifically, it ranks the potential candidates for entering the basis according to their coefficients in the last row of the dictionary: variables with larger coefficients appear to be more promising. But appearances are misleading and the ranking order is easily upset by changes in the scale on which each candidate is measured. For instance, the substitution

$$\bar{x}_1 = x_1, \quad \bar{x}_2 = 0.01x_2, \quad \bar{x}_3 = 0.0001x_3$$

converts the Klee–Minty problem (4.3) into the form

$$\begin{array}{lrcl}
\text{maximize} & 100\bar{x}_1 + 1{,}000\bar{x}_2 + 10{,}000\bar{x}_3 & & \\
\text{subject to} & \bar{x}_1 & \leq & 1 \\
& 20\bar{x}_1 + 100\bar{x}_2 & \leq & 100 \\
& 200\bar{x}_1 + 2{,}000\bar{x}_2 + 10{,}000\bar{x}_3 & \leq & 10{,}000 \\
& \bar{x}_1, \bar{x}_2, \bar{x}_3 & \geq & 0.
\end{array}$$

In the first dictionary associated with this new version of (4.3), the nonbasic variable $\bar{x}_3$ appears most attractive, and so the simplex method reaches the optimal solution in only one iteration.

ALTERNATIVE PIVOTING RULES

Thus we are led to ranking the candidates x_j for entering the basis according to criteria that are independent of changes of scale. One criterion of this kind is the increase in the objective function obtained when x_j actually enters the basis. The resulting rule (always choose that candidate whose entrance into the basis brings about the largest increase in the objective function) is referred to as the *largest-increase* rule. On the Klee–Minty examples (4.2), the largest-increase rule leads the simplex method to the optimal solution in only one iteration, as opposed to the $2^n - 1$ iterations required by the previously used largest-coefficient rule. However, the new rule does not always lead to a small number of iterations: R. G. Jeroslow (1973) constructed LP problems that are to the largest-increase rule what the Klee–Minty problems are to the largest-coefficient rule. (More precisely, the number of iterations required by the largest-increase rule grows exponentially with m and n.) Again, these examples exploit the myopia inherent in the simplex method. It is conceivable that every easily implemented rule for choosing the entering variable can be tricked in a similar way into requiring very large numbers of iterations.

Which of the two rules is better? On problems arising from applications, the number of iterations required by the largest increase is *usually* smaller than the number of iterations required by the largest coefficient. Simulation experiments lead to a similar outcome (see Table 4.2).

Table 4.2 Average Numbers of Iterations Required by the Largest-Increase Rule

m \ n	10	20	30	40	50
10	7.02	9.17	10.8	12.1	12.6
20		16.2	20.2	24.2	27.3
30			28.7	34.5	39.4
40				43.3	39.9
50					58.9

Source: D. Avis and V. Chvátal (1978).

Nevertheless, as the largest-coefficient rule takes less time to execute than the largest increase, it is the former that usually wins in terms of total computing time. More generally, the number of iterations is a poor criterion for assessing the efficiency of a rule for choosing the entering variable. It is the total computing time that counts, and rules that tend to reduce the number of iterations often take too much time to execute. In this light, even the largest-coefficient rule is found too time-consuming and therefore rarely, if ever, used in practice. The choice of entering variables in efficient implementations of the simplex method is influenced by the logistics of handling large problems on a computer; this matter will be studied in Chapter 7.

A systematic rule that always leads to an unambiguous choice of the entering variable, and to an unambiguous choice of the leaving variable in case of a tie, is called a *pivoting rule.* The largest-coefficient rule and the largest-increase rule, amended by unambiguous instructions for tie-breaking, are two examples of pivoting rules; the smallest-subscript rule of Chapter 3 is another.

EFFICIENCY OF ALGORITHMS IN THEORY AND PRACTICE

As noted in Chapter 1, the theoretical and the practical criteria for judging the efficiency of algorithms are radically different. From the theoretical point of view, an algorithm is satisfactory if its running time increases only slowly with the size of the problem. This is a vague definition as it stands; we are going to make it precise.

Let us consider a fixed class of problems (such as linear programming problems in the standard form) and a fixed algorithm (such as the simplex method) for solving problems in this class. A fair interpretation of the "size of the problem" is the time required to transmit the data. To put it differently, the size of a problem is the number of times you have to hit the keyboard of your typewriter in order to write down the data. For instance, the size of the Klee–Minty problems (4.2) is roughly $n^3/3$ when n gets very large (each of the $i - j + 1$ digits in each coefficient $2 \cdot 10^{i-j}$ has to be written down). A fair interpretation of "running time" is the total number of elementary steps (such as adding up, multiplying, or comparing two one-digit numbers; executing a "go to" instruction in a computer program; and so on) that have to be executed. (Thus it is implicitly assumed that each elementary step requires one unit of time.) Now for each s, there may be many (but only finitely many) different problems of size s in our class, and our algorithm may require different amounts of time $t_1, t_2, \ldots, t_M$ for different problems $P_1, P_2, \ldots, P_M$ of this size. Only the largest of these numbers t_i matters in the theoretical context. Of course, this largest t_i depends on s; we shall denote it by $t(s)$. Thus, our algorithm solves every problem of size s within $t(s)$ units of time and actually uses up these $t(s)$ units of time in the worst case. Finally, the algorithm is considered satisfactory if $t(s)$ grows only slowly with s. More precisely, the algorithm is satisfactory if there is a polynomial p such that $t(s) \leq p(s)$ for all s.

This definition, proposed by J. Edmonds (1965), is one of the most fruitful and stimulating concepts in theoretical computer science. [Those wishing for more information on this subject are referred to Garey and Johnson (1979).] Nevertheless, even though this concept does reflect to some extent the reasons why practitioners are satisfied by some algorithms and unsatisfied by others, it fails to capture these reasons fully. Two of the features that make it unrealistic from a practical point of view are:

(i) The worst-case criterion.
(ii) The asymptotic point of view.

The inadequacy of the worst-case criterion is demonstrated most dramatically on the case of the simplex method itself: even eminently useful algorithms may be labeled unsatisfactory on the basis of a few isolated examples of a kind that might never come up in practice. The average running time ($\sum t_i/M$) might provide a more realistic criterion than the worst running time ($\max t_i$); unfortunately, a rigorous analysis of the average performance is often much more difficult than an analysis of the worst performance. The inadequacy of the second feature may manifest itself even when the running time depends only on the size of the problem, so that the average performance and the worst performance coincide. The point is that the actual values $t(s)$, with s restricted to a finite range, do not matter at all; the only thing that counts is the rate of growth of $t(s)$ as s increases beyond every bound. Thus, a hypothetical algorithm with a running time $t = 10^{s/1{,}000{,}000{,}000}$ (rounded up to the nearest integer) would be found theoretically unsatisfactory even though $t(s) \leq 10$ whenever $s \leq 10^9$; on the other hand, an algorithm with a running time $t(s) = 10^{1{,}000}s$ would be found theoretically satisfactory even though $t(s) \geq 10^{1{,}000}$ for all s. Theorists judge algorithms by their worst performance on problems of sizes outside the range of practical interest, whereas practitioners judge algorithms by their typical performance on problems whose sizes are limited to a finite range. (In all fairness, it should be admitted that the theoretical definition is not all that bad. As it turns out, the polynomials bounding the running time of theoretically satisfactory algorithms often assume reasonably small values for reasonably small values of s and, on the other hand, even the typical running time of theoretically unsatisfactory algorithms will often get out of hand already for small values of s.)

For many years, while practitioners were trying to reduce the typical running time of the simplex method by yet another 10% or 20%, theorists were trying to answer a fundamental question: Is there a theoretically satisfactory algorithm for solving linear programming problems? Eventually, L. G. Khachian (1979) provided an affirmative answer by presenting such an algorithm. This "ellipsoid method" is surprisingly simple and elegant; we shall describe its details in the appendix. Will this beautiful gem of pure mathematics ever become a serious challenger of the simplex method's supremacy in solving practical LP problems? That remains to be seen; at the time of this writing, it seems very likely that the answer is no. ☐

PROBLEMS

△ **4.1** Compare the performance of the three pivoting rules discussed in this chapter on the following examples:

a. maximize $4x_1 + 5x_2$

subject to
$$\begin{aligned} 2x_1 + x_2 &\leq 9 \\ x_1 &\leq 4 \\ x_2 &\leq 3 \\ x_1, x_2 &\geq 0 \end{aligned}$$

b. maximize $2x_1 + x_2$

subject to
$$\begin{aligned} 3x_1 + x_2 &\leq 3 \\ x_1, x_2 &\geq 0 \end{aligned}$$

c. maximize $3x_1 + 5x_2$

subject to
$$\begin{aligned} x_1 + 2x_2 &\leq 5 \\ x_1 &\leq 3 \\ x_2 &\leq 2 \\ x_1, x_2 &\geq 0. \end{aligned}$$

4.2 In the Klee–Minty problem (4.2), denote the slack variables by $s_1, s_2, \ldots, s_n$ rather than by $x_{n+1}, x_{n+2}, \ldots, x_{2n}$. Prove that in every feasible dictionary, precisely one of the two variables x_i, s_i is basic.

4.3 Use the result of problem 4.2 and induction on n to prove that, when the simplex method with the largest coefficient rule is applied to (4.2), the resulting dictionaries have the following properties:

(i) After $2^{n-1} - 1$ iterations, the last row reads

$$z = 10\left(100^{n-2} - \sum_{j=1}^{n-2} 10^{n-1-j}x_j - s_{n-1}\right) + x_n.$$

(ii) After 2^{n-1} iterations, the last row reads

$$z = 90 \cdot 100^{n-2} + 10\left(\sum_{j=1}^{n-2} 10^{n-1-j}x_j + s_{n-1}\right) - s_n.$$

(iii) After $2^n - 1$ iterations, the last row reads

$$z = 100^{n-1} - \sum_{j=1}^{n-1} 10^{n-j}x_j - s_n.$$

(iv) After each iteration, all the coefficients in the last row are integers.

5

The Duality Theorem

Every maximization LP problem in the standard form gives rise to a minimization LP problem called the dual problem. The two problems are linked in an interesting way. Every feasible solution in one yields a bound on the optimal value of the other. In fact, if one of the two problems has an optimal solution, then so does the other, and the two optimal values coincide. This fact, known as the Duality Theorem, is the subject of the present chapter. We shall also note that, in managerial applications, the variables featured in the dual problem can be interpreted in a very useful way.

MOTIVATION: FINDING UPPER BOUNDS ON THE OPTIMAL VALUE

We shall begin this chapter with the following LP problem:

$$\begin{array}{lrcl}
\text{maximize} & 4x_1 + x_2 + 5x_3 + 3x_4 & & \\
\text{subject to} & x_1 - x_2 - x_3 + 3x_4 & \leq & 1 \\
 & 5x_1 + x_2 + 3x_3 + 8x_4 & \leq & 55 \\
 & -x_1 + 2x_2 + 3x_3 - 5x_4 & \leq & 3 \\
 & x_1, x_2, x_3, x_4 & \geq & 0.
\end{array}$$

Rather than *solving* it, we shall try to get a quick *estimate* of the optimal value z^* of its objective function. To get a reasonably good lower bound on z^*, we need only come up with a reasonably good feasible solution. For example, the bound $z^* \geq 5$ comes from considering the feasible solution (0, 0, 1, 0). The feasible solution $(2, 1, 1, \frac{1}{3})$ shows that $z^* \geq 15$. Better yet, the feasible solution (3, 0, 2, 0) yields $z^* \geq 22$. Needless to say, such guesswork is vastly inferior to the systematic attack by the simplex method: even if we were lucky enough to hit on the optimal solution, our guess would provide no *proof* that the solution is indeed optimal.

We shall not pursue this line any further: the subject of this chapter stems from a similar quest for *upper* bounds on z^*. For example, a glance at the data suggests that $z^* \leq \frac{275}{3}$. Indeed, multiplying the second constraint by $\frac{5}{3}$ we obtain the inequality

$$\frac{25}{3}x_1 + \frac{5}{3}x_2 + 5x_3 + \frac{40}{3}x_4 \leq \frac{275}{3}.$$

Hence every feasible solution (x_1, x_2, x_3, x_4) satisfies the inequality

$$4x_1 + x_2 + 5x_3 + 3x_4 \leq \frac{25}{3}x_1 + \frac{5}{3}x_2 + 5x_3 + \frac{40}{3}x_4 \leq \frac{275}{3}.$$

In particular, this inequality holds for the optimal solution and so $z^* \leq \frac{275}{3}$. With a little inspiration, we can improve this bound considerably. For instance, the sum of the second and third constraints reads

$$4x_1 + 3x_2 + 6x_3 + 3x_4 \leq 58.$$

Therefore, $z^* \leq 58$. Rather than searching for further improvements in a haphazard way, we shall now describe the strategy in precise and general terms.

We construct *linear combinations* of the constraints. That is, we multiply the first constraint by some number y_1, the second by y_2, the third by y_3, and then we add them up. (In the first case, we had $y_1 = 0$, $y_2 = \frac{5}{3}$, $y_3 = 0$; in the second case, we had $y_1 = 0$, $y_2 = y_3 = 1$.) The resulting inequality reads

$$\begin{aligned}(y_1 + 5y_2 - y_3)x_1 + (-y_1 + y_2 + 2y_3)x_2 + (-y_1 + 3y_2 + 3y_3)x_3 + (3y_1 + 8y_2 - 5y_3)x_4 \\ \leq y_1 + 55y_2 + 3y_3. \qquad (5.1)\end{aligned}$$

Of course, each of the three multipliers y_i must be nonnegative: otherwise the corresponding inequality would reverse its direction. Next, we want to use the left-hand side of (5.1) as an upper bound on $z = 4x_1 + x_2 + 5x_3 + 3x_4$. This can be justified only if in (5.1), the coefficient at each x_j is at least as big as the corresponding coefficient in z. More explicitly, we want

$$\begin{aligned} y_1 + 5y_2 - y_3 &\geq 4 \\ -y_1 + y_2 + 2y_3 &\geq 1 \\ -y_1 + 3y_2 + 3y_3 &\geq 5 \\ 3y_1 + 8y_2 - 5y_3 &\geq 3. \end{aligned}$$

If the multipliers y_i are nonnegative and if they satisfy these four inequalities, then we may safely conclude that every feasible solution (x_1, x_2, x_3, x_4) satisfies the inequality

$$4x_1 + x_2 + 5x_3 + 3x_4 \leq y_1 + 55y_2 + 3y_3.$$

In particular, this inequality is satisfied by the optimal solution; therefore

$$z^* \leq y_1 + 55y_2 + 3y_3.$$

Of course, we want as small an upper bound on z^* as we can possibly get. Thus, we are led to the following LP problem:

$$\begin{array}{ll} \text{minimize} & y_1 + 55y_2 + 3y_3 \\ \text{subject to} & y_1 + 5y_2 - y_3 \geq 4 \\ & -y_1 + y_2 + 2y_3 \geq 1 \\ & -y_1 + 3y_2 + 3y_3 \geq 5 \\ & 3y_1 + 8y_2 - 5y_3 \geq 3 \\ & y_1, y_2, y_3 \geq 0. \end{array}$$

THE DUAL PROBLEM

This problem is called the *dual* of the original one; the original problem is called the *primal* problem. In general, the dual of the problem

$$\begin{array}{lll} \text{maximize} & \sum_{j=1}^{n} c_j x_j & \\ \text{subject to} & \sum_{j=1}^{n} a_{ij} x_j \leq b_i & (i = 1, 2, \ldots, m) \\ & x_j \geq 0 & (j = 1, 2, \ldots, n) \end{array} \tag{5.2}$$

is defined to be the problem

$$\begin{array}{lll} \text{minimize} & \sum_{i=1}^{m} b_i y_i & \\ \text{subject to} & \sum_{i=1}^{m} a_{ij} y_i \geq c_j & (j = 1, 2, \ldots, n) \\ & y_i \geq 0 & (i = 1, 2, \ldots, m). \end{array} \tag{5.3}$$

(Note that the dual of a maximization problem is a minimization problem. Furthermore, the m primal constraints $\sum a_{ij}x_j \leq b_i$ are in a one-to-one correspondence with the m dual variables y_i; conversely, the n dual constraints $\sum a_{ij}y_i \geq c_j$ are in a one-to-one correspondence with the n primal variables x_j. The coefficient at each variable in the objective function, primal or dual, appears in the other problem as the right-hand side of the corresponding constraint.)

As in our example, every feasible solution of the dual yields an upper bound on the optimal value of the primal. More explicitly, for every primal feasible solution $(x_1, x_2, \ldots, x_n)$ and for every dual feasible solution $(y_1, y_2, \ldots, y_m)$ we have

$$\sum_{j=1}^{n} c_j x_j \le \sum_{i=1}^{m} b_i y_i. \tag{5.4}$$

The proof of (5.4), which was illustrated at the beginning of this section, can be written down succinctly as

$$\sum_{j=1}^{n} c_j x_j \le \sum_{j=1}^{n} \left(\sum_{i=1}^{m} a_{ij} y_i \right) x_j = \sum_{i=1}^{m} \left(\sum_{j=1}^{n} a_{ij} x_j \right) y_i \le \sum_{i=1}^{m} b_i y_i.$$

Inequality (5.4) is extremely useful: if we happen to stumble across a primal feasible solution $(x_1^*, x_2^*, \ldots, x_n^*)$ and a dual feasible solution $(y_1^*, y_2^*, \ldots, y_m^*)$ such that

$$\sum_{j=1}^{n} c_j x_j^* = \sum_{i=1}^{m} b_i y_i^*$$

then we may conclude that both of these solutions are optimal. Indeed, (5.4) implies that every primal feasible solution $(x_1, x_2, \ldots, x_n)$ satisfies

$$\sum_{j=1}^{n} c_j x_j \le \sum_{i=1}^{m} b_i y_i^* = \sum_{j=1}^{n} c_j x_j^*$$

and that every dual feasible solution $(y_1, y_2, \ldots, y_m)$ satisfies

$$\sum_{i=1}^{m} b_i y_i \ge \sum_{j=1}^{n} c_j x_j^* = \sum_{i=1}^{m} b_i y_i^*.$$

For instance, we have an easy way of showing that the primal feasible solution $x_1 = 0$, $x_2 = 14$, $x_3 = 0$, $x_4 = 5$ of our original example is optimal: just consider the dual feasible solution $y_1 = 11$, $y_2 = 0$, $y_3 = 6$. It is not at all obvious, however, that an analogous proof of optimality can be given for *every* LP problem that has an optimal solution; this fact is the central theorem of linear programming.

THE DUALITY THEOREM AND ITS PROOF

The explicit version of the theorem comes from D. Gale, H. W. Kuhn, and A. W. Tucker (1951); its notions originated in conversations between G. B. Dantzig and J. von Neumann in the fall of 1947.

THEOREM 5.1 (The Duality Theorem). If the primal (5.2) has an optimal solution $(x_1^*, x_2^*, \ldots, x_n^*)$, then the dual (5.3) has an optimal solution $(y_1^*, y_2^*, \ldots, y_m^*)$ such that

$$\sum_{j=1}^{n} c_j x_j^* = \sum_{i=1}^{m} b_i y_i^*. \tag{5.5}$$

Before presenting the proof, let us briefly illustrate its crucial point: the optimal solution of the *dual* problem can be read off the z-row of the final dictionary for the *primal* problem. In the example that we used to motivate the concept of the dual problem, the final dictionary reads

$$\begin{array}{rcl} x_2 & = & 14 - 2x_1 - 4x_3 - 5x_5 - 3x_7 \\ x_4 & = & 5 - x_1 - x_3 - 2x_5 - x_7 \\ x_6 & = & 1 + 5x_1 + 9x_3 + 21x_5 + 11x_7 \\ \hline z & = & 29 - x_1 - 2x_3 - 11x_5 - 6x_7. \end{array}$$

Note that the slack variables x_5, x_6, x_7 can be matched up with the dual variables y_1, y_2, y_3 in a natural way: for instance, x_5 is the slack variable in the first constraint, whereas y_1 represents the multiplier for the same constraint. By the same logic, x_6 goes with y_2 and x_7 goes with y_3. In the *z-row* of the dictionary, the coefficients at the slack variables are

$$-11 \text{ at } x_5, \quad 0 \text{ at } x_6, \quad -6 \text{ at } x_7.$$

Assigning these values with reversed signs to the corresponding dual variables, we obtain the desired optimal solution of the dual:

$$y_1 = 11, \quad y_2 = 0, \quad y_3 = 6.$$

At first, this may seem like pulling a rabbit out of a hat; however, the following general argument explains the magic.

PROOF OF THEOREM 5.1. We need only find a *feasible* solution $(y_1^*, y_2^*, \ldots, y_m^*)$ satisfying (5.5); indeed, such a solution will be *optimal* by virtue of the remarks following (5.4). In order to find that solution, we solve the primal problem by the simplex method; having introduced the slack variables

$$x_{n+i} = b_i - \sum_{j=1}^{n} a_{ij} x_j \qquad (i = 1, 2, \ldots, m) \tag{5.6}$$

we eventually arrive at the final dictionary. For the sake of definiteness, let us say that the last row of that dictionary reads

$$z = z^* + \sum_{k=1}^{n+m} \bar{c}_k x_k. \tag{5.7}$$

In (5.7), each $\bar{c}_k$ is a nonpositive number (in fact, $\bar{c}_k = 0$ whenever x_k is a basic variable). In addition, z^* is the optimal value of the objective function, and so

$$z^* = \sum_{j=1}^{n} c_j x_j^*. \tag{5.8}$$

Defining

$$y_i^* = -\bar{c}_{n+i} \qquad (i = 1, 2, \ldots, m) \tag{5.9}$$

we claim that $(y_1^*, y_2^*, \ldots, y_m^*)$ is a dual feasible solution satisfying (5.5); the rest of the proof consists of a straightforward verification of our claim. Substituting $\sum c_j x_j$ for z and substituting from (5.6) for the slack variables in (5.7) we obtain the identity

$$\sum_{j=1}^{n} c_j x_j = z^* + \sum_{j=1}^{n} \bar{c}_j x_j - \sum_{i=1}^{m} y_i^* \left(b_i - \sum_{j=1}^{n} a_{ij} x_j \right)$$

which may be written as

$$\sum_{j=1}^{n} c_j x_j = \left(z^* - \sum_{i=1}^{m} b_i y_i^* \right) + \sum_{j=1}^{n} \left(\bar{c}_j + \sum_{i=1}^{m} a_{ij} y_i^* \right) x_j.$$

This identity, having been obtained by algebraic manipulations from the definitions of the slack variables and the objective function, must hold for every choice of values of $x_1, x_2, \ldots, x_n$. Hence we have

$$z^* = \sum_{i=1}^{m} b_i y_i^* \tag{5.10}$$

and

$$c_j = \bar{c}_j + \sum_{i=1}^{m} a_{ij} y_i^* \qquad (j = 1, 2, \ldots, n). \tag{5.11}$$

Since $\bar{c}_k \leq 0$ for every $k = 1, 2, \ldots, n + m$, (5.11) and (5.9) imply

$$\sum_{i=1}^{m} a_{ij} y_i^* \geq c_j \qquad (j = 1, 2, \ldots, n)$$

$$y_i^* \geq 0 \qquad (i = 1, 2, \ldots, m).$$

Finally, (5.10) and (5.8) imply (5.5). ∎

RELATIONSHIP BETWEEN THE PRIMAL AND DUAL PROBLEMS

Next, let us point out that the dual of the dual is always the primal problem. Indeed, the dual problem may be written as

$$\begin{array}{lll} \text{maximize} & \sum_{i=1}^{m} (-b_i)y_i & \\ \text{subject to} & \sum_{i=1}^{m} (-a_{ij})y_i \leq -c_j & (j = 1, 2, \ldots, n) \\ & y_i \geq 0 & (i = 1, 2, \ldots, m). \end{array}$$

The dual of this problem is

$$\begin{array}{lll} \text{minimize} & \sum_{j=1}^{n} (-c_j)x_j & \\ \text{subject to} & \sum_{j=1}^{n} (-a_{ij})x_j \geq -b_i & (i = 1, 2, \ldots, m) \\ & x_j \geq 0 & (j = 1, 2, \ldots, n) \end{array}$$

which is clearly equivalent to the original problem. A nice corollary to this observation and to the duality theorem is that the primal problem has an optimal solution *if and only if* the dual problem has an optimal solution. Note also that if the primal is unbounded, then the dual must be infeasible [this follows directly from (5.4)]. By the same argument, if the dual is unbounded then the primal must be infeasible. However, both primal and dual may be infeasible at the same time. For example, both the problem

$$\begin{array}{ll} \text{maximize} & 2x_1 - x_2 \\ \text{subject to} & x_1 - x_2 \leq 1 \\ & -x_1 + x_2 \leq -2 \\ & x_1, x_2 \geq 0 \end{array}$$

and its dual are infeasible. These conclusions are summarized in Table 5.1.

Table 5.1 Primal–Dual Combinations

		Dual		
		Optimal	Infeasible	Unbounded
Primal	Optimal	Possible	Impossible	Impossible
	Infeasible	Impossible	Possible	Possible
	Unbounded	Impossible	Possible	Impossible

In particular, if the primal problem has a feasible solution *and* if the dual problem has a feasible solution, then both problems have optimal solutions.

Duality has important practical implications. In certain cases, we may find it advantageous to apply the simplex method to the dual of the problem that we are really interested in. (Of course, the optimal solution of the primal problem can then be read directly off the final dictionary for the dual.) For example, if $m = 99$ and $n = 9$, then dictionaries will have 100 rows in the primal problem but only 10 rows in the dual. Since the typical number of simplex iterations is proportional to the number of rows in a dictionary and relatively insensitive to the number of variables, we shall most likely be better off solving the dual problem.

From a theoretical point of view, duality is important because it points out an elegant and succinct way of proving optimality of solutions of LP problems: as we have observed, an optimal solution of the dual problem provides a "certificate of optimality" for an optimal solution of the primal problem, and vice versa. Furthermore, the duality theorem asserts that for *every* optimal solution there is a certificate of optimality. To appreciate the impact of this fact, consider a student who is supposed to solve the problem

$$\begin{aligned} \text{maximize} \quad & \sum_{j=1}^{n} c_j x_j \\ \text{subject to} \quad & \sum_{j=1}^{n} a_{ij} x_j \le b_i \qquad (i = 1, 2, \ldots, m) \\ & x_j \ge 0 \qquad (j = 1, 2, \ldots, n). \end{aligned} \tag{5.12}$$

Applying the simplex method to (5.12), the student finds simultaneously an optimal solution $x_1^*, x_2^*, \ldots, x_n^*$ of (5.12) and an optimal solution $y_1^*, y_2^*, \ldots, y_m^*$ of the dual problem

$$\begin{aligned} \text{minimize} \quad & \sum_{i=1}^{m} b_i y_i \\ \text{subject to} \quad & \sum_{i=1}^{m} a_{ij} y_i \ge c_j \qquad (j = 1, 2, \ldots, n) \\ & y_i \ge 0 \qquad (i = 1, 2, \ldots, m). \end{aligned} \tag{5.13}$$

Then he shows *both* solutions to his supervisor. The supervisor has an easy way of checking the correctness of the answer. To check the *feasibility* of the allegedly optimal solution, she has to verify the inequalities

$$\begin{aligned} & \sum_{j=1}^{n} a_{ij} x_j^* \le b_i \qquad (i = 1, 2, \ldots, m) \\ & x_j^* \ge 0 \qquad (j = 1, 2, \ldots, n). \end{aligned} \tag{5.14}$$

To check its *optimality*, she has to verify the inequalities

$$\begin{aligned} \sum_{i=1}^{m} a_{ij} y_i^* &\geq c_j \qquad (j = 1, 2, \ldots, n) \\ y_i^* &\geq 0 \qquad (i = 1, 2, \ldots, m) \end{aligned} \tag{5.15}$$

and the equation

$$\sum_{j=1}^{n} c_j x_j^* = \sum_{i=1}^{m} b_i y_i^*. \tag{5.16}$$

Of course, the computational effort involved in these *verifications* is much smaller than the computational effort required to *solve* (5.12) from scratch by the simplex method.

COMPLEMENTARY SLACKNESS

Now we shall show how the supervisor can often recover the certificate of optimality $y_1^*, y_2^*, \ldots, y_m^*$ from the optimal solution $x_1^*, x_2^*, \ldots, x_n^*$ alone. The key to the procedure is a convenient way of breaking down equation (5.16) into simple constituents.

THEOREM 5.2. Let $x_1^*, x_2^*, \ldots, x_n^*$ be a feasible solution of (5.12) and let $y_1^*, y_2^*, \ldots, y_m^*$ be a feasible solution of (5.13). Necessary and sufficient conditions for simultaneous optimality of $x_1^*, x_2^*, \ldots, x_n^*$ and $y_1^*, y_2^*, \ldots, y_m^*$ are

$$\sum_{i=1}^{m} a_{ij} y_i^* = c_j \quad \text{or} \quad x_j^* = 0 \quad \text{(or both)} \quad \text{for every} \quad j = 1, 2, \ldots, n \tag{5.17}$$

and

$$\sum_{j=1}^{n} a_{ij} x_j^* = b_i \quad \text{or} \quad y_i^* = 0 \quad \text{(or both)} \quad \text{for every} \quad i = 1, 2, \ldots, m. \tag{5.18}$$

PROOF. Assumptions (5.14) and (5.15) imply

$$c_j x_j^* \leq \left(\sum_{i=1}^{m} a_{ij} y_i^* \right) x_j^* \qquad (j = 1, 2, \ldots, n) \tag{5.19}$$

$$\left(\sum_{j=1}^{n} a_{ij} x_j^* \right) y_i^* \leq b_i y_i^* \qquad (i = 1, 2, \ldots, m) \tag{5.20}$$

and so

$$\sum_{j=1}^{n} c_j x_j^* \leq \sum_{j=1}^{n} \left(\sum_{i=1}^{m} a_{ij} y_i^* \right) x_j^* = \sum_{i=1}^{m} \left(\sum_{j=1}^{n} a_{ij} x_j^* \right) y_i^* \leq \sum_{i=1}^{m} b_i y_i^*. \tag{5.21}$$

Hence, (5.21) holds with equalities throughout if and only if equalities hold in (5.19) and (5.20). One way to guarantee the equality $c_j x_j^* = (\sum a_{ij} y_i^*) x_j^*$ is to insist that $x_j^* = 0$; failing that, we must require $c_j = \sum a_{ij} y_i^*$. Hence, equalities hold in (5.19) if and only if condition (5.17) is satisfied. Similarly, equalities hold in (5.20) if and only if condition (5.18) is satisfied.

To summarize, conditions (5.17) and (5.18) are necessary and sufficient for (5.16) to hold. On the other hand, the Duality Theorem shows that (5.16) is necessary and sufficient for simultaneous optimality of $x_1^*, x_2^*, \ldots, x_n^*$ and $y_1^*, y_2^*, \ldots, y_m^*$. The proof is completed. ∎

Conditions (5.17) and (5.18) gain simplicity as soon as we introduce the slack variables

$$x_{n+i} = b_i - \sum_{j=1}^{n} a_{ij} x_j \qquad (i = 1, 2, \ldots, m)$$

$$y_{m+j} = -c_j + \sum_{i=1}^{m} a_{ij} y_i \qquad (j = 1, 2, \ldots, n).$$

As we observed once before, the primal slack variables $x_{n+1}, x_{n+2}, \ldots, x_{n+m}$ are naturally matched up with the dual decision variables $y_1, y_2, \ldots, y_m$: each variable x_{n+i} denotes the slack in the ith primal constraint, whereas the corresponding y_i represents the multiplier at the same constraint. Similarly, each primal decision variable x_j is matched with the dual slack y_{m+j}. Conditions (5.17) and (5.18) require that in each of the $m + n$ matching pairs, at least one variable must have value zero. These conditions are usually called the *complementary slackness* conditions; Theorem 5.2 itself is referred to as the Complementary Slackness Theorem.

It is an easy task to convert Theorem 5.2 into a form in which its applicability becomes evident.

THEOREM 5.3. A feasible solution $x_1^*, x_2^*, \ldots, x_n^*$ of (5.12) is optimal if and only if there are numbers $y_1^*, y_2^*, \ldots, y_m^*$ such that

$$\begin{aligned} \sum_{i=1}^{m} a_{ij} y_i^* &= c_j \quad \text{whenever} \quad x_j^* > 0 \\ y_i^* &= 0 \quad \text{whenever} \quad \sum_{j=1}^{n} a_{ij} x_j^* < b_i \end{aligned} \qquad (5.22)$$

and such that

$$\begin{aligned} \sum_{i=1}^{m} a_{ij} y_i^* &\ge c_j \quad \text{for all} \quad j = 1, 2, \ldots, n \\ y_i^* &\ge 0 \quad \text{for all} \quad i = 1, 2, \ldots, m. \end{aligned} \qquad (5.23)$$

PROOF. If $x_1^*, x_2^*, \ldots, x_n^*$ is optimal, then, by Theorem 5.1, there is an optimal solution $y_1^*, y_2^*, \ldots, y_m^*$ of (5.13). That solution, being feasible, satisfies (5.23). By Theorem 5.2, the two optimal solutions satisfy the complementary slackness conditions (5.22).

Conversely, if $y_1^*, y_2^*, \ldots, y_m^*$ satisfy (5.23), then they constitute a feasible solution of (5.13). If they satisfy (5.22) as well, then, by Theorem 5.2, $x_1^*, x_2^*, \ldots, x_n^*$ is an optimal solution of (5.12) and $y_1^*, y_2^*, \ldots, y_m^*$ is an optimal solution of (5.13). ■

Theorem 5.3 is often useful in checking the optimality of allegedly optimal solutions when no certificate of optimality is provided. Confronted with an allegedly optimal solution $x_1^*, x_2^*, \ldots, x_n^*$ of (5.12), we first set up the system of linear equations (5.22) and solve for $y_1^*, y_2^*, \ldots, y_m^*$. If the solution $y_1^*, y_2^*, \ldots, y_m^*$ is *unique*, then we are in business: $x_1^*, x_2^*, \ldots, x_n^*$ is optimal if and only if (5.23) holds. We shall illustrate this situation with two examples.

First, let us consider the claim that

$$x_1^* = 2, \quad x_2^* = 4, \quad x_3^* = 0, \quad x_4^* = 0, \quad x_5^* = 7, \quad x_6^* = 0$$

is an optimal solution of the problem

$$\begin{array}{lrcrcrcrcrcrcr}
\text{maximize} & 18x_1 & - & 7x_2 & + & 12x_3 & + & 5x_4 & & & + & 8x_6 & & \\
\text{subject to} & 2x_1 & - & 6x_2 & + & 2x_3 & + & 7x_4 & + & 3x_5 & + & 8x_6 & \leq & 1 \\
& -3x_1 & - & x_2 & + & 4x_3 & - & 3x_4 & + & x_5 & + & 2x_6 & \leq & -2 \\
& 8x_1 & - & 3x_2 & + & 5x_3 & - & 2x_4 & & & + & 2x_6 & \leq & 4 \\
& 4x_1 & & & + & 8x_3 & + & 7x_4 & - & x_5 & + & 3x_6 & \leq & 1 \\
& 5x_1 & + & 2x_2 & - & 3x_3 & + & 6x_4 & - & 2x_5 & - & x_6 & \leq & 5 \\
& & & & & & & \multicolumn{5}{r}{x_1, x_2, x_3, x_4, x_5, x_6} & \geq & 0.
\end{array}$$

In this case, (5.22) reads

$$\begin{array}{rcrcrcrcrcr}
2y_1^* & - & 3y_2^* & + & 8y_3^* & + & 4y_4^* & + & 5y_5^* & = & 18 \\
-6y_1^* & - & y_2^* & - & 3y_3^* & & & + & 2y_5^* & = & -7 \\
3y_1^* & + & y_2^* & & & - & y_4^* & - & 2y_5^* & = & 0 \\
& & y_2^* & & & & & & & = & 0 \\
& & & & & & & & y_5^* & = & 0.
\end{array}$$

Since its solution $(\frac{1}{3}, 0, \frac{5}{3}, 1, 0)$ satisfies (5.23), the proposed solution $x_1^*, x_2^*, \ldots, x_6^*$ is optimal.

Second, let us consider the claim that

$$x_1^* = 0, \quad x_2^* = 2, \quad x_3^* = 0, \quad x_4^* = 7, \quad x_5^* = 0$$

is an optimal solution of the problem

$$\begin{array}{ll} \text{maximize} & 8x_1 - 9x_2 + 12x_3 + 4x_4 + 11x_5 \\ \text{subject to} & 2x_1 - 3x_2 + 4x_3 + x_4 + 3x_5 \le 1 \\ & x_1 + 7x_2 + 3x_3 - 2x_4 + x_5 \le 1 \\ & 5x_1 + 4x_2 - 6x_3 + 2x_4 + 3x_5 \le 22 \\ & x_1, x_2, x_3, x_4, x_5 \ge 0. \end{array}$$

Here (5.22) becomes

$$\begin{aligned} -3y_1^* + 7y_2^* + 4y_3^* &= -9 \\ y_1^* - 2y_2^* + 2y_3^* &= 4 \\ y_2^* &= 0. \end{aligned}$$

Since its unique solution (3.4, 0, 0.3) violates (5.23), the proposed solution $x_1^*, x_2^*, \ldots, x_5^*$ is not optimal.

Of course, this straightforward strategy for verifying optimality of allegedly optimal solutions is applicable only if the system of equations (5.22) has a unique solution. The following result points out conditions under which this is always the case.

> ***THEOREM 5.4.*** If $x_1^*, x_2^*, \ldots, x_n^*$ is a nondegenerate basic feasible solution of (5.12), then (5.22) has a unique solution.

The proof of this theorem is postponed until the end of Chapter 7, where it will become an exercise (problem 7.3).

ECONOMIC SIGNIFICANCE OF DUAL VARIABLES

For many LP problems

$$\begin{array}{lll} \text{maximize} & \sum_{j=1}^{n} c_j x_j & \\ \text{subject to} & \sum_{j=1}^{n} a_{ij} x_j \le b_i & (i = 1, 2, \ldots, m) \\ & x_j \ge 0 & (j = 1, 2, \ldots, n) \end{array} \qquad (5.24)$$

arising in applications, the variables $y_1, y_2, \ldots, y_m$ in the dual problem can be given a meaningful interpretation. An indication of the way these dual variables should be interpreted follows from a heuristic argument occasionally used in elementary physics and known as "dimension analysis." For instance, suppose that (5.24) is the problem of maximizing profit in a furniture manufacturing firm. Each x_j measures the level of the output of the jth product (such as desks or chairs), and each b_i specifies the available amount of the ith resource (such as wood or metal). Note that *each* a_{ij} *is expressed in units of resource i per unit of product j* (in fact, each a_{ij} is the amount of resource i required in making a unit of product j) and that *each* c_j *is in dollars per unit of product j* (in fact, each c_j is the net profit brought in by a unit of product j). To make the left-hand side, $\sum a_{ij}y_i$, of each dual constraint commensurate with the right-hand side c_j, *we must express each* y_i *in dollars per unit of resource i.* Thus, one is led to suspect that each y_i measures the unit worth of the ith resource. The following theorem will validate this suspicion.

THEOREM 5.5. If (5.24) has at least one nondegenerate basic optimal solution, then there is a positive ε with the following property: If $|t_i| \leq \varepsilon$ for all $i = 1, 2, \ldots, m$, then the problem

$$\begin{aligned} &\text{maximize} && \sum_{j=1}^{n} c_j x_j \\ &\text{subject to} && \sum_{j=1}^{n} a_{ij}x_j \leq b_i + t_i \qquad (i = 1, 2, \ldots, m) \\ &&& x_j \geq 0 \qquad (j = 1, 2, \ldots, n) \end{aligned} \tag{5.25}$$

has an optimal solution and its optimal value equals

$$z^* + \sum_{i=1}^{m} y_i^* t_i$$

with z^* standing for the optimal value of (5.24) and with $y_1^*, y_2^*, \ldots, y_m^*$ standing for the optimal solution of its dual.

We postpone the proof of this theorem also until the end of Chapter 7, where it will become an exercise (problem 7.4). At this moment, let us note only that the uniqueness of $y_1^*, y_2^*, \ldots, y_m^*$ is guaranteed by Theorems 5.4 and 5.2.

Theorem 5.5 reveals the effects of small variations in the supplies of the resources on the total net profit of the firm. With each extra unit of resource i, the profit increases by y_i^* dollars. Hence, y_i^* specifies the maximum amount that the firm should be willing to pay, over and above the present trading price, for each extra unit of resource i. For this reason, y_i^* is often called the *marginal value* of the ith resource, the adjective "marginal" referring to the difference between the trading price and the actual worth of the resource. Another term commonly used for y_i^* in this context is the *shadow price* of the ith resource.

To illustrate these findings, imagine a forester who has 100 acres of hardwood timber. Felling the hardwood and letting the area regenerate would cost \$10 per acre in immediate resources and bring a subsequent return of \$50 per acre. An alternative course of action is to fell the hardwood and plant the area with pine; that would cost \$50 per acre with a subsequent return of \$120 per acre. Hence, the *net* profits resulting from the two treatments are \$40 and \$70 per acre, respectively. Unfortunately, the more profitable second treatment cannot be applied to the entire area since only \$4,000 is available to meet the immediate costs. Clearly, the forester's problem is to

$$\begin{array}{ll} \text{maximize} & 40x_1 + 70x_2 \\ \text{subject to} & x_1 + x_2 \leq 100 \\ & 10x_1 + 50x_2 \leq 4{,}000 \\ & x_1, x_2 \geq 0. \end{array}$$

Its optimal solution is $x_1^* = 25$ and $x_2^* = 75$. Hence, the forester should fell the hardwood throughout the entire area, letting 25 acres regenerate and planting the remaining 75 acres with pine. According to this program, the initial investment of \$4,000 yields the ultimate net profit of \$6,250.

Evidently, the forester's initial capital represents a valuable resource. In fact, the forester might be well advised to increase the level of this resource by taking out a short-term loan; the resulting extra profit might make up even for a drastic interest rate. For example, suppose that she could borrow \$100 now and pay back \$180 later; should she do that? On the other hand, she might be tempted to divert some of her \$4,000 to other lucrative enterprises. For example, suppose she could invest \$100 now and collect \$180 later; should she do that? According to Theorem 5.5, the answers lie hidden in the optimal solution

$$y_1^* = 32.5, \quad y_2^* = 0.75$$

of the dual problem: the forester should take out (limited) loans if and only if the interest is lower than 75 cents per dollar and she should make (small) investments if and only if the profit is greater than 75 cents per dollar.

These claims, whose validity is guaranteed by Theorem 5.5, are easy to justify directly. Having borrowed t dollars, the forester aims to

$$\begin{array}{ll} \text{maximize} & 40x_1 + 70x_2 \\ \text{subject to} & x_1 + x_2 \leq 100 \\ & 10x_1 + 50x_2 \leq 4{,}000 + t \\ & x_1, x_2 \geq 0. \end{array} \tag{5.26}$$

Every feasible solution x_1, x_2 of this problem satisfies the inequalities

$$\begin{aligned} 40x_1 + 70x_2 &= 32.5(x_1 + x_2) + 0.75(10x_1 + 50x_2) \\ &\leq 3{,}250 + 0.75(4{,}000 + t) = 6{,}250 + 0.75t \end{aligned} \tag{5.27}$$

and so the extra profit will never exceed $0.75t$. In fact, if $t \leq 1{,}000$, then the forester can realize the additional profit of $0.75t$ by letting

$$x_1 = 25 - 0.025t, \quad x_2 = 75 + 0.025t. \tag{5.28}$$

Investments in other enterprises give rise to negative values of t in (5.26); as a result of such investments, the net profit from the original enterprise diminishes. If $-t$ dollars are diverted to alternative investments ($-t$ is positive!) then, by (5.27), the profit from the hardwood felling enterprise will drop by $0.75(-t)$ or even more. In fact, if $-t \leq 3{,}000$, then the drop can be limited to only $0.75(-t)$ by choosing x_1 and x_2 according to (5.28).

It should perhaps be emphasized that Theorem 5.5 deals with *small* changes t_i in the resource levels; its conclusions may fail when the t_i's are large. For instance, our forester has no use for loans exceeding \$1,000 and, should she wish to invest all of her \$4,000 in another enterprise, she would be ill advised to demand only 75 cents of profit on each dollar. (A part of Theorem 5.5 can be salvaged even if the t_i's are large; see problem 5.9.)

Now suppose that a previously unavailable opportunity arises for the forester to engage in an activity such as, say, felling the hardwood and planting the area with conifer. For a quick assessment of this activity, the forester may appeal to the marginal values of her resources: \$32.5 per acre of hardwood and \$0.75 per dollar of capital. If the new activity requires a dollars per acre, then the resources consumed by this activity per acre are valued at $\$(32.5 + 0.75a)$ and the activity is worth considering if and only if its net profit per acre exceeds this figure. Further examples of this kind are presented in problems 5.6 and 5.7.

In closing, let us mention that models of economy often fall into the realm of linear programming. In particular, many theorems concerning economic equilibria may be deduced from the Duality Theorem and the Complementary Slackness Theorem. Their discussion exceeds the scope of this text; the interested reader is referred to D. Gale (1960) and R. Dorfman, P. A. Samuelson, and R. M. Solow (1958).

PROBLEMS

5.1 Illustrate Theorem 5.1 on each of the three LP problems in problem 2.1.

△ **5.2** Maximize $-x_1 - 2x_2$

subject to

$$\begin{aligned} -3x_1 + x_2 &\le -1 \\ x_1 - x_2 &\le 1 \\ -2x_1 + 7x_2 &\le 6 \\ 9x_1 - 4x_2 &\le 6 \\ -5x_1 + 2x_2 &\le -3 \\ 7x_1 - 3x_2 &\le 6 \\ x_1, x_2 &\ge 0. \end{aligned}$$

△ **5.3** For each of the two problems below, use Theorem 5.3 to check the optimality of the proposed solution.

a. Maximize $7x_1 + 6x_2 + 5x_3 - 2x_4 + 3x_5$

subject to

$$\begin{aligned} x_1 + 3x_2 + 5x_3 - 2x_4 + 2x_5 &\le 4 \\ 4x_1 + 2x_2 - 2x_3 + x_4 + x_5 &\le 3 \\ 2x_1 + 4x_2 + 4x_3 - 2x_4 + 5x_5 &\le 5 \\ 3x_1 + x_2 + 2x_3 - x_4 - 2x_5 &\le 1 \\ x_1, x_2, x_3, x_4, x_5 &\ge 0. \end{aligned}$$

Proposed solution: $x_1^* = 0,\quad x_2^* = \frac{4}{3},\quad x_3^* = \frac{2}{3},\quad x_4^* = \frac{5}{3},\quad x_5^* = 0.$

b. Maximize $4x_1 + 5x_2 + x_3 + 3x_4 - 5x_5 + 8x_6$

subject to

$$\begin{aligned} x_1 - 4x_3 + 3x_4 + x_5 + x_6 &\le 1 \\ 5x_1 + 3x_2 + x_3 - 5x_5 + 3x_6 &\le 4 \\ 4x_1 + 5x_2 - 3x_3 + 3x_4 - 4x_5 + x_6 &\le 4 \\ - x_2 + 2x_4 + x_5 - 5x_6 &\le 5 \\ -2x_1 + x_2 + x_3 + x_4 + 2x_5 + 2x_6 &\le 7 \\ 2x_1 - 3x_2 + 2x_3 - x_4 + 4x_5 + 5x_6 &\le 5 \\ x_1, x_2, x_3, x_4, x_5, x_6 &\ge 0. \end{aligned}$$

Proposed solution: $x_1 = 0,\quad x_2 = 0,\quad x_3 = \frac{5}{2},\quad x_4 = \frac{7}{2},\quad x_5 = 0,\quad x_6 = \frac{1}{2}.$

△ **5.4** In problem 1.6, one of the possible strategies is as follows:

- Smoke all 400 bellies on regular time.
- Smoke 20 picnics on regular time and 210 on overtime.
- Smoke 40 hams on overtime and sell 440 fresh.

Use Theorem 5.3 to find out whether this strategy is optimal or not.

5.5 In problem 1.7, one of the possible strategies is as follows:

- Blend 3,754 barrels of alkylate, 2,666 barrels of catalytic, 920 barrels of straight-run, and 543 barrels of isopentane into 7,883 barrels of Avgas A.
- Blend 60 barrels of alkylate, 3,096 barrels of straight-run, and 672 barrels of isopentane into 3,828 barrels of Avgas B.
- Sell 85 barrels of isopentane raw.

Use Theorem 5.3 to find out whether this strategy is optimal or not.

5.6 In the optimal solution to problem 1.6, all the bellies and picnics are smoked. However, sufficiently drastic changes in market prices might provide an incentive to change this policy. Assume that the market price of fresh bellies increases by x dollars, while all the other prices remain fixed at their original levels. How large would x have to be in order to make it profitable for the plant to sell fresh bellies? Ask and answer a similar question for picnics. How would the sales of small amounts of fresh bellies and picnics affect the rest of the operation? What precisely do "small amounts" mean in this context?

5.7 In the optimal solution to problem 1.7, 85 barrels of isopentane are sold raw at $4.83 per barrel. Find the break-even selling prices for raw alkylate, catalytic, and straight-run. Next, assume there is a demand for Avgas C with PN at least 80 and RVP at most 7. Find the break-even selling price of this gasoline.

5.8 Can you interpret the complementary slackness conditions in economic terms?

5.9 Let z^* be the optimal value of (5.24) and let $y_1^*, y_2^*, \ldots, y_m^*$ be any optimal solution of the dual problem. Prove that

$$\sum_{j=1}^{n} c_j x_j \leq z^* + \sum_{i=1}^{m} y_i^* t_i$$

for every feasible solution $x_1, x_2, \ldots, x_n$ of (5.25).

5.10 Construct an example showing that the conclusion of Theorem 5.5 may fail if the hypothesis that (5.24) has a nondegenerate basic optimal solution is omitted.

6

Gaussian Elimination and Matrices

This chapter reviews facts about systems of linear equations and matrices that will be needed later. We shall begin with Gaussian elimination and discuss ways of enhancing its accuracy and speed. Then we shall introduce the notion of matrix multiplication and review Gaussian elimination in terms of matrices. Finally, we shall establish a few fundamental facts concerning nonsingular matrices.

GAUSSIAN ELIMINATION

Perhaps the most popular method for solving systems of linear equations is known as *Gaussian elimination.* This method consists of a successive elimination of variables followed by a back substitution. We shall illustrate it on the system

$$\begin{aligned} 5x_1 + 4x_2 + 3x_3 + 9x_4 &= 49 \\ 3x_1 + 2x_2 + x_3 + 2x_4 &= 19 \\ -14x_1 - 8x_2 - 7x_3 + 5x_4 &= -68 \\ 12x_1 + 6x_2 - 25x_3 + 10x_4 &= -38. \end{aligned} \tag{6.1}$$

The first equation in (6.1) may be converted into the formula

$$x_1 = 9.8 - 0.8x_2 - 0.6x_3 - 1.8x_4. \tag{6.2}$$

Substituting from (6.2) into the remaining equations in (6.1) we obtain

$$\begin{aligned} 3(9.8 - 0.8x_2 - 0.6x_3 - 1.8x_4) + 2x_2 + x_3 + 2x_4 &= 19 \\ -14(9.8 - 0.8x_2 - 0.6x_3 - 1.8x_4) - 8x_2 - 7x_3 + 5x_4 &= -68 \\ 12(9.8 - 0.8x_2 - 0.6x_3 - 1.8x_4) + 6x_2 - 25x_3 + 10x_4 &= -38 \end{aligned}$$

and, after simplification,

$$\begin{aligned} -0.4x_2 - 0.8x_3 - 3.4x_4 &= -10.4 \\ 3.2x_2 + 1.4x_3 + 30.2x_4 &= 69.2 \\ -3.6x_2 - 32.2x_3 - 11.6x_4 &= -155.6. \end{aligned} \tag{6.3}$$

Clearly, numbers x_1, x_2, x_3, x_4 satisfy (6.1) *if and only if* they satisfy (6.2), (6.3). In this sense, system (6.1) is *equivalent* to system (6.2), (6.3). Since variable x_1 does not appear in (6.3), we say that it has been *eliminated*. Just as we eliminated x_1 from (6.1), we may now eliminate x_2 from (6.3). The first equation in (6.3) may be written as

$$x_2 = 26 - 2x_3 - 8.5x_4. \tag{6.4}$$

Substituting from (6.4) into the remaining two equations in (6.3) we obtain, after simplification, the system

$$\begin{aligned} -5x_3 + 3x_4 &= -14 \\ -25x_3 + 19x_4 &= -62. \end{aligned} \tag{6.5}$$

To eliminate x_3 from (6.5), we write the first equation of (6.5) in the form

$$x_3 = 2.8 + 0.6x_4. \tag{6.6}$$

Substituting from (6.6) into the remaining equation in (6.5) we obtain

$$4x_4 = 8. \tag{6.7}$$

Now, (6.7) yields

$$x_4 = 2. \tag{6.8}$$

Substituting from (6.8) back into (6.6) we find

$$x_3 = 4. \tag{6.9}$$

Substituting from (6.9) and (6.8) back into (6.4) we find

$$x_2 = 1. \tag{6.10}$$

Finally, substituting from (6.10), (6.9), and (6.8) back into (6.2) we obtain

$$x_1 = 3. \tag{6.11}$$

The same strategy applies in general: we eliminate $x_1, x_2, \ldots$ in this order and then find the values of $x_n, x_{n-1}, \ldots, x_1$ by back substitution. The elimination process merits a comment. When the variables $x_1, x_2, \ldots, x_{k-1}$ have been eliminated, we are confronted with a system of $n - k + 1$ equations in the remaining $n - k + 1$ variables $x_k, x_{k+1}, \ldots, x_n$. Among these $n - k + 1$ equations, we choose an equation

$$\sum_{j=k}^{n} a_j x_j = b \tag{6.12}$$

such that $a_k \neq 0$; this number a_k is called the *pivot*. Substituting from

$$x_k = \frac{b}{a_k} - \sum_{j=k+1}^{n} \frac{a_j}{a_k} x_j$$

into the remaining $n - k$ equations, we eliminate x_k and obtain a system of $n - k$ equations in $n - k$ variables. (However, it may happen that each of the $n - k + 1$ equations of (6.12) has $a_k = 0$. In other words, the system of $n - k + 1$ equations may involve only the $n - k$ variables $x_{k+1}, x_{k+2}, \ldots, x_n$. This circumstance indicates that the original system of n equations in n variables either has no solutions at all or it has infinitely many of them. Since this case is of no importance in our present discussion, we shall not elaborate on the details until Theorem 6.1, which will follow later.)

It is natural to ask how much work is required to solve a system of n equations in n variables by this method. For our example, Table 6.1 gives a detailed account of the work involved.

Table 6.1 Operations Count

To calculate	Additions	Multiplications	Divisions
(6.2)	0	0	4
(6.3)	12	12	0
(6.4)	0	0	3
(6.5)	6	6	0
(6.6)	0	0	2
(6.7)	2	2	0
(6.8)	0	0	1
(6.9)	1	1	0
(6.10)	2	2	0
(6.11)	3	3	0
Total	26	26	10

Note: Additions and subtractions are essentially the same operation; therefore we count both under the heading Additions.

In general, the work required to eliminate x_k amounts to $n - k + 1$ divisions, followed by $(n - k)(n - k + 1)$ multiplications and $(n - k)(n - k + 1)$ additions. Then, evaluating x_k by back substitution takes $n - k$ multiplications and $n - k$ additions. The totals are:

Additions: $\sum_{k=1}^{n} (n - k)(n - k + 1) + \sum_{k=1}^{n} (n - k) = n(n - 1)(2n + 5)/6.$

Multiplications: $\sum_{k=1}^{n} (n - k)(n - k + 1) + \sum_{k=1}^{n} (n - k) = n(n - 1)(2n + 5)/6.$

Divisions: $\sum_{k=1}^{n} (n - k + 1) = n(n + 1)/2.$

Hence, there are about $n^3/3$ additions, $n^3/3$ multiplications, and $n^2/2$ divisions to carry out. Since the additions are relatively easy and the divisions are relatively few, the main computational burden comes from the multiplications. (Incidentally, note that the method applied to $2n$ equations in $2n$ variables takes roughly $8n^3/3$ multiplications, *eight times* the amount required by n equations in n variables.)

Similar methods for solving *small* systems of linear equations have been in use for centuries. An ancient Chinese text, "Nine Chapters on the Mathematical Art," includes examples of three equations in three variables solved by an elimination procedure. This book dates from the Han dynasty of 206 B.C.–A.D. 220 (the period in which the *I-Ching* was written), but some of its contents may be as old as 1100 B.C. Other examples appear in various writings scattered throughout the ages. However, a systematic study of the method as a tool for solving *arbitrary* systems of linear equations was first undertaken by C. F. Gauss (1777–1855) and A. M. Legendre (1752–1833). Additional information on the history of elimination methods may be found in Eves (1964), Smith (1953), and Struik (1967).

Two obvious criteria for assessing the performance of any numerical algorithm are (i) the accuracy of results and (ii) the speed of execution. We shall analyze Gaussian elimination from both points of view, beginning with numerical accuracy.

Accuracy of Gaussian Elimination

Numerical computations are usually done in *floating-point arithmetic*. That is, each nonzero number is represented in the form $x \cdot 10^k$ such that k is an integer and $0.1 \le |x| < 1$. The number x is stored only to a *d-digit precision* (or "d significant digits") for some fixed positive integer d; that is, only the first d digits of x following the decimal point are stored. Since the same convention applies to results of arithmetical operations, floating-point computations are subject to rounding errors. For instance, if $d = 3$, then the result of

$$(0.523 \cdot 10^2) \times (-0.324 \cdot 10^{-5}) = -0.169452 \cdot 10^{-3}$$

is stored as $-0.169 \cdot 10^{-3}$. In long chain calculations, these rounding errors may accumulate to such an extent that the final result is useless.

It would be a grave misconception to assume that only long chain calculations lead to numerically inaccurate results: even small systems of linear equations may cause trouble. To illustrate, we shall solve the system

$$\begin{aligned} 0.0001x_1 + \quad x_2 &= 1 \\ 0.5x_1 + 0.5x_2 &= 1 \end{aligned} \tag{6.13}$$

rounding the intermediate results to three significant digits. Using the first equation to eliminate x_1 we obtain

$$\begin{aligned} x_1 &= 10{,}000 - 10{,}000x_2 \\ -5{,}000x_2 &= -5{,}000 \end{aligned}$$

whereupon back substitution yields $x_2 = 1$ and $x_1 = 0$. But the true solution of (6.13), rounded to three significant digits, is $x_1 = 1$ and $x_2 = 1$. What went wrong? The pivot 0.0001 was too small relative to the remaining coefficients. In our first outline of Gaussian elimination, we observed that *zero* pivots *must* be avoided since division by zero is impossible. Now it seems reasonable to infer that *small* pivots *should* be avoided since division by small numbers is likely to produce serious rounding errors. This precaution may be implemented in various ways. For instance, while eliminating $x_1, x_2, \ldots$ in this order, we might always choose the pivot that has the largest magnitude among all the available candidates. Or, going a step further, we might also change the order in which the variables are eliminated, so that the pivot magnitude in each iteration is as large as possible. The latter strategy is known as *complete pivoting*, whereas the former, with the order of elimination fixed, is known as *partial pivoting*. Our discussion will be restricted to partial pivoting since this strategy is quite adequate in practice. In our example, partial pivoting tells us to eliminate x_1 from the second equation. Rounding the intermediate results to three significant digits, we obtain

$$x_1 = 2 - x_2$$
$$x_2 = 1$$

whereupon back substitution yields $x_2 = 1$ and $x_1 = 1$. Thus, all is well—at least in this case.

Investigation into the numerical accuracy of elimination methods has an interesting history. An early paper by H. Hotelling (1943) established a discouraging fact: Even when $n = 100$, the computed solution $\tilde{x}_1, \tilde{x}_2, \ldots, \tilde{x}_n$ of

$$\sum_{j=1}^{n} a_{ij}x_j = b_i \qquad (i = 1, 2, \ldots, n) \tag{6.14}$$

may differ from the true solution $x_1, x_2, \ldots, x_n$ to such extent that (unless we work to the precision of more than 60 decimal digits) the computations are meaningless. Although comparing the computed solution $\tilde{x}_1, \tilde{x}_2, \ldots, \tilde{x}_n$ with the true solution $x_1, x_2, \ldots, x_n$ seems to be the natural way of assessing the accuracy of the computations, this point of view, nowadays known as the *forward error analysis*, fell out of favor around the 1950s. The computed solution may suddenly look much better, if it turns out to be the precise solution of a system

$$\sum_{j=1}^{n} \tilde{a}_{ij}\tilde{x}_j = b_i \qquad (i = 1, 2, \ldots, n) \tag{6.15}$$

whose coefficients $\tilde{a}_{ij}$ differ from the original coefficients a_{ij} only slightly. This new point of view, known as the *backward error analysis*, makes perfect sense in applications where the coefficients a_{ij} are often obtained by imprecise measurements and/or estimations. The idea of backward error analysis, implicit in papers by J. von Neumann and H. H. Goldstine (1947) and by A. M. Turing (1948), was put forth by W. Givens (1954) and championed by J. H. Wilkinson (1963). Wilkinson subjected elimination methods to backward error analysis and found that Gaussian elimination with partial pivoting yields remarkably accurate results. His arguments guarantee the existence of a system (6.15) that is very close to (6.14). Roughly speaking (the precise statement is a little more intricate), each difference $\tilde{a}_{ij} - a_{ij}$ amounts to at most n of the individual errors made in the process of solving (6.14).

Of course, this guarantee loses much of its appeal when the individual rounding errors are large compared with a_{ij}. This situation may happen when the various coefficients differ from each other so widely that the individual rounding errors resulting from handling large a_{ij} become overwhelming when compared with small a_{ij}. To illustrate, let us solve the system

$$\begin{aligned} x_1 + 10{,}000x_2 &= 10{,}000 \\ 0.5x_1 + \quad 0.5x_2 &= 1 \end{aligned} \tag{6.16}$$

by Gaussian elimination with partial pivoting, working to three significant digits. Eliminating x_1 from the first equation, we obtain

$$x_1 = 10{,}000 - 10{,}000x_2$$
$$-5{,}000x_2 = -5{,}000$$

whereupon back substitution yields $x_2 = 1$, $x_1 = 0$. This result is unsatisfactory even from the point of view of backward error analysis since every system

$$\tilde{a}_{11}\tilde{x}_1 + \tilde{a}_{12}\tilde{x}_2 = 10{,}000$$
$$\tilde{a}_{21}\tilde{x}_1 + \tilde{a}_{22}\tilde{x}_2 = 1$$

satisfied by $\tilde{x}_1 = 0$, $\tilde{x}_2 = 1$ must have $\tilde{a}_{22} = 1$, and therefore differ significantly from (6.16). Fortunately, there is often a way of avoiding such blunders. Individual rounding errors tend to become less significant when the various nonzero coefficients a_{ij} have roughly the same magnitude. Systems with this property are called "well-scaled." Many systems

$$\sum_{j=1}^{n} a_{ij}x_j = b_i \qquad (i = 1, 2, \ldots, n) \tag{6.17}$$

that arise in practice are well-scaled; other systems can often be made well-scaled by simple transformations. Having chosen positive numbers $r_1, r_2, \ldots, r_m$ and $s_1, s_2, \ldots, s_n$, we multiply each row i in (6.17) by r_i and substitute $s_jx'_j$ for each x_j. The resulting system reads $\sum r_ia_{ij}(s_jx'_j) = r_ib_i$, or

$$\sum_{j=1}^{n} (r_ia_{ij}s_j)x'_j = r_ib_i \qquad (i = 1, 2, \ldots, n). \tag{6.18}$$

The transformation of (6.17) into (6.18) is called *scaling*, with the numbers r_i and s_j referred to as *scaling factors*. A suitable choice of these scaling factors helps to smooth out discrepancies among the magnitudes of the various nonzero coefficients and often yields a well-scaled system (6.18). For instance, if (6.17) reads

$$\begin{aligned}
0.5x_1 + 0.002x_2 + 2x_3 + 40x_4 &= 12\\
30x_1 + 0.4x_2 + 500x_3 + 1000x_4 &= 600\\
2x_1 + 0.01x_2 + 30x_3 + 300x_4 &= 90\\
0.1x_1 + 0.005x_2 + 4x_3 + 20x_4 &= 14
\end{aligned}$$

and if we choose

$$\begin{aligned}
&r_1 = 10, \quad r_2 = 0.1, \quad r_3 = 1, \quad r_4 = 10\\
&s_1 = 1, \quad s_2 = 100, \quad s_3 = 0.1, \quad s_4 = 0.01
\end{aligned}$$

then (6.18) assumes the well-scaled form

$$\begin{aligned}
5x'_1 + 2x'_2 + 2x'_3 + 4x'_4 &= 120\\
3x'_1 + 4x'_2 + 5x'_3 + x'_4 &= 60\\
2x'_1 + x'_2 + 3x'_3 + 3x'_4 &= 90\\
x'_1 + 5x'_2 + 4x'_3 + 2x'_4 &= 140.
\end{aligned}$$

Incidentally, note the advantage of choosing powers of 10 for the scaling factors. Since we are working in the decimal notation, each evaluation of $r_ia_{ij}s_j$ amounts to a mere shift of the decimal point. For the same reason, the scaling factors in machine computations (working in binary notation) are chosen to be powers of 2. Given any system (6.17), one can always find scaling factors that are powers of 10 and yield a system (6.18) that is nearly well-scaled or, more precisely, *equilibrated* in the sense that

$$0.1 < \max_j |r_ia_{ij}s_j| \le 1 \qquad \text{for all} \quad i$$

and

$$0.1 < \max_i |r_ia_{ij}s_j| \le 1 \qquad \text{for all} \quad j.$$

(In the binary system, powers of 10 would be replaced by powers of 2, and the lower bounds 0.1 would be replaced by 0.5.) The procedure is easy: first find scaling factors r_i such that

$$0.1 < \max_j |r_i a_{ij}| \le 1$$

for all i, and then complete your choice by suitable values of s_j. For instance, this procedure applied to (6.16) yields first $r_1 = 10^{-4}$, $r_2 = 1$ and then $s_1 = 1$, $s_2 = 1$. Next, the resulting system (6.13) is solved by Gaussian elimination with partial pivoting. As observed previously, the resulting computed solution is $\tilde{x}_1 = 1, \tilde{x}_2 = 1$. Note that this result is satisfactory even from the point of view of backward error analysis: $\tilde{x}_1, \tilde{x}_2$ constitute the precise solution of the system

$$\begin{aligned} \tilde{x}_1 + 9{,}999\tilde{x}_2 &= 10{,}000 \\ 0.5\tilde{x}_1 + \quad 0.5\tilde{x}_2 &= 1 \end{aligned}$$

whose coefficients differ relatively little from those of (6.16).

Scaling is a common practice whose theoretical properties are not completely understood. There is often more than one way of scaling a system (6.17) into an equilibrated system (6.18), and it is not clear, a priori, which choice of scaling factors yields the best results. For instance, setting $r_1 = 1$, $r_2 = 1$, $s_1 = 1$, $s_2 = 10^{-4}$ in (6.16), we obtain the equilibrated system

$$\begin{aligned} x_1 + \qquad\quad x_2' &= 10{,}000 \\ 0.5x_1 + 0.00005x_2' &= 1. \end{aligned}$$

Now Gaussian elimination with partial pivoting, working to three significant digits, delivers $x_2' = 10{,}000$ and $x_1 = 0$. Thus we obtain once again the unsatisfactory solution $x_1 = 0$, $x_2 = 1$. Fortunately, examples such as this illustrate the exception rather than the rule. In practice, Gaussian elimination with partial pivoting, preceded by some form of scaling (such as equilibration), produces results whose numerical accuracy is quite adequate. For more information on this subject, the reader is referred to G. Forsythe and C. B. Moler (1967).

Speed of Gaussian Elimination

Now we turn to an analysis of the speed of Gaussian elimination. As we have already observed, Gaussian elimination requires about $n^3/3$ multiplications to solve a system

$$\sum_{j=1}^{n} a_{ij}x_j = b_i \qquad (i = 1, 2, \ldots, n). \tag{6.19}$$

However, it may happen that many of these $n^3/3$ multiplications are multiplications by zero, in which case the number of genuine multiplications is considerably reduced. Typically (the atypical case is illustrated in problem 6.2), such reductions occur only if many of the coefficients a_{ij} in (6.19) are zeros. Systems with this property are called *sparse*, whereas systems with most of the coefficients a_{ij} nonzero are called *dense*. Large systems (6.19) encountered in practice are often very sparse: even when n runs to thousands or more, each variable x_j may occur with a nonzero coefficient a_{ij} in only five or ten equations. However, sparsity alone does not guarantee noticeable reductions in the running time of Gaussian elimination. For instance, consider the sparse system

$$\begin{aligned} x_1 + x_2 + x_3 + x_4 + x_5 + x_6 &= 4 \\ x_1 + 6x_2 \qquad\qquad\qquad\qquad &= 5 \\ x_1 \qquad + 6x_3 \qquad\qquad\qquad &= 5 \\ x_1 \qquad\qquad + 6x_4 \qquad\quad &= 5 \\ x_1 \qquad\qquad\qquad + 6x_5 \quad &= 5 \\ x_1 \qquad\qquad\qquad\qquad + 6x_6 &= 5. \end{aligned} \tag{6.20}$$

Eliminating x_1 from the first equation, we obtain the completely dense system

$$\begin{aligned}
5x_2 - x_3 - x_4 - x_5 - x_6 &= 1\\
-x_2 + 5x_3 - x_4 - x_5 - x_6 &= 1\\
-x_2 - x_3 + 5x_4 - x_5 - x_6 &= 1\\
-x_2 - x_3 - x_4 + 5x_5 - x_6 &= 1\\
-x_2 - x_3 - x_4 - x_5 + 5x_6 &= 1
\end{aligned}$$

and no time savings can be claimed in the subsequent iterations of Gaussian elimination. This example illustrates a phenomenon known as *fill-in*: when a few variables have been eliminated, the remaining system may become considerably denser than the original one.

Attempts to curb fill-in rely on freedom in the choice of pivots. The variables $x_1, x_2, \ldots, x_n$ can be eliminated in an arbitrary order, and each can be eliminated from an arbitrary equation in which it appears with a nonzero coefficient; a judicious choice of the next variable to be eliminated, and of the equation from which to eliminate it, may reduce fill-in considerably. An early proposal along these lines was made by H. M. Markowitz (1957). It goes as follows: let p_i stand for the number of nonzero coefficients in the ith equation, and let q_j stand for the number of equations involving x_j. In the first iteration, when a variable x_j has to be eliminated from an ith equation, choose i and j so that the product $(p_i - 1)(q_j - 1)$ is as small as possible (and, of course, $a_{ij} \neq 0$). In each subsequent iteration, follow the same rule, with the p's and q's updated. For example, in (6.20) we have

$$p_1 = 6, \quad p_2 = 2, \quad p_3 = 2, \quad p_4 = 2, \quad p_5 = 2, \quad p_6 = 2$$

and

$$q_1 = 6, \quad q_2 = 2, \quad q_3 = 2, \quad q_4 = 2, \quad q_5 = 2, \quad q_6 = 2.$$

Since the product $(p_i - 1)(q_j - 1)$ attains the minimum value 1 if and only if $2 \le i \le 6$ and $2 \le j \le 6$, Markowitz's rule directs us to eliminate a variable x_j with $2 \le j \le 6$ from the jth equation (we have to have $a_{ij} \neq 0$!). Breaking the tie arbitrarily, we eliminate x_2 from the second equation,

$$x_2 = \frac{5}{6} - \frac{1}{6}x_1$$

and obtain the system

$$\begin{aligned}
\frac{5}{6}x_1 + x_3 + x_4 + x_5 + x_6 &= \frac{19}{6}\\
x_1 + 6x_3 \qquad\qquad\qquad &= 5\\
x_1 \qquad + 6x_4 \qquad\qquad &= 5\\
x_1 \qquad\qquad + 6x_5 \qquad &= 5\\
x_1 \qquad\qquad\qquad + 6x_6 &= 5.
\end{aligned}$$

This is a significant improvement over the elimination of x_1 from the first equation in (6.20); now we have created no new nonzero coefficients at all. (Furthermore, this phenomenon persists throughout the remaining iterations as long as the pivots are selected by Markowitz's rule.)

We shall return to ways of curbing fill-in by judicious choice of pivots later in this chapter, after Gaussian elimination has been described in terms of matrices.

Summary of Elimination Methods

Gaussian elimination offers a great deal of flexibility in the choice of pivots; this flexibility may be exploited to enhance the accuracy as well as the speed of the method. When the choice of a pivot dictated by considerations of accuracy (partial pivoting) disagrees with a choice suggested by considerations of speed (fill-in reduction), it is usually the latter that gets preference. Considerations of accuracy take over only in extreme situations, such as when a choice of another equation (and the same variable) would increase the size of the pivot a thousand times or even more. (Of course, when Gaussian elimination is applied to completely dense systems, partial pivoting may be given full rein since all hopes of enhancing speed vanish.)

In many textbooks, Gaussian elimination is presented together with a variant, in which each x_k is eliminated not only from the remaining system of equations in $x_{k+1}, x_{k+2}, \ldots, x_n$ but also from the substitution formulas for $x_1, x_2, \ldots, x_{k-1}$. This method appeared in a posthumous edition of a textbook written by a German geometer, W. Jordan, and became known as *Gauss–Jordan elimination*. It is inferior to Gaussian elimination on at least three counts: it requires $n^3/2$ multiplications on completely dense systems, it is liable to suffer from more fill-in on sparse systems, and its numerical accuracy tends to be worse. For these reasons, we shall not discuss it.

MATRIX MULTIPLICATION

The system of linear equations

$$\begin{aligned} y_1 &= 4x_1 + 3x_2 \\ y_2 &= -x_1 + 4x_2 \\ y_3 &= -3x_1 + 2x_2 \\ y_4 &= 2x_1 + 5x_2 \end{aligned} \tag{6.21}$$

may be thought of as a device transforming inputs x_1, x_2 into outputs y_1, y_2, y_3, y_4:

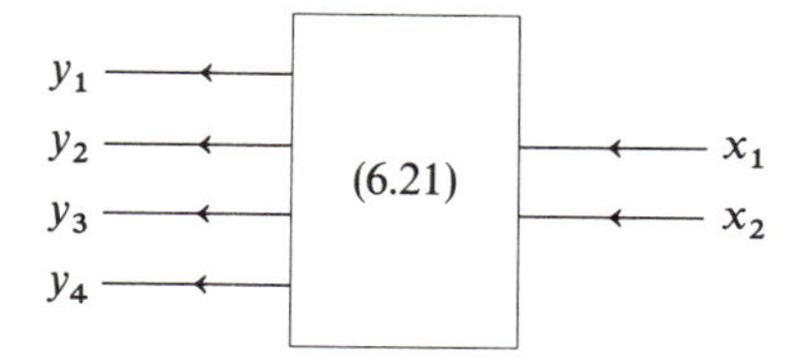

Similarly, the system

$$\begin{aligned} z_1 &= 5y_1 - 2y_2 + 3y_3 + y_4 \\ z_2 &= y_1 + 2y_2 - y_3 + 3y_4 \\ z_3 &= -y_1 - 3y_2 + 3y_3 + 2y_4 \end{aligned} \tag{6.22}$$

may be thought of as a device transforming inputs y_1, y_2, y_3, y_4 into outputs z_1, z_2, z_3:

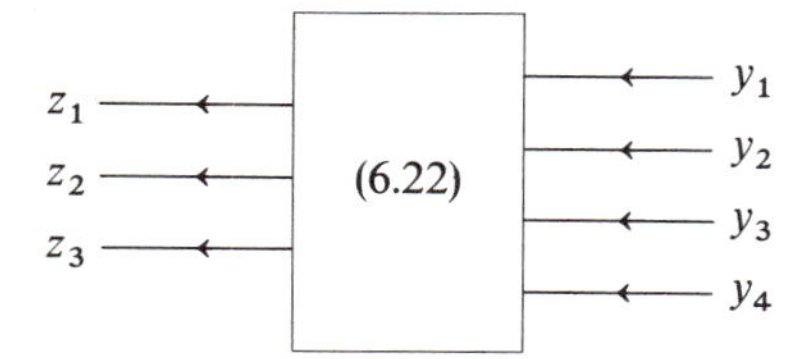

Plugging the outputs of (6.21) into the inputs of (6.22),

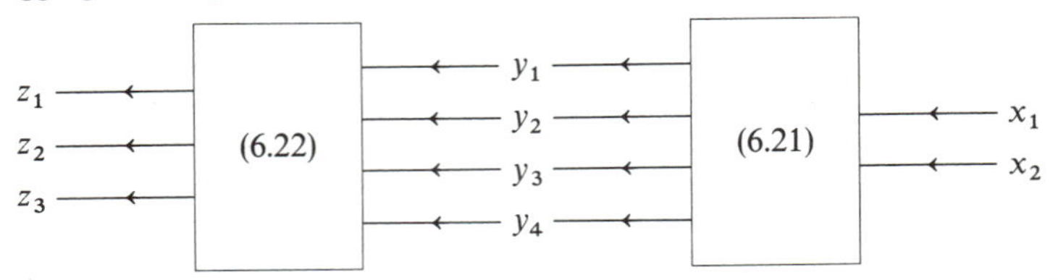

amounts to substituting for y_1, y_2, y_3, y_4 from (6.21) into (6.22): after simplification, we obtain

$$\begin{aligned} z_1 &= 15x_1 + 18x_2 \\ z_2 &= 11x_1 + 24x_2 \\ z_3 &= -6x_1 + x_2. \end{aligned} \tag{6.23}$$

Thus, the concatenation of the two devices (6.21) and (6.22) may be replaced by a single device:

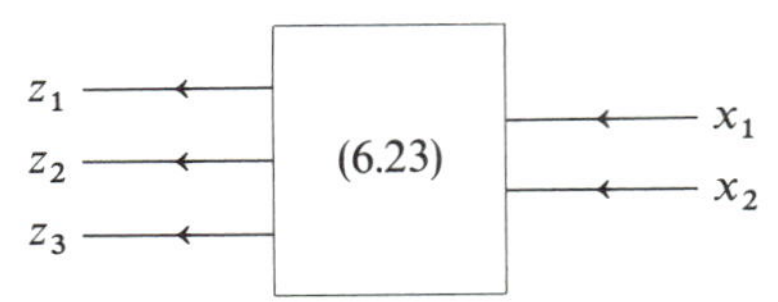

More generally, consider a device that transforms inputs $x_1, x_2, \ldots, x_n$ into outputs

$$y_k = \sum_{j=1}^{n} b_{kj}x_j \qquad (k = 1, 2, \ldots, p) \tag{6.24}$$

and another device that transforms inputs $y_1, y_2, \ldots, y_p$ into outputs

$$z_i = \sum_{k=1}^{p} a_{ik}y_k \qquad (i = 1, 2, \ldots, m). \tag{6.25}$$

Plugging the outputs of the first device into the inputs of the second amounts to substituting for $y_1, y_2, \ldots, y_p$ from (6.24) into (6.25):

$$z_i = \sum_{k=1}^{p} a_{ik}\left(\sum_{j=1}^{n} b_{kj}x_j\right) = \sum_{j=1}^{n}\left(\sum_{k=1}^{p} a_{ik}b_{kj}\right)x_j.$$

Again, the concatenation of the two devices may be replaced by a single device

$$z_i = \sum_{j=1}^{n} c_{ij}x_j \qquad (i = 1, 2, \ldots, m)$$

defined by

$$c_{ij} = \sum_{k=1}^{p} a_{ik}b_{kj}. \tag{6.26}$$

It is customary to represent systems of linear functions by rectangular arrays of the corresponding coefficients. Thus,

$$\mathbf{A} = \begin{bmatrix} 5 & -2 & 3 & 1 \\ 1 & 2 & -1 & 3 \\ -1 & -3 & 3 & 2 \end{bmatrix}, \quad \mathbf{B} = \begin{bmatrix} 4 & 3 \\ -1 & 4 \\ -3 & 2 \\ 2 & 5 \end{bmatrix}, \quad \text{and} \quad \mathbf{C} = \begin{bmatrix} 15 & 18 \\ 11 & 24 \\ -6 & 1 \end{bmatrix}$$

represent (6.22), (6.21), and (6.23), respectively. Such arrays are called *matrices*. [The term *matrix* first appeared in a memoir published by A. Cayley in 1858. For historical details, see T. Hawkins (1974).]

A matrix with m rows and n columns is referred to as a "matrix of size $m \times n$" or simply as an "$m \times n$ matrix." It is convenient to denote the entry in the ith row and in the jth column of a matrix $\mathbf{A}$ by a_{ij}, the corresponding entry in a matrix $\mathbf{B}$ by b_{ij}, and so on. This convention is often stressed in phrases such as "let $\mathbf{A} = (a_{ij})$ be an $m \times n$ matrix." The *product* **AB** of an $m \times p$ matrix **A** and a $p \times n$ matrix **B** is the $m \times n$ matrix **C** defined by (6.26). This definition is motivated by the fact that the matrix $\mathbf{C} = \mathbf{AB}$ represents the device obtained by concatenating the devices represented by **A** and **B**. (Note that the product **AB** is defined only if the number of rows in **B** matches the number of columns in **A**.) The terminology derives from the special case $m = p = n = 1$, when all three of the matrices **A**, **B**, and **AB** reduce to single numbers, and the number **AB** is the ordinary product of the numbers **A** and **B**. However, the familiar properties of number multiplication do not always extend to the more general context of matrix multiplication. For example, if **A** and **B** are matrices, then we do not necessarily have $\mathbf{AB} = \mathbf{BA}$: to begin with, one or both of these products may be undefined; and even if both **AB** and **BA** are defined (which is the case if and only if **A** and **B** are square matrices of the same size), they may be not equal to each other. For instance, if

$$\mathbf{A} = \begin{bmatrix} 1 & 1 \\ 0 & 1 \end{bmatrix}, \quad \mathbf{B} = \begin{bmatrix} 0 & 1 \\ 1 & 0 \end{bmatrix}$$

then

$$\mathbf{AB} = \begin{bmatrix} 1 & 1 \\ 1 & 0 \end{bmatrix}, \quad \mathbf{BA} = \begin{bmatrix} 0 & 1 \\ 1 & 1 \end{bmatrix}.$$

Nevertheless, just as $(ab)c = a(bc)$ for any choice of numbers a, b, c, so we have

$$(\mathbf{AB})\mathbf{C} = \mathbf{A}(\mathbf{BC}) \tag{6.27}$$

for any choice of matrices **A**, **B**, **C** such that the products **AB** and **BC** are defined. This property of matrix multiplication is known as its *associativity*. In view of our motivation of matrix multiplication, (6.27) is intuitively obvious: either of its two sides represents the device obtained by concatenating the three devices represented by **A**, **B**, **C**. Verifying (6.27) directly from the definition of matrix multiplication is easy; we omit the routine details. Once (6.27) is established, we may delete the parentheses on either side and write simply **ABC** without any ambiguity.

In number multiplication, the number "one" plays a special part: we have $1 \cdot x = x \cdot 1 = x$ for all numbers x. In matrix multiplication, this part is played by so-called *identity matrices*, square matrices with ones on the diagonal and zeros off the diagonal. For instance, we have

$$\begin{bmatrix} 1 & 0 & 0 \\ 0 & 1 & 0 \\ 0 & 0 & 1 \end{bmatrix} \cdot \begin{bmatrix} 5 & -2 & 3 & 1 \\ 1 & 2 & -1 & 3 \\ -1 & -3 & 4 & 2 \end{bmatrix} = \begin{bmatrix} 5 & -2 & 3 & 1 \\ 1 & 2 & -1 & 3 \\ -1 & -3 & 4 & 2 \end{bmatrix}$$

and

$$\begin{bmatrix} 5 & -2 & 3 & 1 \\ 1 & 2 & -1 & 3 \\ -1 & -3 & 4 & 2 \end{bmatrix} \cdot \begin{bmatrix} 1 & 0 & 0 & 0 \\ 0 & 1 & 0 & 0 \\ 0 & 0 & 1 & 0 \\ 0 & 0 & 0 & 1 \end{bmatrix} = \begin{bmatrix} 5 & -2 & 3 & 1 \\ 1 & 2 & -1 & 3 \\ -1 & -3 & 4 & 2 \end{bmatrix}.$$

More generally, with $\mathbf{I}_k$ standing for the identity matrix of size $k \times k$, we have

$$\mathbf{I}_m\mathbf{A} = \mathbf{A}\mathbf{I}_n = \mathbf{A}$$

for every $m \times n$ matrix **A**. This fact, apart from being easy to verify mechanically, is also intuitively obvious: the identity matrix $\mathbf{I}_k$ represents a transformation device that transfers the k inputs directly into the k outputs without a change. When there is no danger of confusion, we shall sometimes denote the identity matrix $\mathbf{I}_k$ simply by **I**.

A matrix having only one row is called a *row vector*, and a matrix having only one column is called a *column vector*. The entries in a (row or column) vector are referred to as its *components*, and their number is the vector's *length*. Note that an $m \times n$ matrix may be multiplied by a column vector of length n on the right, and by a row vector of length m on the left. (The following observation concerning matrix products **AB** is often handy: if **a** is the ith row of **A** and if **b** is the jth column of **B**, then **aB** is the ith row of **AB** and **Ab** is the jth column of **AB**.)

The formalism of matrices provides a succinct way of recording systems of linear equations. For instance, the system

$$\begin{aligned} 4x_1 + 3x_2 + 9x_3 &= 3 \\ 2x_1 - x_2 + 2x_3 &= 1 \\ 8x_1 + 7x_2 - 5x_3 &= 4 \end{aligned}$$

may be recorded as $\mathbf{Ax} = \mathbf{b}$ with

$$\mathbf{A} = \begin{bmatrix} 4 & 3 & 9 \\ 2 & -1 & 2 \\ 8 & 7 & -5 \end{bmatrix}, \quad \mathbf{x} = \begin{bmatrix} x_1 \\ x_2 \\ x_3 \end{bmatrix}, \quad \text{and } \mathbf{b} = \begin{bmatrix} 3 \\ 1 \\ 4 \end{bmatrix}.$$

More generally, every system of n linear equations in n variables may be recorded as $\mathbf{Ax} = \mathbf{b}$ such that **A** is an $n \times n$ matrix and **x**, **b** are column vectors of length n.

SPARSE AND SPECIALLY STRUCTURED MATRICES

As we have already observed, matrices **A** appearing in large systems $\mathbf{Ax} = \mathbf{b}$ found in practice are *sparse*: most of their entries a_{ij} are zeros. Such matrices may be multiplied fast and stored compactly. For instance, the matrix

$$\begin{bmatrix} 3 & 2 & 0 & 0 & 0 & 0 & 0 & 0 & 0 & 0 \\ 0 & 0 & 2 & 5 & 0 & 0 & 6 & 0 & 0 & 0 \\ 0 & 0 & 0 & 8 & 2 & 0 & 0 & 0 & 0 & 0 \\ 0 & 9 & 0 & 4 & 0 & 0 & 0 & 0 & 0 & 0 \\ 0 & 0 & 0 & 0 & 5 & 6 & 4 & 0 & 0 & 0 \\ 0 & 0 & 0 & 0 & 0 & 2 & 0 & 0 & 0 & 0 \\ 0 & 0 & 0 & 0 & 0 & 0 & 4 & 0 & 0 & 7 \\ 0 & 0 & 0 & 0 & 0 & 0 & 6 & 5 & 0 & 7 \\ 0 & 0 & 0 & 0 & 0 & 0 & 0 & 2 & 3 & 0 \\ 0 & 0 & 0 & 0 & 0 & 0 & 6 & 0 & 7 & 0 \end{bmatrix}$$

may be stored in the following "packed form":

column 1: 3 in row 1.
column 2: 2 in row 1, 9 in row 4.
column 3: 2 in row 2.
column 4: 5 in row 2, 8 in row 3, 4 in row 4.
column 5: 2 in row 3, 5 in row 5.
column 6: 6 in row 5, 2 in row 6.
column 7: 6 in row 2, 4 in row 5, 4 in row 7, 6 in row 8, 6 in row 10.
column 8: 5 in row 8, 2 in row 9.
column 9: 3 in row 9, 7 in row 10.
column 10: 7 in row 7, 7 in row 8.

The actual technical details involving memory locations and so on would lead us astray; the interested reader is referred to R. P. Tewarson (1973) and J. de Buchet (1971).

We shall sometimes emphasize sparsity by using a suggestive notation. For example, the sparse matrix shown previously might be recorded as

$$\begin{bmatrix} 3 & 2 & & & & & & & & \\ & & 2 & 5 & & & 6 & & & \\ & & & 8 & 2 & & & & & \\ & 9 & & 4 & & & & & & \\ & & & & 5 & 6 & 4 & & & \\ & & & & & 2 & & & & \\ & & & & & & 4 & & & 7 \\ & & & & & & 6 & 5 & & 7 \\ & & & & & & & 2 & 3 & \\ & & & & & & 6 & & 7 & \end{bmatrix}$$

and, more generally, some or all of the zero entries omitted in a presentation of a sparse matrix.

Three special kinds of matrices will play an important part in our discussions: permutation matrices, eta matrices, and triangular matrices. A *permutation matrix*, such as

$$\begin{bmatrix} & & & 1 \\ 1 & & & \\ & & 1 & \\ & 1 & & \end{bmatrix}$$

arises from the identity matrix by permuting its rows and/or columns. To put it differently, a permutation matrix has precisely one nonzero entry in each row and precisely one nonzero entry in each column; all of these nonzero entries are ones. An *eta matrix*, such as

$$\begin{bmatrix} 1 & & 3 & \\ & 1 & 5 & \\ & & 4 & \\ & & 2 & 1 \end{bmatrix}$$

differs from the identity matrix in only one column, referred to as its *eta column.* A matrix $\mathbf{A} = (a_{ij})$ is called *lower triangular* if $a_{ij} = 0$ whenever $j > i$ and *upper triangular* if $a_{ij} = 0$ whenever $j < i$. Thus,

$$\begin{bmatrix} 3 & & & \\ 5 & 6 & & \\ 0 & 9 & 8 & \\ 1 & 7 & 1 & 2 \end{bmatrix} \quad \text{and} \quad \begin{bmatrix} 4 & 1 & 2 & 0 \\ & 3 & 6 & 4 \\ & & 1 & 9 \\ & & & 8 \end{bmatrix}$$

are lower and upper triangular, respectively.

THE ALGEBRA OF MATRICES

Matrices can be not only multiplied; they can also be added. If $\mathbf{A} = (a_{ij})$ and $\mathbf{B} = (b_{ij})$ are matrices of the same size, then their *sum* $\mathbf{A} + \mathbf{B}$ is the matrix $\mathbf{C} = (c_{ij})$, defined by $c_{ij} = a_{ij} + b_{ij}$. This operation generalizes ordinary number addition and obeys several familiar laws of ordinary number arithmetic. For instance,

$$\mathbf{A} + \mathbf{B} = \mathbf{B} + \mathbf{A}$$
$$\mathbf{A} + (\mathbf{B} + \mathbf{C}) = (\mathbf{A} + \mathbf{B}) + \mathbf{C}$$

and

$$(\mathbf{A} + \mathbf{B})\mathbf{C} = \mathbf{AC} + \mathbf{BC}$$
$$\mathbf{A}(\mathbf{B} + \mathbf{C}) = \mathbf{AB} + \mathbf{AC}$$

whenever either side is defined. Finally, matrices may be multiplied by ordinary numbers: if t is a number and if $\mathbf{A} = (a_{ij})$ is a matrix, then $t\mathbf{A}$ denotes the matrix $\mathbf{B} = (b_{ij})$ defined by $b_{ij} = ta_{ij}$. Again, we have

$$t(\mathbf{A} + \mathbf{B}) = t\mathbf{A} + t\mathbf{B}$$
$$(s + t)\mathbf{A} = s\mathbf{A} + t\mathbf{A}$$

and so on.

For every $m \times n$ matrix $\mathbf{A} = (a_{ij})$, the $n \times m$ matrix (a_{ji}) is called the *transpose* of $\mathbf{A}$ and is denoted by $\mathbf{A}^T$. In particular, the transpose of every row vector is a column vector and vice versa. Relying on this convention, we shall often record the column vector with components $x_1, x_2, \ldots, x_n$ as $[x_1, x_2, \ldots, x_n]^T$.

GAUSSIAN ELIMINATION IN TERMS OF MATRICES

Suppose that we have eliminated x_1, x_2, x_3 from a system of six linear equations in $x_1, x_2, \ldots, x_6$. Now we are left with a system of three linear equations in x_4, x_5, x_6 and with back substitution formulas for x_1, x_2, x_3. For definiteness, let us write the formulas as

$$\begin{aligned} x_1 &= d_1 - c_{12}x_2 - c_{13}x_3 - c_{14}x_4 - c_{15}x_5 - c_{16}x_6 \\ x_2 &= d_2 \qquad\quad\; - c_{23}x_3 - c_{24}x_4 - c_{25}x_5 - c_{26}x_6 \\ x_3 &= d_3 \qquad\qquad\qquad\; - c_{34}x_4 - c_{35}x_5 - c_{36}x_6 \end{aligned} \tag{6.28}$$

and the remaining system as

$$\begin{aligned} c_{44}x_4 + c_{45}x_5 + c_{46}x_6 &= d_4 \\ c_{54}x_4 + c_{55}x_5 + c_{56}x_6 &= d_5 \\ c_{64}x_4 + c_{65}x_5 + c_{66}x_6 &= d_6. \end{aligned} \tag{6.29}$$

Now we may convert the first of the last three equations into

$$x_4 = \frac{d_4}{c_{44}} - \frac{c_{45}}{c_{44}} x_5 - \frac{c_{46}}{c_{44}} x_6 \tag{6.30}$$

and, substituting from (6.30) into (6.29), eliminate x_4:

$$\begin{aligned}\left(c_{55} - c_{54}\frac{c_{45}}{c_{44}}\right)x_5 + \left(c_{56} - c_{54}\frac{c_{46}}{c_{44}}\right)x_6 &= d_5 - c_{54}\frac{d_4}{c_{44}}\\ \left(c_{65} - c_{64}\frac{c_{45}}{c_{44}}\right)x_5 + \left(c_{66} - c_{64}\frac{c_{46}}{c_{44}}\right)x_6 &= d_6 - c_{64}\frac{d_4}{c_{44}}.\end{aligned} \tag{6.31}$$

In matrix notation, the intermediate results (6.28), (6.29) *before* the elimination of x_4 may be recorded as

$$\begin{bmatrix} 1 & c_{12} & c_{13} & c_{14} & c_{15} & c_{16} \\ & 1 & c_{23} & c_{24} & c_{25} & c_{26} \\ & & 1 & c_{34} & c_{35} & c_{36} \\ & & & c_{44} & c_{45} & c_{46} \\ & & & c_{54} & c_{55} & c_{56} \\ & & & c_{64} & c_{65} & c_{66} \end{bmatrix} \cdot \begin{bmatrix} x_1 \\ x_2 \\ x_3 \\ x_4 \\ x_5 \\ x_6 \end{bmatrix} = \begin{bmatrix} d_1 \\ d_2 \\ d_3 \\ d_4 \\ d_5 \\ d_6 \end{bmatrix} \tag{6.32}$$

whereas the intermediate results (6.28), (6.30), (6.31) *after* the elimination of x_4 may be recorded as

$$\begin{bmatrix} 1 & c_{12} & c_{13} & c_{14} & c_{15} & c_{16} \\ & 1 & c_{23} & c_{24} & c_{25} & c_{26} \\ & & 1 & c_{34} & c_{35} & c_{36} \\ & & & 1 & \dfrac{c_{45}}{c_{44}} & \dfrac{c_{46}}{c_{44}} \\ & & & & c_{55} - c_{54}\dfrac{c_{45}}{c_{44}} & c_{56} - c_{54}\dfrac{c_{46}}{c_{44}} \\ & & & & c_{65} - c_{64}\dfrac{c_{45}}{c_{44}} & c_{66} - c_{64}\dfrac{c_{46}}{c_{44}} \end{bmatrix} \cdot \begin{bmatrix} x_1 \\ x_2 \\ x_3 \\ x_4 \\ x_5 \\ x_6 \end{bmatrix} = \begin{bmatrix} d_1 \\ d_2 \\ d_3 \\ \dfrac{d_4}{c_{44}} \\ d_5 - c_{54}\dfrac{d_4}{c_{44}} \\ d_6 - c_{64}\dfrac{d_4}{c_{44}} \end{bmatrix} \tag{6.33}$$

Note that the transformation of (6.32) into (6.33) amounts to multiplying both sides of (6.32) by the matrix

$$\mathbf{L} = \begin{bmatrix} 1 & & & & & \\ & 1 & & & & \\ & & 1 & & & \\ & & & \dfrac{1}{c_{44}} & & \\ & & & -\dfrac{c_{54}}{c_{44}} & 1 & \\ & & & -\dfrac{c_{64}}{c_{44}} & & 1 \end{bmatrix}$$

on the left. Of course, if $c_{44} = 0$, then this transformation cannot be carried out and we are forced to use a different row of (6.29) in order to eliminate x_4. For instance, if we use the last row of (6.29), then we obtain

$$x_4 = \frac{d_6}{c_{64}} - \frac{c_{65}}{c_{64}} x_5 - \frac{c_{66}}{c_{64}} x_6$$

and

$$\left(c_{55} - c_{54}\frac{c_{65}}{c_{64}}\right) x_5 + \left(c_{56} - c_{54}\frac{c_{66}}{c_{64}}\right) x_6 = d_5 - c_{54}\frac{d_6}{c_{64}}$$

$$\left(c_{45} - c_{44}\frac{c_{65}}{c_{64}}\right) x_5 + \left(c_{46} - c_{44}\frac{c_{66}}{c_{64}}\right) x_6 = d_4 - c_{44}\frac{d_6}{c_{64}}.$$

In matrix terms again, this operation may be viewed as passing from (6.32) to

$$\begin{bmatrix} 1 & c_{12} & c_{13} & c_{14} & c_{15} & c_{16} \\ & 1 & c_{23} & c_{24} & c_{25} & c_{26} \\ & & 1 & c_{34} & c_{35} & c_{36} \\ & & & c_{64} & c_{65} & c_{66} \\ & & & c_{54} & c_{55} & c_{56} \\ & & & c_{44} & c_{45} & c_{46} \end{bmatrix} \cdot \begin{bmatrix} x_1 \\ x_2 \\ x_3 \\ x_4 \\ x_5 \\ x_6 \end{bmatrix} = \begin{bmatrix} d_1 \\ d_2 \\ d_3 \\ d_6 \\ d_5 \\ d_4 \end{bmatrix} \tag{6.34}$$

and then to

$$\begin{bmatrix} 1 & c_{12} & c_{13} & c_{14} & c_{15} & c_{16} \\ & 1 & c_{23} & c_{24} & c_{25} & c_{26} \\ & & 1 & c_{34} & c_{35} & c_{36} \\ & & & 1 & \dfrac{c_{65}}{c_{64}} & \dfrac{c_{66}}{c_{64}} \\ & & & & c_{55} - c_{54}\dfrac{c_{65}}{c_{64}} & c_{56} - c_{54}\dfrac{c_{66}}{c_{64}} \\ & & & & c_{45} - c_{44}\dfrac{c_{65}}{c_{64}} & c_{46} - c_{44}\dfrac{c_{66}}{c_{64}} \end{bmatrix} \cdot \begin{bmatrix} x_1 \\ x_2 \\ x_3 \\ x_4 \\ x_5 \\ x_6 \end{bmatrix} = \begin{bmatrix} d_1 \\ d_2 \\ d_3 \\ \dfrac{d_6}{c_{64}} \\ d_5 - c_{54}\dfrac{d_6}{c_{64}} \\ d_4 - c_{44}\dfrac{d_6}{c_{64}} \end{bmatrix}. \tag{6.35}$$

Note that the passage from (6.32) to (6.34) amounts to multiplying both sides of (6.32) by the matrix

$$\mathbf{P} = \begin{bmatrix} 1 & & & & & \\ & 1 & & & & \\ & & 1 & & & \\ & & & & & 1 \\ & & & & 1 & \\ & & & 1 & & \end{bmatrix}$$

on the left and that the passage from (6.34) to (6.35) amounts to multiplying both sides of (6.34) by the matrix

$$\mathbf{L} = \begin{bmatrix} 1 & & & & & \\ & 1 & & & & \\ & & 1 & & & \\ & & & \dfrac{1}{c_{64}} & & \\ & & & -\dfrac{c_{54}}{c_{64}} & 1 & \\ & & & -\dfrac{c_{44}}{c_{64}} & & 1 \end{bmatrix}$$

on the left.

More generally, having eliminated $x_1, x_2, \ldots, x_{k-1}$ in this order from a system of n linear equations in $x_1, x_2, \ldots, x_n$, we may record our intermediate results in matrix notation as

$$\mathbf{Cx} = \mathbf{d}. \tag{6.36}$$

The first $k - 1$ rows of this system provide substitution formulas for $x_1, x_2, \ldots, x_{k-1}$, whereas the last $n - k + 1$ rows constitute the residual system of equations in $x_k, x_{k+1}, \ldots, x_n$. Hence, the matrix $\mathbf{C} = (c_{ij})$ has a rather special structure. In its first $k - 1$ columns, all the entries on the diagonal are ones, whereas all the entries below the diagonal are zeros. The subsequent elimination of x_k yields a similar system,

$$\bar{\mathbf{C}}\mathbf{x} = \bar{\mathbf{d}} \tag{6.37}$$

in the first k columns of $\bar{\mathbf{C}}$, all the entries on the diagonal are ones whereas all the entries below the diagonal are zeros. As in our example, the transformation of (6.36) into (6.37) amounts to multiplying both sides of (6.36) on the left first by a permutation matrix $\mathbf{P} = \mathbf{P}_k$ and then by a lower triangular eta matrix $\mathbf{L} = \mathbf{L}_k$. The permutation matrix $\mathbf{P}$ is obtained from the identity matrix by interchanging its row k with some row r such that $r \geq k$ and $c_{rk} \neq 0$. (We shall comment later on the case $c_{ik} = 0$ for all $i = k, k + 1, \ldots, n$.) Thus, the matrix $\mathbf{T} = \mathbf{PC}$ arises from $\mathbf{C}$ by interchanging its rows k and r. In particular, writing $\mathbf{T} = (t_{ij})$ we have $t_{kk} = c_{rk} \neq 0$. The lower triangular eta matrix $\mathbf{L}_k$ differs from the identity matrix only in its kth column whose entries l_{ik} are defined by

$$l_{ik} = \begin{cases} 0 & \text{if } i < k \\ \dfrac{1}{t_{kk}} & \text{if } i = k \\ -\dfrac{t_{ik}}{t_{kk}} & \text{if } i > k \end{cases}.$$

Thus, the new system (6.37) reads $\mathbf{L}_k\mathbf{P}_k\mathbf{Cx} = \mathbf{L}_k\mathbf{P}_k\mathbf{d}$. The elimination process applied to a system $\mathbf{Ax} = \mathbf{b}$ may be seen as moving through a sequence of systems

$$\begin{aligned} \mathbf{Ax} &= \mathbf{b} \\ \mathbf{L}_1\mathbf{P}_1\mathbf{Ax} &= \mathbf{L}_1\mathbf{P}_1\mathbf{b} \\ \mathbf{L}_2\mathbf{P}_2\mathbf{L}_1\mathbf{P}_1\mathbf{Ax} &= \mathbf{L}_2\mathbf{P}_2\mathbf{L}_1\mathbf{P}_1\mathbf{b} \\ &\vdots \\ \mathbf{L}_n\mathbf{P}_n\mathbf{L}_{n-1}\mathbf{P}_{n-1}\cdots\mathbf{L}_1\mathbf{P}_1\mathbf{Ax} &= \mathbf{L}_n\mathbf{P}_n\mathbf{L}_{n-1}\mathbf{P}_{n-1}\cdots\mathbf{L}_1\mathbf{P}_1\mathbf{b}. \end{aligned}$$

Since the last system in this sequence consists entirely of substitution formulas, the matrix

$$\mathbf{U} = \mathbf{L}_n\mathbf{P}_n\mathbf{L}_{n-1}\mathbf{P}_{n-1} \cdots \mathbf{L}_1\mathbf{P}_1\mathbf{A}$$

is upper triangular and all of its diagonal entries are ones.

Triangular Factorization

In Gaussian elimination, as described at the beginning of this chapter, the matrices $\mathbf{P}_1, \mathbf{L}_1, \ldots, \mathbf{P}_n, \mathbf{L}_n$ are not stored explicitly; instead,

$$\mathbf{C} = \mathbf{L}_{k-1}\mathbf{P}_{k-1} \cdots \mathbf{L}_1\mathbf{P}_1\mathbf{A} \quad \text{and} \quad \mathbf{d} = \mathbf{L}_{k-1}\mathbf{P}_{k-1} \cdots \mathbf{L}_1\mathbf{P}_1\mathbf{b}$$

are updated into

$$\bar{\mathbf{C}} = \mathbf{L}_k\mathbf{P}_k\mathbf{C} \quad \text{and} \quad \bar{\mathbf{d}} = \mathbf{L}_k\mathbf{P}_k\mathbf{d}$$

in each iteration. Nevertheless, the description given just now suggests a way of computing $\mathbf{P}_1, \mathbf{L}_1, \ldots, \mathbf{P}_n, \mathbf{L}_n$ and $\mathbf{U} = \mathbf{L}_n\mathbf{P}_n \cdots \mathbf{L}_1\mathbf{P}_1\mathbf{A}$ explicitly and independently of the right-hand side $\mathbf{b}$ in the original system. That is, in each iteration, having just computed $\mathbf{P}_k$ and $\mathbf{L}_k$, we merely have to update $\mathbf{C}$ into $\bar{\mathbf{C}} = \mathbf{L}_k\mathbf{P}_k\mathbf{C}$. Thus the total amount of work involved in computing $\mathbf{P}_1, \mathbf{L}_1, \ldots, \mathbf{P}_n, \mathbf{L}_n$ and $\mathbf{U}$ is roughly equal to the total amount of work involved in solving

$$\mathbf{A}\mathbf{x} = \mathbf{b} \tag{6.38}$$

by Gaussian elimination. Since $\mathbf{U}$ and the eta matrices $\mathbf{L}_1, \mathbf{L}_2, \ldots, \mathbf{L}_n$ are triangular, we shall refer to the matrices $\mathbf{P}_1, \mathbf{L}_1, \ldots, \mathbf{P}_n, \mathbf{L}_n$ and $\mathbf{U}$ as a *triangular factorization* of $\mathbf{A}$. (Strictly speaking, this term is a misnomer, as it is $\mathbf{U}$ rather than $\mathbf{A}$ that is being factorized. Nevertheless, we shall let it stand. A closely related concept is the *LU-decomposition* of $\mathbf{A}$; see problem 6.11.)

Once the triangular factorization of $\mathbf{A}$ is available, system (6.38) may be solved fast for an arbitrary right-hand side $\mathbf{b}$. We first compute

$$\mathbf{v} = \mathbf{L}_n\mathbf{P}_n\mathbf{L}_{n-1}\mathbf{P}_{n-1} \cdots \mathbf{L}_1\mathbf{P}_1\mathbf{b}$$

and then solve the system

$$\mathbf{U}\mathbf{x} = \mathbf{v}.$$

To see that this procedure yields the correct solution $\mathbf{x}$ of (6.38), recall that $\mathbf{U}\mathbf{x} = \mathbf{v}$ is precisely the set of substitution formulas that would have been obtained by Gaussian elimination. In particular, as $\mathbf{U} = (u_{ij})$ is an upper triangular matrix with $u_{ii} = 1$ for all i, solving the system $\mathbf{U}\mathbf{x} = \mathbf{v}$ amounts to a simple back substitution. We obtain the values of $x_n, x_{n-1}, \ldots, x_1$ in this order by setting

$$x_i = v_i - \sum_{j=i+1}^{n} u_{ij}x_j$$

in each iteration. Thus, once $\mathbf{v}$ is known, the total amount of work involved in solving $\mathbf{U}\mathbf{x} = \mathbf{v}$ comes to only $n(n-1)/2$ multiplications and the same number of additions; if the matrix $\mathbf{U}$ is sparse, then this number reduces even further. Similarly, the long chain calculation of $\mathbf{v}$ may be implemented quickly: we first compute $\mathbf{P}_1\mathbf{b}$, then $\mathbf{L}_1(\mathbf{P}_1\mathbf{b})$, then $\mathbf{P}_2(\mathbf{L}_1\mathbf{P}_1\mathbf{b})$, then $\mathbf{L}_2(\mathbf{P}_2\mathbf{L}_1\mathbf{P}_1\mathbf{b})$, and so on. Each multiplication of a $\mathbf{P}_k$ by a column vector on the right is done without any arithmetic; each multiplication of an $\mathbf{L}_k$ by a column vector on the right takes only $n - k + 1$ multiplications and $n - k$ additions. Thus, the total amount of work involved in computing $\mathbf{v}$ comes to only $n(n+1)/2$ multiplications and $n(n-1)/2$ additions; again, if the kth column of each $\mathbf{L}_k$ is very sparse, then these numbers are reduced much further.

The practical impact of these observations is clear: when faced with the task of solving several different systems

$$\mathbf{Ax} = \mathbf{b}, \quad \mathbf{Ax} = \mathbf{b}', \quad \mathbf{Ax} = \mathbf{b}'', \quad \cdots \tag{6.39}$$

with the same left-hand side $\mathbf{Ax}$, one is well-advised to first "preprocess" the matrix $\mathbf{A}$ by finding its triangular factorization. The preprocessing step takes about as much work as solving just one of the systems (6.39) by Gaussian elimination: about $n^3/3$ multiplications and $n^3/3$ additions when $\mathbf{A}$ is completely dense and possibly much less when $\mathbf{A}$ is sparse. As soon as this step is executed, each of the systems (6.39) may be solved relatively quickly: the number of multiplications and the number of additions are roughly equal to the total number of nonzero entries in the matrix $\mathbf{U}$ and in the eta columns of $\mathbf{L}_1, \mathbf{L}_2, \ldots, \mathbf{L}_k$. Even in the completely dense case, n^2 multiplications and $n^2 - n$ additions suffice.

Similarly, every system $\mathbf{yA} = \mathbf{c}$, with $\mathbf{y}$ and $\mathbf{c}$ standing for row rather than column vectors, may be solved by first solving the triangular system $\mathbf{wU} = \mathbf{c}$ and then evaluating $\mathbf{y} = \mathbf{wL}_n\mathbf{P}_n \cdots \mathbf{L}_1\mathbf{P}_1$; now $\mathbf{yA} = \mathbf{wU} = \mathbf{c}$. Again, no more than $n^2/2$ multiplications and additions are required in solving $\mathbf{wU} = \mathbf{c}$ and roughly the same number of operations suffice to evaluate $\mathbf{y} = (((\mathbf{wL}_n)\mathbf{P}_n) \cdots \mathbf{L}_1)\mathbf{P}_1$.

Computing Accurate and Sparse Triangular Factorizations

The tricks used to enhance the accuracy and speed of Gaussian elimination in solving systems of linear equations apply just as well in computing triangular factorizations. In particular, partial pivoting amounts to a judicious choice of the permutation matrix $\mathbf{P}_k$ in each iteration, so that each off-diagonal entry $-t_{ik}/t_{kk}$ in the eta column of $\mathbf{L}_k$ has a size of at most one. A more detailed explanation is called for in the case of tricks such as Markowitz's pivot selection, which change the order in which the variables $x_1, x_2, \ldots, x_n$ are eliminated.

To begin, note that the sequence of pivots selected by Markowitz's strategy may be determined without carrying out the numerical computations. For an illustration, consider the system

$$\begin{aligned}
a_{11}x_1 \qquad\qquad\qquad\qquad\qquad\qquad\qquad\qquad + a_{16}x_6 &= b_1 \\
a_{21}x_1 + a_{22}x_2 \qquad\qquad + a_{24}x_4 + a_{25}x_5 + a_{26}x_6 &= b_2 \\
a_{31}x_1 \qquad\qquad + a_{33}x_3 \qquad\qquad\qquad\qquad\qquad\qquad &= b_3 \\
a_{41}x_1 \qquad\qquad + a_{43}x_3 + a_{44}x_4 + a_{45}x_5 \qquad\qquad &= b_4 \\
a_{54}x_4 + a_{55}x_5 \qquad\qquad &= b_5 \\
a_{62}x_2 + a_{63}x_3 + a_{64}x_4 \qquad\qquad + a_{66}x_6 &= b_6.
\end{aligned} \tag{6.40}$$

Markowitz's selection rule is independent of the actual values of the coefficients a_{ij} shown in (6.40): as long as all of these coefficients are nonzero, the rule tells us to choose a_{16}, a_{33} or a_{55} [all of them having $(p_i - 1)(q_j - 1) = 2$] for the first pivot. Breaking the tie arbitrarily, we choose a_{16}. Unless lucky cancellations (such as those illustrated in problem 6.2) occur, the system that results from the elimination of x_6 from the first equation in (6.40) will read

$$\begin{aligned}
a'_{21}x_1 + a'_{22}x_2 \qquad\qquad + a'_{24}x_4 + a'_{25}x_5 &= b'_2 \\
a'_{31}x_1 \qquad\qquad + a'_{33}x_3 \qquad\qquad\qquad\qquad &= b'_3 \\
a'_{41}x_1 \qquad\qquad + a'_{43}x_3 + a'_{44}x_4 + a'_{45}x_5 &= b'_4 \\
a'_{54}x_4 + a'_{55}x_5 &= b'_5 \\
a'_{61}x_1 + a'_{62}x_2 + a'_{63}x_3 + a'_{64}x_4 \qquad\qquad &= b'_6
\end{aligned}$$

(with the nonzero coefficient $a'_{61} = -a_{11}a_{66}/a_{16}$ created by fill-in). In the second iteration, Markowitz's rule suggests a'_{33} or a'_{55} [both of them having $(p_i - 1)(q_j - 1) = 2$ again] as the pivot; breaking the tie arbitrarily, we choose a'_{33}. Again, the resulting system of equations may be recorded symbolically, the next pivot chosen, and so on. As the reader may verify, a continuation of this process may lead to the sequence of pivots

$$a_{16}, \quad a_{33}, \quad a_{55}, \quad a_{22}, \quad a_{41}, \quad a_{64} \tag{6.41}$$

(where we write a_{33} for a'_{33} and so on).

Gaussian elimination with a prescribed sequence of pivots yields itself to a neat description in terms of matrices. The trick is to permute the variables and the equations of the system $\mathbf{Ax} = \mathbf{b}$ in such a way that, in the resulting system $\tilde{\mathbf{A}}\tilde{\mathbf{x}} = \tilde{\mathbf{b}}$, the prescribed pivots appear on the diagonal in the prescribed order. We leave it to the reader to verify (problem 6.6) that

$$\tilde{\mathbf{A}} = \mathbf{PAQ}, \quad \tilde{\mathbf{x}} = \mathbf{Q}^T\mathbf{x}, \quad \tilde{\mathbf{b}} = \mathbf{Pb}$$

with permutation matrices $\mathbf{P}$ and $\mathbf{Q}$ determined by the prescribed sequence of pivots: if the kth pivot in this sequence is a_{ij}, then the kth row of $\mathbf{P}$ is the ith row of the identity matrix $\mathbf{I}$ and the kth column of $\mathbf{Q}$ is the jth column of $\mathbf{I}$. For instance, if $\mathbf{Ax} = \mathbf{b}$ is (6.40), and the prescribed sequence of pivots is (6.41), then $\tilde{\mathbf{A}}\tilde{\mathbf{x}} = \tilde{\mathbf{b}}$ reads

$$\begin{array}{llllllll}
a_{16}x_6 & & & & + \, a_{11}x_1 & & = b_1 \\
& a_{33}x_3 & & & + \, a_{31}x_1 & & = b_3 \\
& & a_{55}x_5 & & & + \, a_{54}x_4 & = b_5 \\
a_{26}x_6 & & + \, a_{25}x_5 & + \, a_{22}x_2 & + \, a_{21}x_1 & + \, a_{24}x_4 & = b_2 \\
& a_{43}x_3 & + \, a_{45}x_5 & & a_{41}x_1 & + \, a_{44}x_4 & = b_4 \\
a_{66}x_6 & + \, a_{63}x_3 & & + \, a_{62}x_2 & & + \, a_{64}x_4 & = b_6
\end{array}$$

and we have

$$\mathbf{P} = \begin{bmatrix} 1 & & & & & \\ & & 1 & & & \\ & & & & 1 & \\ & 1 & & & & \\ & & & 1 & & \\ & & & & & 1 \end{bmatrix}, \quad \mathbf{Q} = \begin{bmatrix} & & & & 1 & \\ & & & 1 & & \\ & 1 & & & & \\ & & & & & 1 \\ & & 1 & & & \\ 1 & & & & & \end{bmatrix}.$$

Now Gaussian elimination applied to $\mathbf{Ax} = \mathbf{b}$ with the prescribed sequence of pivots is nothing but Gaussian elimination applied to $\tilde{\mathbf{A}}\tilde{\mathbf{x}} = \tilde{\mathbf{b}}$ with the first variable eliminated from the first equation in the first iteration, the second variable eliminated from the second equation in the second iteration, and so on. Thus, it transforms $\tilde{\mathbf{A}}\tilde{\mathbf{x}} = \tilde{\mathbf{b}}$ first into $\tilde{\mathbf{L}}_1\tilde{\mathbf{A}}\tilde{\mathbf{x}} = \tilde{\mathbf{L}}_1\tilde{\mathbf{b}}$, then into $\tilde{\mathbf{L}}_2\tilde{\mathbf{L}}_1\tilde{\mathbf{A}}\tilde{\mathbf{x}} = \tilde{\mathbf{L}}_2\tilde{\mathbf{L}}_1\tilde{\mathbf{b}}$, and so on until an upper triangular system

$$\tilde{\mathbf{L}}_n\tilde{\mathbf{L}}_{n-1} \cdots \tilde{\mathbf{L}}_1\tilde{\mathbf{A}}\tilde{\mathbf{x}} = \tilde{\mathbf{L}}_n\tilde{\mathbf{L}}_{n-1} \cdots \tilde{\mathbf{L}}_1\tilde{\mathbf{b}}$$

is obtained.

When fill-in in Gaussian elimination is curbed by a suitable choice of a pivot sequence, the resulting triangular factorization

$$\tilde{\mathbf{L}}_n\tilde{\mathbf{L}}_{n-1} \cdots \tilde{\mathbf{L}}_1\tilde{\mathbf{A}} = \tilde{\mathbf{U}}$$

is sparser than the triangular factorization

$$\mathbf{L}_n\mathbf{L}_{n-1} \cdots \mathbf{L}_1\mathbf{A} = \mathbf{U}.$$

(This is the case in our example; as the reader may verify by symbolic computations, fill-in creates four extra nonzeros in the triangular factorization of $\mathbf{A}$ and only one in the triangular factorization of $\tilde{\mathbf{A}}$.) Thus Markowitz's pivot selection may be viewed as producing permutation matrices $\mathbf{P}$ and $\mathbf{Q}$ with the aim of reducing the number of nonzeros in the triangular factorization of $\tilde{\mathbf{A}} = \mathbf{PAQ}$.

Why is it important to curb the number of nonzeros in the triangular factorization of $\tilde{\mathbf{A}} = \mathbf{PAQ}$? Sparse triangular factorizations are computed faster than dense ones, but that is not the main point. As we have remarked, every system $\mathbf{Ax} = \mathbf{b}$ may be permuted into $\tilde{\mathbf{A}}\tilde{\mathbf{x}} = \tilde{\mathbf{b}}$ with $\tilde{\mathbf{x}} = \mathbf{Q}^T\mathbf{x}$ and $\tilde{\mathbf{b}} = \mathbf{Pb}$; similarly (problem 6.6 again), every system $\mathbf{yA} = \mathbf{c}$ may be permuted into $\tilde{\mathbf{y}}\tilde{\mathbf{A}} = \tilde{\mathbf{c}}$, with $\tilde{\mathbf{y}} = \mathbf{yP}^T$ and $\tilde{\mathbf{c}} = \mathbf{cQ}$. Hence, $\mathbf{Ax} = \mathbf{b}$ may be solved by first permuting $\mathbf{b}$ into $\tilde{\mathbf{b}}$, then solving $\tilde{\mathbf{A}}\tilde{\mathbf{x}} = \tilde{\mathbf{b}}$, and finally permuting $\tilde{\mathbf{x}}$ into $\mathbf{x}$; similarly, $\mathbf{yA} = \mathbf{c}$ may be solved by first permuting $\mathbf{c}$ into $\tilde{\mathbf{c}}$, then solving $\tilde{\mathbf{y}}\tilde{\mathbf{A}} = \tilde{\mathbf{c}}$, and finally permuting $\tilde{\mathbf{y}}$ into $\mathbf{y}$. In either case, this roundabout procedure requires only one multiplication and one addition per each nonzero in the triangular factorization of $\tilde{\mathbf{A}}$ (not counting the diagonal ones in $\tilde{\mathbf{L}}_1, \tilde{\mathbf{L}}_2, \ldots, \tilde{\mathbf{L}}_n$). Thus, sparser triangular factorizations of $\tilde{\mathbf{A}}$ allow faster solutions of systems $\mathbf{Ax} = \mathbf{b}$ and $\mathbf{yA} = \mathbf{c}$.

The early work of Markowitz was followed by intensive research into reducing the number of nonzeros in the triangular factorization of $\mathbf{PAQ}$ by an appropriate choice of permutation matrices $\mathbf{P}$ and $\mathbf{Q}$. Reducing this number to its absolute minimum seems difficult [in fact, D. J. Rose and R. E. Tarjan (1978) *proved* that it is difficult when $\mathbf{Q}$ is required to be $\mathbf{P}^T$]. Nevertheless, algorithms for finding a good (although not necessarily the best) choice of $\mathbf{P}$ and $\mathbf{Q}$ may be guided by simple observations on the nature of fill-in in computing the triangular factorization

$$\tilde{\mathbf{L}}_n\tilde{\mathbf{L}}_{n-1} \cdots \tilde{\mathbf{L}}_1\tilde{\mathbf{A}} = \tilde{\mathbf{U}}. \tag{6.42}$$

For instance (problem 6.7), if the kth column of $\tilde{\mathbf{A}}$ has no nonzeros above the diagonal, then the pattern of nonzeros in this column is carried intact into the eta column of $\tilde{\mathbf{L}}_k$ and, in addition, the kth column of $\tilde{\mathbf{U}}$ has no nonzeros at all except the diagonal one. To put it differently, fill-in may occur only in those columns where $\tilde{\mathbf{A}}$ has nonzero above the diagonal; such columns are called *spikes*.

A particularly successful procedure, designed by E. Hellerman and D. C. Rarick (1971, 1972), aims to permute $\mathbf{A}$ into an $\tilde{\mathbf{A}} = \mathbf{PAQ}$, with the nonzeros above the diagonal confined to small "bumps", as illustrated in Figure 6.1, and with only a few spikes in each bump. On sparse

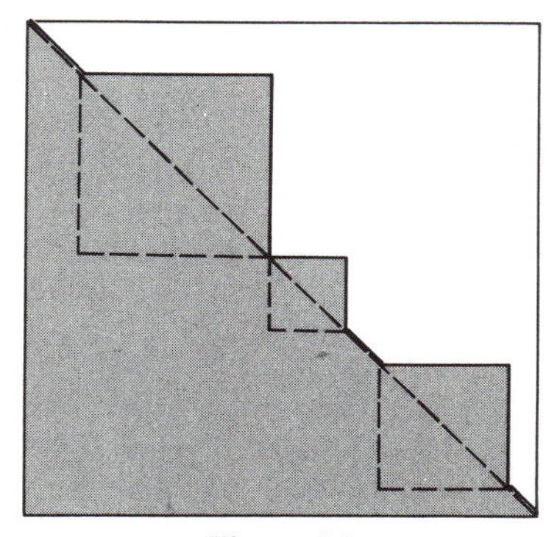

Figure 6.1

matrices **A** that are encountered in practice, typically with up to ten nonzeros per column, this procedure yields excellent results. Usually, the number of nonzeros in the triangular factorization of $\tilde{\mathbf{A}}$ (discounting, as usual, the diagonal ones in $\tilde{\mathbf{L}}_1, \tilde{\mathbf{L}}_2, \ldots, \tilde{\mathbf{L}}_n$) is within 120% of the number of nonzeros in **A**, and the total number of spikes is less than a hundred or so even when **A** has several thousand columns. (Of course, these empirical findings hinge on the special structure of sparse matrices encountered in practice, which allows these matrices to be permuted into the near-triangular form. On randomly generated sparse matrices, even the most sophisticated procedures would fail miserably: since the nonzeros in such matrices **A** form a patternless mosaic, nearly all columns in any **PAQ** are spikes and fill-in is catastrophic.)

The potential conflict between accuracy and speed, discussed earlier in the context of solving systems of linear equations, arises in the same way when triangular factorizations are computed, and may be resolved by the same compromise. More precisely, the kth pivot in the sequence prescribed by **P** and **Q** may vanish or at least become dangerously small in the first $k - 1$ iterations of Gaussian elimination, in which case (6.42) has to be replaced by

$$\tilde{\mathbf{L}}_n\tilde{\mathbf{P}}_n \cdots \tilde{\mathbf{L}}_1\tilde{\mathbf{P}}_1\tilde{\mathbf{A}} = \tilde{\mathbf{U}}$$

with $\tilde{\mathbf{P}}_k \neq \mathbf{I}$. The vague phrase "dangerously small" may be interpreted as "more than c times smaller than another entry in the last $n - k$ rows of the kth column of $\tilde{\mathbf{L}}_{k-1}\tilde{\mathbf{P}}_{k-1} \cdots \tilde{\mathbf{L}}_1\tilde{\mathbf{P}}_1\tilde{\mathbf{A}}$" for some fixed value of c. Setting c equal to 100 or so represents a concession to the rigid rules of partial pivoting (which would require c to be equal to 1) and yet it forces only a few of the matrices $\tilde{\mathbf{P}}_k$ to differ from **I**. As a result, numerical accuracy hardly suffers, while the number of nonzeros in the triangular factorization increases only slightly (typically by a fraction of a percent).

SINGULAR AND NONSINGULAR MATRICES

Gaussian elimination will fail to deliver a triangular factorization of **A** if and only if there will be no available pivots in some iteration. If this happens in the kth iteration, then the matrix

$$\mathbf{C} = \mathbf{L}_{k-1}\mathbf{P}_{k-1} \cdots \mathbf{L}_1\mathbf{P}_1\mathbf{A}$$

has a rather special structure. In its first $k - 1$ columns, all the entries on the diagonal are ones and all the entries below the diagonal are zeros; furthermore, $c_{ik} = 0$ for all $i = k, k + 1, \ldots, n$. For instance, if $n = 6$ and $k = 4$, then

$$\mathbf{C} = \begin{bmatrix} 1 & c_{12} & c_{13} & c_{14} & c_{15} & c_{16} \\ & 1 & c_{23} & c_{24} & c_{25} & c_{26} \\ & & 1 & c_{34} & c_{35} & c_{36} \\ & & & & c_{45} & c_{46} \\ & & & & c_{55} & c_{56} \\ & & & & c_{65} & c_{66} \end{bmatrix}.$$

In this case, defining

$$\begin{aligned} w_3 &= c_{34} \\ w_2 &= c_{24} - c_{23}w_3 \\ w_1 &= c_{14} - c_{12}w_2 - c_{13}w_3 \end{aligned}$$

and $w_4 = -1, w_5 = 0, w_6 = 0$, we obtain

$$\mathbf{C}\cdot\begin{bmatrix} w_1 \\ w_2 \\ w_3 \\ w_4 \\ w_5 \\ w_6 \end{bmatrix} = \begin{bmatrix} 0 \\ 0 \\ 0 \\ 0 \\ 0 \\ 0 \end{bmatrix}.$$

In the last equation, all the components of the right-hand side vector are zeros; such vectors are called *zero vectors* and are denoted by $\mathbf{0}$ when there is no danger of confusion between row and column vectors of different lengths. Our example with $n = 6$ and $k = 4$ illustrates a general procedure for finding a column vector $\mathbf{w}$ such that $\mathbf{w} \neq \mathbf{0}$ but $\mathbf{Cw} = \mathbf{0}$: having defined $w_{k-1}, \ldots, w_1$ in this order by

$$w_i = c_{ik} - \sum_{j=i+1}^{k-1} c_{ij} w_j$$

we set w_k equal to -1 and w_i equal to 0 whenever $i > k$. Now we are in a position to prove a fundamental fact.

THEOREM 6.1. Every square matrix $\mathbf{A}$ has precisely one of the following two properties:

(i) For every right-hand side $\mathbf{b}$, the system $\mathbf{Ax} = \mathbf{b}$ has either no solution at all or infinitely many solutions.

(ii) For every right-hand side $\mathbf{b}$, the system $\mathbf{Ax} = \mathbf{b}$ has precisely one solution.

PROOF. The procedure discussed above either finds those matrices $\mathbf{P}_1, \mathbf{L}_1, \ldots, \mathbf{P}_{k-1}, \mathbf{L}_{k-1}$ and a nonzero vector $\mathbf{w}$ such that

$$\mathbf{L}_{k-1}\mathbf{P}_{k-1}\cdots\mathbf{L}_1\mathbf{P}_1\mathbf{Aw} = \mathbf{0}$$

or else it delivers a triangular factorization of $\mathbf{A}$. In the former case, we recall that, for every right-hand side $\mathbf{b}$, the two systems

$$\mathbf{Ax} = \mathbf{b} \quad \text{and} \quad \mathbf{L}_{k-1}\mathbf{P}_{k-1}\cdots\mathbf{L}_1\mathbf{P}_1\mathbf{Ax} = \mathbf{L}_{k-1}\mathbf{P}_{k-1}\cdots\mathbf{L}_1\mathbf{P}_1\mathbf{b}$$

have the same set of solutions. In particular, setting $\mathbf{b}$ equal to $\mathbf{0}$, we conclude that $\mathbf{Aw} = \mathbf{0}$. But then $\mathbf{A}$ has property (i): as soon as $\mathbf{Ax} = \mathbf{b}$ for some $\mathbf{x}$, we also have $\mathbf{A}(\mathbf{x} + t\mathbf{w}) = \mathbf{b}$ for every real number t. On the other hand, if the procedure delivers a triangular factorization $\mathbf{P}_1, \mathbf{L}_1\mathbf{P}_2, \mathbf{L}_2, \ldots, \mathbf{P}_n, \mathbf{L}_n, \mathbf{U}$ of $\mathbf{A}$, then we recall that, for every right-hand side $\mathbf{b}$, the two systems

$$\mathbf{Ax} = \mathbf{b} \quad \text{and} \quad \mathbf{Ux} = \mathbf{L}_n\mathbf{P}_n\cdots\mathbf{L}_1\mathbf{P}_1\mathbf{b}$$

have the same set of solutions. Since the triangular system $\mathbf{Ux} = \mathbf{v}$ has precisely one solution for every right-hand side $\mathbf{v}$, it follows that $\mathbf{A}$ has property (ii) in this case. ■

Because property (i) is atypical, matrices with this property are called *singular*. Matrices with property (ii) are called *nonsingular*.

The Inverse of a Matrix

Returning to the similarities between matrix multiplication and number multiplication, we observe that the *reciprocal* a^{-1} of a number a is a number b such that $ab = ba = 1$. By analogy, we might say that a matrix **B** is a reciprocal of a matrix **A** if each of the two products **AB**, **BA** is an identity matrix. Nowadays, the term *reciprocal* sounds a little archaic; it has been replaced by the term *inverse*. In order for **B** to be an inverse of **A**, both products **AB** and **BA** must be defined; hence **A** and **B** must be square matrices of the same size.

THEOREM 6.2. A matrix has an inverse if and only if it is nonsingular.

PROOF. To establish the "if" part, consider a nonsingular $n \times n$ matrix **A**. For each of the n columns **b** of the $n \times n$ identity matrix **I**, the system $\mathbf{Ax} = \mathbf{b}$ has a solution; these n solutions may be seen as the n columns of a matrix **X** such that $\mathbf{AX} = \mathbf{I}$. We claim that $\mathbf{XA} = \mathbf{I}$, and so **X** is an inverse of **A**. Our claim may be justified by observing that $\mathbf{A(XA)} = \mathbf{(AX)A} = \mathbf{A}$ and that $\mathbf{AI} = \mathbf{A}$: now if **XA** and **I** differed in their kth columns, then the system $\mathbf{Ax} = \mathbf{b}$, with **b** standing for the kth column of **A**, would have two different solutions (namely, the kth column of **XA** and the kth column of **I**), contradicting the nonsingularity of **A**.

To establish the "only if" part, consider an $n \times n$ matrix **A** along with its inverse **B** and an arbitrary vector **b**. If $\mathbf{Ax} = \mathbf{b}$, then $\mathbf{x} = \mathbf{BAx} = \mathbf{Bb}$; on the other hand, $\mathbf{A(Bb)} = \mathbf{b}$. Hence, **Bb** is the unique solution of $\mathbf{Ax} = \mathbf{b}$ and **A** is nonsingular. ■

Note that an inverse of **A**, if it exists at all, is unique: $\mathbf{BA} = \mathbf{I}$ and $\mathbf{AC} = \mathbf{I}$ imply $\mathbf{B} = \mathbf{BI} = \mathbf{B(AC)} = \mathbf{(BA)C} = \mathbf{IC} = \mathbf{C}$. Hence, we may speak of *the* inverse, rather than *an* inverse, of **A**. Just as the reciprocal of a number a is denoted by a^{-1}, the inverse of a matrix **A** is denoted by $\mathbf{A}^{-1}$.

Inverses will interest us only as theoretical concepts, never as computational devices. The reasons for this are summarized in Box 6.1.

BOX 6.1 Why Inverses Should Not Be Used in Solving Systems of Linear Equations

One might be misled into the belief that the inverse $\mathbf{A}^{-1}$ is a handy device for solving systems $\mathbf{Ax} = \mathbf{b}$: the straight evaluation of $\mathbf{x} = \mathbf{A}^{-1}\mathbf{b}$ takes only n^2 multiplications and $n^2 - n$ additions. However, there are several good reasons for using a triangular factorization of **A** instead.

(i) Computing $\mathbf{A}^{-1}$ takes considerably longer (about three times as long in the completely dense case) than computing the triangular factorization.

(ii) Since extra calculations are involved in computing $\mathbf{A}^{-1}$, the results are more likely to suffer from rounding errors.

(iii) The calculation of $\mathbf{v} = \mathbf{L}_n\mathbf{P}_n \cdots \mathbf{L}_1\mathbf{P}_1\mathbf{b}$ and the solution of $\mathbf{Ux} = \mathbf{v}$ are considerably streamlined when the matrix **U** and the eta columns of $\mathbf{L}_1, \mathbf{L}_2, \ldots, \mathbf{L}_n$ are sparse. (As we have seen, this is the case when **A** is sparse and has only a few spikes.) However, similar time-savings cannot be claimed in the straight evaluation of $\mathbf{x} = \mathbf{A}^{-1}\mathbf{b}$: even very sparse matrices tend to have dense inverses.

PROBLEMS

6.1 Scale the system

$$\begin{aligned}
0.3x_1 + 0.08x_2 + 0.002x_3 + 4x_4 + 0.0002x_5 &= 1.19\\
2{,}000x_1 + 200x_2 + 60x_3 + 20{,}000x_4 + 3x_5 &= 1{,}500\\
60x_1 + 6x_2 + 0.7x_3 + 500x_4 + 0.04x_5 &= 28\\
400x_1 + 40x_2 + 8x_3 + 2{,}000x_4 + 0.3x_5 &= 210\\
5x_1 + 0.2x_2 + 0.06x_3 + 30x_4 + 0.004x_5 &= 2.
\end{aligned}$$

△ **6.2** Solve the system

$$\begin{aligned}
3x_1 - 9x_2 + 6x_3 + 3x_4 - 3x_5 - 6x_6 &= -3\\
2x_1 + x_2 + 4x_3 + 2x_4 - 2x_5 - 4x_6 &= -2\\
x_1 + 2x_2 + 5x_3 + x_4 - x_5 - 2x_6 &= 4\\
3x_1 + 5x_2 + x_3 + 2x_4 - 3x_5 - 6x_6 &= 1\\
x_1 + 4x_2 - 2x_3 + 3x_4 + 2x_5 - 2x_6 &= 8\\
2x_1 + 3x_2 + x_3 - 4x_4 + 5x_5 + x_6 &= 2.
\end{aligned}$$

△ **6.3** Solve the system

$$\begin{aligned}
4x_1 + 3x_4 + 2x_5 &= 14\\
2x_2 + 3x_5 + 4x_6 &= 18\\
3x_1 + 2x_2 + 6x_5 &= 19\\
3x_1 + 2x_3 &= 7\\
6x_1 + 4x_3 + 5x_5 &= 24\\
5x_3 + 6x_4 + 2x_6 &= 26.
\end{aligned}$$

△ **6.4** Find a triangular factorization of the matrix

$$\mathbf{A} = \begin{bmatrix} 1 & 3 & -2 & 4\\ 1 & 5 & -1 & 5\\ 1 & 3 & -3 & 6\\ -1 & -3 & 3 & -8 \end{bmatrix}.$$

Use this result to solve the systems $\mathbf{Ax} = \mathbf{b}$, $\mathbf{Ax} = \mathbf{b}'$, $\mathbf{Ax} = \mathbf{b}''$ with

$$\mathbf{b} = \begin{bmatrix} 6\\ 10\\ 7\\ -1 \end{bmatrix}, \quad \mathbf{b}' = \begin{bmatrix} 8\\ 12\\ 9\\ -11 \end{bmatrix}, \quad \mathbf{b}'' = \begin{bmatrix} 1\\ 4\\ 0\\ 0 \end{bmatrix}$$

and the systems $\mathbf{yA} = \mathbf{c}$, $\mathbf{yA} = \mathbf{c}'$ with

$$\mathbf{c} = [7, 23, -16, 33], \quad \mathbf{c}' = [2, 8, -3, 9].$$

6.5 Prove that $(\mathbf{AB})^T = \mathbf{B}^T\mathbf{A}^T$.

6.6 Let $\mathbf{P}$ and $\mathbf{Q}$ be permutation matrices of the same size as a square matrix $\mathbf{A}$. Prove that $\tilde{\mathbf{A}}\tilde{\mathbf{x}} = \tilde{\mathbf{b}}$ with $\tilde{\mathbf{A}} = \mathbf{PAQ}$, $\tilde{\mathbf{x}} = \mathbf{Q}^T\mathbf{x}$, and $\tilde{\mathbf{b}} = \mathbf{Pb}$ is just another way of recording the system $\mathbf{Ax} = \mathbf{b}$, and that $\tilde{\mathbf{y}}\tilde{\mathbf{A}} = \tilde{\mathbf{c}}$ with $\tilde{\mathbf{A}} = \mathbf{PAQ}$, $\tilde{\mathbf{y}} = \mathbf{yP}^T$, $\tilde{\mathbf{c}} = \mathbf{cQ}$ is just another way of recording the system $\mathbf{yA} = \mathbf{c}$.

6.7 The statement that "in computing the triangular factorization $\mathbf{L}_n\mathbf{L}_{n-1}\cdots\mathbf{L}_1\mathbf{A} = \mathbf{U}$, fill-in occurs in the jth column" means that $\mathbf{L}_j$ has a nonzero entry l_{ij} where $a_{ij} = 0$ or $\mathbf{U}$ has a nonzero entry u_{ij} where $a_{ij} = 0$. Prove that fill-in is restricted to the spikes of $\mathbf{A}$.

6.8 Prove that the product of two lower triangular matrices is lower triangular and that the inverse of a nonsingular lower triangular matrix is lower triangular.

6.9 Prove that the product $\mathbf{AB}$ of two square matrices is nonsingular if and only if both $\mathbf{A}$ and $\mathbf{B}$ are nonsingular.

6.10 Let $\mathbf{L}_n\mathbf{P}_n \cdots \mathbf{L}_1\mathbf{P}_1\mathbf{A} = \mathbf{U}$ be a triangular factorization of $\mathbf{A}$; let matrices $\mathbf{Q}_n, \mathbf{Q}_{n-1}, \ldots, \mathbf{Q}_0$ be defined by $\mathbf{Q}_n = \mathbf{I}$ and $\mathbf{Q}_{k-1} = \mathbf{Q}_k\mathbf{P}_k$ $(k = n, n-1, \ldots, 1)$ and let $\mathbf{L}_k^*$ stand for the matrix $\mathbf{Q}_k\mathbf{L}_k\mathbf{Q}_k^T$. Prove that:

(i) Each $\mathbf{Q}_k$ is a permutation matrix agreeing with $\mathbf{I}$ in the first k rows and columns.

(ii) $\mathbf{L}_n\mathbf{P}_n \cdots \mathbf{L}_1\mathbf{P}_1 = \mathbf{L}_n^* \cdots \mathbf{L}_1^*\mathbf{Q}_0$.

(iii) Each $\mathbf{L}_k^*$ is a lower triangular eta matrix whose eta column is the kth column.

6.11 An *LU-decomposition* of a matrix $\mathbf{A}$ consists of a lower triangular matrix $\mathbf{L}$, an upper triangular matrix $\mathbf{U}$, and a permutation matrix $\mathbf{P}$ such that $\mathbf{PA} = \mathbf{LU}$. Use the results of problems 6.8, 6.9, and 6.10 to prove that every nonsingular matrix has an LU-decomposition.

6.12 At the end of this chapter, the claim was made that "even very sparse matrices tend to have dense inverses." Artificially constructed examples do not support empirical claims such as this one very convincingly, but here is one anyway:

$$\text{if}\quad \mathbf{A} = \begin{bmatrix} 1 & & & & 1 \\ 1 & 1 & & & \\ & 1 & 1 & & \\ & & 1 & 1 & \\ & & & 1 & 1 \end{bmatrix} \quad \text{then}\quad \mathbf{A}^{-1} = \begin{bmatrix} 0.5 & 0.5 & -0.5 & 0.5 & -0.5 \\ -0.5 & 0.5 & 0.5 & -0.5 & 0.5 \\ 0.5 & -0.5 & 0.5 & 0.5 & -0.5 \\ -0.5 & 0.5 & -0.5 & 0.5 & 0.5 \\ 0.5 & -0.5 & 0.5 & -0.5 & 0.5 \end{bmatrix}.$$

Compute a triangular factorization of $\mathbf{A}$ and compare its density with the density of $\mathbf{A}^{-1}$. Generalize this example to arbitrary sizes $n \times n$ such that n is odd. (Why is there no generalization with n even?)

7

The Revised Simplex Method

In each iteration of the simplex method, one basic feasible solution is replaced by another. When the old solution is represented by a dictionary, the new solution is easy to find, and only a small part of the dictionary is actually used for that purpose. This part may be reconstructed directly from the original data and the new solution found without any reference to dictionaries. The resulting implementations of the simplex method are known under the generic name of the *revised simplex method*; the implementation of the simplex method that updates a dictionary in each iteration is known as the *standard simplex method*. We shall see that each iteration of the revised simplex method requires solving two systems of linear equations. Typically, these systems are not solved from scratch; instead, some device facilitating their solution is used and updated in each iteration. The device presented in this chapter is essentially the same as the popular "product form of the inverse" developed by G. B. Dantzig and W. Orchard-Hays (1954).

Each iteration of the revised simplex method may or may not take less time than the corresponding iteration of the standard simplex method. The outcome of this comparison depends not only on the particular implementation of the revised simplex method but also on the nature of the data. We shall see that, on the typical large and sparse LP problems solved in applications, the revised simplex method works faster than the standard simplex method. This is the reason why modern computer programs for solving LP problems always use some form of the revised simplex method.

MATRIX DESCRIPTION OF DICTIONARIES

Our preliminary task is to develop an understanding of the relationship between dictionaries and the original data. For illustration, we shall consider the dictionary

$$\begin{array}{rcl} x_1 &=& 54 - 0.5x_2 - 0.5x_4 - 0.5x_5 + 0.5x_6 \\ x_3 &=& 63 - 0.5x_2 - 0.5x_4 + 0.5x_5 - 1.5x_6 \\ x_7 &=& 15 + 0.5x_2 - 0.5x_4 + 0.5x_5 + 2.5x_6 \\ \hline z &=& 1782 - 2.5x_2 + 1.5x_4 - 3.5x_5 - 8.5x_6 \end{array} \tag{7.1}$$

arising from the problem

$$\begin{array}{ll} \text{maximize} & 19x_1 + 13x_2 + 12x_3 + 17x_4 \\ \text{subject to} & 3x_1 + 2x_2 + x_3 + 2x_4 \leq 225 \\ & x_1 + x_2 + x_3 + x_4 \leq 117 \\ & 4x_1 + 3x_2 + 3x_3 + 4x_4 \leq 420 \\ & x_1, x_2, x_3, x_4 \geq 0 \end{array} \tag{7.2}$$

after two iterations of the standard simplex method. To relate the coefficients in (7.1) to the data in (7.2), we first recall that the top three equations in the dictionary are equivalent to the three equations

$$\begin{array}{rcl} 3x_1 + 2x_2 + x_3 + 2x_4 + x_5 & & = 225 \\ x_1 + x_2 + x_3 + x_4 \quad + x_6 & & = 117 \\ 4x_1 + 3x_2 + 3x_3 + 4x_4 \quad\quad + x_7 & & = 420. \end{array} \tag{7.3}$$

Hence they arise by solving (7.3) for x_1, x_3, and x_7. In matrix terms, this solution may be described quite compactly. First, we record system (7.3) as $\mathbf{Ax} = \mathbf{b}$ with

$$\mathbf{A} = \begin{bmatrix} 3 & 2 & 1 & 2 & 1 & 0 & 0 \\ 1 & 1 & 1 & 1 & 0 & 1 & 0 \\ 4 & 3 & 3 & 4 & 0 & 0 & 1 \end{bmatrix}, \quad \mathbf{b} = \begin{bmatrix} 225 \\ 117 \\ 420 \end{bmatrix}, \quad \mathbf{x} = \begin{bmatrix} x_1 \\ x_2 \\ x_3 \\ x_4 \\ x_5 \\ x_6 \\ x_7 \end{bmatrix}.$$

To emphasize the fact that only the basic variables x_1, x_3, x_7 are treated as unknowns, we write $\mathbf{Ax}$ as $\mathbf{A}_B\mathbf{x}_B + \mathbf{A}_N\mathbf{x}_N$ with

$$\mathbf{A}_B = \begin{bmatrix} 3 & 1 & 0 \\ 1 & 1 & 0 \\ 4 & 3 & 1 \end{bmatrix}, \qquad \mathbf{A}_N = \begin{bmatrix} 2 & 2 & 1 & 0 \\ 1 & 1 & 0 & 1 \\ 3 & 4 & 0 & 0 \end{bmatrix}, \qquad \mathbf{x}_B = \begin{bmatrix} x_1 \\ x_3 \\ x_7 \end{bmatrix}, \qquad \mathbf{x}_N = \begin{bmatrix} x_2 \\ x_4 \\ x_5 \\ x_6 \end{bmatrix}$$

and then cast the system $\mathbf{Ax} = \mathbf{b}$ in the form

$$\mathbf{A}_B\mathbf{x}_B = \mathbf{b} - \mathbf{A}_N\mathbf{x}_N. \tag{7.4}$$

Since the square matrix $\mathbf{A}_B$ happens to be nonsingular, both sides of (7.4) may be multiplied by $\mathbf{A}_B^{-1}$ on the left. Thus we obtain

$$\mathbf{x}_B = \mathbf{A}_B^{-1}\mathbf{b} - \mathbf{A}_B^{-1}\mathbf{A}_N\mathbf{x}_N \tag{7.5}$$

which is a compact record of the top three equations in (7.1). To obtain the fourth equation, we record the objective function z as $\mathbf{cx}$ with

$$\mathbf{c} = [19, 13, 12, 17, 0, 0, 0]$$

or, more suggestively, as $\mathbf{c}_B\mathbf{x}_B + \mathbf{c}_N\mathbf{x}_N$, with

$$\mathbf{c}_B = [19, 12, 0] \quad \text{and} \quad \mathbf{c}_N = [13, 17, 0, 0].$$

Substituting for $\mathbf{x}_B$ from (7.5) we obtain

$$z = \mathbf{c}_B(\mathbf{A}_B^{-1}\mathbf{b} - \mathbf{A}_B^{-1}\mathbf{A}_N\mathbf{x}_N) + \mathbf{c}_N\mathbf{x}_N = \mathbf{c}_B\mathbf{A}_B^{-1}\mathbf{b} + (\mathbf{c}_N - \mathbf{c}_B\mathbf{A}_B^{-1}\mathbf{A}_N)\mathbf{x}_N.$$

Thus, dictionary (7.1) may be recorded in matrix terms as

$$\begin{array}{l} \mathbf{x}_B = \mathbf{A}_B^{-1}\mathbf{b} - \mathbf{A}_B^{-1}\mathbf{A}_N\mathbf{x}_N \\ \hline z \;\;= \mathbf{c}_B\mathbf{A}_B^{-1}\mathbf{b} + (\mathbf{c}_N - \mathbf{c}_B\mathbf{A}_B^{-1}\mathbf{A}_N)\mathbf{x}_N \end{array}. \tag{7.6}$$

More generally, consider an arbitrary LP problem in the standard form

$$\begin{array}{lll} \text{maximize} & \sum_{j=1}^{n} c_j x_j & \\ \text{subject to} & \sum_{j=1}^{n} a_{ij}x_j \le b_i & (i = 1, 2, \ldots, m) \\ & x_j \ge 0 & (j = 1, 2, \ldots, n). \end{array}$$

After the introduction of the slack variables $x_{n+1}, x_{n+2}, \ldots, x_{n+m}$, this problem may be recorded as

$$\begin{array}{ll} \text{maximize} & \mathbf{cx} \\ \text{subject to} & \mathbf{Ax} = \mathbf{b} \\ & \mathbf{x} \ge \mathbf{0}. \end{array}$$

(The matrix $\mathbf{A}$ has m rows and $n + m$ columns, of which the last m form the identity matrix. The column vector $\mathbf{x}$ has length $n + m$ and the column vector $\mathbf{b}$ has length m. The row vector $\mathbf{c}$ has length $n + m$ and its last m components are zeros.) Each basic feasible solution $\mathbf{x}^*$ of this problem partitions $x_1, x_2, \ldots, x_{n+m}$ into m basic and n nonbasic variables. As in our example, this partition induces a partition of $\mathbf{A}$ into $\mathbf{A}_B$ and $\mathbf{A}_N$, a partition of $\mathbf{x}$ into $\mathbf{x}_B$ and $\mathbf{x}_N$, and a partition of $\mathbf{c}$ into $\mathbf{c}_B$ and $\mathbf{c}_N$. We propose to show that

$$\text{matrix } \mathbf{A}_B \text{ is nonsingular} \tag{7.7}$$

by showing that the system $\mathbf{A}_B\mathbf{x}_B = \mathbf{b}$ has precisely one solution. The existence of a solution is evident: since the basic feasible solution $\mathbf{x}^*$ satisfies $\mathbf{A}\mathbf{x}^* = \mathbf{b}$ and $\mathbf{x}_N^* = \mathbf{0}$, it satisfies $\mathbf{A}_B\mathbf{x}_B^* = \mathbf{A}\mathbf{x}^* - \mathbf{A}_N\mathbf{x}_N^* = \mathbf{b}$. To verify that there are no other solutions, consider an arbitrary vector $\tilde{\mathbf{x}}_B$ such that $\mathbf{A}_B\tilde{\mathbf{x}}_B = \mathbf{b}$ and set $\tilde{\mathbf{x}}_N = \mathbf{0}$. Since the resulting vector $\tilde{\mathbf{x}}$ satisfies $\mathbf{A}\tilde{\mathbf{x}} = \mathbf{A}_B\tilde{\mathbf{x}}_B + \mathbf{A}_N\tilde{\mathbf{x}}_N = \mathbf{b}$, it must satisfy the top m equations in the dictionary representing $\mathbf{x}^*$. But then $\tilde{\mathbf{x}}_N = \mathbf{0}$ implies $\tilde{\mathbf{x}}_B = \mathbf{x}_B^*$. Thus the proof of (7.7) is completed. Now the arguments given above show that the dictionary representing $\mathbf{x}^*$ has the form (7.6). The matrix $\mathbf{A}_B$ is called the *basis matrix* or (when there is no danger of confusion with the set of basic variables) simply the *basis*. It is customary to denote the basis matrix by $\mathbf{B}$ rather than $\mathbf{A}_B$. We shall bow to this convention and record the dictionary as

$$\begin{array}{rl} \mathbf{x}_B & = \mathbf{B}^{-1}\mathbf{b} - \mathbf{B}^{-1}\mathbf{A}_N\mathbf{x}_N \\ \hline z & = \mathbf{c}_B\mathbf{B}^{-1}\mathbf{b} + (\mathbf{c}_N - \mathbf{c}_B\mathbf{B}^{-1}\mathbf{A}_N)\mathbf{x}_N. \end{array}$$

Of course, $\mathbf{B}^{-1}\mathbf{b}$ is nothing but the vector $\mathbf{x}_B^*$ specifying the current values of the basic variables.

THE REVISED SIMPLEX METHOD

In each iteration of the simplex method, we first choose the entering variable, then find the leaving variable, and finally update the current basic feasible solution. An examination of the way these tasks are carried out in the standard simplex method will lead us to the alternative, the revised simplex method. For illustration, we shall consider the update of the feasible dictionary (7.1) in the standard simplex method. The corresponding iteration of the revised simplex method begins with

$$\mathbf{x}_B^* = \begin{bmatrix} x_1^* \\ x_3^* \\ x_7^* \end{bmatrix} = \begin{bmatrix} 54 \\ 63 \\ 15 \end{bmatrix} \quad \text{and} \quad \mathbf{B} = \begin{bmatrix} 3 & 1 & 0 \\ 1 & 1 & 0 \\ 4 & 3 & 1 \end{bmatrix}.$$

The entering variable may be any nonbasic variable with a positive coefficient in the last row of the dictionary. As previously observed, the coefficients in this row form the vector $\mathbf{c}_N - \mathbf{c}_B\mathbf{B}^{-1}\mathbf{A}_N$. If the standard simplex method is used, then this vector is readily available as part of the dictionary; in our example, we have

$$z = \cdots - 2.5x_2 + 1.5x_4 - 3.5x_5 - 8.5x_6. \tag{7.8}$$

If the revised simplex method is used, then the vector $\mathbf{c}_N - \mathbf{c}_B\mathbf{B}^{-1}\mathbf{A}_N$ is computed in two steps: first we find $\mathbf{y} = \mathbf{c}_B\mathbf{B}^{-1}$ by solving the system $\mathbf{yB} = \mathbf{c}_B$ and then we calculate $\mathbf{c}_N - \mathbf{yA}_N$. In our example, we would first solve the system

$$[y_1, y_2, y_3] \cdot \begin{bmatrix} 3 & 1 & 0 \\ 1 & 1 & 0 \\ 4 & 3 & 1 \end{bmatrix} = [19, 12, 0]$$

to find $\mathbf{y} = [y_1, y_2, y_3] = [3.5, 8.5, 0]$ and then we would calculate

$$[13, 17, 0, 0] - [3.5, 8.5, 0] \cdot \begin{bmatrix} 2 & 2 & 1 & 0 \\ 1 & 1 & 0 & 1 \\ 3 & 4 & 0 & 0 \end{bmatrix} = [-2.5, 1.5, -3.5, -8.5]$$

to find the vector featured in (7.8). As the only positive component of this vector is its second component, the second component x_4 of the vector $\mathbf{x}_N = [x_2, x_4, x_5, x_6]^T$ enters the basis. Incidentally, note that the components of $\mathbf{c}_N - \mathbf{yA}_N$ may be calculated individually; if a nonbasic variable x_j corresponds to a component c_j of $\mathbf{c}_N$ and to a column $\mathbf{a}$ of $\mathbf{A}_N$, then the corresponding component of $\mathbf{c}_N - \mathbf{yA}_N$ equals $c_j - \mathbf{ya}$. Thus, the entering variable may be any nonbasic variable x_j for which $\mathbf{ya} < c_j$. The corresponding column $\mathbf{a}$ of $\mathbf{A}$ is called the *entering column.*

To determine the leaving variable, we increase the value t of the entering variable from zero to some positive level, maintaining the values of the remaining nonbasic variables at their zero levels and adjusting the values of the basic variables so as to preserve the constraints $\mathbf{Ax} = \mathbf{b}$. As t increases, the values of the basic variables change until a variable whose value is the first to drop to zero leaves the basis. To find the leaving variable and the largest admissible value of t, we have to know how precisely the values of the basic variables change with the changes of t. If the standard simplex method is used, then this information is readily available as part of the dictionary; in our example, we have

$$\begin{aligned} x_1 &= 54 \cdots - 0.5x_4 \cdots & \qquad & x_1 = 54 - 0.5t \\ x_3 &= 63 \cdots - 0.5x_4 \cdots, & \text{and so} \quad & x_3 = 63 - 0.5t. \\ x_7 &= 15 \cdots - 0.5x_4 \cdots & & x_7 = 15 - 0.5t \end{aligned} \tag{7.9}$$

More generally, the top m equations of the dictionary read $\mathbf{x}_B = \mathbf{x}_B^* - \mathbf{B}^{-1}\mathbf{A}_N\mathbf{x}_N$, and so $\mathbf{x}_B$ changes from $\mathbf{x}_B^*$ to $\mathbf{x}_B^* - t\mathbf{d}$, with $\mathbf{d}$ standing for the column of $\mathbf{B}^{-1}\mathbf{A}_N$ that corresponds to the entering variable. Note that $\mathbf{d} = \mathbf{B}^{-1}\mathbf{a}$, with $\mathbf{a}$ standing for the entering column. If the revised simplex method is used, then only $\mathbf{x}_B^*$ is readily available, whereas **d** is obtained by solving the system $\mathbf{Bd} = \mathbf{a}$. In our example, we would solve the system

$$\begin{bmatrix} 3 & 1 & 0 \\ 1 & 1 & 0 \\ 4 & 3 & 1 \end{bmatrix} \cdot \mathbf{d} = \begin{bmatrix} 2 \\ 1 \\ 4 \end{bmatrix} \quad \text{to find the vector} \quad \mathbf{d} = \begin{bmatrix} 0.5 \\ 0.5 \\ 0.5 \end{bmatrix}$$

featured in (7.9). We find easily that t can be increased all the way to 30, at which point $54 - 0.5t = 39$, $63 - 0.5t = 48$, $15 - 0.5t = 0$, and x_7 leaves the basis.

So far, the revised simplex method has been requiring computations not needed in the standard simplex method. This trend gets reversed at the end of the iteration: whereas the standard simplex method requires a laborious update of the entire dictionary, no such computations are needed in the revised simplex method. In our example, the revised simplex method merely enters the next iteration with

$$\mathbf{x}_B^* = \begin{bmatrix} x_1^* \\ x_3^* \\ x_4^* \end{bmatrix} = \begin{bmatrix} 39 \\ 48 \\ 30 \end{bmatrix} \quad \text{and} \quad \mathbf{B} = \begin{bmatrix} 3 & 1 & 2 \\ 1 & 1 & 1 \\ 4 & 3 & 4 \end{bmatrix}.$$

Incidentally, note that the order of the columns of **B** is unimportant as long as it matches the order of the components of $\mathbf{x}_B^*$: the next iteration could just as well be entered with

$$\mathbf{x}_B^* = \begin{bmatrix} x_3^* \\ x_4^* \\ x_1^* \end{bmatrix} = \begin{bmatrix} 48 \\ 30 \\ 39 \end{bmatrix} \quad \text{and} \quad \mathbf{B} = \begin{bmatrix} 1 & 2 & 3 \\ 1 & 1 & 1 \\ 3 & 4 & 4 \end{bmatrix}.$$

To put it differently, the fact that the variables $x_1, x_2, \ldots, x_{n+m}$ happen to be ordered by their subscripts is just coincidental; the columns of **B** may be presented in any other order. An ordered list of the basic variables that specifies the actual order of the m columns of **B** is called the *basis heading*. We shall find it convenient to replace the leaving variable by the entering variable in each update of the basis heading: the corresponding update of **B** amounts to a replacement of the *leaving column* by the entering column.

Our development of the revised simplex method is summarized in Box 7.1.

An Economic Interpretation of the Revised Simplex Method

The revised simplex method is intimately related to two subjects presented in Chapter 5: the Complementary Slackness Theorem and the economic interpretation of dual variables. To illustrate the relationship, we shall consider a hypothetical furniture-manufacturing company.

- A bookcase requires three hours of work, one unit of metal, and four units of wood, and it brings in a net profit of \$19.

BOX 7.1 An Iteration of the Revised Simplex Method

Step 1. Solve the system $\mathbf{yB} = \mathbf{c}_B$.

Step 2. Choose an entering column. This may be any column $\mathbf{a}$ of $\mathbf{A}_N$ such that $\mathbf{ya}$ is less than the corresponding component of $\mathbf{c}_N$. If there is no such column, then the current solution is optimal.

Step 3. Solve the system $\mathbf{Bd} = \mathbf{a}$.

Step 4. Find the largest t such that $\mathbf{x}_B^* - t\mathbf{d} \geq \mathbf{0}$. If there is no such t, then the problem is unbounded; otherwise, at least one component of $\mathbf{x}_B^* - t\mathbf{d}$ equals zero and the corresponding variable is leaving the basis.

Step 5. Set the value of the entering variable at t and replace the values $\mathbf{x}_B^*$ of the basic variables by $\mathbf{x}_B^* - t\mathbf{d}$. Replace the leaving column of $\mathbf{B}$ by the entering column and, in the basis heading, replace the leaving variable by the entering variable.

- A desk requires two hours of work, one unit of metal, and three units of wood, and it brings in a net profit of \$13.
- A chair requires one hour of work, one unit of metal, and three units of wood, and it brings in a net profit of \$12.
- A bedframe requires two hours of work, one unit of metal, and four units of wood, and it brings in a net profit of \$17.
- Only 225 hours of labor, 117 units of metal, and 420 units of wood are available per day.

Note that the problem of maximizing the total net profit of the company, under the assumption that all the furniture can be sold, is nothing but our old example (7.2).

Now suppose that a program of making 54 bookcases and 63 chairs per day has been proposed to the company. To find out if this program is optimal, we may appeal to the Complementary Slackness Theorem (the version presented as Theorem 5.3): a feasible solution $\mathbf{x}^*$ is optimal if and only if there are numbers $y_1, y_2, \ldots, y_m$ that satisfy a certain system of equations and a certain system of inequalities. In this particular example, $\mathbf{x}^*$ is the basic feasible solution with

$$\mathbf{x}_B^* = \begin{bmatrix} x_1^* \\ x_3^* \\ x_7^* \end{bmatrix} = \begin{bmatrix} 54 \\ 63 \\ 15 \end{bmatrix}$$

the system of equations is

$$\begin{aligned} 3y_1 + y_2 + 4y_3 &= 19 \\ y_1 + y_2 + 3y_3 &= 12 \\ y_3 &= 0 \end{aligned} \tag{7.10}$$

and the system of inequalities is

$$\begin{aligned} 2y_1 + y_2 + 3y_3 &\geq 13 \\ 2y_1 + y_2 + 4y_3 &\geq 17 \\ y_1 \qquad\qquad &\geq 0 \\ y_2 \qquad &\geq 0. \end{aligned} \tag{7.11}$$

Note that (7.10) is nothing but $\mathbf{yB} = \mathbf{c}_B$, that (7.11) is nothing but $\mathbf{yA}_N \geq \mathbf{c}_N$, and that this observation generalizes: if $\mathbf{x}^*$ is a nondegenerate basic feasible solution, then the system of equations featured in the Complementary Slackness Theorem is nothing but $\mathbf{yB} = \mathbf{c}_B$ and the system of inequalities is nothing but $\mathbf{yA}_N \geq \mathbf{c}_N$. But $\mathbf{yB} = \mathbf{c}_B$ is precisely the system of equations solved in step 1 of an iteration of the revised simplex method and $\mathbf{yA}_N \geq \mathbf{c}_N$ is the system of inequalities considered in step 2. Thus the first two steps in each iteration of the revised simplex method may be seen as checking the current feasible solution $\mathbf{x}^*$ for optimality by the Complementary Slackness Theorem. (Actually, this statement is not quite correct: if $\mathbf{x}^*$ happens to be degenerate, then the system of equations featured in the Complementary Slackness Theorem consists of fewer than m equations and forms a proper subsystem of $\mathbf{yB} = \mathbf{c}_B$.)

Furthermore, these first two steps may be given an economic interpretation along the lines described in Chapter 5. Solving system (7.10), or $\mathbf{yB} = \mathbf{c}_B$ in general, may be interpreted as assigning temporary *shadow prices* to the *resources* (time, metal, and wood) in such a way that the total shadow price of the resources consumed by each of the three *basic activities* (making bookcases, making chairs, and leaving wood unused) matches the net profit returned by this activity. Thus, the solution $y_1 = 3.5$, $y_2 = 8.5$, $y_3 = 0$ of (7.10) appraises time at \$3.50/hour, metal at \$8.50/unit, and wood at \$0/unit. Evaluating the left-hand side of (7.11), or $\mathbf{yA}_N$ in general, may be interpreted as finding the total shadow price of the resources consumed by each of the *nonbasic activities* (making desks, making bedframes, leaving working time unused, and leaving metal unused); this operation is sometimes referred to as *pricing out* the nonbasic activities. If none of these activities pays back more than it consumes (that is, if $\mathbf{c}_N \leq \mathbf{yA}_N$), then the current program is optimal. (A converse of this implication, although guaranteed by the Complementary Slackness Theorem whenever $\mathbf{x}^*$ is nondegenerate, does not hold in general: a degenerate $\mathbf{x}^*$ may be optimal even if some of the inequalities in $\mathbf{c}_N \leq \mathbf{yA}_N$ are violated. In that case, the simplex method will go through a few degenerate iterations without changing $\mathbf{x}^*$ until it comes up with a basis that yields a vector $\mathbf{y}$ with $\mathbf{yA}_N \geq \mathbf{c}_N$.) In our example, making bedframes does pay back more (\$17) than it consumes (time worth \$7, metal worth \$8.50, and wood worth nothing under the current pricing scheme).

Where the Complementary Slackness Theorem leaves off, the revised simplex method continues: it attempts to construct an improved program by *substituting* the profitable *entering activity* (making bedframes) for a suitable mix of the basic activities. The mix, d_i units of each basic activity i per unit of the entering activity, must consume resources at the same rate as the entering activity itself. In our example, this requirement gives rise to the system

$$\begin{aligned} 3d_1 + d_3 &= 2 \\ d_1 + d_3 &= 1 \\ 4d_1 + 3d_3 + d_7 &= 4 \end{aligned} \tag{7.12}$$

and the solution of this system, $d_1 = 0.5$, $d_3 = 0.5$, $d_7 = 0.5$, specifies the concentrations d_i of the constituents i in the mix: each bedframe will be substituted for half a bookcase plus half a chair plus half a unit of unused wood. [Of course, (7.12) is nothing but the system $\mathbf{Bd} = \mathbf{a}$ solved in step 3 of the iteration.] Since the substitution raises the company's profit (by \$1.50 per bedframe), the largest admissible number t of chairs should be substituted for $0.5t$ bookcases plus $0.5t$ chairs plus $0.5t$ units of unused wood; since only 15 units of unused wood are available, the value of t is limited to 30. (This is step 4 of the iteration, where the largest admissible value of t is determined.) The resulting improved program calls for 39 bookcases, 48 chairs, and 30 bedframes to be made every day. The three new basic activities are making bookcases, making chairs, and making bedframes; the old basic activity of leaving wood unused has just become nonbasic. (This is step 5, where the substitution is actually carried out.)

Along similar lines, each iteration of the revised simplex method may be interpreted in economic terms of pricing and substitution. The interpretation becomes a little less intuitive when some of the numbers $y_1, y_2, \ldots, y_m$ or $d_1, d_2, \ldots, d_m$ come out negative, but it may be justified even in those cases.

Eta Factorization of the Basis

The efficiency of the revised simplex method hinges on the ease of implementing steps 1 and 3 of each iteration. Typically, the systems $\mathbf{yB} = \mathbf{c}_B$ and $\mathbf{Bd} = \mathbf{a}$ are not solved from scratch; instead, some device is used to facilitate their solutions and is updated at the end of each iteration. Thus our description of the revised simplex method encompasses a whole class of implementations, each depending on the choice of device that facilitates solutions of the two systems. We are about to describe the simplest of these devices, almost the same (see problem 7.13) as the popular "product form of the inverse" developed by G. B. Dantzig and W. Orchard-Hays (1954). A class of devices that are more efficient, but also more complicated, will be presented in Chapter 24.

Let $\mathbf{B}_k$ denote the basis matrix obtained after k iterations of the simplex method, so that each $\mathbf{B}_k$ differs from the preceding $\mathbf{B}_{k-1}$ in only one column. Consider a fixed

k and say that it is the pth column in which $\mathbf{B}_k$ differs from $\mathbf{B}_{k-1}$. Now the pth column of $\mathbf{B}_k$ is the entering column $\mathbf{a}$ selected in step 2 of the kth iteration and appearing as the right-hand side in the system $\mathbf{B}_{k-1}\mathbf{d} = \mathbf{a}$, which is solved in step 3 of the same iteration. Hence

$$\mathbf{B}_k = \mathbf{B}_{k-1}\mathbf{E}_k \tag{7.13}$$

with $\mathbf{E}_k$ standing for the identity matrix whose pth column is replaced by $\mathbf{d}$. (To verify this matrix equation, we need only compare its two sides column by column, keeping in mind that the jth column of $\mathbf{B}_{k-1}\mathbf{E}_k$ equals $\mathbf{B}_{k-1}$ multiplied by the jth column of $\mathbf{E}_k$ on the right.) For instance,

$$\begin{bmatrix} 3 & 1 & 2 \\ 1 & 1 & 1 \\ 4 & 3 & 4 \end{bmatrix} = \begin{bmatrix} 3 & 1 & 0 \\ 1 & 1 & 0 \\ 4 & 3 & 1 \end{bmatrix} \cdot \begin{bmatrix} 1 & & 0.5 \\ & 1 & 0.5 \\ & & 0.5 \end{bmatrix}$$

in the example just used. The importance of equation (7.13) for the revised simplex method is paramount: no matter what device is used to solve the two systems $\mathbf{yB}_{k-1} = \mathbf{c}_B$ and $\mathbf{B}_{k-1}\mathbf{d} = \mathbf{a}$, its update invariably relies on the fact that $\mathbf{B}_k = \mathbf{B}_{k-1}\mathbf{E}_k$ with the eta matrix $\mathbf{E}_k$ readily available.

When the initial basis consists of the slack variables, we have $\mathbf{B}_0 = \mathbf{I}$ and successive applications of (7.13) yield $\mathbf{B}_1 = \mathbf{E}_1$, $\mathbf{B}_2 = \mathbf{E}_1\mathbf{E}_2$, $\mathbf{B}_3 = \mathbf{E}_1\mathbf{E}_2\mathbf{E}_3$, and so on. Thus we have

$$\mathbf{B}_k = \mathbf{E}_1\mathbf{E}_2 \cdots \mathbf{E}_k.$$

This *eta factorization* of $\mathbf{B}_k$ suggests a convenient way of solving the two systems of equations: the system $\mathbf{yB}_k = \mathbf{c}_B$ may be seen as

$$(((\mathbf{yE}_1)\mathbf{E}_2) \cdots)\mathbf{E}_k = \mathbf{c}_B$$

and the system $\mathbf{B}_k\mathbf{d} = \mathbf{a}$ may be seen as

$$\mathbf{E}_1(\mathbf{E}_2(\cdots(\mathbf{E}_k\mathbf{d}))) = \mathbf{a}.$$

For instance, $\mathbf{yB}_4 = \mathbf{c}_B$ may be solved by solving the sequence of systems

$$\mathbf{uE}_4 = \mathbf{c}_B, \quad \mathbf{vE}_3 = \mathbf{u}, \quad \mathbf{wE}_2 = \mathbf{v}, \quad \text{and} \quad \mathbf{yE}_1 = \mathbf{w}$$

(so that $\mathbf{yB}_4 = \mathbf{yE}_1\mathbf{E}_2\mathbf{E}_3\mathbf{E}_4 = \mathbf{wE}_2\mathbf{E}_3\mathbf{E}_4 = \mathbf{vE}_3\mathbf{E}_4 = \mathbf{uE}_4 = \mathbf{c}_B$ as desired) and $\mathbf{B}_4\mathbf{d} = \mathbf{a}$ may be solved by solving the sequence of systems

$$\mathbf{E}_1\mathbf{u} = \mathbf{a}, \quad \mathbf{E}_2\mathbf{v} = \mathbf{u}, \quad \mathbf{E}_3\mathbf{w} = \mathbf{v}, \quad \text{and} \quad \mathbf{E}_4\mathbf{d} = \mathbf{w}$$

(so that $\mathbf{B}_4\mathbf{d} = \mathbf{E}_1\mathbf{E}_2\mathbf{E}_3\mathbf{E}_4\mathbf{d} = \mathbf{E}_1\mathbf{E}_2\mathbf{E}_3\mathbf{w} = \mathbf{E}_1\mathbf{E}_2\mathbf{v} = \mathbf{E}_1\mathbf{u} = \mathbf{a}$ as desired). At first, this way of solving $\mathbf{yB}_k = \mathbf{c}_B$ and $\mathbf{B}_k\mathbf{d} = \mathbf{a}$ may seem rather awkward: in order to solve one system of linear equations, we resort to solving k systems. Note, however, that systems such as $\mathbf{vE}_i = \mathbf{u}$ or $\mathbf{E}_i\mathbf{v} = \mathbf{u}$ are extremely easy to solve: if the eta column of $\mathbf{E}_i$ has s nonzero entries, then only $s - 1$ multiplications, $s - 1$ additions, and one division are required. Before discussing the efficiency of this scheme any further, let us illustrate it with an example.

The example is again problem (7.2),

$$\text{maximize} \quad \mathbf{cx} \qquad \text{subject to} \quad \mathbf{Ax} = \mathbf{b}, \quad \mathbf{x} \geq \mathbf{0}$$

with

$$\mathbf{A} = \begin{bmatrix} 3 & 2 & 1 & 2 & 1 & 0 & 0 \\ 1 & 1 & 1 & 1 & 0 & 1 & 0 \\ 4 & 3 & 3 & 4 & 0 & 0 & 1 \end{bmatrix}, \quad \mathbf{b} = \begin{bmatrix} 225 \\ 117 \\ 420 \end{bmatrix}, \quad \mathbf{c} = [19, 13, 12, 17, 0, 0, 0].$$

As usual, we let the slack variables form the initial basis, so that $\mathbf{B}_0 = \mathbf{I}$ and

$$\mathbf{x}_B^* = \begin{bmatrix} x_5^* \\ x_6^* \\ x_7^* \end{bmatrix} = \begin{bmatrix} 225 \\ 117 \\ 420 \end{bmatrix}.$$

The first iteration of the revised simplex method begins.

Step 1. The system $\mathbf{yB}_0 = \mathbf{c}_B$ reduces to $\mathbf{y} = [0, 0, 0]$.

Step 2. Since $c_3 - \mathbf{y}\begin{bmatrix} 1 \\ 1 \\ 3 \end{bmatrix} = 12$, we may let x_3 enter the basis.

Step 3. The system $\mathbf{B}_0\mathbf{d} = \mathbf{a}$ reduces to

$$\mathbf{d} = \begin{bmatrix} 1 \\ 1 \\ 3 \end{bmatrix}.$$

Step 4. The largest t such that $225 - t \geq 0$, $117 - t \geq 0$, $420 - 3t \geq 0$ is $t = 117$. Since $117 - t = 0$, the leaving variable is x_6.

Step 5. Now we have

$$\begin{bmatrix} x_5^* \\ x_3^* \\ x_7^* \end{bmatrix} = \begin{bmatrix} 225 - t \\ t \\ 420 - 3t \end{bmatrix} = \begin{bmatrix} 108 \\ 117 \\ 69 \end{bmatrix} \quad \text{and} \quad \mathbf{B}_1 = \mathbf{E}_1 = \begin{bmatrix} 1 & 1 & \\ & 1 & \\ & 3 & 1 \end{bmatrix}.$$

The second iteration begins.

Step 1. Solving the system $\mathbf{yB}_1 = \mathbf{c}_B$, which reads

$$\mathbf{y}\begin{bmatrix} 1 & 1 & \\ & 1 & \\ & 3 & 1 \end{bmatrix} = [0, 12, 0], \quad \text{we find } \mathbf{y} = [0, 12, 0].$$

Step 2. Since $c_1 - \mathbf{y}\begin{bmatrix} 3 \\ 1 \\ 4 \end{bmatrix} = 7$, we may let x_1 enter the basis.

Step 3. Solving the system $\mathbf{B}_1\mathbf{d} = \mathbf{a}$, which reads

$$\begin{bmatrix} 1 & 1 & \\ & 1 & \\ & 3 & 1 \end{bmatrix} \cdot \mathbf{d} = \begin{bmatrix} 3 \\ 1 \\ 4 \end{bmatrix}, \quad \text{we find} \quad \mathbf{d} = \begin{bmatrix} 2 \\ 1 \\ 1 \end{bmatrix}.$$

Step 4. The largest t such that $108 - 2t \geq 0$, $117 - t \geq 0$, $69 - t \geq 0$ is $t = 54$. Since $108 - 2t = 0$, the leaving variable is x_5.

Step 5. Now we have

$$\begin{bmatrix} x_1^* \\ x_3^* \\ x_7^* \end{bmatrix} = \begin{bmatrix} t \\ 117 - t \\ 69 - t \end{bmatrix} = \begin{bmatrix} 54 \\ 63 \\ 15 \end{bmatrix} \quad \text{and} \quad \mathbf{B}_2 = \mathbf{E}_1\mathbf{E}_2 \quad \text{with} \quad \mathbf{E}_2 = \begin{bmatrix} 2 & & \\ 1 & 1 & \\ 1 & & 1 \end{bmatrix}.$$

The third iteration begins.

Step 1. We shall solve the system $\mathbf{y}\mathbf{B}_2 = \mathbf{c}_B$ as $(\mathbf{y}\mathbf{E}_1)\mathbf{E}_2 = \mathbf{c}_B$. Solving the system $\mathbf{u}\mathbf{E}_2 = \mathbf{c}_B$, which reads

$$\mathbf{u}\begin{bmatrix} 2 & & \\ 1 & 1 & \\ 1 & & 1 \end{bmatrix} = [19, 12, 0], \quad \text{we find} \quad \mathbf{u} = [3.5, 12, 0].$$

Solving the system $\mathbf{y}\mathbf{E}_1 = \mathbf{u}$, which reads

$$\mathbf{y}\begin{bmatrix} 1 & 1 & \\ & 1 & \\ & 3 & 1 \end{bmatrix} = [3.5, 12, 0], \quad \text{we find} \quad \mathbf{y} = [3.5, 8.5, 0].$$

Step 2. Since $c_4 - \mathbf{y}\begin{bmatrix} 2 \\ 1 \\ 4 \end{bmatrix} = 1.5$, we may let x_4 enter the basis.

Step 3. We shall solve the system $\mathbf{B}_2\mathbf{d} = \mathbf{a}$ as $\mathbf{E}_1(\mathbf{E}_2\mathbf{d}) = \mathbf{a}$. Solving the system $\mathbf{E}_1\mathbf{u} = \mathbf{a}$, which reads

$$\begin{bmatrix} 1 & 1 & \\ & 1 & \\ & 3 & 1 \end{bmatrix} \cdot \mathbf{u} = \begin{bmatrix} 2 \\ 1 \\ 4 \end{bmatrix}, \quad \text{we find} \quad \mathbf{u} = \begin{bmatrix} 1 \\ 1 \\ 1 \end{bmatrix}.$$

Solving the system $\mathbf{E}_2\mathbf{d} = \mathbf{u}$, which reads

$$\begin{bmatrix} 2 & & \\ 1 & 1 & \\ 1 & & 1 \end{bmatrix} \cdot \mathbf{d} = \begin{bmatrix} 1 \\ 1 \\ 1 \end{bmatrix}, \quad \text{we find} \quad \mathbf{d} = \begin{bmatrix} 0.5 \\ 0.5 \\ 0.5 \end{bmatrix}.$$

Step 4. The largest t such that $54 - 0.5t \geq 0$, $63 - 0.5t \geq 0$, $15 - 0.5t \geq 0$ is $t = 30$. Since $15 - 0.5t = 0$, the leaving variable is x_7.

Step 5. Now we have

$$\begin{bmatrix} x_1^* \\ x_3^* \\ x_4^* \end{bmatrix} = \begin{bmatrix} 54 - 0.5t \\ 63 - 0.5t \\ t \end{bmatrix} = \begin{bmatrix} 39 \\ 48 \\ 30 \end{bmatrix} \quad \text{and} \quad \mathbf{B}_3 = \mathbf{E}_1\mathbf{E}_2\mathbf{E}_3 \quad \text{with} \quad \mathbf{E}_3 = \begin{bmatrix} 1 & & 0.5 \\ & 1 & 0.5 \\ & & 0.5 \end{bmatrix}.$$

The fourth iteration begins.

Step 1. We shall solve the system $\mathbf{yB}_3 = \mathbf{c}_B$ as $((\mathbf{yE}_1)\mathbf{E}_2)\mathbf{E}_3 = \mathbf{c}_B$. Solving the system $\mathbf{uE}_3 = \mathbf{c}_B$, which reads

$$\mathbf{u}\begin{bmatrix} 1 & & 0.5 \\ & 1 & 0.5 \\ & & 0.5 \end{bmatrix} = [19, 12, 17], \quad \text{we find} \quad \mathbf{u} = [19, 12, 3].$$

Solving the system $\mathbf{vE}_2 = \mathbf{u}$, which reads

$$\mathbf{v}\begin{bmatrix} 2 & & \\ 1 & 1 & \\ 1 & & 1 \end{bmatrix} = [19, 12, 3], \quad \text{we find} \quad \mathbf{v} = [2, 12, 3].$$

Solving the system $\mathbf{yE}_1 = \mathbf{v}$, which reads

$$\mathbf{y}\begin{bmatrix} 1 & 1 & \\ & 1 & \\ & 3 & 1 \end{bmatrix} = [2, 12, 3], \quad \text{we find} \quad \mathbf{y} = [2, 1, 3].$$

Step 2. Since

$$\mathbf{c}_N - \mathbf{yA}_N = [13, 0, 0, 0] - [2, 1, 3] \cdot \begin{bmatrix} 2 & 1 & 0 & 0 \\ 1 & 0 & 1 & 0 \\ 3 & 0 & 0 & 1 \end{bmatrix} = [-1, -2, -1, -3]$$

we find no candidate for entering the basis. Hence the current solution is optimal.

Even though $\mathbf{B}_0 = \mathbf{I}$ whenever the initial basis consists of the slack variables, the case of an arbitrary $\mathbf{B}_0$ is worth considering. In this case, the identity $\mathbf{B}_k = \mathbf{E}_1\mathbf{E}_2 \cdots \mathbf{E}_k$ generalizes into

$$\mathbf{B}_k = \mathbf{B}_0\mathbf{E}_1\mathbf{E}_2 \cdots \mathbf{E}_k$$

and the two systems $\mathbf{yB}_k = \mathbf{c}_B$, $\mathbf{B}_k\mathbf{d} = \mathbf{a}$ may be solved as $(((\mathbf{yB}_0)\mathbf{E}_1) \cdots)\mathbf{E}_k = \mathbf{c}_B$ and $\mathbf{B}_0(\mathbf{E}_1(\cdots(\mathbf{E}_k\mathbf{d}))) = \mathbf{a}$, respectively. Now a triangular factorization

$$\mathbf{L}_m\mathbf{P}_m \cdots \mathbf{L}_1\mathbf{P}_1\mathbf{B}_0 = \mathbf{U}$$

of the initial basis $\mathbf{B}_0$ may be computed before the first iteration and then used again and again in conjunction with the growing sequence $\mathbf{E}_1, \mathbf{E}_2, \ldots, \mathbf{E}_k$. Note that

$$\mathbf{U} = \mathbf{U}_m\mathbf{U}_{m-1} \cdots \mathbf{U}_1$$

with each $\mathbf{U}_j$ standing for the eta matrix obtained when the jth column of $\mathbf{I}$ is replaced by the jth column of $\mathbf{U}$ (a verification of this claim is left for problem 7.6), and so

$$\mathbf{L}_m\mathbf{P}_m \cdots \mathbf{L}_1\mathbf{P}_1\mathbf{B}_k = \mathbf{U}_m\mathbf{U}_{m-1} \cdots \mathbf{U}_1\mathbf{E}_1\mathbf{E}_2 \cdots \mathbf{E}_k.$$

In this notation, the system $\mathbf{y}\mathbf{B}_k = \mathbf{c}_B$ may be solved by first solving $(((\mathbf{y}\mathbf{U}_m)\mathbf{U}_{m-1}) \cdots)\mathbf{E}_k = \mathbf{c}_B$ and then replacing $\mathbf{y}$ by $((\mathbf{y}\mathbf{L}_m\mathbf{P}_m) \cdots)\mathbf{L}_1\mathbf{P}_1$. The details of this procedure may be spelled out as follows.

1. Set $i = k$ and $\mathbf{y} = \mathbf{c}_B$.
2. If $i \geq 1$, then set $\mathbf{v} = \mathbf{y}$, replace $\mathbf{y}$ by the solution of $\mathbf{y}\mathbf{E}_i = \mathbf{v}$, replace i by $i - 1$, and repeat this step.
3. Set $j = 1$.
4. If $j \leq m$, then set $\mathbf{v} = \mathbf{y}$, replace $\mathbf{y}$ by the solution of $\mathbf{y}\mathbf{U}_j = \mathbf{v}$, replace j by $j + 1$, and repeat this step.
5. Set $j = m$.
6. If $j \geq 1$, then replace $\mathbf{y}$ by $\mathbf{y}\mathbf{L}_j\mathbf{P}_j$, replace j by $j - 1$, and repeat this step.

Similarly, the system $\mathbf{B}_k\mathbf{d} = \mathbf{a}$ may be solved as $\mathbf{U}_m(\mathbf{U}_{m-1}(\cdots(\mathbf{E}_k\mathbf{d}))) = (\mathbf{L}_m\mathbf{P}_m(\cdots(\mathbf{L}_1\mathbf{P}_1\mathbf{a})))$; the details of this procedure may be spelled out as follows.

1. Set $j = 1$ and $\mathbf{d} = \mathbf{a}$.
2. If $j \leq m$, then replace $\mathbf{d}$ by $\mathbf{L}_j\mathbf{P}_j\mathbf{d}$, replace j by $j + 1$, and repeat this step.
3. Set $j = m$.
4. If $j \geq 1$, then set $\mathbf{v} = \mathbf{d}$, replace $\mathbf{d}$ by the solution of $\mathbf{U}_j\mathbf{d} = \mathbf{v}$, replace j by $j - 1$, and repeat this step.
5. Set $i = 1$.
6. If $i \leq k$, then set $\mathbf{v} = \mathbf{d}$, replace $\mathbf{d}$ by the solution of $\mathbf{E}_i\mathbf{d} = \mathbf{v}$, replace i by $i + 1$, and repeat this step.

To store each $\mathbf{E}_i$, we need only store its eta column and record the position of this column in the matrix. Furthermore, if the eta columns are sufficiently sparse, then they may be stored in the "packed form" mentioned in Chapter 6, so that only the nonzero entries are stored and their positions in the column recorded. The same remark applies to the triangular eta matrices $\mathbf{L}_j$ and $\mathbf{U}_j$. Each of the permutation matrices $\mathbf{P}_j$, obtained by interchanging some row of $\mathbf{I}$ with the jth row, may be represented by a single pointer specifying the interchanged row. A sequential file storing the matrices

$$\mathbf{P}_1, \mathbf{L}_1, \mathbf{P}_2, \mathbf{L}_2, \ldots, \mathbf{P}_m, \mathbf{L}_m, \mathbf{U}_m, \mathbf{U}_{m-1}, \ldots, \mathbf{U}_1, \mathbf{E}_1, \mathbf{E}_2, \ldots, \mathbf{E}_k$$

in this fashion is called the *eta file*. This file is scanned backward, from $\mathbf{E}_k$ to $\mathbf{P}_1$, in solving the system $\mathbf{y}\mathbf{B}_k = \mathbf{c}_B$, and it is solved forward, from $\mathbf{P}_1$ to $\mathbf{E}_k$, in solving the system $\mathbf{B}_k\mathbf{d} = \mathbf{a}$. For this reason, the procedure for solving $\mathbf{y}\mathbf{B}_k = \mathbf{c}_B$ is sometimes referred to as the *backward transformation*, or BTRAN, and the procedure for solving $\mathbf{B}_k\mathbf{d} = \mathbf{a}$ is referred to as the *forward transformation*, or FTRAN. Note that the backward and the forward scans alternate and that each new item $\mathbf{E}_{k+1}$ is added to the open end of the file after the file has been scanned forward all the way to $\mathbf{E}_k$ and before the next scan backward to $\mathbf{P}_1$ begins. (The reader should be warned that the term *eta file* is usually employed in connection with the "product form of the inverse," in which case it refers to a different file; see problem 7.13.)

Refactorizations

Since the eta file grows with each iteration, BTRAN and FTRAN become progressively more and more laborious; eventually, they could even take longer than solving the two systems $\mathbf{y}\mathbf{B}_k = \mathbf{c}_B$ and $\mathbf{B}_k\mathbf{d} = \mathbf{a}$ from scratch. Such counterproductive uses of the eta file may be avoided by discarding the whole file from time to time and treating the current $\mathbf{B}_k$ as a new $\mathbf{B}_0$: compute a fresh triangular factorization of this matrix, and let a new sequence $\mathbf{E}_1, \mathbf{E}_2, \mathbf{E}_3, \ldots$ grow from that point on. These periodic *refactorizations* of the basis keep the overall time spent on executions of steps 1 and 3 within acceptable limits. (For historical reasons, refactorizations are sometimes referred to as "reinversions.")

How often should the basis be refactorized? If T_0 stands for the time spent on the refactorization, if T_k stands for the time spent on BTRAN and FTRAN in the kth iteration after refactorization, and if the basis is refactorized after r iterations, then the average time per execution of steps 1 and 3, including an appropriate share of the overhead T_0, comes to

$$T_r^* = \frac{1}{r}\sum_{k=0}^{r} T_k. \tag{7.14}$$

Obviously, r should be chosen so as to minimize T_r^*. A trivial way of doing so relies on the observation that $T_1^*, T_2^*, T_3^*, \ldots$ first decrease (as the overhead T_0 gets distributed over more and more iterations) and then they begin to grow (as the length of the eta file begins to take over). Thus, we need only keep track of the cumulative total $T_0 + T_1 + \cdots + T_k$ and refactorize as soon as this quantity divided by k stops decreasing. (A rigorous proof of this claim, relying only on the natural assumption that $T_1 \le T_2 \le T_3 \le \cdots$ is left for problem 7.7.)

In solving large sparse problems arising from applications, the basis is refactorized quite frequently, often after every twenty iterations or so. An exact analysis of the reasons behind these frequent refactorizations is both impossible and unnecessary: impossible, since the relevant statistics vary unpredictably from one problem to the next and unnecessary, since there is no point in a theoretical justification of a policy whose practical success has been firmly established. All the same, the insight provided by an inexact analysis is better than no insight at all. For this reason, we are going to present a few observations concerning the behavior of the large sparse problems encountered in practice.

The matrices featured in these problems are very sparse, typically with at most ten nonzeros per column. Furthermore, they have an implicit special structure allowing each basis matrix to be permuted into a near-triangular form, as discussed in Chapter 6. The resulting triangular factorizations of $\mathbf{B}_0$ are very sparse, typically with up to $12m$ nonzeros or so (discounting, as usual, the diagonal ones in $\mathbf{L}_1, \mathbf{L}_2, \ldots, \mathbf{L}_m$). If $\mathbf{B}_0 = \mathbf{I}$, then the eta columns of $\mathbf{E}_1, \mathbf{E}_2, \mathbf{E}_3, \ldots$ are initially very sparse: in particular, the eta column of $\mathbf{E}_1$ is nothing but some column of $\mathbf{A}$. However, as more and more slack variables are pivoted out of the basis, the density of these eta columns increases until it reaches a steady state that persists after subsequent refactorizations. Typically, the eta columns are 25%–50% dense in this steady state, although densities outside this range do occur.

These statistics permit us to estimate the time required to execute BTRAN and FTRAN in each iteration of the revised simplex method. Taking the time required to execute one multiplication and one addition as our basic unit, we observe that the time required to execute BTRAN equals the length of the eta file; assuming that the eta columns are 50% dense, we may estimate the length of the eta file in the kth iteration by $12m + 0.5m(k - 1)$. The same quantity provides an upper bound on the time required to execute FTRAN. However, many of the multiplications in FTRAN are multiplications by zero, and so considerable time-savings can be claimed in this case. In short, the point is that the right-hand side $\mathbf{a}$ of the system $\mathbf{B}_k\mathbf{d} = \mathbf{a}$ will be very sparse, but the right-hand side $\mathbf{c}_B$ of the system $\mathbf{y}\mathbf{B}_k = \mathbf{c}_B$ may be completely dense. More precisely, consider the last stage of FTRAN, where systems $\mathbf{E}_i\mathbf{d} = \mathbf{v}$ are solved. Since the right-hand side $\mathbf{v}$ is a solution of $\mathbf{B}_{i-1}\mathbf{v} = \mathbf{a}$, about $m/2$ or more of its m components may be expected to be zeros: $\mathbf{v}$ would have been the eta column in the $(i - 1)$th iteration after the last refactorization if $\mathbf{a}$ were the entering column then. Thus there is at least an even chance that $\mathbf{v}$ will have a zero in that position in which the eta column of $\mathbf{E}_i$ occurs; if that is the case, then the system $\mathbf{E}_i\mathbf{d} = \mathbf{v}$ reduces to $\mathbf{d} = \mathbf{v}$. Since similar, if not better, time-savings occur in the earlier stages of FTRAN, we estimate the time required to execute FTRAN by a half of the time required to execute BTRAN. Thus, we are led to estimate the total time required to execute BTRAN and FTRAN in the kth iteration by

$$T_k \approx 18m + 0.75m(k - 1).$$

Estimating the time T_0 required to refactorize $\mathbf{B}_0$ is much trickier: permuting $\mathbf{B}_0$ into the near-triangular form may take just about as much time as the subsequent computation of the triangular factorization. For the sake of argument, we make an educated guess at

$$T_0 \approx 100m.$$

Substituting these estimations into (7.14), we obtain

$$T_r^* \approx \frac{m}{r}\left(100 + \sum_{k=1}^{r} (18 + 0.75(k - 1))\right) \approx m\left(\frac{100}{r} + 18 + 0.375r\right).$$

Now a routine exercise in calculus (left for problem 7.8) shows that right-hand side is minimized by $r \approx 16$, a value within the range of the typical refactorization frequencies used in practice. For future reference, note that refactorizations with this frequency (i) keep the average time required to execute steps 1 and 3 (including the refactorization overhead) down to an estimated $30m$ or so, and (ii) keep the average length of the eta file down to an estimated $16m$ or so.

One flaw in this approximate analysis is the assumption that the density of the eta columns has reached its steady state. This steady state is reached only after most of the slack variables have been pivoted out of the basis, which takes about m iterations. But the simplex method may terminate after just about m iterations, in which case the steady state is simply never reached. Nevertheless, the results of the analysis are not completely misleading. True, the density of the eta columns will fall far below the estimated 50% in the early iterations. However, $\mathbf{B}_0$ will share many columns with $\mathbf{I}$ in this case, and so the time required to find its triangular factorization and the number of nonzeros in this factorization will be reduced as well. Even though $30m$ will become a gross overestimate of the average time required to execute steps 1 and 3, the optimal refactorization frequency will remain in the neighborhood of twenty or so. □

THE REVISED SIMPLEX METHOD VERSUS THE STANDARD SIMPLEX METHOD

On the large sparse problems encountered in practice, an iteration of the revised simplex method takes less time than an iteration of the standard simplex method.

□

To gain an insight into this phenomenon, let us extend the approximate analysis just presented. We have estimated the average time required to execute steps 1 and 3 by $30m$ or so; the time required to execute step 2 does not exceed the total number of nonzeros in $\mathbf{A}$, estimated at no more than $10n$. Since step 4 requires up to m divisions and step 5 requires up to m multiplications and additions, we are led to estimate the total time per iteration of the revised simplex method at about

$$32m + 10n.$$

On the other hand, the time required to update the dictionary in an iteration of the revised simplex method equals the number of nonzeros in the pivot row times the number of nonzeros in the pivot column. Since each eta column is nothing but a column of the corresponding dictionary, and since we have estimated the density of the eta columns at 50%, we shall estimate the density of the dictionaries by the same figure. Now about $n/2$ nonzeros may be expected in each row and about $m/2$ nonzeros in each column of the dictionary; thus, we are led to estimate the total time per iteration of the standard simplex method at

$$mn/4.$$

Note that we may assume $n \geq m$, for otherwise it would be easier to solve the problem by solving its dual, as noted in Chapter 5. In fact, n is considerably bigger than m in many of the problems encountered in practice; even $n \geq 2m$ is quite usual. Thus, we are led to conclude that, on the typical sparse problems encountered in practice, the revised simplex method beats the standard simplex method as soon as the number of rows exceeds 100. (Problems with 100 rows are considered small by present-day standards; some people call a problem large only if it has at least 2,000 rows.)

Furthermore, when the simplex method is applied to very large problems, its running time is influenced by another factor, the nature of which is more engineering than mathematical. This factor arises from the way computers are constructed: data may be stored either in central core memory, which can be accessed fast but whose capacity is relatively small, or in peripheral memory (on disks, drums, or tapes), which has practically unlimited capacity but is slow to access. The time required to retrieve data from peripheral memory may be considerable compared to the time required to execute the various arithmetical operations. When the length of the eta file (estimated before at about $16m$ on the average) exceeds the capacity of the central memory, the file has to be retrieved from peripheral memory twice in each iteration of the revised simplex method: first in step 1 (BTRAN) and then in step 3 (FTRAN). In addition, when the number of nonzeros in $\mathbf{A}$ (estimated before at no more than $10n$) exceeds the capacity of central memory, then these nonzeros, or at least some of them, have to be retrieved from the peripheral memory in step 2. Thus we are led to estimate the average number of items retrieved from the peripheral memory in each iteration of the revised simplex method at

$$32m + 10n.$$

On the other hand, even if the dictionaries used in the standard simplex method remain reasonably sparse, new nonzeros in these dictionaries are created in unpredictable places with each iteration. Thus storing the dictionaries in a packed form would create awkward difficulties. If all their entries, including the zeros, are stored explicitly, then the total number of items retrieved from the peripheral memory in each iteration of the standard simplex method comes to

$$mn.$$

We conclude that considerations of the retrieval time favor the revised simplex method even more decisively than considerations of the time required to carry out the various arithmetical operations. □

For these reasons, it is the revised simplex method rather than the standard simplex method that is used in computer programs for solving LP problems. The sparsity and the special structure of the data make the revised simplex method superior and constitute the two leitmotifs of large-scale linear programming.

However, it would be wrong to believe that the revised simplex method is always faster than the standard simplex method: an easy argument, left for problem 7.9, shows that an iteration of the revised simplex method takes longer than an iteration of the standard simplex method whenever the bases $\mathbf{B}_k$ are completely dense and $n < 2m$.

Incidentally, it may be a good policy to use the revised simplex method even in solving small LP problems by hand. In addition to its obvious didactic value, this practice makes it possible to spot numerical mistakes immediately by comparing $\mathbf{yB}$ with $\mathbf{c}_B$, $\mathbf{Bd}$ with $\mathbf{a}$, and $\mathbf{Bx}_B^*$ with $\mathbf{b}$.

PRICING IN THE REVISED SIMPLEX METHOD

The fact that the components of $\mathbf{c}_N - \mathbf{yA}_N$ may be computed individually in the revised simplex method has important practical implications. One may discover quite a few candidates for entering the basis by computing only a part of the vector $\mathbf{c}_N - \mathbf{yA}_N$. This strategy is known as *partial pricing*; the generic term *pricing* refers to any method of computing the vector $\mathbf{c}_N - \mathbf{y}\mathbf{A}_N$, or a part of it, for the purpose of selecting an entering variable. Variations on partial pricing are unlimited. For instance, the heuristic assumption that a current candidate for entering the basis is likely to remain eligible during the next few iterations suggests the following strategy. Compute the first, for instance, $n/3$ components of $\mathbf{c}_N - \mathbf{y}\mathbf{A}_N$ and set aside perhaps 40 of the most promising candidates for entering the basis. In the subsequent iterations, limit your choice of entering variables to these 40 candidates. When only about 20 of these candidates remain eligible, compute the next $n/3$ components of $\mathbf{c}_N - \mathbf{y}\mathbf{A}_N$ and repeat the whole process. Another popular alternative is known as *multiple pricing*: having selected, say, 10 promising candidates for entering the basis, consider the restricted problem that involves only these 10 nonbasic variables along with the m current basic variables. Only after an optimal solution of the restricted problem is found, go on to select the next 10 candidates and to repeat the whole process. The effect of these strategies on the overall number of iterations is hard to predict; the motivation behind their development is the desire to minimize the overall computing time by reducing the time per iteration.

The merits of partial and multiple pricing become even more pronounced when the problem is so large that the matrix $\mathbf{A}$ has to be stored in peripheral memory and its columns retrieved in step 2 of each iteration. Partial pricing reduces the retrieval time by retrieving fewer columns; multiple pricing transfers the 10 or so columns into central memory and solves the restricted problem there. (Because of its special format, the restricted problem is solved by the *standard* simplex method; each dictionary, having $m + 1$ rows and only 10 or so columns, is small enough to fit in core.)

Incidentally, judging candidates for entering the basis simply by the magnitude of the components of $\mathbf{c}_N - \mathbf{y}\mathbf{A}_N$ is not the best policy: typically, the number of iterations may be reduced by the use of another criterion. Two such criteria, closely related to each other, have been developed by P. M. J. Harris (1973) and D. Goldfarb and J. K. Reid (1977); they are known as *Devex* and *steepest edge*, respectively. The extra computations involved in these alternatives do pay off: the overall running time decreases in most cases and sometimes it drops to only 50% or less of the original figure.

ZERO TOLERANCES

The number zero plays a special role in the selection of an entering variable: a nonbasic variable is eligible for entering the basis if and only if the corresponding component of $\mathbf{c}_N - \mathbf{y}\mathbf{A}_N$ is greater than zero. Rounding errors can make zeros or small negative numbers appear as small positive numbers; making the corresponding nonbasic variables enter the basis would be a blunder. To safeguard against this trap, a small positive number ε_1, called a *zero tolerance*, is chosen in advance; then a component of $\mathbf{c}_N - \mathbf{y}\mathbf{A}_N$ is considered positive if and only if its computed value exceeds ε_1. Another zero tolerance, ε_2, safeguards against divisions by extremely small numbers (which tend to produce the most dangerous rounding errors): if some choice of the entering variable leads to an eta matrix whose diagonal element in the eta column is less than ε_2 in magnitude, then this choice is rejected and another nonbasic variable is made to enter the basis. Yet another zero tolerance, ε_3, is used in monitoring the accuracy of computations by comparing $\mathbf{B}\mathbf{x}_B^*$ with $\mathbf{b}$: if these two vectors differ in any component by more than ε_3, then the basis is refactorized immediately. Setting the actual values of the various zero tolerances is a delicate matter requiring some expertise; in particular, an inappropriate choice of ε_1 might lead the computations astray. B. A. Murtagh (1981) gives $\varepsilon_1 = 10^{-5}$, $\varepsilon_2 = 10^{-8}$, $\varepsilon_3 = 10^{-6}$ as typical values for computations with 15 decimal digit precision. □

PROBLEMS

7.1 Solve problem 2.1 by the revised simplex method.

7.2 Solve problem 1.6 by the revised simplex method and interpret each iteration in economic terms.

7.3 Prove Theorem 5.4.

7.4 Prove Theorem 5.5.

△ **7.5** Solve the systems $\mathbf{y}\mathbf{E}_1\mathbf{E}_2\mathbf{E}_3\mathbf{E}_4 = [1, 2, 3]$ and $\mathbf{E}_1\mathbf{E}_2\mathbf{E}_3\mathbf{E}_4\mathbf{d} = [1, 2, 3]^T$ with

$$\mathbf{E}_1 = \begin{bmatrix} 1 & 3 & \\ & 0.5 & \\ & 4 & 1 \end{bmatrix}, \quad \mathbf{E}_2 = \begin{bmatrix} 2 & & \\ 1 & 1 & \\ 4 & & 1 \end{bmatrix}, \quad \mathbf{E}_3 = \begin{bmatrix} 1 & & 1 \\ & 1 & 3 \\ & & 1 \end{bmatrix}, \quad \mathbf{E}_4 = \begin{bmatrix} -0.5 & & \\ 3 & 1 & \\ 1 & & 1 \end{bmatrix}.$$

7.6 Let $\mathbf{U}$ be an $m \times m$ upper triangular matrix and let $\mathbf{U}_j$ be the eta matrix obtained when the jth column of $\mathbf{I}$ is replaced by the jth column of $\mathbf{U}$. Prove that $\mathbf{U} = \mathbf{U}_m\mathbf{U}_{m-1} \cdots \mathbf{U}_1$.

7.7 Let $T_0, T_1, T_2, \ldots$ be a sequence of numbers such that $T_{k+1} \geq T_k$ whenever $k \geq 1$. Prove that the sequence $T_1^*, T_2^*, T_3^* \cdots$ defined by

$$T_r^* = \frac{1}{r} \sum_{k=0}^{r} T_k$$

has the following property: if $T_r^* \leq T_{r+1}^*$, then $T_{r+1}^* \leq T_{r+2}^*$.

7.8 Given positive numbers a, b, c, minimize $a/r + b + cr$.

7.9 Show that an iteration of the revised simplex method takes longer than an iteration of the standard simplex method whenever the bases $\mathbf{B}_k$ are completely dense and $n < 2m$.

7.10 What is the optimal refactorization frequency in the revised simplex method when all the bases are completely dense?

7.11 Show that, if all the bases are completely dense, an iteration of the revised simplex method takes less time on the average than an iteration of the standard simplex method as long as (i) the basis is refactorized with an appropriate frequency, (ii) $n > 5m$, and (iii) only $n/3$ out of the n nonbasic variables are priced out in each iteration. (Note, however, that assumption (iii) makes the comparison unfair. First, the number of iterations tends to increase when partial pricing is used. Secondly, (iii) is untenable when there are only a few candidates for entering the basis, which is the typical situation in the last iterations.)

7.12 Design a procedure that, given an $m \times m$ nonsingular matrix $\mathbf{B}_0$, finds eta matrices $\mathbf{E}_1^*, \mathbf{E}_2^*, \ldots, \mathbf{E}_m^*$ such that

$$\mathbf{B}_0 = \mathbf{E}_1^*\mathbf{E}_2^* \cdots \mathbf{E}_m^*.$$

(*Hint*: Assuming that you have found $\mathbf{E}_1^*, \mathbf{E}_2^*, \ldots, \mathbf{E}_j^*$, whose product agrees with $\mathbf{B}_0$ in the first j columns, show how to find $\mathbf{E}_{j+1}^*$.) In the revised simplex method, this eta factorization of $\mathbf{B}_0$ may be updated into

$$\mathbf{B}_k = \mathbf{E}_1^*\mathbf{E}_2^* \cdots \mathbf{E}_{m+k}^*$$

(by setting $\mathbf{E}_{m+i}^* = \mathbf{E}_i$ in the ith iteration) and used to solve the two systems $\mathbf{y}\mathbf{B}_k = \mathbf{c}_B$ and $\mathbf{B}_k\mathbf{d} = \mathbf{a}$. However, it is better to use a triangular factorization of $\mathbf{B}_0$ instead (as we have done in the text): triangular factorizations are sparser and easier to compute than eta factorizations.

7.13 The device commonly presented in place of our "eta factorization of the basis" is the "product form of the inverse," which consists of a sequence of eta matrices whose product is $\mathbf{B}_k^{-1}$. How can this device be used and updated in each iteration? (*Hint*: Consult problem 7.12.)

7.14 Tradition seems to be the reason for clinging to the product form of the inverse: early versions of the revised simplex method used $\mathbf{B}_k^{-1}$, represented by its individual entries, to solve the systems $\mathbf{y}\mathbf{B}_k = \mathbf{c}_B$ and $\mathbf{B}_k\mathbf{d} = \mathbf{a}$. How can this device be used and updated in each iteration? Why is it unsuitable for solving the large sparse problems typically encountered in practice? (*Hint*: Consult problem 6.12.)

8

General LP Problems: Solutions by the Simplex Method

Many linear programming problems involve explicit upper bounds on individual variables. For instance, one might be confronted with the problem

$$\begin{array}{lll} \text{maximize} & \sum_{j=1}^{n} c_j x_j & \\ \text{subject to} & \sum_{j=1}^{n} a_{ij} x_j \leq b_i & (i = 1, 2, \ldots, m) \\ & 0 \leq x_j \leq u_j & (j = 1, 2, \ldots, n) \end{array} \tag{8.1}$$

such that each u_j is a positive number. This problem may be cast in the standard form

$$\begin{array}{lll} \text{maximize} & \sum_{j=1}^{n} c_j x_j & \\ \text{subject to} & \sum_{j=1}^{n} a_{ij} x_j \leq b_i & (i = 1, 2, \ldots, m) \\ & x_j \leq u_j & (j = 1, 2, \ldots, n) \\ & x_j \geq 0 & (j = 1, 2, \ldots, n) \end{array}$$

and then solved by the simplex method; when the revised simplex method is used, the size of each basis matrix is $(m + n) \times (m + n)$. Alternatively, the simplex method can be made to work on (8.1) directly in such a way that, when the revised simplex method is used, the size of each basis matrix is only $m \times m$. This technique, suggested by G. B. Dantzig (1955) and known as *upper bounding*, can be applied in the more general context of problems

$$\begin{aligned} &\text{maximize} && \sum_{j=1}^{n} c_j x_j \\ &\text{subject to} && \sum_{j=1}^{n} a_{ij} x_j = b_i \qquad (i = 1, 2, \ldots, m) \\ & && l_j \leq x_j \leq u_j \qquad (j = 1, 2, \ldots, n). \end{aligned} \tag{8.2}$$

Here, each l_j is either a number or the symbol $-\infty$, meaning that no lower bound is imposed on x_j, and each u_j is either a number or the symbol $+\infty$, meaning that no upper bound is imposed on x_j. [To cast (8.1) in the form of (8.2), we need only add the slack variables $x_{n+1}, x_{n+2}, \ldots, x_{n+m}$, with $l_{n+i} = 0$ and $u_{n+i} = +\infty$ for all i.] No generality is lost in assuming that $l_j \leq u_j$ for all j [otherwise (8.2) is trivially infeasible] but, for technical reasons, it is convenient to admit variables x_j with $l_j = u_j$ (even though such variables, whose values are fixed, can be deleted at once and the right-hand side adjusted appropriately).

The modifications that enable the simplex method to handle (8.2) directly are the subject of this chapter. Thus, we shall learn to solve even the most general LP problems, for *every LP problem is easily presentable in the form of (8.2)*. To justify this claim, we need only observe that minimizing $\sum c_j x_j$ is tantamount to maximizing $\sum(-c_j)x_j$ and that every inequality constraint can be converted into an equation by the introduction of an appropriate slack variable. (The only inequalities not converted into equations are the explicit bounds on individual variables; each of these is simply recorded as $x_j \geq l_j$ or $x_j \leq u_j$.)

HOW TO HANDLE EXPLICIT BOUNDS ON INDIVIDUAL VARIABLES

In matrix terms, (8.2) may be recorded as

$$\text{maximize} \quad \mathbf{cx} \qquad \text{subject to} \quad \mathbf{Ax} = \mathbf{b}, \quad \mathbf{l} \leq \mathbf{x} \leq \mathbf{u}. \tag{8.3}$$

We shall say that the matrix $\mathbf{A}$ has m rows and n columns; then $\mathbf{b}$ is a column vector of length m, $\mathbf{c}$ is a row vector of length n, and $\mathbf{x}$, $\mathbf{l}$, $\mathbf{u}$ are column vectors of length n. Note that we allow variables x_j with $l_j = -\infty$ and $u_j = +\infty$. Such variables are called *free* or *unrestricted*, whereas the remaining variables are called *bounded* or *restricted*. We shall say that a solution $\mathbf{x}^*$ of $\mathbf{Ax} = \mathbf{b}$ is a *basic solution* of (8.3) if the n components of $\mathbf{x}$ can be partitioned into m "basic" and $n - m$ "nonbasic" variables in such a way that (i) the m columns of $\mathbf{A}$ corresponding to the basic variables form a nonsingular matrix and (ii) the value x_j^* of each bounded nonbasic variable x_j is l_j or u_j. A basic solution $\mathbf{x}^*$ is called *feasible* if $\mathbf{l} \le \mathbf{x}^* \le \mathbf{u}$. [This definition is consistent with our original definition of a basic feasible solution in the special case when (8.3) arises from an LP problem in the standard form.] For instance, if

$$\mathbf{A} = \begin{bmatrix} 3 & 1 & 5 & 6 & 9 & 4 & 3 & 4 & 7 & 6 & 4 & 5 \\ 1 & 0 & 9 & 5 & 8 & 1 & 2 & 7 & 8 & 7 & 9 & 1 \end{bmatrix}, \quad \mathbf{b} = \begin{bmatrix} 72 \\ 62 \end{bmatrix},$$

$$\mathbf{c} = [2, \ 1, \ -2, \ -2, \ 3, \ 2, \ 3, \ -4, \ 0, \ -2, \ -3, \ 3]$$

and

$$\mathbf{l} = [-5, \ -\infty, \ -4, \ -2, \ 2, \ 0, \ 0, \ 3, \ -\infty, \ -\infty, \ -\infty, \ -\infty]^T$$
$$\mathbf{u} = [+\infty, \ 3, \ -2, \ 3, \ 5, \ 1, \ +\infty, \ +\infty, \ 0, \ 5, \ +\infty, \ +\infty]^T$$

then

$$\mathbf{x}^* = [1, \ 0, \ -2, \ 3, \ 2, \ 0, \ 0, \ 3, \ 0, \ 5, \ -1, \ 1]^T$$

is a basic feasible solution; the variables x_1, x_2 are basic, whereas $x_3, x_4, \ldots, x_{12}$ are nonbasic. (Note that, in the present context, a basic solution is not always determined by the choice of basic variables alone. For instance,

$$\mathbf{x}^* = [2, \ 2, \ -2, \ 3, \ 2, \ 0, \ 0, \ 3, \ 0, \ 5, \ -1, \ 0]^T$$

and

$$\mathbf{x}^* = [1, \ 1, \ -2, \ 3, \ 2, \ 1, \ 0, \ 3, \ 0, \ 5, \ -1, \ 0]^T$$

are other basic feasible solutions with x_1, x_2 basic and $x_3, x_4, \ldots, x_{12}$ nonbasic in our example.)

To solve problem (8.3), the simplex method replaces one basic feasible solution by another in every (nondegenerate) iteration. Again, the idea is to change the value of only one of the nonbasic variables (the entering variable) in such a way that, when the constraints $\mathbf{Ax} = \mathbf{b}$ are maintained by adjusting the values of the basic variables, feasibility is preserved and the value of the objective function increases. For instance, we might change the value of x_7 in our example from zero to some positive t. When the values of the basic variables are adjusted to $x_1 = 1 - 2t$ and $x_2 = 3t$, feasibility is preserved (as long as t is sufficiently small) and the value of the objective function increases from -10 to $-10 + 2t$. Alternatively, we might change the value of x_{12} from 1 to $1 - t$ for some positive t. When the values of the basic variables are adjusted to $x_1 = 1 + t$ and $x_2 = 2t$, feasibility is preserved (as long as t is sufficiently small) and the value of the objective function increases from -10 to $-10 + t$.

General changes of this kind may be investigated along the lines of Chapter 7: having defined $\mathbf{B}$, $\mathbf{A}_N$, $\mathbf{x}_B$, $\mathbf{x}_N$ and $\mathbf{c}_B$, $\mathbf{c}_N$, we observe that the constraints $\mathbf{Ax} = \mathbf{b}$ are equivalent to

$$\mathbf{x}_B = \mathbf{B}^{-1}\mathbf{b} - \mathbf{B}^{-1}\mathbf{A}_N\mathbf{x}_N \tag{8.4}$$

and that every vector $\mathbf{x}$ satisfying these constraints satisfies

$$\mathbf{cx} = \mathbf{yb} + (\mathbf{c}_N - \mathbf{yA}_N)\mathbf{x}_N \tag{8.5}$$

with $\mathbf{y} = \mathbf{c}_B\mathbf{B}^{-1}$. Now consider a basic feasible solution $\mathbf{x}^*$; let $\mathbf{a}$ stand for the column of $\mathbf{A}_N$ corresponding to some nonbasic x_j, and write $\mathbf{d} = \mathbf{B}^{-1}\mathbf{a}$. If the value of x_j changes from x_j^* to $x_j^* + t$ but the remaining nonbasic variables stay fixed at their current values, then the right-hand side of (8.4) changes from $\mathbf{B}^{-1}\mathbf{b} - \mathbf{B}^{-1}\mathbf{A}_N\mathbf{x}_N^*$ (which equals $\mathbf{x}_B^*$) to $\mathbf{B}^{-1}\mathbf{b} - \mathbf{B}^{-1}\mathbf{A}_N\mathbf{x}_N^* - t\mathbf{B}^{-1}\mathbf{a}$ (which equals $\mathbf{x}_B^* - t\mathbf{d}$). Therefore, to keep the constraints (8.4) satisfied, we must change the values of the basic variables from $\mathbf{x}_B^*$ to $\mathbf{x}_B^* - t\mathbf{d}$. The effect of this change on the value of the objective function can be seen in (8.5): since the component of $\mathbf{c}_N - \mathbf{yA}_N$ corresponding to x_j is $c_j - \mathbf{ya}$, the right-hand side of (8.5) changes from $\mathbf{yb} + (\mathbf{c}_N - \mathbf{yA}_N)\mathbf{x}_N^*$ (which equals $\mathbf{cx}^*$) to $\mathbf{yb} + (\mathbf{c}_N - \mathbf{yA}_N)\mathbf{x}_N^* + (c_j - \mathbf{ya})t$ (which equals $\mathbf{cx}^* + (c_j - \mathbf{ya})t$). To summarize:

> If x_j changes from x_j^* to $x_j^* + t$, then $\mathbf{x}_B$ has to change from $\mathbf{x}_B^*$ to $\mathbf{x}_B^* - t\mathbf{d}$ and $\mathbf{cx}$ changes from $\mathbf{cx}^*$ to $\mathbf{cx}^* + (c_j - \mathbf{ya})t$.

Replacing t with $-t$, we see the following:

> If x_j changes from x_j^* to $x_j^* - t$, then $\mathbf{x}_B$ has to change from $\mathbf{x}_B^*$ to $\mathbf{x}_B^* + t\mathbf{d}$ and $\mathbf{cx}$ changes from $\mathbf{cx}^*$ to $\mathbf{cx}^* - (c_j - \mathbf{ya})t$.

We wish to increase the value of the objective function $\mathbf{cx}$ by changing the value of the entering nonbasic variable from x_j^* to $x_j^* + t$ or $x_j^* - t$ for some positive t. The increase to $x_j^* + t$ is allowed only if $x_j^* < u_j$ and brings about the desired increase in $\mathbf{cx}$ only if $c_j - \mathbf{ya} > 0$; the decrease to $x_j^* - t$ is allowed only if $x_j^* > l_j$ and brings about the desired increase in $\mathbf{cx}$ only if $c_j - \mathbf{ya} < 0$. Hence, the entering variable x_j must satisfy either

$$c_j - \mathbf{ya} > 0 \quad \text{and} \quad x_j^* < u_j \tag{8.6}$$

or

$$c_j - \mathbf{ya} < 0 \quad \text{and} \quad x_j^* > l_j. \tag{8.7}$$

The absence of nonbasic variables x_j satisfying (8.6) or (8.7) implies optimality of our current solution $\mathbf{x}^*$. To justify this claim, denote the row vector $\mathbf{c}_N - \mathbf{yA}_N$ by $\bar{\mathbf{c}}_N$. If no nonbasic variable x_j satisfies (8.6), then $x_j^* = u_j$ whenever $\bar{c}_j > 0$; if no nonbasic variable x_j satisfies (8.7), then $x_j^* = l_j$ whenever $\bar{c}_j < 0$. But then $\bar{c}_j x_j \leq \bar{c}_j x_j^*$ whenever $l_j \leq x_j \leq u_j$. Thus, every feasible solution $\mathbf{x}$ of (8.3) satisfies $\bar{\mathbf{c}}_N \mathbf{x}_N \leq \bar{\mathbf{c}}_N \mathbf{x}_N^*$ and, by virtue of (8.5), it satisfies $\mathbf{cx} \leq \mathbf{cx}^*$. Hence, $\mathbf{x}^*$ is an optimal solution of (8.3).

As in Chapter 7, the vector $\mathbf{y} = \mathbf{c}_B \mathbf{B}^{-1}$ may be found by solving the system $\mathbf{yB} = \mathbf{c}_B$. In our example, this system reads

$$\mathbf{y}\begin{bmatrix} 3 & 1 \\ 1 & 0 \end{bmatrix} = [2, 1], \quad \text{and so we have} \quad \mathbf{y} = [1, -1].$$

It is easy to check that the only nonbasic variables satisfying (8.6) are x_5, x_7, and x_{11}, whereas the only nonbasic variables satisfying (8.7) are x_4, x_{10}, and x_{12}. Hence, the entering variable may be any of the nonbasic variables $x_4, x_5, x_7, x_{10}, x_{11}, x_{12}$.

It will be convenient to write

$$x_j(t) = x_j^* + t, \quad \mathbf{x}_B(t) = \mathbf{x}_B^* - t\mathbf{d}$$

if the entering variable x_j satisfies (8.6), and

$$x_j(t) = x_j^* - t, \quad \mathbf{x}_B(t) = \mathbf{x}_B^* + t\mathbf{d}$$

if the entering variable x_j satisfies (8.7). As in Chapter 7, the vector $\mathbf{d} = \mathbf{B}^{-1}\mathbf{a}$ featured here may be found by solving the system $\mathbf{Bd} = \mathbf{a}$. We intend to replace $x_j^* = x_j(0)$ by $x_j(t)$ and $\mathbf{x}_B^* = \mathbf{x}_B(0)$ by $\mathbf{x}_B(t)$ for some positive t. The resulting solution of $\mathbf{Ax} = \mathbf{b}$ will remain feasible as long as

$$l_j \leq x_j(t) \leq u_j \quad \text{and} \quad \mathbf{l}_B \leq \mathbf{x}_B(t) \leq \mathbf{u}_B. \tag{8.8}$$

Since the value of the objective function increases with t, we wish to make t as large as these constraints allow Let us distinguish among three cases:

(i) The upper bound on t imposed by $\mathbf{l}_B \leq \mathbf{x}_B(t) \leq \mathbf{u}_B$ is stricter than the upper bound (if any) imposed by $l_j \leq x_j(t) \leq u_j$.

(ii) The upper bound on t imposed by $l_j \leq x_j(t) \leq u_j$ is at least as strict as the upper bound (if any) imposed by $\mathbf{l}_B \leq \mathbf{x}_B(t) \leq \mathbf{u}_B$.

(iii) Neither $l_j \leq x_j(t) \leq u_j$ nor $\mathbf{l}_B \leq \mathbf{x}_B(t) \leq \mathbf{u}_B$ impose any upper bounds on t.

Each of these three cases will be illustrated on our example.

If the entering variable is x_5, then we solve the system

$$\begin{bmatrix} 3 & 1 \\ 1 & 0 \end{bmatrix} \cdot \mathbf{d} = \begin{bmatrix} 9 \\ 8 \end{bmatrix} \quad \text{and find} \quad \mathbf{d} = \begin{bmatrix} 8 \\ -15 \end{bmatrix}.$$

Now the constraints (8.8) assume the form

$$2 \leq 2 + t \leq 5 \quad \text{and} \quad -5 \leq 1 - 8t, \quad 15t \leq 3$$

which reduces to $t \leq 0.2$. This is an instance of (i): when t is set at its largest admissible value, one of the basic variables reaches its lower or upper bound. This variable leaves the basis (whereas the entering variable enters). In our case, the new basic feasible solution is

$$\mathbf{x}^* = [-0.6, 3, -2, 3, 2.2, 0, 0, 3, 0, 5, -1, 1]^T$$

with the variables x_1 and x_5 basic.

If the entering variable is x_4, then we solve the system

$$\begin{bmatrix} 3 & 1 \\ 1 & 0 \end{bmatrix} \cdot \mathbf{d} = \begin{bmatrix} 6 \\ 5 \end{bmatrix} \quad \text{and find} \quad \mathbf{d} = \begin{bmatrix} 5 \\ -9 \end{bmatrix}.$$

Now the constraints (8.8) assume the form

$$-2 \leq 3 - t \leq 3 \quad \text{and} \quad -5 \leq 1 + 5t, \quad -9t \leq 3$$

which reduces to $t \leq 5$. This is an instance of (ii): when t is set at its largest admissible value, the entering variable switches from one of its bounds to the other, and the basis remains unchanged. (In particular, the entering variable does *not* enter the basis.) In our case, the new basic feasible solution is

$$\mathbf{x}^* = [26, -45, -2, -2, 2, 0, 0, 3, 0, 5, -1, 1]^T$$

with the old basic variables x_1, x_2 remaining basic.

If the entering variable is x_{10}, then we solve the system

$$\begin{bmatrix} 3 & 1 \\ 1 & 0 \end{bmatrix} \cdot \mathbf{d} = \begin{bmatrix} 6 \\ 7 \end{bmatrix} \quad \text{and find} \quad \mathbf{d} = \begin{bmatrix} 7 \\ -15 \end{bmatrix}.$$

Now the constraints (8.8) assume the form

$$5 - t \leq 5 \quad \text{and} \quad -5 \leq 1 + 7t, \quad -15t \leq 3$$

imposing no upper bound on t whatsoever. This is an instance of (iii): as t can be made arbitrarily large, the problem is unbounded. In our case, every

$$\mathbf{x}^* = [1 + 7t, -15t, -2, 3, 2, 0, 0, 3, 0, 5 - t, -1, 1]^T$$

with $t \geq 0$ constitutes a feasible solution with $\mathbf{cx}^* = -10 + t$.

To summarize, when the revised simplex method is made to work directly on (8.3), each of its iterations can be described as in Box 8.1.

BOX 8.1 An Iteration of the Revised Simplex Method

Step 1. Solve the system $\mathbf{yB} = \mathbf{c}_B$.

Step 2. Choose an entering variable x_j. This may be any nonbasic variable x_j such that, with $\mathbf{a}$ standing for the corresponding column of $\mathbf{A}$, we have either $\mathbf{ya} < c_j$, $x_j^* < u_j$, or $\mathbf{ya} > c_j$, $x_j^* > l_j$. If there is no such variable then stop; the current solution $\mathbf{x}^*$ is optimal.

Step 3. Solve the system $\mathbf{Bd} = \mathbf{a}$.

Step 4. Define $x_j(t) = x_j^* + t$ and $\mathbf{x}_B(t) = \mathbf{x}_B^* - t\mathbf{d}$ in case $\mathbf{ya} < c_j$ and $x_j(t) = x_j^* - t$, $\mathbf{x}_B(t) = \mathbf{x}_B^* + t\mathbf{d}$ in case $\mathbf{ya} > c_j$. If the constraints

$$l_j \leq x_j(t) \leq u_j, \quad \mathbf{l}_B \leq \mathbf{x}_B(t) \leq \mathbf{u}_B$$

are satisfied for all positive t then stop; the problem is unbounded. Otherwise set t at the largest value allowed by these constraints. If the upper bound imposed on t by the constraints $\mathbf{l}_B \leq \mathbf{x}_B(t) \leq \mathbf{u}_B$ is stricter than the upper bound imposed by $l_j \leq x_j(t) \leq u_j$, then determine the leaving variable. This may be any basic variable x_i such that the upper bound imposed on t by $l_i \leq x_i(t) \leq u_i$ alone is as strict as the upper bound imposed by all the constraints in $\mathbf{l}_B \leq \mathbf{x}_B(t) \leq \mathbf{u}_B$.

Step 5. Replace x_j^* by $x_j(t)$ and $\mathbf{x}_B^*$ by $\mathbf{x}_B(t)$. If the value of the entering variable x_j has just switched from one of its bounds to the other, then proceed directly to step 2 of the next iteration. Otherwise, replace the leaving variable x_i by the entering variable x_j in the basis heading, and replace the leaving column of $\mathbf{B}$ by the entering column $\mathbf{a}$.

Degeneracy and Termination

Our definition of a basic feasible solution admits basic feasible solutions $\mathbf{x}^*$ with $x_i^* = l_i$ or $x_i^* = u_i$ for one or more basic variables x_i. Such basic feasible solutions are called *degenerate.* When a simplex iteration begins with a degenerate basic feasible solution $\mathbf{x}^*$, the constraints $\mathbf{l}_B \leq \mathbf{x}_B(t) \leq \mathbf{u}_B$ may force t equal to 0 in step 4. In that case, the entering variable enters the basis and the leaving variable leaves, but the solution $\mathbf{x}^*$ remains unchanged. Such iterations are called *degenerate.* The value of the objective function remains unchanged after each degenerate iteration and increases with each nondegenerate one. Hence only a sequence of degenerate iterations may create *cycling,* defined as the appearance of the same basic feasible solution with the same set of basic variables in two different iterations. [Note that, in the special case when (8.3) arises from an LP problem in the standard form, this definition of cycling is equivalent to the definition used in Chapter 3.] Theorem 3.1 told us that, on LP problems in the standard form cycling is the only reason why the simplex method may fail to terminate. Now we exter. result to arbitrary LP problems.

THEOREM 8.1. If the simplex method, applied to a problem

$$\text{maximize} \quad \mathbf{cx} \quad \text{subject to} \quad \mathbf{Ax} = \mathbf{b}, \quad \mathbf{l} \le \mathbf{x} \le \mathbf{u}$$

fails to terminate, then it must cycle.

PROOF. Assume that the simplex method fails to terminate and so goes through an endless sequence of iterations. At the beginning of each iteration, ask (i) which variables are basic, (ii) which nonbasic restricted variables are at their lower bounds, and (iii) which nonbasic restricted variables are at their upper bounds. Since there are only a finite number of possible answers to these three questions, one answer must come up in two different iterations. It remains to be proved that the basic feasible solutions $\mathbf{x}'$ and $\mathbf{x}''$ featured in these two iterations are also the same. Since

$$\mathbf{x}'_B = \mathbf{B}^{-1}\mathbf{b} - \mathbf{B}^{-1}\mathbf{A}_N\mathbf{x}'_N \quad \text{and} \quad \mathbf{x}''_B = \mathbf{B}^{-1}\mathbf{b} - \mathbf{B}^{-1}\mathbf{A}_N\mathbf{x}''_N$$

proving that $\mathbf{x}' = \mathbf{x}''$ reduces to proving that $\mathbf{x}'_N = \mathbf{x}''_N$. Since the answers to questions (ii) and (iii) were the same in the two iterations, we have $x'_j = x''_j$ whenever x_j is a nonbasic restricted variable. Now it only remains to be proved that $x'_j = x''_j$ whenever x_j is a nonbasic free variable. But this conclusion follows at once from the observations that (a) if a nonbasic free variable changes its value in some iteration, then it enters the basis in that iteration and (b) a basic free variable never leaves the basis. ■

As noted in Chapter 3, cycling is very rare in practice, and its remote possibility is disregarded in most computer implementations of the simplex method. Furthermore, Theorem 3.3 may be extended as follows.

THEOREM 8.2. The simplex method, applied to a problem

$$\text{maximize} \quad \mathbf{cx} \quad \text{subject to} \quad \mathbf{Ax} = \mathbf{b}, \quad \mathbf{l} \le \mathbf{x} \le \mathbf{u}$$

terminates as long as the entering and the leaving variables are selected by the smallest-subscript rule in each iteration. ■

The proof follows the proof of Theorem 3.3 almost completely, except that now some of the inequalities in the argument may get reversed, depending on which of the three variables x_t, x_s, x_r are at their lower bounds and which of them are at their upper bounds. We leave the tedious details for problem 8.3. □

The Two-Phase Simplex Method

Now the only remaining difficulty in applying the simplex method to problems with explicit bounds on individual variables occurs in initialization: How do we get hold of a basic feasible solution to begin with? This difficulty may be overcome by a number of variations on the two-phase simplex method of Chapter 3. The idea remains the same: first the constraints $\mathbf{Ax} = \mathbf{b}$ are extended by means of *artificial variables*, making a basic feasible solution readily available, and then the values of these variables are driven to zero by solving an *auxiliary problem*. This strategy can be implemented in a variety of ways; one of them goes as follows.

When confronted with a problem

$$\begin{array}{lll} \text{maximize} & \sum_{j=1}^{n} c_j x_j & \\ \text{subject to} & \sum_{j=1}^{n} a_{ij} x_j = b_i & (i = 1, 2, \ldots, m) \\ & l_j \le x_j \le u_j & (j = 1, 2, \ldots, n) \end{array} \tag{8.9}$$

introduce the artificial variables $x_{n+1}, x_{n+2}, \ldots, x_{n+m}$ and consider the constraints

$$\begin{array}{ll} \sum_{j=1}^{n} a_{ij} x_j + x_{n+i} = b_i & (i = 1, 2, \ldots, m) \\ l_j \le x_j \le u_j & (j = 1, 2, \ldots, n + m) \end{array} \tag{8.10}$$

with $l_{n+1}, l_{n+2}, \ldots, l_{n+m}$ and $u_{n+1}, u_{n+2}, \ldots, u_{n+m}$ defined as follows. Choose numbers $\tilde{x}_1, \tilde{x}_2, \ldots, \tilde{x}_n$ such that $\tilde{x}_j = l_j$ or $\tilde{x}_j = u_j$ for every bounded variable x_j, and set $\tilde{x}_{n+i} = b_i - \sum a_{ij}\tilde{x}_j$ for all $i = 1, 2, \ldots, m$. If $\tilde{x}_{n+i} \ge 0$, then set $l_{n+i} = 0$ and $u_{n+i} = +\infty$; if $\tilde{x}_{n+i} < 0$, then set $l_{n+i} = -\infty$ and $u_{n+i} = 0$.

Now (8.9) has a feasible solution if and only if (8.10) has a solution with all the artificial variables at zero. In a sense, the degree to which a solution of (8.10) fails to provide a feasible solution of (8.9) can be measured by the sum of the magnitudes of the artificial variables. This sum equals $\sum w_{n+i} x_{n+i}$, with $w_{n+i} = 1$ when $l_{n+i} = 0$ and $w_{n+i} = -1$ when $u_{n+i} = 0$; its value is nonnegative for every solution of (8.10), and it equals zero if and only if all the artificial variables equal zero. Thus, the issue of existence of a feasible solution of (8.9) may be settled by solving the auxiliary problem,

$$\text{minimize} \quad \sum_{i=1}^{m} w_{n+i} x_{n+i} \qquad \text{subject to (8.10)}$$

or, in the maximization form,

$$\text{maximize} \quad \sum_{i=1}^{m} (-w_{n+i}) x_{n+i} \qquad \text{subject to (8.10).}$$

The first phase of the two-phase simplex method consists of solving the auxiliary problem by the simplex method; this phase is initialized by the basic feasible solution $\tilde{\mathbf{x}}$ just described, with the m artificial variables being basic. At the end of the first phase, we either discover that the optimal value of the auxiliary problem is not zero, and so the original problem (8.9) has no feasible solutions at all, or else we find an optimal solution $\mathbf{x}^*$ of the auxiliary problem such that $x^*_{n+1} = x^*_{n+2} = \cdots = x^*_{n+m} = 0$. In the latter case, the basic feasible solution $\mathbf{x}^*$ is used to initialize the second phase of the two-phase simplex method. The second phase consists of applying the simplex method to an equivalent version of the original problem (8.9),

$$\begin{array}{lll} \text{maximize} & \sum_{j=1}^{n} c_j x_j & \\ \text{subject to} & \sum_{j=1}^{n} a_{ij} x_j + x_{n+i} = b_i & (i = 1, 2, \ldots, m) \\ & l_j \le x_j \le u_j & (j = 1, 2, \ldots, n) \\ & 0 \le x_{n+i} \le 0 & (i = 1, 2, \ldots, m). \end{array} \tag{8.11}$$

The only reason for retaining the artificial variables x_{n+i} in (8.11) is the fact that some of them may be featured as basic in the initial basic feasible solution $\mathbf{x}^*$. Each artifical variable that is nonbasic in $\mathbf{x}^*$ may be deleted at once; each artificial variable that is basic in $\mathbf{x}^*$ but becomes nonbasic later may be deleted as soon as it leaves the basis.

It is often the case that at least some of the m artificial variables x_{n+i} involved in the auxiliary problem may be omitted from the auxiliary problem from the very start and that the roles these variables play in the solution $\tilde{\mathbf{x}}$ initializing the first phase may be played by suitably chosen variables x_j of the original problem (8.9). In particular, this is the case for those artificial variables that simply duplicate slack variables in the constraints of the auxiliary problem. For instance, if (8.9) reads

$$\begin{array}{lrcl} \text{maximize} & 3x_1 + x_2 + x_3 + 2x_4 & & \\ \text{subject to} & 3x_1 + 2x_2 + 4x_3 + 2x_4 & = & 2 \\ & 4x_1 + x_2 - 3x_3 + x_4 + x_5 & = & -1 \\ & x_1 - 3x_2 + x_3 + 3x_4 \quad + x_6 & = & 3 \\ & 2x_1 + x_2 + x_3 + 3x_4 \quad\quad + x_7 & = & 5 \\ & x_1, x_2, \ldots, x_7 & \ge & 0 \end{array}$$

then the auxiliary problem assumes the form

$$\begin{array}{lrcl} \text{maximize} & -x_8 + x_9 - x_{10} - x_{11} & & \\ \text{subject to} & 3x_1 + 2x_2 + 4x_3 + 2x_4 \quad + x_8 & = & 2 \\ & 4x_1 + x_2 - 3x_3 + x_4 + x_5 \quad + x_9 & = & -1 \\ & x_1 - 3x_2 + x_3 + 3x_4 \quad + x_6 \quad + x_{10} & = & 3 \\ & 2x_1 + x_2 + x_3 + 3x_4 \quad + x_7 \quad + x_{11} & = & 5 \\ & x_1, x_2, \ldots, x_7 \ge 0,\ x_8 \ge 0,\ x_9 \le 0,\ x_{10} \ge 0,\ x_{11} & \ge & 0. \end{array}$$

In the constraints of this auxiliary problem, the artificial variables x_{10} and x_{11} duplicate the slack variables x_6 and x_7, respectively. Having made this observation, we may replace the auxiliary problem by

$$\begin{array}{llr}
\text{maximize} & & -x_8 + x_9 \\
\text{subject to} & 3x_1 + 2x_2 + 4x_3 + 2x_4 \qquad\qquad + x_8 & = 2 \\
& 4x_1 + x_2 - 3x_3 + x_4 + x_5 \qquad\qquad + x_9 & = -1 \\
& x_1 - 3x_2 + x_3 + 3x_4 \qquad + x_6 & = 3 \\
& 2x_1 + x_2 + x_3 + 3x_4 \qquad\qquad + x_7 & = 5 \\
& x_1, x_2, \ldots, x_7 \geq 0, x_8 \geq 0, x_9 \leq & 0
\end{array}$$

which can be initialized by $\tilde{\mathbf{x}}_B = [\tilde{x}_8, \tilde{x}_9, \tilde{x}_6, \tilde{x}_7]^T = [2, -1, 3, 5]^T$, $\tilde{\mathbf{x}}_N = \mathbf{0}$. Thus, we bypass two iterations in the first phase: the first iteration with x_6 entering and x_{10} leaving, the second iteration with x_7 entering and x_{11} leaving. (In fact, some people prefer not to use the term *artificial variable* at all. They would think of x_9 as the slack variable x_5, whose bounds $l_5 = 0, u_5 = +\infty$ have been temporarily changed to $l_5 = -\infty, u_5 = 0$, and refer to x_8 as "the slack variable of the equation $3x_1 + 2x_2 + 4x_3 + 2x_4 = 2$," whose bounds $l_8 = 0, u_8 = 0$ have been temporarily changed to $l_8 = 0, u_8 = +\infty$.)

Incidentally, observe that the original objective function $\sum c_j x_j$ does not affect the auxiliary problem at all. This is only fair: the issue to be settled in the first phase is that of feasibility, and the objective function is quite irrelevant there. Nevertheless, a complete disregard for the original objective may have unpleasant side effects: if the basic feasible solution delivered at the end of the first phase gives the original objective function an extremely low value, then the second phase may require a relatively large number of iterations to reach an optimal solution. (Of course, these statements are just intuitive observations; even though they tend to be true in general, they are not true under all circumstances.) It seems sensible, therefore, to keep an eye on the original objective function while the artificial variables are being driven out of the basis, so as to avoid unnecessarily sharp drops in its value. This idea is sometimes implemented by maximizing

$$\sum_{j=1}^{n} c_j x_j - M \sum_{i=1}^{m} w_{n+i} x_{n+i}$$

rather than $\sum(-w_{n+i})x_{n+i}$ in the first phase. When the coefficient M is very large, minimizing $\sum w_{n+i}x_{n+i}$ is still the primary concern, but sharp drops in the value of $\sum c_j x_j$ do become undesirable to some extent. This variation on the two-phase method is referred to as the "big M method." (There is no evidence to support the belief that the "big M" stands for "big mother.")

Other popular variations involve maximizing

$$r \sum_{j=1}^{n} c_j x_j - \sum_{i=1}^{m} s_i w_{n+i} x_{n+i}$$

for suitably chosen "scaling factors" r and $s_1, s_2, \ldots, s_m$. The actual choice of these numbers may depend on the current values of the artificial variables x_{n+i}, and so it may change from one iteration to the next. Then the auxiliary problem is no longer a linear programming problem (the objective function ceases to be linear), but all is well as long as all the artificial variables vanish in the end. The simplest example arises when $r = 0$ and

$$s_i = \begin{cases} 1 & \text{if} \quad x_{n+i} \neq 0 \\ 0 & \text{if} \quad x_{n+i} = 0 \end{cases}$$

with each artificial variable x_{n+i} deleted as soon as it leaves the basis.

Finally, the two-phase simplex method may often be implemented without the use of artificial variables. To do so on a problem

$$\text{maximize} \quad \mathbf{cx} \quad \text{subject to} \quad \mathbf{Ax} = \mathbf{b}, \quad \mathbf{l} \leq \mathbf{x} \leq \mathbf{u}$$

with $\mathbf{A}$ having size $m \times n$ as usual, we require only a nonsingular matrix $\mathbf{B}$ formed by some m columns of $\mathbf{A}$. Now we may choose the values $\mathbf{x}_N^*$ of the nonbasic variables in such a way that $x_i^* = l_i$ or $x_i^* = u_i$ for each bounded nonbasic x_i. Having computed the values $\mathbf{x}_B^*$ of the basic variables by solving the system $\mathbf{Bx}_B^* = \mathbf{b} - \mathbf{A}_N \mathbf{x}_N^*$, we obtain a basic solution $\mathbf{x}^*$ that, in general, is infeasible: for each basic variable x_i, we may have $x_i^* < l_i$ or $x_i^* > u_i$. Writing $i \in P$ in the former case and $i \in Q$ in the latter, we may measure the "infeasibility" of $\mathbf{x}^*$ by the quantity

$$\sum_{i \in P} (l_i - x_i^*) + \sum_{i \in Q} (x_i^* - u_i).$$

Since we wish to drive the basic variables inside their bounds, our momentary aim is to increase the value of the objective function

$$\sum_{i \in P} x_i - \sum_{i \in Q} x_i$$

subject to the original constraints $\mathbf{Ax} = \mathbf{b}$, $\mathbf{l} \leq \mathbf{x} \leq \mathbf{u}$ with l_i replaced by $-\infty$ whenever $i \in P$ and u_i replaced by $+\infty$ whenever $i \in Q$. For this purpose, we may use the simplex method as usual. However, as the value of our objective function increases, the value of some x_i with $i \in P$ may rise to l_i, or the value of some x_i with $i \in Q$ may drop to u_i. At this very moment, the objective function must be updated and the original bounds l_i, u_i restored (otherwise further changes in the value of the entering variable might actually increase the infeasibility of the current solution); the variable x_i leaves the basis at the same time. Thus the first phase produces a sequence of basic solutions whose infeasibility decreases with each nondegenerate iteration until a basic feasible solution is found or its nonexistence established. We shall not go into the details any further.

Getting Rid of Artificial Basic Variables

Unless the first phase of the two-phase simplex method discovers that (8.9) is infeasible, it delivers a basic feasible solution of the equivalent problem (8.11). This solution is not necessarily basic in the original problem (8.9), for some of its basic variables may be the artificial variables x_{n+i}. As previously observed, the presence of artificial variables in the basis creates no problem in the second phase of the two-phase simplex method. Nevertheless, it is only natural to question the role of these variables. Our investigations begin with a procedure which, given a basic feasible solution of (8.11), attempts to drive all artificial variables out of the basis (see Box 8.2). Without loss of generality, let us assume that each artificial basic variable x_{n+k} appears in the kth position of the basis heading. Note that each column of $\mathbf{B}$ is either a column of $\mathbf{A}$ or a column of $\mathbf{I}$.

BOX 8.2 Driving Artificial Variables Out of the Basis

Step 0. Let S denote the set of all the subscripts i such that x_{n+i} is a basic variable.

Step 1. If S is empty then stop. Otherwise delete some subscript k from S.

Step 2. Solve the system $\mathbf{rB} = \mathbf{e}$ with $\mathbf{e}$ standing for the kth row of the identity matrix. If $\mathbf{rA}$ is the zero vector, then return to step 1. Otherwise, there is a nonbasic variable x_j such that $\mathbf{ra} \neq 0$ for the corresponding column $\mathbf{a}$ of $\mathbf{A}$. Replace the kth column of $\mathbf{B}$ by $\mathbf{a}$; replace x_{n+k} by x_j in the basis heading and return to step 1.

To verify that the updated matrix $\mathbf{B}$ remains nonsingular after each execution of step 2, let $\mathbf{B}$ stand for the matrix entering step 2 and note that its updated version equals $\mathbf{BE}$ with $\mathbf{E}$ standing for the eta matrix whose kth column is $\mathbf{B}^{-1}\mathbf{a}$. Thus we need only verify that $\mathbf{E}$ is nonsingular. But $\mathbf{E}$ is nonsingular if and only if the diagonal entry in its eta column is nonzero; now it remains to be observed only that this diagonal entry is $\mathbf{e}(\mathbf{B}^{-1}\mathbf{a})$ and that $(\mathbf{eB}^{-1})\mathbf{a} = \mathbf{ra} \neq 0$.

Upon termination of this procedure, some artificial variables may persist in the basis. This apparent failure is actually a success in its own way: persistence of x_{n+k} in the basis means that the kth equation in (8.9) is redundant and may be deleted altogether. More precisely, let J denote the set of subscripts k such that x_{n+k} persists in the basis, and let I denote the set of subscripts $1, 2, \ldots, m$ that do not belong to J. We claim that every solution of

$$\sum_{j=1}^{n} a_{ij}x_j = b_i \qquad (i \in I) \tag{8.12}$$

satisfies all the equations

$$\sum_{j=1}^{n} a_{ij}x_j = b_i \qquad (i = 1, 2, \ldots, m).$$

To justify this claim, consider an arbitrary subscript k in J, along with the vector $\mathbf{r}$ computed immediately after the deletion of k from S. Since $\mathbf{rA}$ is the zero vector, we have

$$\sum_{i=1}^{m} r_i a_{ij} = 0 \qquad \text{for all } j = 1, 2, \ldots, n. \tag{8.13}$$

Since $\mathbf{rB} = \mathbf{e}$, we have $r_k = 1$ and $r_j = 0$ for all the remaining subscripts j in J. Hence (8.13) may be recorded as

$$a_{kj} = -\sum_{i \in I} r_i a_{ij}$$

and every solution of (8.12) must satisfy

$$\sum_{j=1}^{n} a_{kj} x_j = \sum_{j=1}^{n} \left(-\sum_{i \in I} r_i a_{ij}\right) x_j = \sum_{i \in I} \left(-r_i \sum_{j=1}^{n} a_{ij} x_j\right) = \sum_{i \in I} (-r_i b_i). \tag{8.14}$$

In particular, (8.14) must be satisfied by the feasible solution $x_1^*, x_2^*, \ldots, x_n^*$ of (8.9), and so

$$\sum_{i \in I} (-r_i b_i) = \sum_{j=1}^{n} a_{kj} x_j^* = b_k.$$

We conclude that every solution of (8.12) must satisfy

$$\sum_{j=1}^{n} a_{kj} x_j = b_k$$

which is the desired result.

Thus (8.9) and the problem

$$\begin{aligned} \text{maximize} \quad & \sum_{j=1}^{n} c_j x_j \\ \text{subject to} \quad & \sum_{j=1}^{n} a_{ij} x_j = b_i \qquad (i \in I) \\ & l_j \le x_j \le u_j \qquad (j = 1, 2, \ldots, n) \end{aligned} \tag{8.15}$$

are equivalent in the sense that they have precisely the same set of feasible solutions. In particular, the feasible solution of (8.9) delivered by the first phase of the two-phase simplex method is a feasible solution of (8.15). We claim that this solution is basic in (8.15), with the appropriate set of basic variables having been delivered by the procedure described earlier. To justify this claim, denote by B the set of subscripts j with $1 \le j \le n$ for which x_j appears in the basis heading upon termination of the procedure. We have to verify only that the system

$$\sum_{j \in B} a_{ij} x_j = b_i \qquad (i \in I)$$

has a unique solution. But this claim follows at once from the fact that the system

$$\begin{aligned} \sum_{j \in B} a_{ij} x_j &= b_i \qquad (i \in I) \\ \sum_{j \in B} a_{ij} x_j + x_{n+i} &= b_i \qquad (i \in J) \end{aligned}$$

has a unique solution.

Our findings may be summarized as follows.

THEOREM 8.3. If (8.9) has a feasible solution, then some set I of subscripts $1, 2, \ldots, m$ has the following two properties: (i) problems (8.9) and (8.15) have precisely the same set of feasible solutions and (ii) problem (8.15) has a basic feasible solution.

Of course, more important than the mere statement of Theorem 8.3 is the fact that we can actually find the set I, and that we can do so reasonably quickly. If our procedure fails to drive all the artificial variables out of the basis, then we can apply the second phase of the two-phase simplex method to the smaller problem (8.15), where each iteration requires less work than it would have required in the larger problem (8.11). □

Treatment of Free Variables

As the reader may have observed, each free variable x_j may be eliminated from the problem

$$\text{maximize } \mathbf{cx} \quad \text{subject to} \quad \mathbf{Ax} = \mathbf{b}, \quad \mathbf{l} \le \mathbf{x} \le \mathbf{u}$$

by first converting one of the equations into a formula for x_j and then substituting from this formula into the remaining equations. For illustration, consider the problem

$$\begin{aligned}
\text{maximize} \quad & x_1 \\
\text{subject to} \quad & x_1 + x_2 + x_3 + x_4 + x_5 + x_6 - x_7 = 10 \\
& x_1 + 2x_2 + x_7 = 2 \\
& x_3 + 2x_4 + x_7 = 5 \\
& x_5 + 2x_6 + x_7 = 8 \\
& x_1, x_2, x_3, x_4, x_5, x_6 \ge 0, \quad -\infty \le x_7 \le +\infty.
\end{aligned}$$

Having converted the first equation into the formula

$$x_7 = -10 + x_1 + x_2 + x_3 + x_4 + x_5 + x_6$$

we substitute into the remaining three equations and obtain an equivalent version of our problem,

$$\begin{aligned}
\text{maximize} \quad & x_1 \\
\text{subject to} \quad & 2x_1 + 3x_2 + x_3 + x_4 + x_5 + x_6 = 12 \\
& x_1 + x_2 + 2x_3 + 3x_4 + x_5 + x_6 = 15 \\
& x_1 + x_2 + x_3 + x_4 + 2x_5 + 3x_6 = 18 \\
& x_1, x_2, x_3, x_4, x_5, x_6 \ge 0.
\end{aligned}$$

More generally, whenever all the free variables are eliminated in this manner, not only do these variables disappear but the number of equations is reduced as well. Now the following strategy suggests itself: first eliminate all the free variables, then solve the new version of the problem by the simplex method, and finally determine the values of the free variables by going back to the substitution formulas. Since the new version is smaller than the original, this strategy has at least two apparent advantages: (i) each of the systems $\mathbf{yB} = \mathbf{c}_B$ and $\mathbf{Bd} = \mathbf{a}$, being smaller in the new version, is easier to solve, (ii) the number of iterations is likely to be smaller in the new version. However, if the original problem is sparse, then the elimination of the free variables may yield a problem that, although smaller, is also much denser. In that case, the first advantage is lost: relatively small but dense systems may be harder to solve than relatively large but sparse ones. On the other hand, the effects of the second advantage may be felt even without the elimination of free variables: when a free variable has entered the basis, it will never leave again. Thus, a problem with k free variables in the basis is like a problem with k equations deleted. For this reason, it is good policy to pivot all the free variables into the basis as soon as possible.

THE FUNDAMENTAL THEOREM OF LINEAR PROGRAMMING REVIEWED

In Chapter 3, we established a fundamental property of LP problems in the standard form: whenever such a problem has a feasible solution, it has a basic feasible solution. This property is not shared by every problem

$$\text{maximize} \quad \mathbf{cx} \qquad \text{subject to} \quad \mathbf{Ax} = \mathbf{b}, \quad \mathbf{l} \le \mathbf{x} \le \mathbf{u}. \tag{8.16}$$

For instance, the problem

$$\begin{array}{ll} \text{maximize} & x_1 + x_2 + x_3 + x_4 + x_5 \\ \text{subject to} & 3x_1 + 4x_2 + 2x_3 + 5x_4 + x_5 = 6 \\ & 2x_1 + x_2 + 3x_3 \qquad\quad - x_5 = -1 \\ & x_1 + 2x_2 \qquad\quad + 3x_4 + x_5 = 4 \\ & x_1, x_2, x_3, x_4, x_5 \ge 0 \end{array} \tag{8.17}$$

has feasible solutions ($x_1^* = x_2^* = x_3^* = 0$, $x_4^* = x_5^* = 1$ is one) but it has no basic feasible solution. To justify this claim, recall that (8.16) can have a basic feasible solution only if some three columns of

$$\mathbf{A} = \begin{bmatrix} 3 & 4 & 2 & 5 & 1 \\ 2 & 1 & 3 & 0 & -1 \\ 1 & 2 & 0 & 3 & 1 \end{bmatrix}$$

form a nonsingular matrix $\mathbf{B}$. But we have

$$[3, -2, -5] \cdot \mathbf{A} = [0, 0, 0, 0, 0]$$

and so $[3, -2, -5] \cdot \mathbf{B} = [0, 0, 0]$ for each matrix $\mathbf{B}$ formed by three columns of $\mathbf{A}$. Now $\mathbf{B}$ must be singular, for otherwise

$$[3, -2, -5] = [3, -2, -5] \cdot \mathbf{BB}^{-1} = [0, 0, 0] \cdot \mathbf{B}^{-1} = [0, 0, 0]$$

which is a contradiction.

More generally, (8.16) can have a basic solution only if some m columns of the $m \times n$ matrix $\mathbf{A}$ form a nonsingular matrix. Not all matrices $\mathbf{A}$ have this property; those that do are said to have *full row rank*. (If $\mathbf{rA}$ is the zero vector of length n for some nonzero vector $\mathbf{r}$ of length m, then, by the argument used in the previous example, **A** does not have full row rank. See also problem 8.4.) Thus, Theorem 3.4 can be generalized only to problems (8.16) whose matrices have full row rank.

Furthermore, a little care must be taken, so that the actual impact of Theorem 3.4, as well as its statement, is generalized. An important consequence of Theorem 3.4 is the fact that the search for an optimal solution of an LP problem in the standard form may be restricted to a finite set, namely, the set of basic feasible solutions. Confining the search for an optimal solution of (8.16) to the set of its basic feasible solutions would not hold the same attraction: problems involving free variables may have infinitely many basic feasible solutions. For instance, if

$$\mathbf{A} = \begin{bmatrix} 2 & 1 & 1 & 1 \\ 1 & 0 & 3 & 1 \end{bmatrix}, \quad \mathbf{b} = \begin{bmatrix} 2 \\ 1 \end{bmatrix}$$

and $\mathbf{l} = [-\infty, 0, 0, -\infty]^T$, $\mathbf{u} = [1, +\infty, 5, +\infty]^T$, then $[1 - t, t, 0, t]^T$ is a basic feasible solution of (8.16), with x_1 and x_2 basic, whenever $t \geq 0$. To remove this blemish, we introduce the notion of a *normal* basic solution: this is any basic solution in which each nonbasic free variable is set at zero. Clearly, every problem (8.16) has only a finitely many normal basic solutions.

THEOREM 8.4. If (8.16) has no optimal solution, then it is either infeasible or unbounded. Furthermore, if **A** has full row rank, then (8.16) has the following two properties:

(i) If it has a feasible solution, then it has a normal basic feasible solution.

(ii) If it has an optimal solution, then it has a normal basic optimal solution.

PROOF. The first statement follows easily from an analysis of the two-phase simplex method applied to (8.9). The first phase terminates either by discovering that (8.9) is infeasible or by producing a basic feasible solution of the equivalent problem (8.11). In the latter case, the second phase takes place. This phase terminates either by discovering that (8.11), and therefore also (8.9), is unbounded, or by producing a basic optimal solution of (8.11) that yields an optimal solution of (8.9). The termination of each phase is guaranteed by Theorem 8.2 as long as the smallest-subscript pivoting rule is used.

To prove (i) and (ii), assume that **A** has full row rank. If (8.9) has a feasible solution, then the first phase terminates by producing a basic feasible solution $x_1^*, x_2^*, \ldots, x_{n+m}^*$ of (8.11). The procedure used in the proof of Theorem 8.3 will drive all the artificial variables out of the basis; otherwise, it would find a nonzero row vector **r** of length m such that **rA** is the zero vector of length n, contradicting the assumption that **A** has full row rank. Hence $x_1^*, x_2^*, \ldots, x_n^*$ is a basic feasible solution of (8.9) in this case. To ensure that this solution is normal, we need only choose the numbers $\tilde{x}_1, \tilde{x}_2, \ldots, \tilde{x}_n$ used to construct (8.10) in such a way that $\tilde{x}_j = 0$ whenever x_j is free. Indeed, as soon as the simplex method is initialized by a normal feasible solution, it will construct only normal basic feasible solutions; if a nonbasic free variable actually changes its value in some iteration, then it enters the basis in the same iteration and it remains basic from that point on. Finally, if (8.9) has an optimal solution, then the second phase, initialized by the normal basic feasible solution $x_1^*, x_2^*, \ldots, x_n^*$, finds a normal basic optimal solution. ■

As observed earlier, the assumption that **A** has full row rank cannot be dropped in proving (i) and (ii). Nevertheless, this assumption is satisfied whenever (8.16) arises from a problem in the standard form. In this case, the last m columns of **A** form the identity matrix. Hence Theorem 8.4 does generalize Theorem 3.4. □

PROBLEMS

△ **8.1** Solve the following problems:

a.
$$\begin{array}{ll} \text{maximize} & -3x_1 - x_2 - x_3 + 2x_4 - x_5 + x_6 + x_7 - 4x_8 \\ \text{subject to} & x_1 + 3x_3 + x_4 - 5x_5 - 2x_6 + 4x_7 - 6x_8 = 7 \\ & x_2 - 2x_3 - x_4 + 4x_5 + x_6 - 3x_7 + 5x_8 = -3 \\ & 0 \le x_1 \le 8 \\ & 0 \le x_2 \le 6 \\ & 0 \le x_3 \le 4 \\ & 0 \le x_4 \le 15 \\ & 0 \le x_5 \le 2 \\ & 0 \le x_6 \le 10 \\ & 0 \le x_7 \le 10 \\ & 0 \le x_8 \le 3. \end{array}$$

b.
$$\begin{array}{ll} \text{maximize} & 3x_1 + x_2 + 4x_3 + 2x_4 \\ \text{subject to} & x_1 + 4x_2 + 3x_3 + 3x_4 \le 2 \\ & x_1 + 3x_2 - x_3 + x_4 \le -2 \\ & x_1 + 2x_2 + 3x_3 + 2x_4 \le 3 \\ & x_1 + 3x_2 - 2x_3 + x_4 \le -3 \\ & x_3, x_4 \ge 0. \end{array}$$

c.
$$\begin{array}{ll} \text{maximize} & 5x_1 + 2x_2 - 3x_3 + 3x_4 + 6x_5 + x_6 \\ \text{subject to} & 3x_1 + x_2 - 4x_3 + 2x_4 + 5x_5 + x_6 \le 3 \\ & -5x_1 + 4x_2 + 2x_3 - 3x_4 + 2x_5 + 3x_6 \le 25 \\ & x_1 + x_2 + 2x_3 + x_4 + x_5 + 2x_6 = 4 \\ & x_1 \ge 0, 2 \le x_2 \le 10, x_3 \le 0, -3 \le x_4 \le 3. \end{array}$$

d.
$$\begin{array}{ll} \text{maximize} & 8x_1 + 6x_2 - 2x_3 + 6x_4 - 3x_5 \\ \text{subject to} & 3x_1 + 2x_2 - x_3 + 3x_4 + 2x_5 = 7 \\ & -x_1 + 2x_2 - 2x_3 = 10 \\ & x_1 - 3x_4 + 2x_5 = 7 \\ & 2x_1 - 3x_2 + 2x_3 + 3x_4 = -10 \\ & 2x_1 + x_3 - x_5 \le -6 \\ & x_1, x_2 \ge 0. \end{array}$$

e.
$$\begin{array}{ll} \text{Maximize} & x_1 \\ \text{subject to} & 3x_1 + 5x_2 + 3x_3 + 2x_4 + x_5 + 2x_6 + x_7 = 6 \\ & 4x_1 + 6x_2 + 5x_3 + 3x_4 + 5x_5 + 5x_6 + 6x_7 = 11 \\ & 2x_1 + 4x_2 + x_3 + x_4 - 3x_5 - x_6 - 4x_7 = 1 \\ & x_1 + x_2 + 2x_3 + x_4 + 4x_5 + 3x_6 + 5x_7 = 5 \\ & x_1, x_2, \ldots, x_7 \ge 0. \end{array}$$

8.2 Solve Polly's diet problem from Chapter 1 and illustrate each iteration in economic terms.

8.3 Give a detailed proof of Theorem 8.2.

8.4 Prove that a matrix $\mathbf{A}$ does not have full row rank *if and only if* $\mathbf{rA} = \mathbf{0}$ for some nonzero row vector $\mathbf{r}$.

8.5 Illustrate Theorem 8.3 on problem 8.1.e.

8.6 Construct an example of a problem

$$\text{maximize} \quad \mathbf{cx} \qquad \text{subject to} \quad \mathbf{Ax} = \mathbf{b}, \quad \mathbf{l} \le \mathbf{x} \le \mathbf{u}$$

such that (i) there is a feasible solution, (ii) the $m \times n$ matrix $\mathbf{A}$ has a full row rank, (iii) in every basic feasible solution, at least one free variable is nonbasic, and yet (iv) there are at most m free variables.

8.7 Prove the following theorem. If the problem

$$\begin{aligned} \text{maximize} \quad & \sum_{j=1}^{n} c_j x_j \\ \text{subject to} \quad & \sum_{j=1}^{n} a_{ij} x_j = b_i \qquad (i = 1, 2, \ldots, m) \\ & l_j \le x_j \le u_j \qquad (j = 1, 2, \ldots, n) \end{aligned} \tag{8.18}$$

has at least one nondegenerate basic optimal solution, then there are uniquely defined numbers $y_1^*, y_2^*, \ldots, y_m^*$ and a positive ε with the following property: If $|t_i| \le \varepsilon$ for all $i = 1, 2, \ldots, m$, then the problem

$$\begin{aligned} \text{maximize} \quad & \sum_{j=1}^{n} c_j x_j \\ \text{subject to} \quad & \sum_{j=1}^{n} a_{ij} x_j = b_i + t_i \qquad (i = 1, 2, \ldots, m) \\ & l_j \le x_j \le u_j \qquad (j = 1, 2, \ldots, n) \end{aligned}$$

has an optimal solution and its optimal value equals

$$z^* + \sum_{i=1}^{m} y_i^* t_i$$

with z^* standing for the optimal value of (8.18).

8.8 Illustrate the economic significance of the preceding result on the example of problem 1.8.

8.9 State and prove a variation on problem 8.7, involving changes in the bounds l_j and u_j rather than changes in the right-hand sides b_i.

9

General LP Problems: Theorems on Duality and Infeasibility

In Chapter 8, we learned to solve general LP problems by the simplex method. The focus of this chapter is more theoretical: we shall extend the duality theorem, proved in Chapter 5 for LP problems in the standard form, to the general context of arbitrary LP problems. Furthermore, we shall prove a theorem characterizing infeasible LP problems: in other words, we shall characterize unsolvable systems of linear inequalities and equations.

THE DUALITY THEOREM

In Chapter 8, we observed that every LP problem can be converted into the form

$$\text{maximize} \quad \mathbf{cx} \quad \text{subject to} \quad \mathbf{Ax} = \mathbf{b}, \quad \mathbf{l} \le \mathbf{x} \le \mathbf{u}.$$

In the present context, it will be convenient to view LP problems from a different angle: (i) the inequalities $\sum a_{ij}x_j \le b_i$ will be left alone rather than converted into equations with slack variables and (ii) except for the nonnegativity constraints $x_j \ge 0$, explicit bounds on individual variables will be regarded as instances of $\sum a_{ij}x_j \le b_i$. Thus, every LP problem will be presented as

$$\begin{aligned} \text{maximize} \quad & \sum_{j=1}^{n} c_j x_j \\ \text{subject to} \quad & \sum_{j=1}^{n} a_{ij} x_j \leq b_i \qquad (i \in I) \\ & \sum_{j=1}^{n} a_{ij} x_j = b_i \qquad (i \in E) \\ & x_j \geq 0 \qquad (j \in R). \end{aligned} \tag{9.1}$$

Here the set of the constraint subscripts $1, 2, \ldots, m$ splits into subsets I and E corresponding to inequalities and equations, respectively. In the set of the variable subscripts $1, 2, \ldots, n$, the subset R marks out the variables x_j that are explicitly *restricted* to the nonnegative range. Variables x_j with $j \notin R$ will be called *free* (even though some of the constraints $\sum a_{ij} x_j \leq b_i$ may reduce to explicit bounds on these variables) and the corresponding set of variable subscripts j will be denoted by F. In particular, (9.1) is in the standard form if and only if $E = \varnothing$ and $F = \varnothing$.

By a *linear combination* of the constraints

$$\begin{aligned} & \sum_{j=1}^{n} a_{ij} x_j \leq b_i \qquad (i \in I) \\ & \sum_{j=1}^{n} a_{ij} x_j = b_i \qquad (i \in E) \end{aligned} \tag{9.2}$$

we mean any linear inequality

$$\sum_{j=1}^{n} \left(\sum_{i=1}^{m} a_{ij} y_i \right) x_j \leq \sum_{i=1}^{m} b_i y_i \tag{9.3}$$

such that $y_1, y_2, \ldots, y_m$ are numbers satisfying

$$y_i \geq 0 \quad \text{whenever} \quad i \in I. \tag{9.4}$$

Every linear combination of (9.2) arises by first multiplying both sides of each constraint by the appropriate y_i, so that

$$\begin{aligned} & y_i \left(\sum_{j=1}^{n} a_{ij} x_j \right) \leq b_i y_i \qquad (i \in I) \\ & y_i \left(\sum_{j=1}^{n} a_{ij} x_j \right) = b_i y_i \qquad (i \in E) \end{aligned}$$

is obtained, and then adding up the results, so that

$$\sum_{i=1}^{m} y_i \left(\sum_{j=1}^{n} a_{ij} x_j \right) \leq \sum_{i=1}^{m} b_i y_i$$

is obtained, and finally changing the order of summation in the left-hand side of the last inequality. Hence, every solution $x_1, x_2, \ldots, x_n$ of the inequalities and equations (9.2) must satisfy all their linear combinations (9.3).

Next, if numbers $y_1, y_2, \ldots, y_m$ are chosen in such a way that

$$\sum_{i=1}^{m} a_{ij}y_i \geq c_j \quad \text{whenever} \quad j \in R \tag{9.5}$$

and

$$\sum_{i=1}^{m} a_{ij}y_i = c_j \quad \text{whenever} \quad j \in F \tag{9.6}$$

then every feasible solution $x_1, x_2, \ldots, x_n$ of (9.1) satisfies

$$c_j x_j \leq \left(\sum_{i=1}^{m} a_{ij}y_i \right) x_j \quad \text{for all} \quad j = 1, 2, \ldots, n$$

(with the sign of equality whenever $j \in F$), and so

$$\sum_{j=1}^{n} c_j x_j \leq \sum_{j=1}^{n} \left(\sum_{i=1}^{m} a_{ij}y_i \right) x_j. \tag{9.7}$$

Finally, if numbers $x_1, x_2, \ldots, x_n$ satisfy both (9.3) and (9.7), then they satisfy the inequality

$$\sum_{j=1}^{n} c_j x_j \leq \sum_{i=1}^{m} b_i y_i. \tag{9.8}$$

To summarize, the number $\sum b_i y_i$ provides an upper bound on the optimal value of (9.1) whenever the numbers $y_1, y_2, \ldots, y_m$ satisfy (9.4), (9.5), and (9.6). Naturally, we are interested in finding as good an upper bound as we can. Thus, we are led to

$$\begin{aligned}
\text{minimize} \quad & \sum_{i=1}^{m} b_i y_i \\
\text{subject to} \quad & \sum_{i=1}^{m} a_{ij}y_i \geq c_j \qquad (j \in R) \\
& \sum_{i=1}^{m} a_{ij}y_i = c_j \qquad (j \in F) \\
& y_i \geq 0 \qquad (i \in I).
\end{aligned} \tag{9.9}$$

This linear programming problem is called the *dual problem* or simply the *dual* of the linear programming problem (9.1); in this context, the original problem (9.1) is sometimes referred to as the *primal problem* or simply the *primal*. [Note that, in the special case when (9.1) is in the standard form, the present definition reduces to that given in Chapter 5.] As we have just observed, every feasible solution $x_1, x_2, \ldots, x_n$ of the primal problem (9.1) and every feasible solution $y_1, y_2, \ldots, y_m$ of the dual problem (9.9) satisfy the inequality (9.8). This inequality is sometimes referred to as "the weak duality theorem" or "the easy part of the duality theorem." (The more difficult part, which we shall prove, asserts that the gap between the two sides of (9.8) can always be closed by a suitable choice of $x_1, x_2, \ldots, x_n$ and $y_1, y_2, \ldots, y_m$.)

Incidentally, note that the variables of the dual are in a one-to-one correspondence with the constraints of the primal and that the constraints of the dual are in a one-to-one correspondence with the variables of the primal.

Table 9.1 Primal–Dual Correspondence

In the dual	In the primal
Restricted variables	Inequality constraints
Free variables	Equation constraints
Inequality constraints	Restricted variables
Equation constraints	Free variables

Note also that the dual of the dual is the primal: problem (9.9) may be presented as

$$\begin{aligned} \text{maximize} \quad & \sum_{i=1}^{m} (-b_i) y_i \\ \text{subject to} \quad & \sum_{i=1}^{m} (-a_{ij}) y_i \leq -c_j \qquad (j \in R) \\ & \sum_{i=1}^{m} (-a_{ij}) y_i = -c_j \qquad (j \in F) \\ & y_i \geq 0 \qquad (i \in I) \end{aligned}$$

and its dual problem

$$\begin{aligned} \text{minimize} \quad & \sum_{j=1}^{n} (-c_j) x_j \\ \text{subject to} \quad & \sum_{j=1}^{n} (-a_{ij}) x_j \geq -b_i \qquad (i \in I) \\ & \sum_{j=1}^{n} (-a_{ij}) x_j = -b_i \qquad (i \in E) \\ & x_j \geq 0 \qquad (j \in R) \end{aligned}$$

is just another presentation of (9.1).

Occasionally, students get confused when asked to construct dual problems of LP problems such as

$$\begin{array}{ll} \text{maximize} & 3x_1 + 2x_2 + 5x_3 \\ \text{subject to} & 5x_1 + 3x_2 + x_3 = -8 \\ & 4x_1 + 2x_2 + 8x_3 \le 23 \\ & 6x_1 + 7x_2 + 3x_3 \ge 1 \\ & \qquad x_1 \le 4,\ x_3 \ge 0. \end{array}$$

The routine way of accomplishing this task consists of presenting the maximization problem in the form of (9.1) and then writing (9.9). Thus, our example would be presented as

$$\begin{array}{ll} \text{maximize} & 3x_1 + 2x_2 + 5x_3 \\ \text{subject to} & 5x_1 + 3x_2 + x_3 = -8 \\ & 4x_1 + 2x_2 + 8x_3 \le 23 \\ & -6x_1 - 7x_2 - 3x_3 \le -1 \\ & x_1 \qquad\qquad\qquad \le 4 \\ & \qquad\qquad\quad x_3 \ge 0 \end{array}$$

and its dual found to be

$$\begin{array}{ll} \text{minimize} & -8y_1 + 23y_2 - y_3 + 4y_4 \\ \text{subject to} & 5y_1 + 4y_2 - 6y_3 + y_4 = 3 \\ & 3y_1 + 2y_2 - 7y_3 \qquad\ \ = 2 \\ & y_1 + 8y_2 - 3y_3 \qquad\ \ \ge 5 \\ & \qquad\qquad\quad y_2, y_3, y_4 \ge 0. \end{array}$$

Similarly, the routine way of finding the dual of a minimization problem consists of presenting the problem as (9.9) and then writing (9.1). When this convention is observed, the dual of a maximization problem is always a minimization problem and vice versa.

***THEOREM 9.1* (*The Duality Theorem*).** If a linear programming problem has an optimal solution, then its dual has an optimal solution and the optimal values of the two problems coincide.

PROOF. We shall rely on an analysis of the simplex method as presented in Chapter 8. Every LP problem may be presented in the form (9.1), whereupon its dual assumes the form (9.9). If (9.1) has an optimal solution, then the simplex method, applied to its equivalent version

$$\begin{aligned}
\text{maximize} \quad & \sum_{j=1}^{n} c_j x_j \\
& \sum_{j=1}^{n} a_{ij} x_j + x_{n+i} = b_i \qquad (i \in I) \\
& \sum_{j=1}^{n} a_{ij} x_j = b_i \qquad (i \in E) \\
& x_j \geq 0 \qquad (j \in R) \\
& x_{n+i} \geq 0 \qquad (i \in I)
\end{aligned}$$

finds an optimal solution $x_1^*, x_2^*, \ldots, x_n^*$. Consider the vector $\mathbf{y} = [y_1^*, y_2^*, \ldots, y_m^*]$ featured in the final iteration of the simplex method. In step 1, this vector is computed by solving the system $\mathbf{yB} = \mathbf{c}_B$; hence

$$\sum_{i=1}^{m} a_{ij} y_i^* = c_j \quad \text{whenever} \quad 1 \leq j \leq n \text{ and } x_j \text{ is basic,}$$

$$y_i^* = 0 \quad \text{whenever} \quad i \in I \text{ and } x_{n+i} \text{ is basic.}$$

In step 2, no candidate for entering the basis is found; hence

$$\sum_{i=1}^{m} a_{ij} y_i^* = c_j \quad \text{whenever} \quad j \in F \text{ and } x_j \text{ is nonbasic}$$

$$\sum_{i=1}^{m} a_{ij} y_i^* \geq c_j \quad \text{whenever} \quad j \in R \text{ and } x_j \text{ is nonbasic}$$

$$y_i^* \geq 0 \quad \text{whenever} \quad i \in I \text{ and } x_{n+i} \text{ is nonbasic.}$$

Now it follows that

$$\left(\sum_{i=1}^{m} a_{ij} y_i^*\right) x_j^* = c_j x_j^* \quad \text{for all} \quad j = 1, 2, \ldots, n: \tag{9.10}$$

the inequality $\sum a_{ij} y_i^* \geq c_j$, valid for all $j = 1, 2, \ldots, n$, can hold with the sharp inequality sign only if $j \in R$ and x_j is nonbasic, in which case $x_j^* = 0$. Similarly, it follows that

$$\left(\sum_{j=1}^{n} a_{ij} x_j^*\right) y_i^* = b_i y_i^* \quad \text{for all} \quad i = 1, 2, \ldots, m: \tag{9.11}$$

the inequality $\sum a_{ij} x_j^* \leq b_i$, valid for all $i = 1, 2, \ldots, m$, can hold with the sharp inequality sign only if $i \in I$ and x_{n+i} is basic, in which case $y_i^* = 0$. Note that (9.10) and (9.11) imply

$$\sum_{j=1}^{n} c_j x_j^* = \sum_{j=1}^{n} \left(\sum_{i=1}^{m} a_{ij} y_i^*\right) x_j^* = \sum_{i=1}^{m} \left(\sum_{j=1}^{n} a_{ij} x_j^*\right) y_i^* = \sum_{i=1}^{m} b_i y_i^*.$$

To summarize, $y_1^*, y_2^*, \ldots, y_m^*$ is a feasible solution of the dual (9.9) and

$$\sum_{i=1}^{m} b_i y_i^* = \sum_{j=1}^{n} c_j x_j^*. \tag{9.12}$$

Now it only remains to be shown that every feasible solution $y_1, y_2, \ldots, y_m$ of (9.9) has

$$\sum_{i=1}^{m} b_i y_i \geq \sum_{i=1}^{m} b_i y_i^*$$

for then $y_1^*, y_2^*, \ldots, y_m^*$ is an optimal solution of (9.9). But this inequality follows directly from (9.12) and (9.8) with $x_1, x_2, \ldots, x_n$ replaced by $x_1^*, x_2^*, \ldots, x_n^*$. ■

Our proof shows that the simplex method, applied to an arbitrary LP problem that has an optimal solution, finds an optimal solution of this problem simultaneously with an optimal solution of its dual. Additional results involving arbitrary LP problems and their duals can be developed along the lines of Chapter 5.

UNSOLVABLE SYSTEMS OF LINEAR INEQUALITIES AND EQUATIONS

In the remainder of this chapter, we shall characterize infeasible linear programs. Now it is convenient to present every LP problem in the form

$$\begin{aligned} \text{maximize} \quad & \sum_{j=1}^{n} c_j x_j \\ \text{subject to} \quad & \sum_{j=1}^{n} a_{ij} x_j \leq b_i \qquad (i \in I) \\ & \sum_{j=1}^{n} a_{ij} x_j = b_i \qquad (i \in E) \end{aligned}$$

with the nonnegativity constraints $x_j \geq 0$, if any, viewed as special instances of $\sum a_{ij} x_j \leq b_i$. Since we are concerned only with the existence of feasible solutions, the objective function $\sum c_j x_j$ is irrelevant; our task amounts to characterizing unsolvable systems

$$\begin{aligned} & \sum_{j=1}^{n} a_{ij} x_j \leq b_i \qquad (i \in I) \\ & \sum_{j=1}^{n} a_{ij} x_j = b_i \qquad (i \in E). \end{aligned} \tag{9.13}$$

At the beginning of this chapter, we defined a linear combination of the constraints (9.13) as any linear inequality

$$\sum_{j=1}^{n}\left(\sum_{i=1}^{m} a_{ij}y_i\right)x_j \le \sum_{i=1}^{m} b_i y_i$$

such that $y_1, y_2, \ldots, y_m$ are numbers satisfying $y_i \ge 0$ whenever $i \in I$. In addition, we observed that every solution of a system of linear constraints must satisfy all linear combinations of this system. This observation may be used to demonstrate the unsolvability of systems such as

$$\begin{aligned} x_1 + 3x_2 + 2x_3 + 4x_4 &\le 5 \\ 3x_1 + x_2 + 2x_3 + x_4 &\le 4 \\ 5x_1 + 3x_2 + 3x_3 + 3x_4 &= 9 \\ -x_3 &\le 0 \\ -x_4 &\le 0 \end{aligned} \tag{9.14}$$

in a simple and convincing manner. In this particular example, we need only observe that the multipliers

$$y_1 = 1, \quad y_2 = 3, \quad y_3 = -2, \quad y_4 = 2, \quad y_5 = 1$$

yield a linear combination of (9.14) that reads

$$0 \cdot x_1 + 0 \cdot x_2 + 0 \cdot x_3 + 0 \cdot x_4 \le -1. \tag{9.15}$$

Now it becomes obvious that (9.14) is unsolvable: any solution of this system would have to satisfy the clearly unsolvable inequality (9.15).

More generally, we shall follow H. W. Kuhn (1956) in saying that a system (9.13) is *inconsistent* if there are numbers $y_1, y_2, \ldots, y_m$ such that

$$\left.\begin{aligned} & y_i \ge 0 \quad \text{whenever} \quad i \in I \\ & \sum_{i=1}^{m} a_{ij}y_i = 0 \qquad \text{for all} \quad j = 1, 2, \ldots, n \\ & \sum_{i=1}^{m} b_i y_i < 0 \end{aligned}\right\}. \tag{9.16}$$

Thus, (9.13) is inconsistent if and only if one of its linear combinations reads

$$0 \cdot x_1 + 0 \cdot x_2 + \cdots + 0 \cdot x_n \le c \tag{9.17}$$

for some negative number c. As soon as the multipliers $y_1, y_2, \ldots, y_m$ satisfying (9.16) are exhibited, it becomes obvious that (9.13) is unsolvable: any solution of this system would have to satisfy the clearly unsatisfiable inequality (9.17). Hence every inconsistent system is unsolvable; now we propose to show the converse.

THEOREM 9.2. A system of linear inequalities and equations is unsolvable if and only if it is inconsistent.

PROOF. Since the "if" part has just been proved, we have to prove only the "only if" part. For this purpose, we shall appeal to the relationship between system (9.13) and the problem

$$\begin{aligned}
\text{maximize} \quad & \sum_{i=1}^{m} (-x_{n+i}) \\
\text{subject to} \quad & \sum_{j=1}^{n} a_{ij}x_j + w_i x_{n+i} \le b_i \qquad (i \in I) \\
& \sum_{j=1}^{n} a_{ij}x_j + w_i x_{n+i} = b_i \qquad (i \in E) \\
& x_{n+i} \ge 0 \qquad (i = 1, 2, \ldots, m)
\end{aligned} \tag{9.18}$$

with $w_i = 1$ if $b_i \ge 0$ and $w_i = -1$ if $b_i < 0$. Theorem 8.4 guarantees that (9.18) always has an optimal solution; obviously, the optimal value of (9.18) is zero if and only if (9.13) is solvable. In particular, if (9.13) is unsolvable, then the optimal value of (9.18) is negative. In that case, Theorem 9.1 guarantees that the dual of (9.18),

$$\begin{aligned}
\text{minimize} \quad & \sum_{i=1}^{m} b_i y_i \\
\text{subject to} \quad & \sum_{i=1}^{m} a_{ij} y_i = 0 \qquad (j = 1, 2, \ldots, n) \\
& w_i y_i \ge -1 \qquad (i = 1, 2, \ldots, m) \\
& y_i \ge 0 \qquad (i \in I)
\end{aligned}$$

has an optimal solution and that its optimal value is negative. But then the optimal solution $y_1, y_2, \ldots, y_m$ satisfies (9.16), and so (9.13) is inconsistent. ■

Before proving a supplementary result on unsolvable systems of linear inequalities, let us make note of an immediate corollary of Theorem 8.3.

THEOREM 9.3. If a system of m linear equations has a nonnegative solution, then it has a solution with at most m variables positive.

PROOF. If the system

$$\begin{aligned}
& \sum_{j=1}^{n} a_{ij}x_j = b_i \qquad (i = 1, 2, \ldots, m) \\
& x_j \ge 0 \qquad (j = 1, 2, \ldots, n)
\end{aligned} \tag{9.19}$$

has a solution, then, by Theorem 8.3, there is some set I of subscripts $1, 2, \ldots, m$ such that (i) systems (9.19) and

$$\begin{aligned} \sum_{j=1}^{n} a_{ij}x_j &= b_i \qquad (i \in I) \\ x_j &\geq 0 \qquad (j = 1, 2, \ldots, n) \end{aligned} \tag{9.20}$$

have precisely the same set of solutions and (ii) system (9.20) has a basic feasible solution $x_1^*, x_2^*, \ldots, x_n^*$. Now it only remains to be observed that $x_j^* = 0$ whenever x_j is nonbasic and that at most $|I|$ variables x_j are basic. ■

> ***THEOREM 9.4.*** Every unsolvable system of linear inequalities in n variables contains an unsolvable subsystem of at most $n + 1$ inequalities.

PROOF. If $\mathbf{Ax} \leq \mathbf{b}$ is unsolvable, then, by Theorem 9.2, it is inconsistent. Hence, some row vector $\mathbf{y}^*$ satisfies $\mathbf{y}^* \geq \mathbf{0}$, $\mathbf{y}^*\mathbf{A} = \mathbf{0}$, $\mathbf{y}^*\mathbf{b} < 0$. Denote $\mathbf{y}^*\mathbf{b}$ by c, and consider the system $\mathbf{yA} = \mathbf{0}$, $\mathbf{yb} = c$ consisting of $n + 1$ equations. Since $\mathbf{y}^*$ is a nonnegative solution, Theorem 9.3 guarantees the existence of a nonnegative solution $\bar{\mathbf{y}}$ with at most $n + 1$ positive components $\bar{y}_i$. The desired subsystem consists of those inequalities $\sum a_{ij}x_j \leq b_i$ for which $\bar{y}_i > 0$; since $\sum \bar{y}_i a_{ij} = 0$ for all j but $\sum \bar{y}_i b_i = c < 0$, this subsystem is inconsistent and therefore unsolvable. ■ □

PROBLEMS

9.1 Construct the dual problems of

$$\begin{array}{ll} \text{maximize} & x_1 - x_2 \\ \text{subject to} & 2x_1 + 3x_2 - x_3 + x_4 \leq 0 \\ & 3x_1 + x_2 + 4x_3 - 2x_4 \geq 3 \\ & -x_1 - x_2 + 2x_3 + x_4 = 1 \\ & x_2, x_3 \geq 0 \end{array}$$

and of

$$\begin{array}{ll} \text{minimize} & x_1 - x_2 \\ \text{subject to} & 2x_1 + 3x_2 - x_3 + x_4 \leq 0 \\ & 3x_1 + x_2 + 4x_3 - 2x_4 \geq 3 \\ & -x_1 - x_2 + 2x_3 + x_4 = 1 \\ & x_2, x_3 \geq 0. \end{array}$$

9.2 In matrix notation, an LP in the standard form reads

$$\text{maximize } \mathbf{cx} \quad \text{subject to } \mathbf{Ax} \leq \mathbf{b}, \ \mathbf{x} \geq \mathbf{0}$$

and its dual reads

$$\text{minimize } \mathbf{yb} \quad \text{subject to } \mathbf{yA} \geq \mathbf{c}, \ \mathbf{y} \geq \mathbf{0}.$$

What is the dual of: maximize $\mathbf{cx}$ subject to $\mathbf{Ax} \leq \mathbf{b}$? The dual of: maximize $\mathbf{cx}$ subject to $\mathbf{Ax} = \mathbf{b}, \mathbf{x} \geq \mathbf{0}$?

9.3 Construct the dual problem of

$$\begin{aligned} &\text{maximize} \quad z \\ &\text{subject to} \quad z - \sum_{i=1}^{m} a_{ij}x_i \leq 0 \qquad (j = 1, 2, \ldots, n) \\ &\qquad\qquad \sum_{i=1}^{m} x_i = 1 \\ &\qquad\qquad x_i \geq 0 \qquad (i = 1, 2, \ldots, m). \end{aligned}$$

△**9.4** Maximize $x_1 + x_2 + x_3$

subject to

$$\begin{aligned} 2x_1 - 4x_2 + x_3 &\leq -1 \\ x_1 + 5x_2 + x_3 &\leq 16 \\ x_1 + x_3 &\leq 5 \\ 2x_1 + 4x_2 - x_3 &\leq 8 \\ x_1 - 3x_2 + x_3 &\leq 0 \\ -4x_1 + 3x_2 &\leq 4 \\ 4x_1 - 3x_2 + 5x_3 &\leq 10 \\ x_1 + 2x_2 + x_3 &\leq 9. \end{aligned}$$

9.5 Generalize Theorems 5.2 and 5.3 to the context of general LP problems. Use your result to find out if

$$x_1^* = 3, \quad x_2^* = -1, \quad x_3^* = 0, \quad x_4^* = 2$$

is an optimal solution of the problem

$$\begin{aligned} &\text{maximize} \quad 6x_1 + x_2 - x_3 - x_4 \\ &\text{subject to} \quad x_1 + 2x_2 + x_3 + x_4 \leq 5 \\ &\qquad\qquad 3x_1 + x_2 - x_3 \leq 8 \\ &\qquad\qquad x_2 + x_3 + x_4 = 1 \\ &\qquad\qquad x_3, x_4 \geq 0. \end{aligned}$$

9.6 One feasible solution of problem 1.8 calls for 10 assemblers and 30 instructors in the first week, followed by 130 assemblers in each of the remaining three weeks. Use the result of problem 9.5 to find out if this solution is optimal or not.

9.7 Illustrate Theorems 9.2 and 9.4 on the system

$$\begin{aligned} 2x + 6y - z &\leq -6 \\ 4x - 4y + 3z &\leq 0 \\ 3x + y - 4z &\leq 3 \\ 7x - 3y - 6z &\leq 10 \\ x - 5y + 2z &\leq 2 \\ 7x + 13y - 6z &\leq -7 \\ -6x - 2y + 3z &\leq 0. \end{aligned}$$

10

Sensitivity Analysis

In linear programming problems arising from applications, the numerical data often represent only rough estimates of quantities that are inherently difficult to measure or predict. Market prices may fluctuate, supplies of raw materials and demands for finished products are often unknown in advance, and production may be affected by a variety of accidental events such as machine breakdowns. In such cases, solving the initial LP problem is only the starting point for further analysis of the situation.

By replacing the original data by more pessimistic or optimistic estimates of the unknown quantities, we may create a number of variations on the original theme. Since each of these new LP problems might possibly represent the actual situation, it is useful to find out how the optimal solutions vary with the changes in data. It may be, for example, that the optimal solution is particularly sensitive to changes in only a small set of parameters; if possible, these parameters should then be estimated with greater accuracy. Other variations on the original theme arise when new variables and/or constraints are introduced. New variables may appear when new products are developed. Similarly, new constraints may result from new production policies. It may also happen that a new constraint actually had to be satisfied from the beginning but was simply forgotten in the original formulation. Investigations of such changes are referred to as *sensitivity analysis* or *postoptimality analysis.*

Sensitivity analysis could be conducted in a straightforward way by solving each of the modified problems from scratch. However, similar problems are likely to have similar solutions, and so it is often easier to exploit the results obtained in solving the original problem. This alternative strategy is the subject of the present chapter.

DUAL-FEASIBLE DICTIONARIES

In Chapter 2, we introduced the notion of a *dictionary* associated with a problem

$$\begin{aligned} \text{maximize} \quad & \sum_{j=1}^{n} c_j x_j \\ \text{subject to} \quad & \sum_{j=1}^{n} a_{ij} x_j \le b_i \qquad (i = 1, 2, \ldots, m) \\ & x_j \ge 0 \qquad (j = 1, 2, \ldots, n). \end{aligned} \tag{10.1}$$

The dictionary is any system of equations

$$\begin{aligned} x_r &= \bar{b}_r - \sum_{s \notin B} \bar{a}_{rs} x_s \qquad (r \in B) \\ \hline z &= \bar{d} + \sum_{s \notin B} \bar{c}_s x_s \end{aligned} \tag{10.2}$$

such that every solution $x_1, x_2, \ldots, x_{n+m}, z$ of the system

$$\begin{aligned} x_{n+i} &= b_i - \sum_{j=1}^{n} a_{ij} x_j \qquad (i = 1, 2, \ldots, m) \\ \hline z &= \sum_{j=1}^{n} c_j x_j \end{aligned}$$

is a solution of (10.2) and vice versa. Note that (10.2) may be a dictionary even if $\bar{b}_r < 0$ for some $r \in B$. That is, not every dictionary is *feasible* in the sense defined in Chapter 2.

Our immediate objective is to point out a certain one-to-one correspondence between the dictionaries associated with (10.1) and the dictionaries associated with the dual problem. In order to make this correspondence more transparent, we shall change the way in which the variables x_k are subscripted; instead of denoting the decision variables by $x_1, x_2, \ldots, x_n$ and the slack variables by $x_{n+1}, x_{n+2}, \ldots, x_{n+m}$ as usual, we shall denote the decision variables by $x_{m+1}, x_{m+2}, \ldots, x_{m+n}$ and the slack variables by $x_1, x_2, \ldots, x_m$. Thus, (10.1) will be recorded as

$$\begin{aligned} \text{maximize} \quad & \sum_{j=m+1}^{m+n} c_j x_j \\ \text{subject to} \quad & \sum_{j=m+1}^{m+n} a_{ij} x_j \leq b_i \qquad (i = 1, 2, \ldots, m) \\ & x_j \geq 0 \qquad (j = m+1, m+2, \ldots, m+n) \end{aligned} \tag{10.3}$$

and the corresponding initial dictionary will be recorded as

$$\begin{array}{l} x_i = b_i - \sum_{j=m+1}^{m+n} a_{ij} x_j \qquad (i = 1, 2, \ldots, m) \\ \hline z = \quad \sum_{j=m+1}^{m+n} c_j x_j \end{array} \tag{10.4}$$

The dual problem of (10.3) is

$$\begin{aligned} \text{minimize} \quad & \sum_{i=1}^{m} b_i y_i \\ \text{subject to} \quad & \sum_{i=1}^{m} a_{ij} y_i \geq c_j \qquad (j = m+1, m+2, \ldots, m+n) \\ & y_i \geq 0 \qquad (i = 1, 2, \ldots, m) \end{aligned} \tag{10.5}$$

and the corresponding initial dictionary reads

$$\begin{array}{l} y_j = -c_j + \sum_{i=1}^{m} a_{ij} y_i \qquad (j = m+1, m+2, \ldots, m+n) \\ \hline -w = \quad - \sum_{i=1}^{m} b_i y_i \,. \end{array} \tag{10.6}$$

Note that dictionaries (10.4) and (10.6) are mirror images of each other in the sense that the actual coefficients appearing in a row of (10.4) are found with reversed signs in a column of (10.6) and vice versa. For instance, if (10.3) is the problem

$$\begin{aligned} \text{maximize} \quad & 4x_3 - 13x_4 + 7x_5 \\ \text{subject to} \quad & 3x_3 + 2x_4 + 5x_5 \leq 5 \\ & x_3 - 3x_4 + 2x_5 \leq 3 \\ & x_3, x_4, x_5 \geq 0 \end{aligned}$$

then the dual problem (10.5) reads

$$\text{minimize} \quad 5y_1 + 3y_2$$

$$\begin{aligned} \text{subject to} \quad 3y_1 + y_2 &\geq 4 \\ 2y_1 - 3y_2 &\geq -13 \\ 5y_1 + 2y_2 &\geq 7 \\ y_1, y_2 &\geq 0 \end{aligned}$$

and the two dictionaries (10.4) and (10.6) are

$$\begin{array}{l} x_1 = 5 - 3x_3 - 2x_4 - 5x_5 \\ x_2 = 3 - x_3 + 3x_4 - 2x_5 \\ \hline z = 4x_3 - 13x_4 + 7x_5 \end{array} \quad \text{and} \quad \begin{array}{l} y_3 = -4 + 3y_1 + y_2 \\ y_4 = 13 + 2y_1 - 3y_2 \\ y_5 = -7 + 5y_1 + 2y_2 \\ \hline -w = - 5y_1 - 3y_2 \end{array}$$

respectively. A further examination of this particular example reveals that the primal and the dual dictionaries come in matching pairs. For instance,

$$\begin{array}{l} x_1 = -4 + 3x_2 - 11x_4 + x_5 \\ x_3 = 3 - x_2 + 3x_4 - 2x_5 \\ \hline z = 12 - 4x_2 - x_4 - x_5 \end{array} \quad \text{goes with} \quad \begin{array}{l} y_2 = 4 - 3y_1 + y_3 \\ y_4 = 1 + 11y_1 - 3y_3 \\ y_5 = 1 - y_1 + 2y_3 \\ \hline -w = -12 + 4y_1 - 3y_3 \end{array}$$

$$\begin{array}{l} x_3 = -5 - 2x_1 + 5x_2 - 19x_4 \\ x_5 = 4 + x_1 - 3x_2 + 11x_4 \\ \hline z = 8 - x_1 - x_2 - 12x_4 \end{array} \quad \text{goes with} \quad \begin{array}{l} y_1 = 1 + 2y_3 - y_5 \\ y_2 = 1 - 5y_3 + 3y_5 \\ y_4 = 12 + 19y_3 - 11y_5 \\ \hline -w = -8 + 5y_3 - 4y_5 \end{array}$$

and so on. In each of these pairs, the primal dictionary is a mirror image of the dual dictionary in the same way that (10.4) is the mirror image of (10.6). This situation is no accident: if

$$\begin{array}{l} x_r = \bar{b}_r - \sum_{s \in N} \bar{a}_{rs} x_s \qquad (r \in B) \\ \hline z = \bar{d} + \sum_{s \in N} \bar{c}_s x_s \end{array} \tag{10.7}$$

is a primal dictionary arising from (10.4), with $x_r (r \in B)$ basic and $x_s (s \in N)$ nonbasic, then

$$\begin{array}{l} y_s = -\bar{c}_s + \sum_{r \in B} \bar{a}_{rs} y_r \qquad (s \in N) \\ \hline -w = -\bar{d} - \sum_{r \in B} \bar{b}_r y_r \end{array} \tag{10.8}$$

is a dual dictionary arising from (10.6), with $y_s (s \in N)$ basic and $y_r (r \in B)$ nonbasic.

A verification of this claim, left for problem 10.1, amounts to tedious plugging and grinding.

Since Chapter 2, we have been referring to dictionaries (10.7) as *feasible* if $\bar{b}_r \geq 0$ for all $r \in B$. Now we shall refer to (10.7) as *dual-feasible* if the corresponding dual dictionary (10.8) is feasible. Thus, (10.7) is dual-feasible if and only if $\bar{c}_s \leq 0$ for all $s \in N$.

THE DUAL SIMPLEX METHOD

As observed in Chapter 5, every LP problem in the standard form may be solved by applying the simplex method to its dual; that is, an optimal solution of the primal problem may be read off the optimal dictionary for the dual problem. The observations made in the last section imply that this strategy may be implemented without any reference to the dual problem: the sequence of feasible dictionaries created by the simplex method working on the dual problem may be represented by a sequence of dual-feasible dictionaries associated with the primal problem. The resulting algorithm, designed by C. E. Lemke (1954) and known as the *dual simplex method*, constitutes a valuable tool of sensitivity analysis.

An explicit description of the dual simplex method follows mechanically from the fact that a dual variable y_k is basic in (10.8) if and only if the corresponding primal variable x_k is nonbasic in (10.7). Thus, a dual dictionary arising from (10.8) by a single pivot, with y_i entering and y_j leaving the basis, will correspond to the primal dictionary arising from (10.7) by a single pivot, with x_i leaving and x_j entering the basis. In particular, if (10.8) is one of the feasible dictionaries created by the simplex method, then y_i and y_j are determined by the familiar rules: the choice of y_i is motivated by the desire to increase $-w$ and the choice of y_j is dictated by the need to preserve feasibility when y_i increases. Formally, i may be any subscript $i \in B$ with

$$\bar{b}_i < 0 \tag{10.9}$$

and j must be a subscript $j \in N$ that has

$$\bar{a}_{ij} < 0 \quad \text{and} \quad \bar{c}_j/\bar{a}_{ij} \leq \bar{c}_s/\bar{a}_{is} \qquad \text{for all} \quad s \in N \quad \text{with} \quad \bar{a}_{is} < 0. \tag{10.10}$$

Hence an iteration of the dual simplex method, beginning with a dual-feasible dictionary (10.7), consists of first choosing a subscript $i \in B$ that satisfies (10.9), then finding a subscript $j \in N$ that satisfies (10.10), and finally pivoting, with x_i leaving and x_j entering the basis.

For illustration, suppose that (10.7) reads

$$\begin{array}{rcrrrr}
x_1 &=& -4 &+ 3x_2 &- 11x_4 &+ x_5 \\
x_3 &=& 3 &- x_2 &+ 3x_4 &- 2x_5 \\
\hline
z &=& 12 &- 4x_2 &- x_4 &- x_5.
\end{array}$$

To choose the leaving variable, we examine $\bar{b}_1 = -4$ and $\bar{b}_2 = 3$; the leaving variable must be x_1. To find the entering variable, we examine the ratios $\bar{c}_2/\bar{a}_{12} = 4/3$ and $\bar{c}_5/\bar{a}_{15} = 1$ (ignoring $\bar{c}_4/\bar{a}_{14}$ since $\bar{a}_{14} \geq 0$); the entering variable must be x_5. The pivot with x_5 entering and x_1 leaving the basis yields the dual-feasible dictionary

$$\begin{array}{rcl} x_3 &=& -5 - 2x_1 + 5x_2 - 19x_4 \\ x_5 &=& 4 + x_1 - 3x_2 + 11x_4 \\ \hline z &=& 8 - x_1 - x_2 - 12x_4. \end{array}$$

In general, each iteration of the dual simplex method is nothing but a disguised version of an iteration of the simplex method working on the dual problem. This particular example disguises the iteration that leads from

$$\begin{array}{rcl} y_2 &=& 4 - 3y_1 + y_3 \\ y_4 &=& 1 + 11y_1 - 3y_3 \\ y_5 &=& 1 - y_1 + 2y_3 \\ \hline -w &=& -12 + 4y_1 - 3y_3 \end{array} \quad \text{to} \quad \begin{array}{rcl} y_1 &=& 1 + 2y_3 - y_5 \\ y_2 &=& 1 - 5y_3 + 3y_5 \\ y_4 &=& 12 + 19y_3 - 11y_5 \\ \hline -w &=& -8 + 5y_3 - 4y_5. \end{array}$$

If no $i \in B$ satisfies (10.9), then the computations terminate: since dictionary (10.7) is not only dual-feasible but also feasible, it describes an optimal solution $x_1^*, x_2^*, \ldots, x_{n+m}^*$ by $x_s^* = 0$ for all $s \in N$ and $x_r^* = \bar{b}_r$ for all $r \in B$. Similarly, if no $j \in N$ satisfies (10.10), then the computations terminate: since the dual problem is unbounded, the primal problem is infeasible. The latter conclusion may be also reached directly, without any reference to the dual problem: if no $j \in N$ satisfies (10.10), then $\bar{a}_{is} \geq 0$ for all $s \in N$, and so the right-hand side of the equation

$$x_i = \bar{b}_i - \sum_{s \in N} \bar{a}_{is} x_s,$$

with $\bar{b}_i < 0$, assumes a negative value whenever the values of all $x_s (s \in N)$ are nonnegative.

In Chapter 7, we saw how the simplex method may be implemented in the revised format without the use of dictionaries. The dual simplex method may be implemented similarly, with only the numbers $\bar{b}_r (r \in B)$ and $\bar{c}_s (s \in N)$ stored and updated in each iteration. Working out the details amounts to answering the following four questions:

(i) How is the leaving variable x_i found?

(ii) How is the entering variable x_j found?

(iii) How are the numbers $\bar{b}_r$ updated?

(iv) How are the numbers $\bar{c}_s$ updated?

The first question does not present any difficulty since the numbers $\bar{b}_r (r \in B)$ featured in (10.9) are readily available.

To answer the second question, we need only find a way of recovering the numbers $\bar{a}_{is} (s \in N)$ featured in (10.10) from the original data. For this purpose, let us recall that the first m rows of dictionary (10.7) may be recorded (in the notation of Chapter 7) as

$$\mathbf{x}_B = \mathbf{B}^{-1}\mathbf{b} - \mathbf{B}^{-1}\mathbf{A}_N\mathbf{x}_N. \tag{10.11}$$

If the leaving variable x_i appears in the pth position of the basis heading, then it is the pth equation in (10.11) that expresses x_i in terms of the nonbasic variables, and so it is the pth row of the matrix $\mathbf{B}^{-1}\mathbf{A}_N$ that consists of the desired numbers $\bar{a}_{is}$. But the pth row of $\mathbf{B}^{-1}\mathbf{A}_N$ equals $\mathbf{v}\mathbf{A}_N$ with $\mathbf{v}$ standing for the pth row of $\mathbf{B}^{-1}$; in turn, $\mathbf{v}$ itself may be found by solving the system $\mathbf{v}\mathbf{B} = \mathbf{e}$ with $\mathbf{e}$ standing for the pth row of the $m \times m$ identity matrix. Hence the numbers $\bar{a}_{is}(s \in N)$ may be computed by first solving the system $\mathbf{v}\mathbf{B} = \mathbf{e}$ and then computing the row vector $\mathbf{w}_N = \mathbf{v}\mathbf{A}_N$; each $\bar{a}_{is}$ is a component w_s of this vector $\mathbf{w}_N$.

To answer the third question, we need only recall that the numbers $\bar{b}_r(r \in B)$ are nothing but the components $x_r^*(r \in B)$ of the vector $\mathbf{x}^* = [x_1^*, x_2^*, \ldots, x_{n+m}^*]^T$ associated with dictionary (10.7); in fact, (10.11) may be written as

$$\mathbf{x}_B = \mathbf{x}_B^* - \mathbf{B}^{-1}\mathbf{A}_N\mathbf{x}_N.$$

These numbers may be updated as in Chapter 7: having determined the entering variable x_j, and therefore the entering column $\mathbf{a}$, we solve the system $\mathbf{Bd} = \mathbf{a}$ and replace $\mathbf{x}_B^*$ by $\mathbf{x}_B^* - t\mathbf{d}$, with t standing for the new value of the entering variable. This value t equals $\bar{b}_i/\bar{a}_{ij}$, which may be written as x_i^*/w_j.

To answer the fourth question, let us record the last row of (10.7) as

$$z = \bar{d} + \bar{c}_j x_j + \sum_{s \in R} \bar{c}_s x_s$$

with $s \in R$ if and only if $s \in N$ and $s \neq j$. Since pivoting amounts to substituting for x_j from

$$x_j = \left(\bar{b}_i - x_i - \sum_{s \in R} \bar{a}_{is} x_s\right) \Big/ \bar{a}_{ij}$$

the formula for z gets updated into

$$z = \bar{d} + \bar{c}_j\left(\bar{b}_i - x_i - \sum_{s \in R} \bar{a}_{is} x_s\right) \Big/ \bar{a}_{ij} + \sum_{s \in R} \bar{c}_s x_s.$$

After simplification, and writing w_s for each $\bar{a}_{is}(s \in N)$, we obtain

$$z = \left(\bar{d} + \bar{b}_i \frac{\bar{c}_j}{w_j}\right) - (\bar{c}_j/w_j)x_i + \sum_{s \in R}\left(\bar{c}_s - w_s \frac{\bar{c}_j}{w_j}\right)x_s.$$

Thus, the new coefficient $\bar{c}_i$ at x_i equals $-\bar{c}_j/w_j$, and the coefficient $\bar{c}_s$ at each x_s with $s \in R$ gets replaced by $\bar{c}_s - \bar{c}_j w_s/w_j = \bar{c}_s + \bar{c}_i w_s$.

These findings are summarized in Box 10.1. Rather than beginning with a dual-feasible dictionary (10.7), the iteration begins just with a vector $\mathbf{x}_B^*$ whose components $x_r^*(r \in B)$ are the numbers $\bar{b}_r$ and with a vector $\bar{\mathbf{c}}_N$ whose components are the numbers $\bar{c}_s(s \in N)$.

BOX 10.1 Iteration of the Revised Dual Simplex Method

Step 1. If $\mathbf{x}_B^* \geq 0$ then stop: $\mathbf{x}^*$ is an optimal solution. Otherwise, choose the leaving variable; this may be any basic variable x_i with $x_i^* < 0$.

Step 2. Solve the system $\mathbf{vB} = \mathbf{e}$ with $\mathbf{e}$ standing for the pth row of the identity matrix and with p such that x_i appears in the pth position of the basis heading. Compute $\mathbf{w}_N = \mathbf{vA}_N$.

Step 3. Let J be the set of those nonbasic variables x_j for which $w_j < 0$. If J is empty then stop: the problem is infeasible. Otherwise, find the x_j in J that minimizes $\bar{c}_j/w_j$ and let it be the entering variable.

Step 4. Solve the system $\mathbf{Bd} = \mathbf{a}$ with $\mathbf{a}$ standing for the entering column.

Step 5. Set the value x_j^* of the entering variable at $t = x_i^*/w_j$ and replace the values $\mathbf{x}_B^*$ of the basic variables by $\mathbf{x}_B^* - t\mathbf{d}$. Replace the leaving column of $\mathbf{B}$ by the entering column and, in the basis heading, replace the leaving variable by the entering variable. Set $\bar{c}_i = -\bar{c}_j/w_j$ and add $\bar{c}_i w_s$ to each $\bar{c}_s$ with $s \neq i$.

For illustration, we shall apply the revised dual simplex method to the problem

$$\text{maximize} \quad \mathbf{cx} \quad \text{subject to} \quad \mathbf{Ax} = \mathbf{b}, \quad \mathbf{x} \geq \mathbf{0}$$

with

$$\mathbf{A} = \begin{bmatrix} -6 & 1 & 2 & 4 & 1 & 0 & 0 \\ 3 & -2 & -1 & -5 & 0 & 1 & 0 \\ -2 & 1 & 0 & 2 & 0 & 0 & 1 \end{bmatrix}, \qquad \mathbf{b} = \begin{bmatrix} 14 \\ -25 \\ 14 \end{bmatrix},$$

$$\mathbf{c} = [-5 \quad -3 \quad -3 \quad -6 \quad 0 \quad 0 \quad 0].$$

We may initialize by

$$\mathbf{x}_B^* = \begin{bmatrix} x_5^* \\ x_6^* \\ x_7^* \end{bmatrix} = \begin{bmatrix} 14 \\ -25 \\ 14 \end{bmatrix}$$

and $\bar{\mathbf{c}}_N = [\bar{c}_1, \bar{c}_2, \bar{c}_3, \bar{c}_4] = [-5, -3, -3, -6]$.
The first iteration begins.

Step 1. The leaving variable is x_6.

Step 2. Since $\mathbf{B} = \mathbf{I}$, the system $\mathbf{vB} = [0, 1, 0]$ reduces to $\mathbf{v} = [0, 1, 0]$. We have

$$[w_1, w_2, w_3, w_4] = \mathbf{v}\begin{bmatrix} -6 & 1 & 2 & 4 \\ 3 & -2 & -1 & -5 \\ -2 & 1 & 0 & 2 \end{bmatrix} = [3, -2, -1, -5].$$

Step 3. The set J consists of x_2, x_3, x_4. Comparing the ratios 3/2, 3/1, and 6/5, we find that x_4 has to enter the basis.

Step 4. Since $\mathbf{B} = \mathbf{I}$, the system

$$\mathbf{Bd} = \begin{bmatrix} 4 \\ -5 \\ 2 \end{bmatrix} \quad \text{reduces to} \quad \mathbf{d} = \begin{bmatrix} 4 \\ -5 \\ 2 \end{bmatrix}.$$

Step 5. We have $t = x_6^*/w_4 = 5$,

$$\mathbf{x}_B^* = \begin{bmatrix} x_5^* \\ x_4^* \\ x_7^* \end{bmatrix} = \begin{bmatrix} 14 - 4t \\ t \\ 14 - 2t \end{bmatrix} = \begin{bmatrix} -6 \\ 5 \\ 4 \end{bmatrix}, \qquad \mathbf{B} = \mathbf{E}_1 \quad \text{with} \quad \mathbf{E}_1 = \begin{bmatrix} 1 & 4 & \\ & -5 & \\ & 2 & 1 \end{bmatrix},$$

$$\begin{aligned}[\bar{c}_1, \bar{c}_2, \bar{c}_3, \bar{c}_6] &= [-5 + 3\bar{c}_6, -3 - 2\bar{c}_6, -3 - \bar{c}_6, \bar{c}_6] \\ &= [-8.6, -0.6, -1.8, -1.2].\end{aligned}$$

The second iteration begins.

Step 1. The leaving variable is x_5.

Step 2. Solving the system $\mathbf{vB} = [1, 0, 0]$ we find $\mathbf{v} = [1, 0.8, 0]$. Hence

$$[w_1, w_2, w_3, w_6] = \mathbf{v}\begin{bmatrix} -6 & 1 & 2 & 0 \\ 3 & -2 & -1 & 1 \\ -2 & 1 & 0 & 0 \end{bmatrix} = [-3.6, -0.6, 1.2, 0.8].$$

Step 3. The set J consists of x_1 and x_2. Comparing the ratios 8.6/3.6 and 0.6/0.6, we find that x_2 has to enter the basis.

Step 4. Solving the system

$$\mathbf{Bd} = \begin{bmatrix} 1 \\ -2 \\ 1 \end{bmatrix} \quad \text{we find} \quad \mathbf{d} = \begin{bmatrix} -0.6 \\ 0.4 \\ 0.2 \end{bmatrix}.$$

Step 5. We have $t = x_5^*/w_2 = 10$,

$$x_B^* = \begin{bmatrix} x_2^* \\ x_4^* \\ x_7^* \end{bmatrix} = \begin{bmatrix} t \\ 5 - 0.4t \\ 4 - 0.2t \end{bmatrix} = \begin{bmatrix} 10 \\ 1 \\ 2 \end{bmatrix}, \qquad \mathbf{B} = \mathbf{E}_1\mathbf{E}_2 \quad \text{with} \quad \mathbf{E}_2 = \begin{bmatrix} -0.6 & & \\ 0.4 & 1 & \\ 0.2 & & 1 \end{bmatrix},$$

$$\begin{aligned}[\bar{c}_1, \bar{c}_3, \bar{c}_5, \bar{c}_6] &= [-8.6 - 3.6\bar{c}_5, -1.8 + 1.2\bar{c}_5, \bar{c}_5, -1.2 + 0.8\bar{c}_5] \\ &= [-5, -3, -1, -2].\end{aligned}$$

The third iteration begins.

Step 1. Since $\mathbf{x}_B^* \geq \mathbf{0}$, the current solution is optimal. (This observation could have been made in step 5 of the previous iteration; the update of $\bar{\mathbf{c}}_N$ was not strictly necessary.)

Extensions to the general setting of problems

$$\text{maximize} \quad \mathbf{cx} \qquad \text{subject to} \quad \mathbf{Ax} = \mathbf{b}, \quad \mathbf{l} \leq \mathbf{x} \leq \mathbf{u} \tag{10.12}$$

are as follows. For every basic solution of (10.12), as defined in Chapter 9, we have

$$\mathbf{cx} = \mathbf{c}_B\mathbf{x}_B + \mathbf{c}_N\mathbf{x}_N = \mathbf{c}_B(\mathbf{B}^{-1}\mathbf{b} - \mathbf{B}^{-1}\mathbf{A}_N\mathbf{x}_N) + \mathbf{c}_N\mathbf{x}_N = \mathbf{c}_B\mathbf{B}^{-1}\mathbf{b} + (\mathbf{c}_N - \mathbf{c}_B\mathbf{B}^{-1}\mathbf{A}_N)\mathbf{x}_N.$$

We shall write $\bar{\mathbf{c}}_N = \mathbf{c}_N - \mathbf{c}_B\mathbf{B}^{-1}\mathbf{A}_N$. A basic solution $\mathbf{x}^*$ is *dual-feasible* if each nonbasic x_j with $\bar{c}_j > 0$ has $x_j^* = u_j$ and if each nonbasic x_j with $\bar{c}_j < 0$ has $x_j^* = l_j$. Each iteration of the revised dual simplex method, beginning with some dual-feasible basic solution $\mathbf{x}^*$ and with the corresponding vector $\mathbf{c}_N$, is as described in Box 10.2.

BOX 10.2 An Iteration of Revised Dual Simplex Method Extended

Step 1. If $\mathbf{l}_B \leq \mathbf{x}_B^* \leq \mathbf{u}_B$ then stop: $\mathbf{x}^*$ is an optimal solution. Otherwise, choose the leaving variable: this may be any basic variable x_i with $x_i^* < l_i$ or $x_i^* > u_i$.

Step 2. Solve the system $\mathbf{vB} = \mathbf{e}$ with $\mathbf{e}$ standing for the pth row of the identity matrix and with p such that x_i appears in the pth position of the basis heading. Compute $\mathbf{w}_N = \mathbf{vA}_N$.

Step 3. If $x_i^* < l_i$, then let J be the set of those nonbasic variables x_j for which $w_j < 0$, $x_j^* < u_j$ or $w_j > 0$, $x_j^* > l_j$. If $x_i^* > u_i$, then let J be the set of those nonbasic variables x_j for which $w_j > 0$, $x_j^* < u_j$ or $w_j < 0$, $x_j^* > l_j$. If J is empty then stop: the problem is infeasible. Otherwise, find the x_j in J that minimizes $|\bar{c}_j/w_j|$ and let it be the entering variable.

Step 4. Solve the system $\mathbf{Bd} = \mathbf{a}$ with $\mathbf{a}$ standing for the entering column.

Step 5. Set $t = (x_i^* - l_i)/w_j$ in case $x_i^* < l_i$ and $t = (x_i^* - u_i)/w_j$ in case $x_i^* > u_i$. Replace the value x_j^* of x_j by $x_j^* + t$ and replace the values $\mathbf{x}_B^*$ of the basic variables by $\mathbf{x}_B^* - t\mathbf{d}$. Replace the leaving column of $\mathbf{B}$ by the entering column and, in the basis heading, replace the leaving variable by the entering variable. Set $\bar{c}_i = -\bar{c}_j/w_j$ and add $\bar{c}_i w_s$ to each $\bar{c}_s$ with $s \neq i$.

We shall not worry about initializing the dual simplex method: whenever an application of this method is called for in sensitivity analysis, a dual-feasible basic solution is readily available.

SENSITIVITY ANALYSIS

Now we shall explain how the results found by the simplex method in solving a problem

$$\text{maximize} \quad \mathbf{cx} \qquad \text{subject to} \quad \mathbf{Ax} = \mathbf{b}, \quad \mathbf{l} \leq \mathbf{x} \leq \mathbf{u} \tag{10.13}$$

can be exploited in solving modified versions of (10.13).

- If the changes leading from (10.13) to its modified version

$$\text{maximize} \quad \tilde{\mathbf{c}}\mathbf{x} \qquad \text{subject to} \quad \mathbf{Ax} = \mathbf{b}, \quad \mathbf{l} \leq \mathbf{x} \leq \mathbf{u} \tag{10.14}$$

are restricted to the objective function $\mathbf{cx}$, then the previously found optimal basic solution $\mathbf{x}^*$ of (10.13) remains feasible in (10.14), and so it can be used to initialize the simplex method on (10.14). The intuitive advantage of this strategy over solving (10.14) from scratch is obvious: if $\tilde{\mathbf{c}}$ differs from $\mathbf{c}$ only slightly, then an optimal solution of (10.14) is likely to differ from the optimal solution $\mathbf{x}^*$ of (10.13) only slightly, and the number of simplex iterations leading from $\mathbf{x}^*$ to an optimal solution of (10.14) is likely to be relatively small. In fact, $\mathbf{x}^*$ may even be dual-feasible, and therefore optimal, in (10.14).

- If the changes leading from (10.13) to its modified version

$$\text{maximize} \quad \mathbf{cx} \qquad \text{subject to} \quad \mathbf{Ax} = \tilde{\mathbf{b}}, \quad \mathbf{l} \leq \mathbf{x} \leq \mathbf{u} \tag{10.15}$$

are restricted to the right-hand side $\mathbf{b}$, then the previously found optimal basic solution $\mathbf{x}^*$ of (10.13) defines a basic solution $\bar{\mathbf{x}}$ of (10.15) by $\bar{\mathbf{x}}_N = \mathbf{x}_N^*$ and $\mathbf{B}\bar{\mathbf{x}}_B = \tilde{\mathbf{b}} - \mathbf{A}_N\bar{\mathbf{x}}_N$. It is easy to see that $\bar{\mathbf{x}}$ is dual-feasible, and so it can be used to initialize the dual simplex method on (10.15). Again, the intuitive advantage of this strategy over solving (10.15) from scratch is obvious: if $\tilde{\mathbf{b}}$ differs from $\mathbf{b}$ only slightly, then an optimal solution of (10.15) is likely to differ from $\mathbf{x}^*$ and $\bar{\mathbf{x}}$ only slightly, and the number of dual simplex iterations leading from $\bar{\mathbf{x}}$ to an optimal solution of (10.15) is likely to be relatively small. In fact, $\bar{\mathbf{x}}$ may even be feasible, and therefore optimal, in (10.15).

- The technique for handling changes in the objective function can handle even more extensive changes as long as a feasible basic solution remains available for initializing the simplex method on the resulting problem. In particular, this is the case when new variables x_j with $l_j = 0$ or $u_j = 0$ are introduced. For example, if $[x_1^*, x_2^*, \ldots, x_6^*]^T$ is an optimal basic solution of the problem

$$\begin{aligned} &\text{maximize} && \sum_{j=1}^{6} c_j x_j \\ &\text{subject to} && \sum_{j=1}^{6} a_{ij} x_j = b_i \qquad (i = 1, 2, 3) \\ & && l_j \leq x_j \leq u_j \qquad (j = 1, 2, \ldots, 6) \end{aligned} \tag{10.16}$$

then $[x_1^*, x_2^*, \ldots, x_6^*, 0]^T$ is a feasible basic solution of the problem

$$\begin{aligned} \text{maximize} \quad & \sum_{j=1}^{7} c_j x_j \\ \text{subject to} \quad & \sum_{j=1}^{7} a_{ij} x_j = b_i \qquad (i = 1, 2, 3) \\ & l_j \leq x_j \leq u_j \qquad (j = 1, 2, \ldots, 7) \end{aligned}$$

as long as $l_7 = 0$ or $u_7 = 0$. (The new variable x_7 is nonbasic.)

• Similarly, the technique for handling changes in the right-hand side can handle even more extensive changes as long as a dual-feasible solution remains available for initializing the dual simplex method on the resulting problem. In particular, this is the case when new constraints are introduced. For illustration, let us add two new constraints,

$$\sum_{j=1}^{6} a_{4j} x_j \leq b_4 \quad \text{and} \quad \sum_{j=1}^{6} a_{5j} x_j = b_5$$

to problem (10.16). The resulting problem may be recorded as

$$\begin{aligned} \text{maximize} \quad & \sum_{j=1}^{6} c_j x_j \\ \text{subject to} \quad & \sum_{j=1}^{6} a_{ij} x_j = b_i \qquad (i = 1, 2, 3) \\ & \sum_{j=1}^{6} a_{4j} x_j + x_7 = b_4 \\ & \sum_{j=1}^{6} a_{5j} x_j + x_8 = b_5 \\ & l_j \leq x_j \leq u_j \qquad (j = 1, 2, \ldots, 6) \\ & 0 \leq x_7 \leq +\infty \\ & 0 \leq x_8 \leq 0. \end{aligned} \tag{10.17}$$

It is a routine matter to verify that a dual-feasible basic solution $[x_1^*, x_2^*, \ldots, x_8^*]^T$ of (10.17) can be obtained from the previously found optimal solution $[x_1^*, x_2^*, \ldots, x_6^*]^T$ of (10.16) by setting

$$x_7^* = b_4 - \sum_{j=1}^{6} a_{4j} x_j^*, \qquad x_8^* = b_5 - \sum_{j=1}^{6} a_{5j} x_j^*$$

and by letting the new set of five basic variables consist of the three old basic variables along with the two new slack variables x_7, x_8. The details are left to the reader as an exercise.

More generally, if k constraints are added to a problem with m rows and n columns, then the resulting problem has $m + k$ rows and $n + k$ columns. The previously found optimal solution of the original problem extends into a dual-feasible basic solution of the new problem; the new set of $m + k$ basic variables consists of the old m basic variables along with the k new slack variables.

- Now consider a completely general transformation of a problem

$$\text{maximize} \quad \mathbf{cx} \qquad \text{subject to} \quad \mathbf{Ax} = \mathbf{b}, \quad \mathbf{l} \leq \mathbf{x} \leq \mathbf{u} \tag{10.18}$$

into a problem

$$\text{maximize} \quad \tilde{\mathbf{c}}\mathbf{x} \qquad \text{subject to} \quad \tilde{\mathbf{A}}\mathbf{x} = \tilde{\mathbf{b}}, \quad \tilde{\mathbf{l}} \leq \mathbf{x} \leq \tilde{\mathbf{u}}. \tag{10.19}$$

If the changes leading from (10.18) to (10.19) are only slight, then an optimal solution of (10.19) is likely to differ from the previously found optimal solution $\mathbf{x}^*$ of (10.18) only slightly. Rather than solving (10.19) from scratch, we may convert $\mathbf{x}^*$ into a basic solution $\bar{\mathbf{x}}$ of (10.19) and, if $\bar{\mathbf{x}}$ turns out to be feasible or dual-feasible, use it to initialize the simplex method or the dual simplex method, respectively. As long as the entries a_{ij} of $\mathbf{A}$ remain unchanged for all variables x_j basic in $\mathbf{x}^*$, the vector $\bar{\mathbf{x}}$ may be obtained by first converting $\mathbf{x}_N^*$ into $\bar{\mathbf{x}}_N$ and then solving the system $\mathbf{B}\bar{\mathbf{x}}_B = \tilde{\mathbf{b}} - \tilde{\mathbf{A}}_N\bar{\mathbf{x}}_N$. (Changes of a_{ij} at basic variables x_j might ruin this plan by making $\mathbf{B}$ into a singular $\tilde{\mathbf{B}}$. A way of bypassing this difficulty is suggested later in this section.)

How do we convert $\mathbf{x}_N^*$ into $\bar{\mathbf{x}}_N$? Since $\bar{\mathbf{x}}$ must be a basic solution, each restricted nonbasic variable must be set at one of its bounds. For nonbasic variables with one-sided restrictions, this requirement leaves us no choice; we must have

$$\begin{aligned} \bar{x}_j &= \tilde{l}_j \qquad \text{whenever} \quad \tilde{l}_j \quad \text{is finite and} \quad \tilde{u}_j = +\infty \\ \bar{x}_j &= \tilde{u}_j \qquad \text{whenever} \quad \tilde{u}_j \quad \text{is finite and} \quad \tilde{l}_j = -\infty. \end{aligned}$$

On the other hand, unrestricted nonbasic variables may assume arbitrary values; to keep $\bar{\mathbf{x}}_N$ close to $\mathbf{x}_N^*$, we may set

$$\bar{x}_j = x_j^* \qquad \text{whenever} \quad \tilde{l}_j = -\infty \quad \text{and} \quad \tilde{u}_j = +\infty.$$

Now there remain only the nonbasic variables that are restricted from both sides in (10.19). For each of these variables x_j, we must have $\bar{x}_j = \tilde{l}_j$ or $\bar{x}_j = \tilde{u}_j$. It will be convenient to set

$$\begin{aligned} \bar{x}_j &= \tilde{l}_j \qquad \text{if} \quad x_j^* = l_j \\ \bar{x}_j &= \tilde{u}_j \qquad \text{if} \quad x_j^* = u_j. \end{aligned}$$

(Actually, these instructions may be self-contradictory if $l_j = u_j$ but $\tilde{l}_j < \tilde{u}_j$. If such is the case, we may choose between $\bar{x}_j = \tilde{l}_j$ and $\bar{x}_j = \tilde{u}_j$. A similar choice can be made in the only remaining case where $\tilde{l}_j$ and $\tilde{u}_j$ are both finite but $l_j = -\infty$ and $u_j = +\infty$.)

In particular, note that $\bar{\mathbf{x}}$ *is dual-feasible in (10.19) if the changes leading from (10.18) to (10.19) are restricted to the lower bounds* l_j *and the upper bounds* u_j, *with each finite* l_j *replaced by a finite* $\tilde{l}_j$ *and each finite* u_j *replaced by a finite* $\tilde{u}_j$.

• Even when $\bar{\mathbf{x}}$ turns out to be neither feasible nor dual-feasible in (10.19), the transformation of (10.18) into (10.19) can still be handled by a combination of the simplex method and the dual simplex method. The trick is to interpose an intermediate problem between (10.18) and (10.19) so that the transformation of (10.18) into the intermediate problem can be handled by the simplex method, and the transformation of the intermediate problem into (10.19) can be handled by the dual simplex method. The problem

$$\text{maximize} \quad \tilde{\mathbf{c}}\mathbf{x} \qquad \text{subject to} \quad \tilde{\mathbf{A}}\mathbf{x} = \tilde{\mathbf{b}}, \quad \mathbf{l}^* \le \mathbf{x} \le \mathbf{u}^* \tag{10.20}$$

with

$$\begin{aligned} l_j^* &= \bar{x}_j, \quad u_j^* = \tilde{u}_j \qquad \text{whenever} \quad \bar{x}_j < \tilde{l}_j \\ l_j^* &= \tilde{l}_j, \quad u_j^* = \bar{x}_j \qquad \text{whenever} \quad \bar{x}_j > \tilde{u}_j \\ l_j^* &= \tilde{l}_j, \quad u_j^* = \tilde{u}_j \qquad \text{whenever} \quad \tilde{l}_j \le \bar{x}_j \le \tilde{u}_j \end{aligned}$$

can always serve as the intermediate problem. [In this case, $\bar{\mathbf{x}}$ is a feasible basic solution of (10.20), and each optimal basic solution of (10.20) can be converted into a dual-feasible basic solution of (10.19) as described earlier.]

• To fit changes of the coefficients a_{ij} at variables x_j basic in $\mathbf{x}^*$ into this scheme, we need only think of each such change as an introduction of a new variable combined with a replacement of l_j and u_j by zeros. For instance, consider the problem

$$\text{maximize} \quad \mathbf{c}\mathbf{x} \qquad \text{subject to} \quad \mathbf{A}\mathbf{x} = \mathbf{b}, \quad \mathbf{l} \le \mathbf{x} \le \mathbf{u}$$

with

$$\mathbf{A} = \begin{bmatrix} a_{11} & a_{12} & a_{13} & a_{14} & a_{15} \\ a_{21} & a_{22} & a_{23} & a_{24} & a_{25} \\ a_{31} & a_{32} & a_{33} & a_{34} & a_{35} \end{bmatrix}$$

$$\begin{aligned} \mathbf{c} &= [c_1 \quad c_2 \quad c_3 \quad c_4 \quad c_5] \\ \mathbf{l} &= [l_1 \quad l_2 \quad l_3 \quad l_4 \quad l_5]^T \\ \mathbf{u} &= [u_1 \quad u_2 \quad u_3 \quad u_4 \quad u_5]^T. \end{aligned}$$

If x_2 is a basic variable in the previously found optimal solution $\mathbf{x}^*$ and if the coefficients a_{12}, a_{22}, a_{32} are replaced by $\tilde{a}_{12}, \tilde{a}_{22}, \tilde{a}_{32}$, respectively, then the resulting problem may be presented as (10.19) with $\tilde{\mathbf{b}} = \mathbf{b}$ and

$$\tilde{\mathbf{A}} = \begin{bmatrix} a_{11} & a_{12} & a_{13} & a_{14} & a_{15} & \tilde{a}_{12} \\ a_{21} & a_{22} & a_{23} & a_{24} & a_{25} & \tilde{a}_{22} \\ a_{31} & a_{32} & a_{33} & a_{34} & a_{35} & \tilde{a}_{32} \end{bmatrix}$$

$$\begin{aligned} \tilde{\mathbf{c}} &= [c_1 \quad c_2 \quad c_3 \quad c_4 \quad c_5 \quad c_2] \\ \tilde{\mathbf{l}} &= [l_1 \quad 0 \quad l_3 \quad l_5 \quad l_5 \quad l_2]^T \\ \tilde{\mathbf{u}} &= [u_1 \quad 0 \quad u_3 \quad u_4 \quad u_5 \quad u_2]^T. \end{aligned}$$

Parametric Linear Programming

In linear programming problems arising from applications, the coefficients in the objective function may be linear functions of some parameter p rather than constants. For instance, net profits may depend on a variable price p of some raw material. In other situations, it may be unclear which of the two objective functions should be maximized. In such cases, a compromise between maximizing $\mathbf{cx}$ and maximizing $\tilde{\mathbf{c}}\mathbf{x}$ is to maximize $p\mathbf{cx} + (1 - p)\tilde{\mathbf{c}}\mathbf{x} = (\tilde{\mathbf{c}} + p(\mathbf{c} - \tilde{\mathbf{c}}))\mathbf{x}$ for some fixed value of p between zero and one. Here, the parameter p measures the relative importance of $\mathbf{cx}$ as opposed to that of $\tilde{\mathbf{c}}\mathbf{x}$. Since the choice of its actual value may be a highly subjective matter, it may be enlightening to find out how the optimal solution changes with changing p. Such investigations are known as *parametric linear programming*.

What is called an LP problem with a *parametric objective function* is actually an infinite family of LP problems

$$\text{maximize} \quad (\mathbf{c}' + p\mathbf{c}'')\mathbf{x} \qquad \text{subject to} \quad \mathbf{Ax} = \mathbf{b}, \quad \mathbf{l} \le \mathbf{x} \le \mathbf{u} \tag{10.21}$$

one for every value of p. We are going to show that the dependence of the optimal value of (10.21) on the value of the parameter p always follows the same simple pattern. For illustration, let us consider the problem

$$\begin{array}{lrl} \text{maximize} & (3 - 2p)x_1 + (-3 + p)x_2 + x_3 & \\ \text{subject to} & x_1 + 2x_2 + 3x_3 \le 5 & \\ & 2x_1 + x_2 + 4x_3 \le 7 & \\ & x_1 \ge 0, 0 \le x_2 \le 2, x_3 \le 0 & \end{array} \tag{10.22}$$

which, after the usual introduction of slack variables, may be presented in the format (10.21) with $\mathbf{c}' = [3, -3, 1, 0, 0]$, $\mathbf{c}'' = [-2, 1, 0, 0, 0]$, and so on. It is a routine, if tedious, matter to verify that (10.22) has precisely six basic feasible solutions,

$$\begin{aligned} \mathbf{x}^* &= [\,3, \quad 1, \quad 0, \quad 0, \quad 0]^T \\ \mathbf{x}^* &= [\,0, \quad 0, \quad 0, \quad 5, \quad 7]^T \\ \mathbf{x}^* &= [3.5, \quad 0, \quad 0, \quad 1.5, \quad 0]^T \\ \mathbf{x}^* &= [\,0, \quad 2, \quad 0, \quad 1, \quad 5]^T \\ \mathbf{x}^* &= [\,1, \quad 2, \quad 0, \quad 0, \quad 3]^T \\ \mathbf{x}^* &= [5.5, \quad 2, \quad -1.5, \quad 0, \quad 0]^T. \end{aligned}$$

For every fixed value of p, one of these six solutions must be optimal or else the problem must be unbounded. Hence the optimal value of (10.22), if it exists at all, equals the largest of the six numbers $6 - 5p$, 0, $10.5 - 7p$, $-6 + 2p$, -3, $9 - 9p$, specifying the respective values of the objective function at our six solutions. More generally, we have the following result.

THEOREM 10.1. If (10.21) has an optimal solution for at least one value of p, then there are a finite number of linear functions $r_1 + s_1 p, r_2 + s_2 p, \ldots, r_k + s_k p$ and an interval I with the following properties:

(i) If $p \notin I$ then (10.21) is unbounded.

(ii) If $p \in I$ then (10.21) has an optimal solution and its optimal value $z^* = z^*(p)$ equals the largest of the k numbers $r_1 + s_1 p, r_2 + s_2 p, \ldots, r_k + s_k p$.

PROOF. Let I denote the set of those values of p for which (10.21) is not unbounded. Since (10.21) has an optimal solution for at least one value of p, it has a feasible solution for every value of p, and so it has an optimal solution whenever $p \in I$. Showing that I is an interval amounts to showing that $p \in I$ whenever $p_1 < p < p_2$ and $p_1, p_2 \in I$. To establish this fact, observe that

$$(\mathbf{c}' + p\mathbf{c}'')\mathbf{x} = \frac{p_2 - p}{p_2 - p_1}(\mathbf{c}' + p_1\mathbf{c}'')\mathbf{x} + \frac{p - p_1}{p_2 - p_1}(\mathbf{c}' + p_2\mathbf{c}'')\mathbf{x}$$

and so every feasible solution $\mathbf{x}$ of (10.21) satisfies

$$(\mathbf{c}' + p\mathbf{c}'')\mathbf{x} \le \frac{p_2 - p}{p_2 - p_1} z^*(p_1) + \frac{p - p_1}{p_2 - p_1} z^*(p_2).$$

To establish (ii), we rely on the results of Chapter 8. Theorems 8.3 and 8.4 guarantee that (10.21) has a finite number k of feasible solutions such that for every $p \in I$, one of these k solutions is an optimal solution of (10.21). Hence, $z^*(p)$ is the largest of the k numbers $z_1(p), z_2(p), \ldots, z_k(p)$, the ith of which stands for the value of the objective function at the ith of our k solutions. To complete the proof, write $r_i = \mathbf{c}'\mathbf{x}^*$ and $s_i = \mathbf{c}''\mathbf{x}^*$ for the ith solution $\mathbf{x}^*$, and observe that $z_i(p) = (\mathbf{c}' + p\mathbf{c}'')\mathbf{x}^* = r_i + s_i p$. ■

Now we turn to the algorithmic problem of solving the infinite family of problems (10.21) in only a finite time. The procedure implicit in our proof of Theorem 10.1 is unsatisfactory on at least two counts: it does not provide a constructive way of finding I and, in general, it enumerates far too many candidates for optimal solutions of (10.21). To illustrate the second point, we observe that the first and fifth of the six basic feasible solutions in our example are never optimal; in fact,

$$\max(6 - 5p, 0, 10.5 - 7p, -6 + 2p, -3, 9 - 9p) = \begin{cases} 9 - 9p & \text{for } p \le -0.75 \\ 10.5 - 7p & \text{for } -0.75 \le p \le 1.5 \\ 0 & \text{for } 1.5 \le p \le 3 \\ -6 + 2p & \text{for } p \ge 3. \end{cases}$$

A procedure designed by S. I. Gass and T. Saaty (1955) bypasses such hopeless candidates automatically by continuously varying the value of p and monitoring the attendant changes in the optimal solution of (10.21). We are going to illustrate this procedure on our example (10.22).

As a preliminary step, the parameter p is set at some arbitrarily chosen level, and the resulting LP problem is solved by the simplex method. Setting $p = 0$ in (10.22), we obtain the problem

$$\begin{array}{ll} \text{maximize} & 3x_1 - 3x_2 + x_3 \\ \text{subject to} & x_1 + 2x_2 + 3x_3 \leq 5 \\ & 2x_1 + x_2 + 4x_3 \leq 7 \\ & x_1 \geq 0, 0 \leq x_2 \leq 2, x_3 \leq 0. \end{array} \tag{10.23}$$

Solving this problem by the simplex method, we arrive at the basic feasible solution $[3.5, 0, 0, 1.5, 0]^T$, compute the vector $\mathbf{y} = [0, 1.5]$, find that x_3 may enter the basis, and discover that there is no candidate for leaving the basis. In short, we discover that (10.23) is unbounded: its solution

$$[3.5 + 2t, 0, -t, 1.5 + t, 0]^T \tag{10.24}$$

is feasible whenever $t \geq 0$, and the corresponding value $10.5 + 5t$ of the objective function can be made arbitrarily large by making t sufficiently large.

The next step consists of finding all the other values of p for which this status quo persists. The value of the parametric objective function $(3 - 2p)x_1 + (-3 + p)x_2 + x_3$ at (10.24) equals $(10.5 - 7p) + (5 - 4p)t$. This quantity can be made arbitrarily large as long as $5 - 4p > 0$. Thus, we conclude that (10.22) is unbounded whenever $p < 1.25$.

To find out what happens when $p \geq 1.25$, we raise p to 1.25 and discover that the basic feasible solution $[3.5, 0, 0, 1.5, 0]^T$ is optimal in the resulting problem. Again, the next step consists of finding all the other values of p for which this status quo persists. Toward this end, we first solve the system $\mathbf{yB} = \mathbf{c}_B$ with $\mathbf{c}_B = \mathbf{c}'_B + p\mathbf{c}''_B$ and then compare the vectors $\mathbf{yA}_N$ and $\mathbf{c}_N = \mathbf{c}'_N + p\mathbf{c}''_N$. Solving the system

$$\mathbf{y}\begin{bmatrix} 1 & 1 \\ 2 & \end{bmatrix} = [3 - 2p, 0] \quad \text{we obtain} \quad \mathbf{y} = [0, 1.5 - p].$$

Comparing the vector $\mathbf{yA}_N = [1.5 - p, 6 - 4p, 1.5 - p]$ with $\mathbf{c}_N = [-3 + p, 1, 0]$, we find that $[3.5, 0, 0, 1.5, 0]^T$ remains optimal as long as

$$1.5 - p \geq -3 + p, \quad 6 - 4p \leq 1, \quad \text{and} \quad 1.5 - p \geq 0. \tag{10.25}$$

In short, $[3.5, 0, 0, 1.5, 0]^T$ is an optimal solution of (10.22) whenever $1.25 \leq p \leq 1.5$.

To find out what happens when $p > 1.5$, we think of p as being raised just slightly above 1.5 and observe that the inequality $1.5 - p \geq 0$ in (10.25) becomes violated. Now x_5 becomes eligible for entering the basis; the resulting simplex iteration brings us to the basic feasible solution $[0, 0, 0, 5, 7]^T$. For which values of p is this solution optimal? Solving the system

$$\mathbf{y}\begin{bmatrix} 1 & \\ & 1 \end{bmatrix} = [0, 0] \quad \text{we obtain} \quad \mathbf{y} = [0, 0].$$

Comparing the vector $\mathbf{yA}_N = [0, 0, 0]$ with $\mathbf{c}_N = [3 - 2p, -3 + p, 1]$ we find that $[0, 0, 0, 5, 7]^T$ is optimal whenever

$$0 \geq 3 - 2p, \quad 0 \geq -3 + p, \quad \text{and} \quad 0 \leq 1. \tag{10.26}$$

In short, $[0, 0, 0, 5, 7]^T$ is an optimal solution of (10.22) whenever $1.5 \leq p \leq 3$.

To find out what happens when $p > 3$, we think of p as being raised just slightly above 3 and observe that the inequality $0 \geq -3 + p$ in (10.26) becomes violated. Now x_2 becomes eligible for entering the basis; the resulting simplex iteration brings us to the basic feasible solution

$[0, 2, 0, 1, 5]^T$. For which values of p is this solution optimal? Comparing the vector $\mathbf{y}\mathbf{A}_N = [0, 0, 0]$ with $\mathbf{c}_N = [3 - 2p, -3 + p, 1]$ we find that the answer is affirmative whenever

$$0 \geq 3 - 2p, \quad 0 \leq -3 + p, \quad \text{and} \quad 0 \leq 1.$$

In short, $[0, 2, 0, 1, 5]^T$ is an optimal solution of (10.22) whenever $p \geq 3$. Now (10.22) is solved completely; note that the useless basic feasible solutions $[3, 1, 0, 0, 0,]^T$, $[1, 2, 0, 0, 3]^T$, and $[5.5, 2, -1.5, 0, 0]^T$ have never even come under consideration.

Incidentally, even this procedure may take a considerable time on small but sufficiently nasty examples such as

$$\begin{aligned} &\text{maximize} && \sum_{j=1}^{n} (p - 2^{n-j})x_j \\ &\text{subject to} && x_i + 2 \sum_{j=i+1}^{n} x_j \leq 4^{n-i} \qquad (i = 1, 2, \ldots, n) \\ & && x_j \geq 0 \qquad (j = 1, 2, \ldots, n). \end{aligned} \tag{10.27}$$

These examples, constructed by K. G. Murty (1980), are closely related to the Klee–Minty examples of Chapter 4. They have 2^n basic feasible solutions, each of which is a unique optimal solution of (10.27) for a suitably chosen value of p. It follows that any procedure for solving (10.27) must enumerate 2^n basic feasible solutions in this case; already when n is as small as ten, there are more than a thousand basic feasible solutions to be enumerated.

Before leaving the subject of parametric objective function, let us note that Theorem 10.1 applies just as well to minimization problems

$$\text{minimize} \quad (\mathbf{c}' + p\mathbf{c}'')\mathbf{x} \qquad \text{subject to} \quad \mathbf{A}\mathbf{x} = \mathbf{b}, \quad \mathbf{l} \leq \mathbf{x} \leq \mathbf{u}$$

as soon as "the largest" in (ii) is replaced by "the smallest." After this observation, we are ready to tackle LP problems with a *parametric right-hand side*,

$$\text{maximize} \quad \mathbf{c}\mathbf{x} \qquad \text{subject to} \quad \mathbf{A}\mathbf{x} = \mathbf{b}' + p\mathbf{b}'', \quad \mathbf{l} \leq \mathbf{x} \leq \mathbf{u}$$

and, more generally, parametric LP problems

$$\text{maximize} \quad \mathbf{c}\mathbf{x} \qquad \text{subject to} \quad \mathbf{A}\mathbf{x} = \mathbf{b}' + p\mathbf{b}'', \quad \mathbf{l}' + p\mathbf{l}'' \leq \mathbf{x} \leq \mathbf{u}' + p\mathbf{u}''. \tag{10.28}$$

THEOREM 10.2. If (10.28) has an optimal solution for at least one value of p, then there are a finite number of linear functions $r_1 + s_1p, r_2 + s_2p, \ldots, r_k + s_kp$ and an interval I with the following properties:

(i) If $p \notin I$ then (10.28) is infeasible.

(ii) If $p \in I$ then (10.28) has an optimal solution and its optimal value $z^* = z^*(p)$ equals the smallest of the k numbers $r_1 + s_1p, r_2 + s_2p, \ldots, r_k + s_kp$.

PROOF. Note that the dual of (10.28) is a minimization problem with a parametric objective function. Now the desired conclusion follows directly from the observation made just before the statement of this theorem, combined with the duality theorem: if the dual is unbounded, then the primal is infeasible, and if the dual has an optimal value z^*, then z^* is also the optimal value of the primal. ■

Every infinite family (10.28) of LP problems can be solved in only a finite time by a procedure analogous to the procedure suggested for solving (10.21), except that now the optimal solutions vary continuously with p and simplex iterations are replaced by dual simplex iterations. For illustration, let us consider the parametric problem

$$\begin{array}{lr} \text{maximize} & x_1 + 4x_2 + x_3 \\ \text{subject to} & 2x_1 + 5x_2 + x_3 \leq 7 - 3p \\ & x_1 + 3x_2 + x_3 \leq 5 - 2p \\ & x_1 \geq p, x_2 \geq 0, 0 \leq x_3 \leq 3 + 2p. \end{array} \tag{10.29}$$

Setting p at zero and applying the simplex method to the resulting problem, we find the optimal solution $[0, 1, 2, 0, 0]^T$. Even though this vector ceases to be a solution of (10.29) as soon as $p \neq 0$, it points out a family of basic solutions

$$[p, 1 - p, 2, 0, 0]^T \tag{10.30}$$

by specifying that x_2, x_3 are basic and that x_1, x_4, x_5 are at their lower bounds. In fact, each of these basic solutions is dual-feasible regardless of the actual value of p; in addition, it is feasible as long as $1 - p \geq 0$ and $2 \leq 3 + 2p$. Thus we conclude that (10.30) is an optimal solution of (10.29) whenever $-0.5 \leq p \leq 1$.

To find out what happens for the remaining values of p, we first think of p as being raised just slightly above 1. Now x_2 becomes eligible for leaving the basis in the dual-feasible basic solution (10.30); the resulting dual simplex iteration brings us to the basic solution

$$[p, 0, 7 - 5p, 0, -2 + 2p]^T. \tag{10.31}$$

Since this dual-feasible solution remains feasible as long as $0 \leq 7 - 5p \leq 3 + 2p$ and $-2 + 2p \geq 0$, we conclude that (10.31) is an optimal solution of (10.29) whenever $1 \leq p \leq 1.4$.

Next, we think of p as being raised above 1.4. Now x_3 becomes eligible for leaving the basis in the dual-feasible basic solution (10.31); as there are no candidates for entering the basis, we conclude that (10.29) is infeasible whenever $p > 1.4$.

Finally, we return to our starting point $-0.5 \leq p \leq 1$ and think of p as being lowered below -0.5. Now x_3 becomes eligible for leaving the basis in the dual-feasible basic solution (10.30); the resulting dual simplex iteration brings us to the basic solution

$$[p, 0.8 - 1.4p, 3 + 2p, 0, -0.4 - 0.8p]^T. \tag{10.32}$$

Since this dual-feasible solution remains feasible as long as $0.8 - 1.4p \geq 0$ and $-0.4 - 0.8p \geq 0$, we conclude that (10.32) is an optimal solution of (10.29) whenever $p \leq -0.5$. □

PROBLEMS

10.1 Writing

$$c_j = 0 \quad \text{and} \quad a_{ij} = \begin{cases} 1 & \text{if} \quad i = j \\ 0 & \text{if} \quad i \neq j \end{cases} \qquad \text{for} \quad j = 1, 2, \ldots, m$$

$$b_i = 0 \quad \text{and} \quad a_{ij} = \begin{cases} -1 & \text{if} \quad i = j \\ 0 & \text{if} \quad i \neq j \end{cases} \qquad \text{for} \quad i = m+1, m+2, \ldots, m+n$$

$$\bar{c}_s = 0 \quad \text{and} \quad \bar{a}_{rs} = \begin{cases} 1 & \text{if} \quad r = s \\ 0 & \text{if} \quad r \neq s \end{cases} \qquad \text{for} \quad s \in B$$

$$\bar{b}_r = 0 \quad \text{and} \quad \bar{a}_{rs} = \begin{cases} -1 & \text{if} \quad r = s \\ 0 & \text{if} \quad r \neq s \end{cases} \qquad \text{for} \quad r \notin B$$

prove the following:

(i) If every solution $x_1, x_2, \ldots, x_{m+n}, z$ of (10.4) satisfies (10.7), then

$$\sum_{i=1}^{m+n} \bar{a}_{ri} a_{ij} = 0, \quad \sum_{i=1}^{m+n} \bar{a}_{ri} b_i = b_r, \quad \sum_{i=1}^{m+n} \bar{c}_i a_{ij} = -c_j, \quad \sum_{i=1}^{m+n} \bar{c}_i b_i = -\bar{d} \tag{10.33}$$

for all $r \in B$ and all $j = m+1, m+2, \ldots, m+n$.

(ii) If every solution $x_1, x_2, \ldots, x_{m+n}, z$ of (10.7) satisfies (10.4), then

$$\sum_{r=1}^{m+n} a_{ir} \bar{a}_{rs} = 0, \quad \sum_{r=1}^{m+n} a_{ir} \bar{b}_r = b_i, \quad \sum_{r=1}^{m+n} c_r \bar{a}_{rs} = -\bar{c}_s, \quad \sum_{r=1}^{m+n} c_r \bar{b}_r = \bar{d} \tag{10.34}$$

for all $s \in N$ and all $i = 1, 2, \ldots, m$.

(iii) If (10.34) holds for all $s \in N$ and all $i = 1, 2, \ldots, m$, then every solution $y_1, y_2, \ldots, y_{m+n}, w$ of (10.6) is a solution of (10.8).

(iv) If (10.33) holds for all $r \in B$ and all $j = m+1, m+2, \ldots, m+n$, then every solution $y_1, y_2, \ldots, y_{m+n}, w$ of (10.8) satisfies (10.6).

△**10.2** The optimal solution of the illustrative problem of Chapter 7 calls for 39 bookcases, 48 chairs, and 30 bedframes to be produced per day. Solve the following variations:

(i) The net profit brought in by each desk increases from \$13 to \$15.

(ii) The availability of metal increases from 117 to 125 units per day.

(iii) The company may also produce coffee tables, each of which requires three hours of work, one unit of metal, two units of wood, and brings in a net profit of \$14.

(iv) The number of chairs produced must be at most five times the numbers of desks.

(v) The demand for wood per chair increases from three to four units.

10.3 The optimal solution of the problem

$$\text{maximize } \mathbf{cx} \quad \text{subject to } \mathbf{Ax} = \mathbf{b}, \quad \mathbf{x} \geq \mathbf{0}$$

with

$$\mathbf{A} = \begin{bmatrix} 8 & 5 & 1 & 0 & 0 \\ 8 & 6 & 0 & 1 & 0 \\ 8 & 7 & 0 & 0 & 1 \end{bmatrix}, \qquad \mathbf{b} = \begin{bmatrix} 32 \\ 33 \\ 35 \end{bmatrix},$$

$$\mathbf{c} = [1 \quad 1 \quad 0 \quad 0 \quad 0]$$

is $[0, 5, 7, 3, 0]^T$. Use this information to solve the problem

$$\text{maximize} \quad \tilde{\mathbf{c}}\mathbf{x} \qquad \text{subject to} \quad \mathbf{A}\mathbf{x} = \mathbf{b}, \quad \mathbf{x} \geq \mathbf{0}$$

with $\tilde{\mathbf{c}} = [2, 1, 0, 0, 0]$. Would it be more difficult to solve the new problem from scratch?

10.4 Solve (10.27) with $n = 3$.

10.5 Solve the variation on problem 1.8 in which the demand for the radios is a parameter unknown in advance.

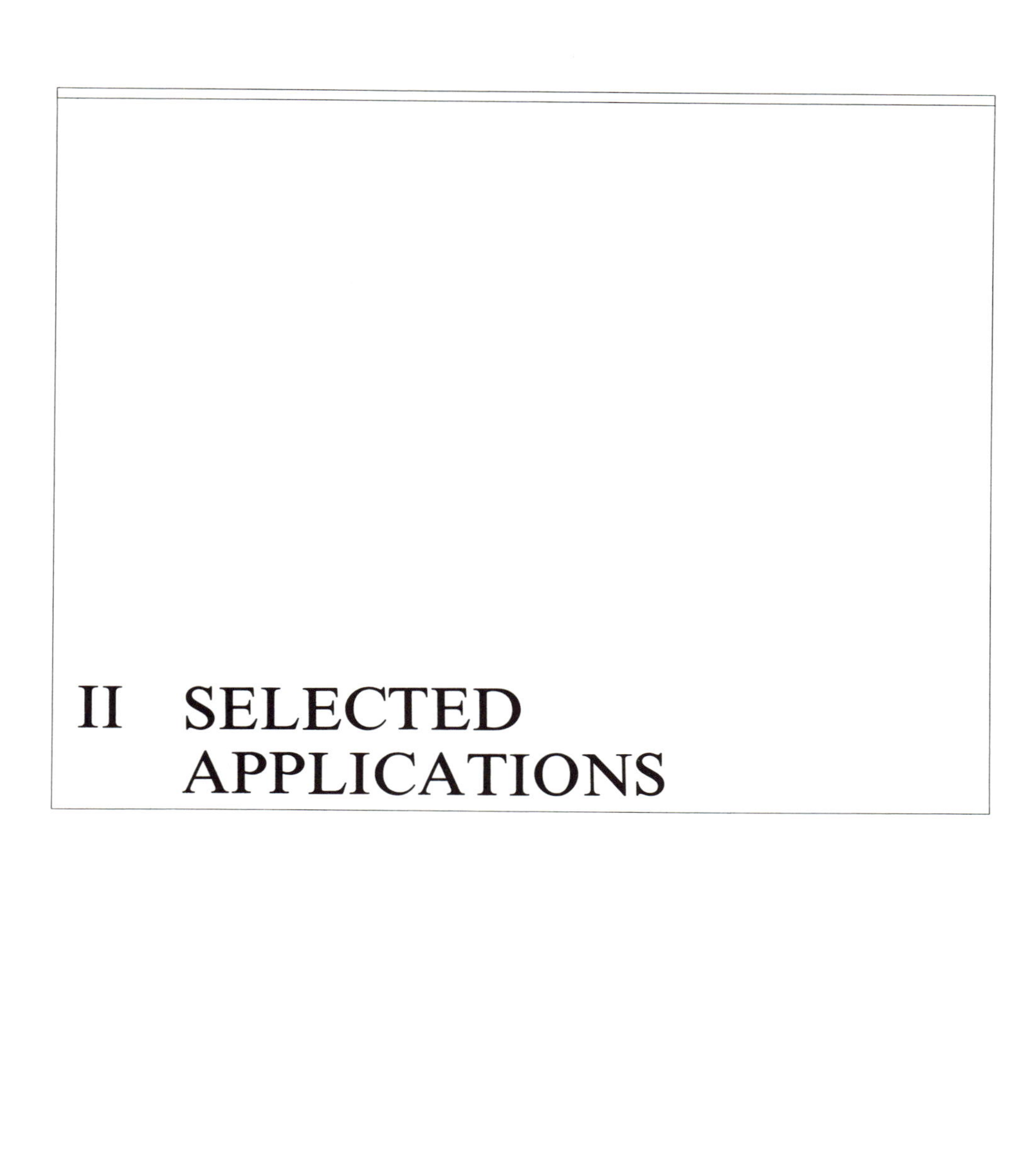

II SELECTED APPLICATIONS

11

Efficient Allocation of Scarce Resources

The problems that come under the general heading of this chapter constitute perhaps the most obvious area of linear programming applications. We shall illustrate these applications on two examples. The first is a case study reported by P. A. Wardle (1965); the second comes from an expository paper by E. R. Swanson and K. Fox (1954). At the end of the chapter, we shall review the diet problem of Chapter 1 from a more realistic point of view. Even though the problems in this chapter are the kind a consultant might come across in practice, our discussion will be confined to the strictly mathematical aspects of the consultant's job. For an introduction to some of the other aspects of the job, the reader is referred to the essays of R. E. D. Woolsey (1972a, 1972b, 1973, and the "Fifth Column" essays in *Interfaces* from 1974 to 1981). We recommend these essays highly for their wit, eloquence, and, above all, the worthy causes they champion: honesty, realism, and common sense.

CASE STUDY IN FORESTRY

The setting, New Forest, is a park and forest district of approximately 145 square miles situated in Hampshire, England. Thus we shall count money in pounds and measure lumber in Hoppus feet. (A Hoppus foot, abbreviated h.ft., is the volume of a board 1 foot square and 1 inch thick; close equivalent American measure is the board foot.) The management of New Forest had to choose a felling program for an area of about 30,000 acres, with the objective of maximizing the net discounted revenue over the next decade. The problem considered here involves only part of that area; some 8,500 acres with six different crop types shown in Table 11.1.

Table 11.1 New Forest Crop Types

Crop type	Description	Acres	Volume if felled (h.ft./acre)
1	High-volume hardwoods	2,754	2,000
2	Medium-volume hardwoods	850	1,200
3	Low-volume hardwoods	855	700
4	Conifer high forest	1,598	4,000
5	Mixed high forest	405	2,500
6	Bare land	1,761	

The hardwood areas are further classified into those with a complete undergrowth, those with a partial undergrowth, and those with no undergrowth. The corresponding acreages are shown in Table 11.2.

Table 11.2 Classification of Hardwood Areas

	Complete undergrowth	Partial undergrowth	No undergrowth	Total
High-volume hardwoods	357	500	1,897	2,754
Medium-volume hardwoods	197	130	523	850
Low-volume hardwoods	39	170	646	855

Any number of acres of any crop type can receive one of two basic treatments: fell and plant conifer (treatment 1A) or fell and plant hardwood (treatment 1B). When applied to bare land, these treatments become "plant conifer" or "plant hardwood." In addition, for hardwood areas with a complete undergrowth, management has the option of felling and retaining the undergrowth (treatment 2); similarly, for hardwood areas with a partial undergrowth, management has the option of felling and enriching the undergrowth (treatment 3). A final option is simply to postpone treatment altogether for any number of acres of any crop type.

The net discounted revenue (NDR) over the next ten years varies with treatment and crop type. These figures, in pounds per acre (£/acre), are estimated in Table 11.3.

Table 11.3 Estimate of Net Discounted Revenue (£/acre)

	Treatment				
Crop type	1A	1B	2	3	No treatment
1	287	215	228	292	204
2	207	135	148	212	148
3	157	85	98	162	112
4	487	415	—	—	371
5	337	265	—	—	264
6	87	15	—	—	61

In abstract jargon, the various treatments would be referred to as *activities* and the various areas with different crop types would be referred to as *resources*; hence the title of the present chapter.

Visual amenity requirements and a limited labor capacity dictate the following four conditions.

(i) The treated area must not exceed 5,000 acres.

(ii) The resulting conifer area must not exceed 3,845 acres.

(iii) The volume of felled hardwood must not exceed 2.44 million h.ft.

(iv) The volume of felled conifer and mixed high forest must not exceed 4.16 million h.ft.

The conifer area in (ii) is the area of newly planted conifer *plus* the untreated area of old conifer. Estimates of the average volume per acre of each of the five crops are listed in Table 11.1.

To formulate the problem in linear programming terms, let us first examine Table 11.3. For each of the six crop types, the NDR resulting from treatment 1A exceeds the NDR resulting from treatment 1B by £72. This is hardly surprising: the NDR brought in by felling and planting equals the sum of two components, the NDR brought in by felling plus the NDR brought in by planting. Actually, the first component (the NDR brought in by felling) comes to £1 per 10 h.ft. felled; the second component (the NDR brought in by planting) is £87 per acre if conifer is planted, and £15 per acre if hardwood is planted. Thus, where only felling is concerned, the distinction between treatments 1A and 1B is irrelevant; we may refer to both 1A and 1B as treatment 1. On the other hand, where only planting is concerned, the crop types formerly occupying the area are irrelevant: all that matters are the total acreages of newly planted conifer and newly planted hardwood.

If x_{ij} acres of crop type i receive treatment j, and if x_0 acres are planted with conifer, then

$$x_{11} + x_{21} + x_{31} + x_{41} + x_{51} + x_{61} - x_0$$

acres get planted with hardwood. If no conifer is planted ($x_0 = 0$), then the resulting NDR equals

$$\begin{aligned}
&215x_{11} + 228x_{12} + 292x_{13} + 204(2{,}754 - x_{11} - x_{12} - x_{13})\\
&+ 135x_{21} + 148x_{22} + 212x_{23} + 148(850 - x_{21} - x_{22} - x_{23})\\
&+ 85x_{31} + 98x_{32} + 162x_{33} + 112(855 - x_{31} - x_{32} - x_{33})\\
&+ 415x_{41} + 371(1{,}598 - x_{41})\\
&+ 265x_{51} + 264(405 - x_{51})\\
&+ 15x_{61} + 61(1{,}761 - x_{61}) = 1{,}590{,}575 + 11x_{11} + 24x_{12} + 88x_{13}\\
&\qquad - 13x_{21} + 64x_{23}\\
&\qquad - 27x_{31} - 14x_{32} + 50x_{33}\\
&\qquad + 44x_{41} + x_{51} - 46x_{61}.
\end{aligned}$$

With each acre where hardwood planting is replaced by conifer planting (an increase of x_0 by one unit), the NDR grows by £72. Hence, in general, the total NDR equals £1,590,575 plus

$$\begin{aligned}
&11x_{11} + 24x_{12} + 88x_{13} - 13x_{21} + 64x_{23} - 27x_{31}\\
&\quad -14x_{32} + 50x_{33} + 44x_{41} + x_{51} - 46x_{61} + 72x_0.
\end{aligned} \tag{11.1}$$

We wish to maximize (11.1) subject to the physical constraints

$$\begin{aligned}
x_{11} + x_{12} + x_{13} &\le 2{,}754\\
x_{21} + x_{22} + x_{23} &\le 850\\
x_{31} + x_{32} + x_{33} &\le 855\\
-x_{11} - x_{21} - x_{31} - x_{41} - x_{51} - x_{61} + x_0 &\le 0\\
x_{11} \ge 0, \quad 0 \le x_{12} \le 357, \quad 0 \le x_{13} &\le 500,\\
x_{21} \ge 0, \quad 0 \le x_{22} \le 197, \quad 0 \le x_{23} &\le 130,\\
x_{31} \ge 0, \quad 0 \le x_{32} \le 39, \quad 0 \le x_{33} &\le 170,\\
0 \le x_{41} \le 1{,}598, \quad 0 \le x_{51} \le 405, \quad 0 \le x_{61} \le 1{,}761, &\quad x_0 \ge 0
\end{aligned} \tag{11.2}$$

combined with the four constraints imposed by the management:

$$\begin{aligned}
\sum_{i,j} x_{ij} &\le 5{,}000\\
x_0 - x_{41} &\le 2{,}247\\
2(x_{11} + x_{12} + x_{13}) + 1.2(x_{21} + x_{22} + x_{23}) + 0.7(x_{31} + x_{32} + x_{33}) &\le 2{,}440\\
4x_{41} + 2.5x_{51} &\le 4{,}160.
\end{aligned} \tag{11.3}$$

[The second of our constraints in (11.3) is an equivalent version of the inequality $x_0 + (1{,}598 - x_{41}) \leq 3{,}845$ that stipulates the condition (ii).]

The simplex method takes only seven iterations to find the optimal solution,

$$\begin{aligned} &x_{11}^* = 582.5, \quad x_{13}^* = 500, \quad x_{23}^* = 130, \quad x_{33}^* = 170, \\ &x_{41}^* = 1{,}040, \quad x_{61}^* = 1{,}664.5, \quad x_0^* = 3{,}287 \end{aligned} \tag{11.4}$$

with all the remaining variables at zero. The managers of New Forest found this proposal unsatisfactory on grounds that they had neglected to verbalize *before* the LP problem was solved. Now they declared that, in order to "conform with requirements of visual amenity," a felling program must meet yet another condition:

(v) At least 500 acres must be planted with hardwood.

[Note that (11.4) calls for no hardwood planting at all.] In algebraic terms, this requirement amounts to replacing the right-hand side 0 in the fourth inequality in (11.2) by -500. Clearly, this situation calls for the dual simplex method. Along the lines of Chapter 10, the optimal solution (11.4) of the original problem may be converted into a dual-feasible basic solution of the modified problem,

$$\begin{aligned} &\bar{x}_{11} = 582.5, \quad \bar{x}_{13} = 500, \quad \bar{x}_{23} = 130, \quad \bar{x}_{33} = 170, \\ &\bar{x}_{41} = 1{,}040, \quad \bar{x}_{61} = 2{,}164.5, \quad \bar{x}_0 = 3{,}287 \end{aligned} \tag{11.5}$$

with all the remaining variables at zero. The dual simplex method, initialized by (11.5), takes only one iteration to find the optimal solution of the modified problem,

$$\begin{aligned} &x_{11}^* = 365.23, \quad x_{13}^* = 500, \quad x_{23}^* = 130, \quad x_{33}^* = 170, \\ &x_{31}^* = 620.77, \quad x_{41}^* = 1{,}040, \quad x_{61}^* = 1{,}761, \quad x_0^* = 3{,}287 \end{aligned}$$

with all the remaining variables at zero.

The information supplied by the simplex method extends beyond the optimal values of the *primal* variables: as explained in Chapter 5, the optimal values y_i^* of the *dual* variables may be interpreted as marginal values of resources. The y_i^*'s associated with the physical constraints (11.2) indicate the extra profits that we would realize if we had a few extra acres of a specified crop type. Since the few extra acres are unlikely to materialize out of the blue, the corresponding y_i^*'s are of little importance. However, the y_i^*'s associated with conditions (i) through (v) are interesting to inspect. Their respective values—0, 24.5, 29.2, 29, 47.5—indicate the rates of penalties paid in lost profits for conforming to the five requirements. (It does not come as much of a surprise that the first of these values is zero: since the optimal solution calls for treatment of only 4,587 acres, condition (i) represents no real restriction.)

Our report to the management of New Forest might go as follows.

- We suggest the following course of action. Fell hardwood with a partial undergrowth, and enrich the undergrowth in the entire 800 acre area. In addition, fell 365 acres of high-volume hardwoods, 621 acres of low-volume hardwoods, and 1,040 acres of conifer high forest. In the resulting area of 2,026 acres, combined with all the bare land, plant conifers on 3,287 acres and hardwood on 500 acres.

  The total treated area will amount to only 4,587 acres, well within the upper limit of 5,000. As stipulated, 500 acres will be planted with hardwood. The 558 acres of untreated conifer high forest will combine with 3,287 acres of newly planted conifer to yield only 3,845 acres. If the estimates of volumes per acre are more or less correct, then the volume felled from hardwood will be about 2.44 million h.ft. and the volume felled from conifer will be about 4.16 million h.ft.

  Finally, if the figures predicting the various net discounted revenues are reasonably accurate, then this felling program will yield a total NDR of some £1,840,000. Under the present circumstances, this is the most that can be hoped for.

- Additional profits could be realized if you relaxed your requirements (ii), (iii), (iv), and (v). Should you be tempted to do so, the following figures may help you in your decision. Each thousand h.ft. felled in addition to the 6,600 thousand h.ft. that may be felled now would bring in an additional NDR of about £29. Similarly, relaxing the upper bound on the conifer area would bring in extra NDR at the rate of £24.5 per acre; relaxing the lower bound on the area planted with hardwood would bring in extra NDR at the rate of £47.5 per acre.

  The upper limit on the volume felled seems to stem mainly from limited workforce capacity. If that is the case, then you should look into the possibility of having your people work overtime or hiring extra labor. The extra £29 per acre might be well worth the overtime bonuses and/or training costs.

  The rates quoted here apply to *moderate* changes in your policy; for drastic changes, the profits may increase at slower rates. Nevertheless, as soon as you decide to consider *specific* changes in the policy, we can easily work out the corresponding optimal felling program. If you are willing to consider, even tentatively, several alternative policies, do not hesitate to tell us about them; having many different scenarios to choose from cannot possibly hurt your final decision.

- How reliable are the NDR predictions? We might be able to help you further if we knew how these figures were calculated and where the data came from. If some members of your group have more faith in a different set of figures, tell us about it. If the corresponding optimal felling program differs from that presented here, then we shall also provide you with a range of in-between felling programs.

This report is an invitation to a continuing dialogue with the managers. Of course, if the managers trust the data and if their policy is absolutely firm, then we simply collect our fee and catch the 5:32 to Victoria Station.

EXAMPLE FROM FARM PLANNING

Unlike the first example, this is not an actual case study. The objective of E. R. Swanson and K. Fox was to promote the use of linear programming by *illustrating* its applicability to farm planning. For that purpose, they considered a fictitious farm somewhere in central Illinois, with 320 acres of land and a total labor force of two workers for the entire year. The data were compiled by averaging from various statistics and estimates.

The farm activities fall into two broad groups: crop enterprises and livestock enterprises. The authors assume that a fixed cropping system has already been adopted; the remaining labor supply available for the livestock enterprises varies with the time of year as follows.

Month	Hours
January	420
February	415
March	355
April	345
May	160
June	95
July	380
August	395
September	270
October	230
November	310
December	420

Five livestock enterprises are common to the corn belt:

(i) Spring litter hog enterprise.
(ii) Fall litter hog enterprise.
(iii) Full-feed drylot feeder cattle enterprise.
(iv) Full-feed pasture feeder cattle enterprise.
(v) Delayed-feeding feeder cattle enterprise.

In the hog enterprises, the pigs are farrowed either in February (spring litter) or in August (fall litter) and sold about six months later. In the cattle enterprises, feeder calves are bought in October and sold about a year later. Estimated labor requirements are shown in Table 11.4.

Table 11.4 Estimated Labor Requirements

	Enterprise				
Month	(i)*	(ii)*	(iii)†	(iv)†	(v)†
January	1.4	1.8	1.5	1.4	1.4
February	9.8	2.4	1.4	1.4	1.4
March	4.0	0.4	1.4	1.4	1.4
April	2.8	0.6	1.3	1.4	1.5
May	2.2	0.4	1.3	1.5	1.2
June	2.2	0.4	1.3	1.3	1.2
July	2.2	0.6	1.3	1.3	1.2
August	2.6	5.8	1.5	1.5	1.2
September	0.6	4.0	1.3	—	—
October	0.6	1.2	1.3	1.3	2.6
November	0.6	1.8	1.2	1.2	1.2
December	0.6	1.8	1.5	1.4	1.4

* Hours per litter.
† Hours per cow.

In addition to labor, each of the five enterprises requires a certain amount of roughage (pasture and/or hay). Pasture is measured in units called *pasture days*; a pasture day is the amount of pasture eaten in one day by a mature horse or a cow receiving no other feed. Pasture is available from April to September; during that period, it can also be converted into hay. The availability of pasture is shown in Table 11.5 together with its demand by each of the five enterprises.

Table 11.5 Pasture Demands

	Demand (pasture days/unit)					
Period	(i)	(ii)	(iii)	(iv)	(v)	Availability (pasture days)
April and May	16	0	0	12	35	5,200
June and July	20	0	0	36	50	5,200
August and September	16	0	0	12	35	3,600
Total	52	0	0	60	120	14,000

Pasture may be converted into hay at an estimated rate of 5.5 hours and 50 pasture days per ton of hay. Except for the spring litter hogs, the remaining four enterprises require hay in the off-season period. The overall demands for hay are

0.1 ton per fall litter of hogs.

0.9 ton per full-feed drylot feeder calf.

0.8 ton per full-feed pasture feeder calf.

2.3 ton per delayed-feeding feeder calf.

Finally, the *net* profits for these five enterprises are estimated as follows:

(i) $139 per litter.
(ii) $88 per litter.
(iii) $133 per cow.
(iv) $137 per cow.
(v) $165 per cow.

(Calculations of these figures involve subtracting the cost of feed, the farmhands' wages, and so on, so we won't have to worry about the corresponding variables.) The objective is to design the most profitable livestock plan.

Needless to say, we are facing great uncertainty. It seems highly unlikely that a farmer would come up with the kind of data presented here. His experience might indicate that 40 spring litters of pigs can just about be taken care of, whereas 50 litters would create a serious bottleneck in February. However, it would be absurd to expect him to say that in February, each spring litter requires 9.8 hours. (Actually, this figure comes from a cost report issued by the Department of Agriculture, University of Illinois.) Given a specific livestock program, the farmer will be able to estimate whether it can be carried out or not. Even though he may not be inclined to formulate his criteria explicitly, we may be able to discern their contours from the reasons for rejecting various programs. Once we have roughly approximated them by linear constraints, we can confront the farmer with a variety of optimal and near-optimal solutions. His reactions will provide the feedback required to refine our simplified description of the farm operations and prepare us for the next round. With enough patience on both sides, the dialogue may eventually terminate in a program the farmer finds acceptable, profitable, and perhaps even clever. In spite of their unreliability, the figures quoted earlier may serve as a starting point.

The cloud of uncertainty, upsetting as it may be, does have a silver lining: it allows us to treat the data with a healthy irreverence. A program that requires 97 hours in June will be just as feasible as a program requiring only the alloted 95 hours. A program that yields a profit of $12,600 will be just as profitable as a program yielding $12,800. Trying to impose the standards of precision found in physics on speculations about livestock would be preposterous. [An amusing example of such an endeavour has been reported by R. E. D. Woolsey and H. S. Swanson (1975): "... an operations research group in an oil company was worried about the third decimal place in a linear programming output, when the data on viscosity of the input stream had been obtained by a refinery worker rubbing it between his fingers."]

Before proceeding, let us point out that this example seems to suffer from a lack of realism on at least two counts. First, trying to *optimize* the livestock program subject to restrictions stemming from a *fixed* cropping program is a little like trying to roller-skate with only one skate on. It would be more sensible to view the farm as a whole, optimizing simultaneously the livestock plan *and* the cropping plan. Striking the most profitable balance between the two is precisely the kind of problem in which the power of linear programming manifests itself most clearly; by keeping the cropping plan fixed, we forfeit potential profits.

Secondly, we still have not asked a fundamental question. What period of time are we planning for? Planning for short periods, such as a year, does not allow much flexibility; durations of the various enterprises overlap and so we may already be committed to some of them. (In addition, pregnancy in swine lasts for about 114 days, so the hog enterprises may have to be planned at least four months in advance.) On the other hand, planning for longer periods requires predicting market prices in a relatively distant future, and that can make the data extremely unreliable.

In spite of these shortcomings, the example remains instructive and we shall carry on its discussion. Assuming a reasonable degree of faith in the predictions of net profits, we shall aim for the most lucrative livestock plan to be repeated over several consecutive years.

Linear programming formulation is straightforward. Decision variables $x_1, x_2, \ldots, x_5$ measure the levels of the five enterprises, whereas $x_6, x_7, \ldots, x_{11}$ denote the amounts of hay to be made in the six months from April to September. Twelve linear inequalities reflect the limited availability of labor; three additional linear inequalities reflect the limited availability of pasture. The last constraint is the law of conservation of hay: the amount of hay produced must equal the amount of hay consumed. Altogether, the problem reads as follows:

$$
\begin{array}{llr}
\text{maximize} & 139x_1 + 88x_2 + 133x_3 + 137x_4 + 165x_5 & \\
\text{subject to} & 1.4x_1 + 1.8x_2 + 1.5x_3 + 1.4x_4 + 1.4x_5 & \leq 420 \\
& 9.8x_1 + 2.4x_2 + 1.4x_3 + 1.4x_4 + 1.4x_5 & \leq 415 \\
& 4.0x_1 + 0.4x_2 + 1.4x_3 + 1.4x_4 + 1.4x_5 & \leq 355 \\
& 2.8x_1 + 0.6x_2 + 1.3x_3 + 1.4x_4 + 1.5x_5 + 5.5x_6 & \leq 345 \\
& 2.2x_1 + 0.4x_2 + 1.3x_3 + 1.5x_4 + 1.2x_5 + 5.5x_7 & \leq 160 \\
& 2.2x_1 + 0.4x_2 + 1.3x_3 + 1.3x_4 + 1.2x_5 + 5.5x_8 & \leq 95 \\
& 2.2x_1 + 0.6x_2 + 1.3x_3 + 1.3x_4 + 1.2x_5 + 5.5x_9 & \leq 380 \\
& 2.6x_1 + 5.8x_2 + 1.5x_3 + 1.5x_4 + 1.2x_5 + 5.5x_{10} & \leq 395 \\
& 0.6x_1 + 4.0x_2 + 1.3x_3 \qquad\qquad\qquad\quad + 5.5x_{11} & \leq 270 \\
& 0.6x_1 + 1.2x_2 + 1.3x_3 + 1.3x_4 + 2.6x_5 & \leq 230 \\
& 0.6x_1 + 1.8x_2 + 1.2x_3 + 1.2x_4 + 1.2x_5 & \leq 310 \\
& 0.6x_1 + 1.8x_2 + 1.5x_3 + 1.4x_4 + 1.4x_5 & \leq 420
\end{array}
$$

and

$$16x_1 + 12x_4 + 35x_5 + 50x_6 + 50x_7 \le 5{,}200$$
$$20x_1 + 36x_4 + 50x_5 + 50x_8 + 50x_9 \le 5{,}200$$
$$16x_1 + 12x_4 + 35x_5 + 50x_{10} + 50x_{11} \le 3{,}600$$

and

$$0.1x_2 + 0.9x_3 + 0.8x_4 + 2.3x_5 - \sum_{j=6}^{11} x_j = 0$$

and

$$x_j \ge 0 \qquad (j = 1, 2, \ldots, 11).$$

The optimal solution $[x_1^*, x_2^*, \ldots, x_{11}^*]$ delivered by the simplex method (after only eleven iterations) reads

$$[0, 54.7, 0, 20.7, 38.4, 41.1, 11.1, 0, 49.9, 0, 9.3].$$

The optimal solution $[y_1^*, y_2^*, \ldots, y_{16}^*]$ of the dual problem, delivered by the simplex method at the same time, reads

$$[0, 0, 0, 4.6, 4.6, 68.7, 4.6, 5.6, 4.6, 0, 0, 0, 0, 0, 0, 25.3].$$

As y_6^* is quite large, it may be illuminating to investigate changes in the June supply of labor by solving at least one modified problem with the right-hand side of the sixth constraint changed from 95 to some larger number. Having done so, we might make the following points in our first conversation with the farmer.

- We suggest that you raise about 55 fall pig litters, 21 full-fed pasture cows, and 38 delayed-fed cows. These animals would require about 110 tons of hay; you could make 41 in April, 11 in May, 49 in July, and 9 in September. According to the market price predictions, you would make about $14,000.

  We do not necessarily expect that you will like this proposal. If you don't, tell us the reasons for your dissatisfaction, so we can try again. If you do, then we have a couple of additional suggestions to make.

- It is clear that June, with its yearly low of only 95 hours, represents a serious bottleneck for the livestock enterprises. Our analysis shows more than just that: it also points out that your profit would increase by nearly $70 with each extra hour available in June. Out of curiosity, we found out what would happen if 145 rather than 95 hours were available in June. In that case, you could raise 43 fall pig litters, 12 full-fed drylot cows, 81 full-fed pasture cows, and 4 delayed-fed cows, with 33 tons of hay made in April, 41 in July, and 15 in September. The resulting profit would be slightly over $17,000.

  We find this conclusion staggering: an extra farmhand available for a mere week in June would make your *annual* profit increase by more than 20 percent!

What are the prospects of, say, hiring a student whose vacation begins in June? This is definitely something to think about.

- We do realize that hay is bulky, awkward to handle, and therefore rarely bought or sold. Should it be otherwise, our analysis tells us that a fair price for you to pay would be about $25/ton. If, by any chance, some of your neighbors were willing to sell hay at a substantially lower price, then you should close the deal. Conversely, if your neighbors were willing to buy hay from you, and pay substantially more than $25/ton, then you should, again, close the deal. If you are at all inclined to look into this possibility and if prospective sellers and/or buyers do materialize, then we can analyze the various options more carefully.

Depending on the farmer's satisfaction, a dialogue may or may not develop from here. Since the new variations, if any, would be handled in much the same way as the original theme, we shall terminate our discussion now.

THE DIET PROBLEM REVIEWED

Finally, let us critically examine the first example in this book: Polly's diet problem of Chapter 1. To begin, let us look at some of the recommended daily allowances (RDA) in Table 11.6.

The table's original caption, "Designed for the maintenance of good nutrition of *practically all* healthy people in the U.S.A." (italics ours) sounds a little suspect. How much can we trust these figures? Where do they come from? Let us quote from *Recommended Dietary Allowances* (1974):

> The ideal method, rarely if ever achieved, to develop an allowance would be to (1) determine the average requirement of a healthy and

Table 11.6 Recommended Daily Dietary Allowances

					Minerals	
Sex	Age	Weight (kg)	Energy kcal	Protein (g)	Calcium (mg)	Iron (mg)
M	19–22	67	3,000	54	800	10
M	23–50	70	2,700	56	800	10
M	51+	70	2,400	56	800	10
F	19–22	58	2,100	46	800	18
F	23–50	58	2,000	46	800	18
F	51+	58	1,800	46	800	10

representative segment of each age group for the nutrient under consideration; (2) assess statistically the variability among the individuals within the group; and (3) calculate from this the amount by which the average requirement must be increased to meet the needs of nearly all healthy individuals.

The starting point for developing nutrient allowances is the scientific evidence of nutrient requirements judged by the Committee on Dietary Allowances to be most reliable. Unfortunately, experiments on man are costly, they must often be of long duration, certain types of experiments are not possible for ethical reasons, and, even under the best conditions, only a small number of subjects can be studied in a single experiment. Thus, requirement estimates must often be derived from limited information. For some nutrients the requirement must be assessed largely from one or two experimental trials on a small number of subjects; for some there are so few experiments on human subjects that requirements must be estimated either from information about the requirements of other mammals or from information about the minimum amount of the nutrient known, from food analyses and dietary surveys, to be consumed by apparently healthy people.

Thus, RDA figures are of necessity only vague estimates. Along the same line, it is interesting to compare U.S. figures with those of other countries. In France, for example, the daily amount of protein recommended for a 143-lb male is 90 g whereas the corresponding figure agreed on by the Food and Agriculture Organization–World Health Organization Expert Committee is only 46 g. Clearly, differences of this magnitude are due to differences in methodology rather than to cultural variations in metabolism. For additional reading on nutritional requirements, we suggest N. S. Scrimshaw and V. R. Young (1976).

Vitamins				
A (IU)	Thiamin (mg)	Riboflavin (mg)	Niacin (mg)	C (mg)
5,000	1.5	1.8	20	45
5,000	1.4	1.6	18	45
5,000	1.2	1.5	16	45
4,000	1.1	1.4	14	45
4,000	1.0	1.2	13	45
4,000	1.0	1.1	12	45

Source: National Academy of Sciences, *Recommended Dietary Allowances* (Washington, D.C., 1974).

The uncertainty inherent in the RDA estimates is matched by that in the data on nutritive values of various foods. Of course, no two pork chops are quite alike and the amount of vitamin C in an apple depends, among many other factors, on its freshness. In addition, the contents of packaged foods vary greatly by brand name. For example, B. K. Watt and A. L. Merrill (1963) estimate that tuna canned in oil contains 288 kcal per 100 g of solids and liquids. However, the corresponding estimates for canned tuna actually on the U.S. market in 1974 ranged from 163 to 290 depending on the brand. [See *Consumer Guide* (1974).] Yet the data are customarily presented as single figures, without any indication of their reliability. In B. K. Watt and A. L. Merrill (1963), this practice is justified as follows.

> The question of listing a range in compositional data along with the single values was considered early in the planning of this edition. Also considered was the inclusion of the number of cases (or samples) on which each figure listed in the tables was based, as well as the number of separate investigations or laboratories represented in the figure. These statistical expressions of data are important to the investigator planning or conducting research on the composition of foods and to a few others who apply the values to particular problems. Most users, however, including research workers, dietitians, teachers, and other professional workers, find single representative values more serviceable.

Is an answer's tidy appearance more important than information on its reliability?

The optimal solution of Polly's diet problem in Chapter 1 calls for 4 servings of oatmeal, $4\frac{1}{2}$ servings of milk, and 2 servings of cherry pie per day. The reader may find this solution unsatisfactory on several counts. For instance, it clearly violates the rule of "four food groups." According to this rule, a balanced diet should include a reasonable amount of food from each of the following groups:

(i) The milk group (milk, cheese, ice cream, and other milk products).
(ii) The meat group (meats, fish, eggs, peanut butter).
(iii) The vegetable–fruit group (all fruits and vegetables, including potatoes).
(iv) The grain group (bread, flour, cereals).

The remedy is clear: enlarge the original list of foods by including enough representatives from each of the four food groups, and refine the mathematical problem by expressing the four-food-groups rule in terms of linear constraints. If the new optimal solution turns out to be unpalatable or unsatisfactory for some other reason, the objections can probably be formalized again in terms of violated linear constraints. In that case, the whole process may be repeated again and again, until a satisfactory menu is found. Of course, in view of the unreliability of the data, the menu should be taken with a grain of salt (the average daily American intake: 10–15 g).

PROBLEMS

△ **11.1** In the farm planning example, the January constraint

$$1.4x_1 + 1.8x_2 + 1.5x_3 + 1.4x_4 + 1.4x_5 \leq 420$$

is more restrictive than the December constraint

$$0.6x_1 + 1.8x_2 + 1.5x_3 + 1.4x_4 + 1.4x_5 \leq 420.$$

The December constraint, therefore, may be deleted without changing the nature of the problem. Can you spot other redundant constraints?

11.2 Design yourself a diet. Ideally, you should consult the sources quoted in the text for the nutritive values of foods, and your local food stores for food prices. An excellent source of information on nutrition is L. J. Bogert, G. M. Briggs, and D. H. Calloway (1973). Of course, the choice of variables, constraints, and objective function is completely up to you.

The following is a list of 29 foods, followed by Table 11.7 that lists approximate nutritive values per serving of each food.

The milk group

Cheddar:	1 oz $\doteq$ 30 g
Cottage cheese:	55 g $\doteq \frac{1}{4}$ c* of uncreamed cottage cheese
Milk:	8 oz $\doteq$ 1 c of whole milk
Yogurt:	246 g $\doteq$ 1 c of plain, low-fat yogurt

The meat group

Chicken:	100 g $\doteq \frac{1}{2}$ breast, fried
Eggs:	50 g $\doteq$ 1 whole egg (raw, hard-cooked, or poached)
Fish:	flounder or sole, 100 g $\doteq$ 4 oz before cooking
Ham:	100 g $\doteq 3\frac{1}{2}$ oz, cooked
Hamburger:	$\frac{1}{4}$ pound $\doteq$ 85 g, market-ground
Liver:	beef, 75 g fried
Peanut butter:	16 g = 1 tbsp
Pork with beans:	canned with tomato sauce, 130 g $\doteq \frac{1}{2}$ c
Steak:	beef, round, 100 g broiled
Tuna:	canned in oil, 100 g drained

The vegetable–fruit group

Apples:	150 g $\doteq$ 1 medium large, fresh
Broccoli:	100 g $\doteq \frac{2}{3}$ c, boiled and drained
Carrots:	50 g $\doteq$ 1 carrot, raw
Oranges:	150 g $\doteq$ 1 medium
Orange juice:	6 oz $\doteq \frac{3}{4}$ c, canned
Potatoes:	100 g $\doteq$ 1 medium, baked
Spinach:	90 g $\doteq \frac{1}{2}$ c, boiled and drained
Tomatoes:	150 g $\doteq$ 1 medium, fresh

* c is the standard abbreviation for cup.

The grain group	
Bread:	American rye, 23 g ≐ 1 slice
Corn flakes:	with added nutrients, 25 g ≐ 1 c
Oatmeal:	120 g ≐ 0.7 c
Pasta:	macaroni or spaghetti, 140 g ≐ 1 c, cooked 14–20 min
Rice:	white, enriched, 100 g ≐ $\frac{2}{3}$ c

Miscellaneous	
Butter:	14 g ≐ 1 tbsp
Sugar:	white, granulated, 100 g ≐ $\frac{1}{2}$ c

Table 11.7 Approximate Nutritive Value per Serving

Food	Energy (kcal)	Protein (g)	Calcium (mg)	Iron (mg)	Vitamin A (IU)	Thiamin (mg)	Riboflavin (mg)	Niacin (mg)	Vitamin C (mg)
Cheddar	120	7.5	225	0.3	390	0.01	0.14	0	0
Cottage cheese	50	9.5	50	0.2	5	0.02	0.15	0	0
Milk	160	8.5	285	0	370	0.07	0.41	0.2	2
Yogurt	125	8.5	295	0	170	0.10	0.44	0.2	2
Chicken	205	32.5	12	1.7	90	0.05	0.22	14.7	0
Eggs	80	6.5	27	1.2	590	0.06	0.15	0.1	0
Fish	200	30.0	23	1.4	0	0.07	0.08	2.5	2
Ham	290	21.0	9	2.6	0	0.47	0.18	3.6	0
Hamburger	245	20.5	9	2.7	30	0.08	0.18	4.6	0
Liver	170	20.0	8	6.6	40,000	0.20	3.14	12.4	20
Peanut butter	90	4.0	10	0.3	0	0.02	0.02	2.4	0
Pork w/beans	160	8.0	70	2.3	170	0.10	0.04	0.8	3
Steak	260	28.5	12	3.5	30	0.08	0.22	5.6	0
Tuna	195	29.0	8	1.9	80	0.05	0.12	11.9	0
Apples	90	0.3	11	0.5	140	0.05	0.03	0.2	6
Broccoli	25	3.0	88	0.8	2,500	0.09	0.20	0.8	90
Carrots	20	0.6	19	0.4	5,500	0.03	0.03	0.3	4
Oranges	75	1.5	62	0.6	300	0.15	0.06	0.6	75
Orange juice	80	1.0	20	0.4	370	0.17	0.06	0.7	93
Potatoes	95	2.5	9	0.7	0	0.10	0.04	1.7	20
Spinach	20	3.0	84	2.0	7,290	0.06	0.13	0.5	25
Tomatoes	35	1.5	20	0.8	1,350	0.09	0.06	1.1	35
Bread	55	2.0	17	0.5	0	0.04	0.02	0.3	0
Corn flakes	95	2.0	4	0.4	0	0.11	0.02	0.5	0
Oatmeal	65	2.0	11	0.7	0	0.10	0.02	0.1	0
Pasta	155	5.0	11	0.6	0	0.01	0.01	0.4	0
Rice	110	2.0	10	0.9	0	0.11	0	1.0	0
Butter	100	0	3	0	460	0	0	0	0
Sugar	385	0	0	0.1	0	0	0	0	0

Source: From *Bogert's Nutrition and Physical Fitness*, 10th edition, by George Briggs and Doris Calloway. Copyright © 1979 by W. B. Saunders Company. Reprinted by permission of Holt, Rinehart and Winston, CBS College Publishing.

12

Scheduling Production and Inventory

In this chapter, we shall explain how certain multistage scheduling problems can be turned into LP problems, and we shall comment on the special structure of these LP problems.

AN EXAMPLE OF A SCHEDULING PROBLEM

Consider a hypothetical factory producing goods to satisfy a fluctuating demand. The best available estimates of the demand over the next twelve months are as follows:

Month	Items	Month	Items	Month	Items
January	5,300	May	4,100	September	7,300
February	5,100	June	4,800	October	7,800
March	4,400	July	6,000	November	7,600
April	2,800	August	7,100	December	6,400

To adjust to these fluctuations, the management can use any combination of the following strategies:

(i) Change the work force level by hiring and firing.

(ii) Cover temporary shortages by overtime work.

(iii) Store some of the present surplus to cover future shortages.

Each of these three strategies has limitations. Changes in the work force level, whether up or down, are limited to at most 40 workers per month. In addition, these changes are expensive: it costs \$300 to hire and \$420 to fire a worker. Next, overtime production is limited: each worker, producing 20 units per month on regular time, will produce no more than 6 units per month on overtime. Furthermore, overtime production costs exceed regular production costs by \$20 per unit. Finally, storage can become quite expensive over prolonged periods of time: it costs \$8 per month to store each unit. (The data are only imprecise estimates: future demands are difficult to forecast, production rate may be affected by machine breakdowns, and so on. Consequently, any results of a mathematical analysis must be interpreted as possible suggestions rather than unquestionable commands.)

At present, there are 290 workers and no inventory; the long-range planning policy dictates a zero inventory by next December. One of the many possible schedules meeting the predicted demand is shown in Table 12.1. The corresponding extra charges incurred in addition to the normal production costs are itemized in Table 12.2.

It may not be immediately obvious that our problem can be seen as a linear programming problem. Hence we shall settle for a general mathematical description as our starting point. In the jth month, we shall denote the demand by d_j, the level of normal production by x_j, the level of overtime production by y_j, and the level of inventory by z_j. Note that we need no special symbol for the work-force level, which is simply $x_j/20$. The upper bound on the monthly changes in the work-force level requires that

$$\left|\frac{x_j}{20} - \frac{x_{j-1}}{20}\right| \leq 40$$

or, equivalently,

$$|x_j - x_{j-1}| \leq 800 \tag{12.1}$$

for every $j = 1, 2, \ldots, 12$. The upper bound on overtime says that the level of overtime production cannot exceed 30% of the level of normal production:

$$y_j \leq 0.3x_j \tag{12.2}$$

for every $j = 1, 2, \ldots, 12$. An additional physical constraint,

$$z_{j-1} + x_j + y_j = d_j + z_j \tag{12.3}$$

for every $j = 1, 2, \ldots, 12$, amounts to a "law of conservation of goods." As for the

Table 12.1 A Possible Schedule

Month	Demand	Work force	Production		Inventory
			Normal	Overtime	
Jan	5,300	265	5,300	—	—
Feb	5,100	255	5,100	—	—
Mar	4,400	220	4,400	—	—
Apr	2,800	220	4,400	—	1,600
May	4,100	220	4,400	—	1,900
Jun	4,800	240	4,800	—	1,900
Jul	6,000	280	5,600	—	1,500
Aug	7,100	320	6,400	—	800
Sep	7,300	360	7,200	—	700
Oct	7,800	320	6,400	700	—
Nov	7,600	320	6,400	1,200	—
Dec	6,400	320	6,400	—	—

Table 12.2 Itemized Extra Charges

Month	Hiring	Firing	Overtime	Storage
Jan	—	10,500	—	—
Feb	—	4,200	—	—
Mar	—	14,700	—	—
Apr	—	—	—	12,800
May	—	—	—	15,200
Jun	6,000	—	—	15,200
Jul	12,000	—	—	12,000
Aug	12,000	—	—	6,400
Sep	12,000	—	—	5,600
Oct	—	16,800	14,000	—
Nov	—	—	24,000	—
Dec	—	—	—	—

expenses incurred during the jth month, overtime costs an extra $20y_j$, storage costs $8z_j$, and the hiring/firing costs t_j are defined by

$$t_j = \begin{cases} 15(x_j - x_{j-1}) & \text{if} \quad x_j \geq x_{j-1} \\ 21(x_{j-1} - x_j) & \text{if} \quad x_{j-1} \geq x_j. \end{cases} \tag{12.4}$$

Hence we wish to minimize

$$\sum_{j=1}^{n} (20y_j + 8z_j + t_j) \tag{12.5}$$

subject to (12.1), (12.2), (12.3), (12.4), the "boundary conditions"

$$x_0 = 5800, \quad z_0 = 0, \quad z_{12} = 0 \tag{12.6}$$

and, of course,

$$x_j, y_j, z_j, t_j \geq 0 \tag{12.7}$$

for all $j = 1, 2, \ldots, 12$.

Except for constraints (12.1) and (12.4), our problem has the linear programming form. Actually, (12.1) can be viewed as a shorthand for two linear inequalities,

$$\begin{aligned} x_j - x_{j-1} &\leq 800 \\ x_{j-1} - x_j &\leq 800. \end{aligned} \tag{12.8}$$

Thus the only nontrivial obstacle comes from (12.4). Note, however, that (12.4) may be stated as

$$t_j = \max \begin{cases} 15(x_j - x_{j-1}) \\ 21(x_{j-1} - x_j) \end{cases} \tag{12.9}$$

and so

$$\begin{aligned} t_j &\geq 15(x_j - x_{j-1}) \\ t_j &\geq 21(x_{j-1} - x_j). \end{aligned} \tag{12.10}$$

Furthermore, if each t_j gets slammed down against the larger of the two lower bounds (12.10), then (12.9) holds. We conclude that our problem is equivalent to the linear programming problem of minimizing (12.5) subject to (12.8), (12.2), (12.3), (12.10), (12.6), and (12.7). Only one small obstacle remains: in order to describe a feasible schedule, the numbers $x_1, x_2, \ldots, x_{12}$ have to be multiples of 20 (the number $x_j/20$ of workers in month j is necessarily an integer). However, we shall ignore this obstacle altogether: having obtained an optimal solution of the linear programming problem, we may simply round each x_j to the nearest multiple of 20 and adjust the levels of overtime and storage accordingly. The resulting schedule will be at least very close to optimal. In view of the uncertainty inherent in the data, this procedure is quite satisfactory.

Table 12.3 Optimal Solution

j	x_j	y_j	z_j	t_j
1	5,300	0	0	10,500
2	5,100	0	0	4,200
3	4,400	0	0	14,700
4	4,016.67	0	1,216.67	8,050
5	4,016.67	0	1,133.33	0
6	4,816.67	0	1,150	12,000
7	5,616.67	0	766.67	12,000
8	6,416.67	0	83.33	12,000
9	7,216.67	0	0	12,000
10	7,216.67	583.33	0	0
11	7,200	400	0	350
12	6,400	0	0	16,800

Thus the optimal solution (Table 12.3) of our linear programming problem yields the schedule shown in Table 12.4. The corresponding charges incurred in addition to the normal production costs come to $157,160. This figure is nearly equal to the optimum value 157,067 of the linear programming problem.

In qualitative terms, the schedule may be described as follows:

- During the period from January to March, no overtime is used and no inventory created. During this period, work-force level is adjusted monthly to meet the predicted demand.
- The period from April to September is characterized by a large initial inventory that gradually dwindles to nearly zero. During this period, no overtime is used. The April work force is kept at the same level in May; from June to September, 40 new workers are hired monthly. The initial work-force level in April is set so that inventory is almost gone by the end of September. Evidently, this initial level is quite sensitive to the overall demand in this period.
- From October to December, no inventory is created; in fact, some overtime production is required to cover the demand in October and November. The movement of the work-force level during this period is much affected by the drop in demand from November to December; this drop is too sharp to be reflected by an instant decrease in the labor force. For this reason, November witnesses the apparently irrational policy of firing and having overtime work at the same time.

Table 12.4 Optimal Schedule

Month	Demand	Work force	Production		Inventory
			Normal	Overtime	
Jan	5,300	265	5,300	—	—
Feb	5,100	255	5,100	—	—
Mar	4,400	220	4,400	—	—
Apr	2,800	201	4,020	—	1,220
May	4,100	201	4,020	—	1,140
Jun	4,800	241	4,820	—	1,160
Jul	6,000	281	5,620	—	780
Aug	7,100	321	6,420	—	100
Sep	7,300	361	7,220	—	20
Oct	7,800	361	7,220	560	—
Nov	7,600	360	7,200	400	—
Dec	6,400	320	6,400	—	—

This discussion followed the lines of F. Hanssmann and S. W. Hess (1960). Some of the earlier works on this subject are A. Charnes et al. (1953), J. F. Magee (1953), H. A. Antosiewicz and A. J. Hoffman (1954), and A. J. Hoffman and W. Jacobs (1954). We shall return to two special classes of multiperiod scheduling problems in Chapter 20.

STAIRCASE STRUCTURE

Linear programming problems arising in multiperiod scheduling have a special structure: the nonzero entries in the constraint matrix are restricted to the shaded area illustrated in Figure 12.1. Such problems sometimes require more iterations of the simplex method than other LP problems of comparable size. [This phenomenon is related to an empirical observation, made by G. B. Dantzig (1955), concerning optimal solutions of multiperiod scheduling problems: an activity, such as overtime or inventory, which is maintained at a positive level during one of the periods, is likely to persist at a positive level in the next period. To put it differently, the periods during which a specified activity is maintained at a positive level tend to come in blocks rather than at haphazard points.] In addition, the simplex method seems more likely to suffer from rounding errors in solving problems of this kind.

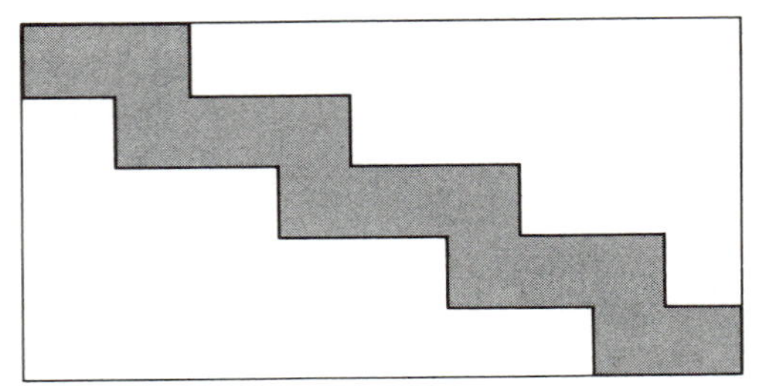

Figure 12.1

Several attempts have been made to facilitate solutions of multiperiod scheduling problems by exploiting their staircase structure. For instance, A. F. Perold and G. B. Dantzig (1978) proposed a factorization of **B** that relies on the characteristic properties of multiperiod scheduling problems to facilitate the solution of $\mathbf{yB} = \mathbf{c}_B$ and $\mathbf{Bd} = \mathbf{a}$ in each iteration of the revised simplex method. A completely different line of attack is represented in an algorithm developed by J. K. Ho and A. S. Manne (1974). In this algorithm, the simplex method is abandoned altogether in favor of an iterative application of the "Dantzig–Wolfe decomposition algorithm." We shall present the Dantzig–Wolfe decomposition algorithm in Chapter 26.

PROBLEMS

△ **12.1** Solve the no-overtime-allowed variant of the problem in text.

13

The Cutting-Stock Problem

Materials such as paper, textiles, cellophane, and metallic foil are manufactured in rolls of large widths. These rolls, referred to as *raws*, are later cut into rolls of small widths, called *finals*. Each manufacturer produces raws of a few standard widths; the widths of the finals are specified by different customers and may vary widely. The cutting is done on machines by knives that slice through the rolls in much the same way as a knife slices bread. For example, a raw that is 100 in. wide may be cut into two finals with 31-in. widths and one final with a 36-in. width, with the 2 in. left over going to waste. When a complicated summary of orders has to be filled, the most economical way of cutting the existing raws into the desired finals is rarely obvious. The problem of finding such a way is known as the *cutting-stock problem*.

THE CUTTING-STOCK PROBLEM AND LINEAR PROGRAMMING

Let us consider an example where the raws are 100 in. wide and we have to fill the following summary of orders:

97 finals of width 45 in.

610 finals of width 36 in.

395 finals of width 31 in.

211 finals of width 14 in.

First, we shall tabulate all the ways of cutting a raw into a_1 finals of width 45, a_2 finals of width 36, a_3 finals of width 31, and a_4 finals of width 14. As it turns out, there are 37 ways of doing this. They are listed in Table 13.1, with the jth pattern specified by a_{1j}, a_{2j}, a_{3j}, and a_{4j}. Now we may speculate about various ways of filling the order summary. If the jth pattern is used x_j times then we must have

$$\sum_{j=1}^{37} a_{ij}x_j = b_i \qquad (i = 1, 2, 3, 4) \tag{13.1}$$

with $b_1 = 97$, $b_2 = 610$, $b_3 = 395$, $b_4 = 211$. Thus we are led to

$$\text{minimize} \quad \sum_{j=1}^{37} x_j \quad \text{subject to} \quad (13.1) \quad \text{and} \quad x_j \geq 0 \quad (j = 1, 2, \ldots, 37). \tag{13.2}$$

Note that there is an obvious catch: an optimal solution of (13.2) may not be integer-valued and yet, in order to describe a realistic plan, our numbers x_j must be nonnegative *integers*. Nevertheless, as K. Eisemann (1957) pointed out, analogues of (13.2) may sometimes describe the actual problem with a sufficient degree of realism. For example, rolls of expensive materials, such as silk, are *not* simply cut through. Instead, while a raw roll is being unwound, it is sliced lengthwise for a certain total length. This process may be stopped at any time and the knife setting altered; the very same raw roll continues to unwind, now being sliced to different widths. In such situations, it does make sense to speak of a cutting pattern used a fractional number of times. In a paper mill, however, each raw roll is either sliced completely or not sliced at all; hence, a working plan is described by *integers* x_j. Even in those situations, though, it is advantageous to solve analogues of (13.2): their optimal solutions point out a variety of working plans that are, at least, close to optimal. For instance, an optimal solution of (13.2) is

$$x_1^* = 48.5, \quad x_{10}^* = 105.5, \quad x_{12}^* = 100.75, \quad x_{13}^* = 197.5.$$

Rounding each x_j^* down to the nearest integer, we obtain a schedule using 450 raws to produce 96 finals of width 45 in., 607 finals of width 36 in., 394 finals of width 31 in., and 210 finals of width 14 in. The residual unsatisfied demand can be easily satisfied by as few as three extra raws. Thus we find a way of satisfying the total demand by using only 453 raws. Since the optimum value of (13.2) is 452.25, our integer-valued solution is clearly optimal.

The same line of attack applies in general. However, there are difficulties, which stem from two radically different sources:

Table 13.1 Thirty-Seven Cutting Patterns

j	1	2	3	4	5	6	7	8	9	10	11	12	13
a_{1j}	2	1	1	1	1	1	1	1	1	0	0	0	0
a_{2j}	0	1	1	0	0	0	0	0	0	2	2	2	1
a_{3j}	0	0	0	1	1	0	0	0	0	0	0	0	2
a_{4j}	0	1	0	1	0	3	2	1	0	2	1	0	0

	14	15	16	17	18	19	20	21	22	23	24	25	26
a_{1j}	0	0	0	0	0	0	0	0	0	0	0	0	0
a_{2j}	1	1	1	1	1	1	1	1	0	0	0	0	0
a_{3j}	1	1	1	0	0	0	0	0	3	2	2	2	1
a_{4j}	2	1	0	4	3	2	1	0	0	2	1	0	4

	27	28	29	30	31	32	33	34	35	36	37
a_{1j}	0	0	0	0	0	0	0	0	0	0	0
a_{2j}	0	0	0	0	0	0	0	0	0	0	0
a_{3j}	1	1	1	1	0	0	0	0	0	0	0
a_{4j}	3	2	1	0	7	6	5	4	3	2	1

Note: If the width of the trimmed-off piece exceeds 14 in., which is the width of the smallest final, then the pattern is clearly wasteful. For instance, it would be foolish to use pattern 9 twice in a row: the resulting two 45-in. finals could just as well be cut from a *single* raw by pattern 1. These wasteful patterns, however, might be called for to take care of various odds and ends. In any event, as we shall see later, the cutting-stock problem is customarily solved *without* tabulating all the patterns in advance.

(i) Problems commonly encountered in the paper industry may involve astronomical numbers of variables. For example, if the raw rolls are 200 in. wide and if the finals are ordered in 40 different lengths ranging from 20 in. to 80 in., then the number of different patterns can easily exceed 10 or even 100 million. The amount of time and space required just to tabulate these patterns (not to mention *solving* the problem afterwards) may be simply unavailable.

(ii) Passing from an optimal *fractional-valued* solution to an optimal *integer-valued* solution is not easy. Rounding the fractional values down and then satisfying the residual demand, as we did in the example, may not yield the optimal working plan. If the finals are ordered in small enough quantities, then the patterns used in the optimal integer-valued solution may be quite different from those used originally in the optimal fractional-valued solutions. In fact, if the order summary is small, then experienced schedulers often find a working plan that is better than any plan obtained by simple adjustments of the fractional optimum; see R. E. D. Woolsey (1972a).

An ingenious way of getting around the first difficulty has been suggested by P. C. Gilmore and R. E. Gomory (1961). The trick, in short, is to work with only a few patterns at a time and to generate new patterns only when they are really needed. We are going to discuss this *delayed column-generation technique* in the next section. (Actually, delayed column generation was used even before Gilmore and Gomory in a different context; it constitutes a crucial part of the Dantzig–Wolfe decomposition algorithm presented in Chapter 26.)

No efficient way of handling the second difficulty is known. Fortunately, rounding the fractional optimum up or down, with subsequent ad hoc adjustments, is quite satisfactory for typical problems arising in the paper industry. If m different lengths of finals are ordered, then the fractional optimum produced by the simplex method involves at most m nonzero variables x_j. Hence the cost difference between the rounded-off solution and the true integer optimum will be *at most* the cost of m raws (and often considerably less). Since a typical value of m is no more than 50, and most final widths are ordered in hundreds or thousands, the possible cost increases due to inefficient rounding are relatively negligible. Problems with wide spectra of final widths are usually referred to as *bin-packing problems*; the term *cutting-stock problem* is normally reserved for bin-packing problems with narrow spectra of final widths and large order summaries. These characteristic features make cutting-stock problems amenable to the linear programming approach described here.

DELAYED COLUMN GENERATION

Consider a cutting-stock problem where the raws are r inches wide and the order summary calls for b_i finals of width w_i ($i = 1, 2, \ldots, m$). As in the previous example, we are led to

$$\text{minimize} \quad \mathbf{cx} \qquad \text{subject to} \quad \mathbf{Ax} = \mathbf{b}, \mathbf{x} \geq \mathbf{0}.$$

Here $\mathbf{b}$ is a column vector with components $b_1, b_2, \ldots, b_m$ and $\mathbf{c}$ is a row vector with components $1, 1, \ldots, 1$. Each column $\mathbf{a} = [a_1, a_2, \ldots, a_m]^T$ of $\mathbf{A}$ specifies a pattern cutting a raw into a_i finals of width w_i ($i = 1, 2, \ldots, m$). Thus $\mathbf{a}$ is a column of $\mathbf{A}$ if and only if $a_1, a_2, \ldots, a_m$ are nonnegative integers such that $\sum w_i a_i \leq r$. The revised simplex method refers to the nonbasic columns of $\mathbf{A}$ only in step 2 of each iteration, when a new entering column $\mathbf{a}$ has to be found; having computed the row vector $\mathbf{y}$, we look for nonnegative integers $a_1, a_2, \ldots, a_m$ such that

$$\sum_{i=1}^{m} w_i a_i \leq r \quad \text{and} \quad \sum_{i=1}^{m} y_i a_i > 1 \tag{13.3}$$

(we are *minimizing* the objective function). If we can always find such integers or prove their nonexistence, then there is no need to tabulate all the columns of **A** in advance. Instead, the entering columns may be generated individually at the rate of one column per iteration.

To illustrate the details, we shall consider an example where the raws are 91 in. wide and the order summary calls for

78 finals of width $25\frac{1}{2}$ in.

40 finals of width $22\frac{1}{2}$ in.

30 finals of width 20 in.

30 finals of width 15 in.

Here, the revised simplex method may be initialized by

$$\mathbf{B} = \begin{bmatrix} 3 & & & \\ & 4 & & \\ & & 4 & \\ & & & 6 \end{bmatrix} \quad \text{and} \quad \mathbf{x}_B^* = \begin{bmatrix} 26 \\ 10 \\ 7.5 \\ 5 \end{bmatrix}.$$

In general, one can always initialize by m cutting patterns such that the ith pattern yields only finals of width w_i. Now the first iteration may begin.

Step 1. Solving the system $\mathbf{yB} = [1, 1, 1, 1]$, we obtain $\mathbf{y} = [\frac{1}{3}, \frac{1}{4}, \frac{1}{4}, \frac{1}{6}]$.

Step 2. Now we are looking for nonnegative integers a_1, a_2, a_3, a_4 such that

$$25.5a_1 + 22.5a_2 + 20a_3 + 15a_4 \leq 91$$

$$\frac{1}{3}a_1 + \frac{1}{4}a_2 + \frac{1}{4}a_3 + \frac{1}{6}a_4 > 1 .$$

A certain method, which will be discussed in the next section, finds $a_1 = 2, a_2 = 0, a_3 = 2, a_4 = 0$. Hence $\mathbf{a} = [2, 0, 2, 0]^T$ is our entering column.

Step 3. Solving the system $\mathbf{Bd} = \mathbf{a}$, we find $\mathbf{d} = [\frac{2}{3}, 0, \frac{1}{2}, 0]^T$.

Step 4. Comparing the ratios $26/\frac{2}{3}$ and $7.5/\frac{1}{2}$, we find that $t = 15$ and that the third column has to leave **B**.

Step 5. Now we have

$$\mathbf{B} = \begin{bmatrix} 3 & & 2 & \\ & 4 & & \\ & & 2 & \\ & & & 6 \end{bmatrix} \quad \text{and} \quad \mathbf{x}_B^* = \begin{bmatrix} 26 - 2t/3 \\ 10 \\ t \\ 5 \end{bmatrix} = \begin{bmatrix} 16 \\ 10 \\ 15 \\ 5 \end{bmatrix}.$$

The second iteration begins.

Step 1. Solving the system $\mathbf{yB} = [1, 1, 1, 1]$, we obtain $\mathbf{y} = [\frac{1}{3}, \frac{1}{4}, \frac{1}{6}, \frac{1}{6}]$.

Step 2. Now we are looking for nonnegative integers a_1, a_2, a_3, a_4 such that

$$25.5a_1 + 22.5a_2 + 20a_3 + 15a_4 \leq 91$$

$$\frac{1}{3}a_1 + \frac{1}{4}a_2 + \frac{1}{6}a_3 + \frac{1}{6}a_4 > 1 .$$

The method to be discussed later finds $a_1 = 2$, $a_2 = 1$, $a_3 = 0$, $a_4 = 1$. Hence, $\mathbf{a} = [2, 1, 0, 1]^T$ is our entering column.

Step 3. Solving the system $\mathbf{Bd} = \mathbf{a}$, we find $\mathbf{d} = [\frac{2}{3}, \frac{1}{4}, 0, \frac{1}{6}]^T$.

Step 4. Comparing the ratios $16/\frac{2}{3}$, $10/\frac{1}{4}$, and $5/\frac{1}{6}$, we find that $t = 24$ and that the first column has to leave **B**.

Step 5. Now we have

$$\mathbf{B} = \begin{bmatrix} 2 & & 2 & \\ 1 & 4 & & \\ & & 2 & \\ 1 & & & 6 \end{bmatrix} \quad \text{and} \quad \mathbf{x}_B^* = \begin{bmatrix} t \\ 10 - t/4 \\ 15 \\ 5 - t/6 \end{bmatrix} = \begin{bmatrix} 24 \\ 4 \\ 15 \\ 1 \end{bmatrix}.$$

The third iteration begins.

Step 1. Solving the system $\mathbf{yB} = [1, 1, 1, 1]$, we obtain $\mathbf{y} = [\frac{7}{24}, \frac{1}{4}, \frac{5}{24}, \frac{1}{6}]$.

Step 2. Now we are looking for nonnegative integers a_1, a_2, a_3, a_4 such that

$$25.5a_1 + 22.5a_2 + 20a_3 + 15a_4 \leq 91$$

$$\frac{7}{24}a_1 + \frac{1}{4}a_2 + \frac{5}{24}a_3 + \frac{1}{6}a_4 > 1 .$$

The method to be discussed later establishes the nonexistence of such integers. We conclude that our current solution is optimal.

Clearly, this strategy is applicable to any cutting-stock problem with a single raw width. Its success hinges on an efficient subroutine for finding nonnegative integers $a_1, a_2, \ldots, a_m$ satisfying (13.3) or establishing their nonexistence.

We are about to present an algorithm that does a little more: it solves the problem:

$$\begin{aligned} &\text{maximize} && \sum_{i=1}^{m} y_i a_i \\ &\text{subject to} && \sum_{i=1}^{m} w_i a_i \leq r \\ &&& a_i = \text{nonnegative integer} \qquad (i = 1, 2, \ldots, m). \end{aligned} \tag{13.4}$$

This problem is known as the *knapsack problem.* (The name derives from a frivolous interpretation where a narrow knapsack of height r has to be filled with food packages, the ith brand having height w_i and value y_i.)

Solving the Knapsack Problem

In the usual notation, the knapsack problem is recorded as

$$\begin{aligned} \text{maximize} \quad & \sum_{i=1}^{m} c_i x_i \\ \text{subject to} \quad & \sum_{i=1}^{m} a_i x_i \leq b \\ & x_i = \text{nonnegative integer} \qquad (i = 1, 2, \ldots, m). \end{aligned} \tag{13.5}$$

Thus the variables are denoted by $x_1, x_2, \ldots, x_m$ rather than $a_1, a_2, \ldots, a_m$. Similarly, the coefficients w_i in (13.4) are replaced by a_i in (13.5); the coefficients y_i in (13.4) are replaced by c_i in (13.5); and the right-hand side is denoted by b rather than r. In typical applications, such as the cutting-stock problem, the numbers $a_1, a_2, \ldots, a_m$ and b are positive. Now we may assume that the numbers $c_1, c_2, \ldots, c_m$ are positive (variables x_i with $c_i \leq 0$ could be deleted at once). In the frivolous interpretation, each a_i is thought of as the height of a food package, whereas c_i represents the corresponding value. Thus the ratio c_i/a_i amounts to the value per inch of the ith brand. In an intuitive sense, the desirability of including the ith brand in the knapsack increases with this ratio. We shall refer to c_i/a_i as the *efficiency* of the variable x_i. Without loss of generality, we may assume that the variables are subscripted in order of decreasing efficiency:

$$c_1/a_1 \geq c_2/a_2 \geq \cdots \geq c_m/a_m. \tag{13.6}$$

Finally, note that every optimal solution of (13.5) satisfies

$$b - \sum_{i=1}^{m} a_i x_i < a_m \tag{13.7}$$

for otherwise x_m could be increased by one unit. We shall refer to feasible solutions satisfying (13.7) as *sensible.*

The key to our method lies in a certain way of enumerating all the sensible solutions. We shall illustrate it on the problem

$$\begin{aligned} \text{maximize} \quad & 4x_1 + 5x_2 + 5x_3 + 2x_4 \\ \text{subject to} \quad & 33x_1 + 49x_2 + 51x_3 + 22x_4 \leq 120 \\ & x_1, x_2, x_3, x_4 = \text{nonnegative integer.} \end{aligned}$$

The reader may easily verify that this problem has precisely 13 sensible solutions; we shall describe them by the diagram in Figure 13.1. Diagrams of this kind are called *enumeration trees.* The small circles are the *nodes* of the tree; the leftmost node is the *root* and each of the 13 rightmost nodes is a *leaf*. (For typographical reasons, this tree grows in a direction seldom found in nature: from left to right.) Proceeding from the root to a leaf in four stages, we specify first a value of x_1, then a value of x_2, and so on. In this sense, each of the 13 leaves corresponds to a sensible solution,

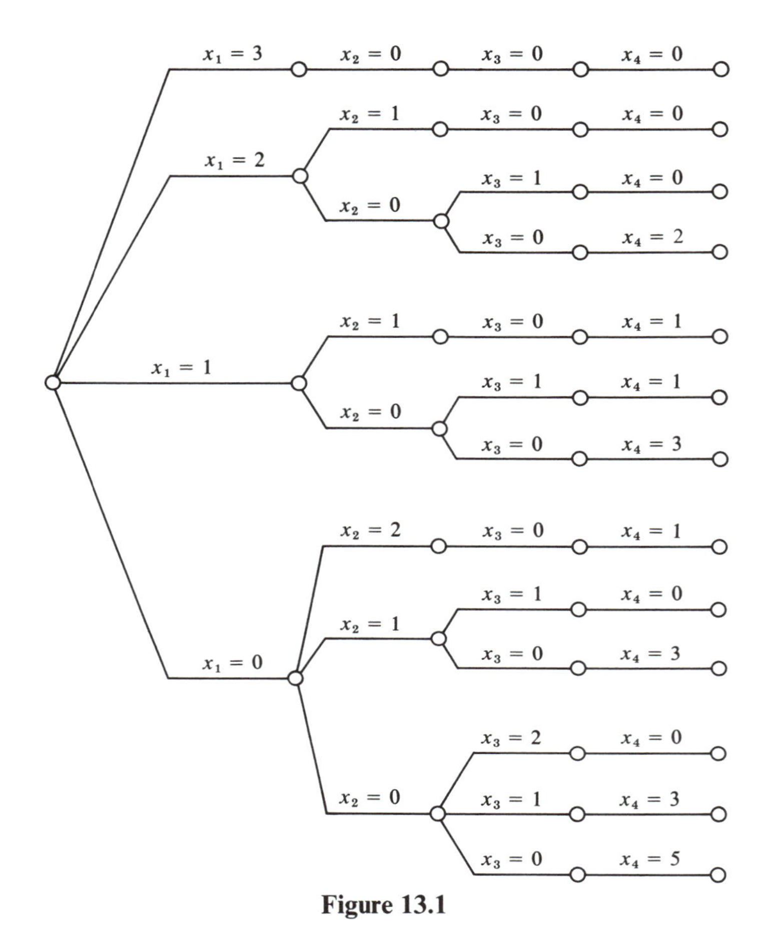

Figure 13.1

and vice versa. Note also that, whenever two or more branches grow from the same node, the higher branches correspond to higher values of x_j. In general, the solution corresponding to the *topmost* leaf is obtained by making x_1 as large as possible and, once x_1 has been fixed, making x_2 as large as possible, and so on. Formally, with $\lfloor z \rfloor$ denoting the integer obtained by rounding down z, this solution is defined by the recursion

$$x_j = \left\lfloor \left(b - \sum_{i=1}^{j-1} a_i x_i \right) \bigg/ a_j \right\rfloor \qquad (j = 1, 2, \ldots, m).$$

In particular, $x_1 = \lfloor b/a_1 \rfloor$. We begin with this solution. If it weren't for certain shortcuts, we would scan through the list of leaves down to the bottom, keeping track of the best solution $x_1^*, x_2^*, \ldots, x_m^*$ found so far, and replacing it with a better solution whenever one comes up.

The scanning instructions make use of the tree structure. From each leaf that has just been examined, we backtrack toward the root step by step, keeping a constant lookout for as yet unexplored branches. When one is found, we reverse direction and proceed toward a new leaf; at each junction, we resolve possible ties by preferring higher branches to lower ones. As the reader may verify by consulting Figure 13.1, an algebraic description of this procedure is as follows. Having just examined a sensible solution $x_1, x_2, \ldots, x_m$, we set k equal to $m - 1$ and, if necessary, keep reducing k until $x_k > 0$ is found. Then we replace x_k by $x_k - 1$ and define the values of $x_{k+1}, x_{k+2}, \ldots, x_m$ recursively by

$$x_j = \left\lfloor \left(b - \sum_{i=1}^{j-1} a_i x_i \right) \Big/ a_j \right\rfloor.$$

The shortcuts will prevent us from moving along branches that are *obviously* hopeless. More precisely, suppose that the current best solution $x_1^*, x_2^*, \ldots, x_m^*$ yields

$$\sum_{i=1}^{m} c_i x_i^* = M$$

and that, backtracking from some $x_1, x_2, \ldots, x_m$ toward the root, we have just found the largest k such that $k \leq m - 1$ and $x_k > 0$. Proceeding as before, we would set

$$\bar{x}_i = x_i \qquad \text{for all} \quad i = 1, 2, \ldots, k - 1$$

$$\bar{x}_k = x_k - 1$$

and begin to examine various choices of $\bar{x}_{k+1}, \bar{x}_{k+2}, \ldots, \bar{x}_m$ leading to sensible solutions $\bar{x}_1, \bar{x}_2, \ldots, \bar{x}_m$. Before doing so, let us ask whether any of these solutions has a fighting chance against $x_1^*, x_2^*, \ldots, x_m^*$. By (13.6), each of the variables x_{k+1}, $x_{k+2}, \ldots, x_m$ has an efficiency of *at most* c_{k+1}/a_{k+1}, and so

$$\sum_{i=k+1}^{m} c_i \bar{x}_i \leq \frac{c_{k+1}}{a_{k+1}} \sum_{i=k+1}^{m} a_i \bar{x}_i.$$

Hence the assumption

$$\sum_{i=1}^{m} a_i \bar{x}_i \leq b$$

implies

$$\sum_{i=1}^{m} c_i \bar{x}_i \leq \sum_{i=1}^{k} c_i \bar{x}_i + \frac{c_{k+1}}{a_{k+1}} \left(b - \sum_{i=1}^{k} a_i \bar{x}_i \right). \tag{13.8}$$

Unless the right-hand side of (13.8) *exceeds* M, none of the solutions $\bar{x}_1, \bar{x}_2, \ldots, \bar{x}_m$ has a chance of improving $x_1^*, x_2^*, \ldots, x_m^*$, and so we might as well forget about them. In terms of the enumeration tree, the inequality

$$\sum_{i=1}^{k} c_i\bar{x}_i + \frac{c_{k+1}}{a_{k+1}}\left(b - \sum_{i=1}^{k} a_i\bar{x}_i\right) \leq M \tag{13.9}$$

makes the path specified by $\bar{x}_1, \bar{x}_2, \ldots, \bar{x}_k$ *not* worth exploring any further. In fact, if all the coefficients $c_1, c_2, \ldots, c_m$ are positive *integers*, then M is an integer and (13.9) may be replaced by the weaker inequality

$$\sum_{i=1}^{k} c_i\bar{x}_i + \frac{c_{k+1}}{a_{k+1}}\left(b - \sum_{i=1}^{k} a_i\bar{x}_i\right) < M + 1. \tag{13.10}$$

Let us see how useful these shortcuts turn out to be in our example. We initialize by

$$\begin{aligned}
x_1 &= \lfloor 120/33 \rfloor = 3\\
x_2 &= \lfloor (120 - 99)/49 \rfloor = 0\\
x_3 &= \lfloor (120 - 99)/51 \rfloor = 0\\
x_4 &= \lfloor (120 - 99)/22 \rfloor = 0
\end{aligned}$$

and file the initial solution as our current best:

$$x_1^* = 3, \quad x_2^* = x_3^* = x_4^* = 0 \quad \text{and} \quad M = 12. \tag{13.11}$$

Beginning with $k = 3$, we reduce k until $k = 1$ with $x_k > 0$ is found. Then we change $x_1 = 3$ into $x_1 = 2$. Before exploring the branch $x_1 = 2$ any further, let us test inequality (13.10) with $k = 1$ and $\bar{x}_1 = 2$. Since the left-hand side equals

$$8 + \frac{5}{49}(120 - 66) \doteq 13.5$$

which is *not* smaller than $M + 1 = 13$, the branch *may* be worth exploring. Computing

$$\begin{aligned}
x_2 &= \lfloor (120 - 66)/49 \rfloor = 1\\
x_3 &= \lfloor (120 - 115)/51 \rfloor = 0\\
x_4 &= \lfloor (120 - 115)/22 \rfloor = 0
\end{aligned}$$

we indeed find an improvement over (13.11). Hence, we replace (13.11) by

$$x_1^* = 2, \quad x_2^* = 1, \quad x_3^* = x_4^* = 0 \quad \text{and} \quad M = 13. \tag{13.12}$$

Again, we begin with $k = 3$ and reduce k until $k = 2$ with $x_k > 0$ is found. Then we change $x_2 = 1$ into $x_2 = 0$. To find out whether the path $x_1 = 2, x_2 = 0$ is worth exploring further, we test inequality (13.10) with $k = 2$ and $\bar{x}_1 = 2, \bar{x}_2 = 0$. The left-hand side equals

$$8 + \frac{5}{51}(120 - 66) \doteq 13.3$$

which is smaller than $M + 1 = 14$. Hence, we reduce k further until the next k with $x_k > 0$. Having found $k = 1$, we replace $x_1 = 2$ by $x_1 = 1$. Is the branch $x_1 = 1$ worth exploring? Testing (13.10) with $k = 1$ and $\bar{x}_1 = 1$ shows that

$$4 + \frac{5}{49}(120 - 33) \doteq 12.9 < 14$$

and so the answer is negative. The branch $x_1 = 0$ would do even worse:

$$\frac{5}{49} \cdot 120 \doteq 12.2 < 14.$$

Hence, the search terminates and we conclude that (13.12) is an optimal solution.

In terms of the enumeration tree, not exploring the obviously hopeless branches may be viewed as pruning these branches off. In effect, the enumeration tree of Figure 13.1 reduces to the much smaller tree shown in Figure 13.2.

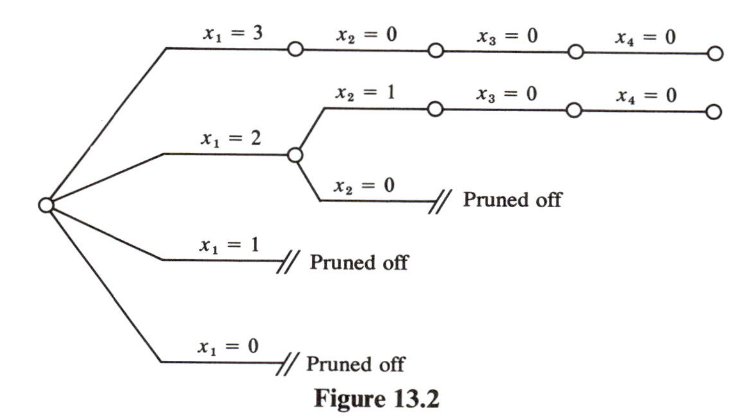

Figure 13.2

Note that having pruned off the branch $\bar{x}_1, \bar{x}_2, \ldots, \bar{x}_k$, we can prune off $\bar{x}_1, \bar{x}_2, \ldots, \bar{x}_k - 1$ at once, without performing the appropriate test. The point is that the relevant inequality,

$$\sum_{i=1}^{k-1} c_i\bar{x}_i + c_k(\bar{x}_k - 1) + \frac{c_{k+1}}{a_{k+1}}\left(b - \sum_{i=1}^{k-1} a_i\bar{x}_i - a_k(\bar{x}_k - 1)\right) < M + 1. \qquad (13.13)$$

is already implied by (13.10). Indeed, the left-hand side of (13.10) exceeds the left-hand side of (13.13) by

$$c_k - \frac{c_{k+1}}{a_{k+1}} a_k$$

which is nonnegative by virtue of (13.6). Of course, this observation brings about further computational savings.

This method, known as *branch-and-bound*, is described succintly in Box 13.1.

BOX 13.1 Branch-and-Bound Method for Solving the Knapsack Problem

Step 1. [Initialize.] Set $M = 0, k = 0$.

Step 2. [Find the most promising extension of the current branch.] For $j = k + 1, k + 2, \ldots, m$, set

$$x_j = \left\lfloor \left(b - \sum_{i=1}^{j-1} a_i x_i\right) \Big/ a_j \right\rfloor.$$

Then replace k by m.

Step 3. [An improved solution obtained?] If $\sum_{i=1}^{m} c_i x_i > M$, then replace M by $\sum_{i=1}^{m} c_i x_i$ and replace $x_1^*, x_2^*, \ldots, x_m^*$ by $x_1, x_2, \ldots, x_m$.

Step 4. [Backtrack to the next branch.]

a. If $k = 1$, then stop; otherwise replace k by $k - 1$.

b. If $x_k = 0$, then return to a; otherwise replace x_k by $x_k - 1$.

Step 5. [Branch worth exploring?] If (13.10) with x_i in place of $\bar{x}_i$ fails, then return to step 2; otherwise return to step 4.

Of course, if some of the coefficients c_i are not integers, then test (13.10) in step 5 must be replaced by (13.9). An alternative, and preferable, course of action is to find a positive d such that each dc_i is an integer, and then replace $c_1, c_2, \ldots, c_m$ by $dc_1, dc_2, \ldots, dc_m$. For instance, the objective function in the knapsack problem

$$\begin{array}{ll} \text{maximize} & \frac{1}{5}x_1 + \frac{1}{7}x_2 + \frac{1}{9}x_3 + \frac{1}{9}x_4 \\ \text{subject to} & 17.25x_1 + 12.75x_2 + 10x_3 + 15x_4 \leq 90 \\ & x_1, x_2, x_3, x_4 = \text{nonnegative integer} \end{array}$$

may be replaced by $63x_1 + 45x_2 + 35x_3 + 35x_4$.

An alternative method is based on a simple idea known as the principle of *dynamic programming*. Dynamic programming algorithms, however, tend to get quite tedious as the right-hand side b increases. In experiments with the cutting-stock problem, P. C. Gilmore and R. E. Gomory (1963) found the branch-and-bound method about five times as fast as dynamic programming. On the examples discussed here, dynamic programming would certainly require far more work than we would care to go through. We leave this line of attack for problems at the end of this chapter (13.4–13.8); for more information, the reader is referred to Chapter 6 of R. S. Garfinkel and G. L. Nemhauser (1972).

Finding a Good Initial Solution

In solving the cutting-stock problem by the revised simplex method with delayed column generation, it is desirable to begin with a near-optimal basic feasible solution: if the initial value of the objective function is close to the optimum value, then the number of simplex iterations required to reach the optimum is likely to be small. We are about to describe a very fast procedure for finding reasonably good initial solutions. This procedure is based on the assumption that it pays to get the wide finals out of the way first, letting the versatile narrow ones compete for the leftovers.

Again, we shall denote the single raw width by r and say that the order summary calls for b_i finals of width w_i ($i = 1, 2, \ldots, m$). Furthermore, we shall assume that the final widths are arranged in decreasing order:

$$w_1 > w_2 > \cdots > w_m.$$

Our procedure constructs the initial $\mathbf{B}$ and $\mathbf{x}$ in m iterations. At the beginning of iteration j, we have a certain set R containing precisely $m - j + 1$ of the subscripts $1, 2, \ldots, m$; for each subscript i in R, we have a nonnegative number b_i', interpreted as the residual demand for finals of width w_i. (To initialize, we set $b_i' = b_i$ for all i.) In the jth iteration, we define the jth column $\mathbf{a} = [a_1, a_2, \ldots, a_m]^T$ of $\mathbf{B}$ recursively by

$$a_i = \begin{cases} 0 & \text{if } \; i \notin R \\ \left\lfloor \left(r - \sum_{k=1}^{i-1} w_k a_k \right) \Big/ w_i \right\rfloor & \text{if } \; i \in R. \end{cases}$$

Our initial solution x will use this pattern until some residual demand b_k' gets fully satisfied. To put it differently, the component x_j corresponding to the jth column of $\mathbf{B}$ will equal the smallest of the ratios b_i'/a_i with $i \in R$ and $a_i > 0$. Thus, $x_j a_i \leq b_i'$ for all $i \in R$ and $x_j a_k = b_k'$ for at least one $k \in R$. We delete this subscript k from R, replace each remaining b_i' by $b_i' - x_j a_i$ and proceed to the $(j + 1)$th iteration.

For illustration, we return to our first cutting-stock example with $r = 100$, $w_1 = 45$, $w_2 = 36$, $w_3 = 31$, $w_4 = 14$, $b_1 = 97$, $b_2 = 610$, $b_3 = 395$, $b_4 = 211$. Here the initialization procedure works as follows.

Iteration 1. We find $a_1 = \lfloor 100/45 \rfloor = 2$, $a_2 = \lfloor 10/36 \rfloor = 0$, $a_3 = \lfloor 10/31 \rfloor = 0$, $a_4 = \lfloor 10/14 \rfloor = 0$. Now $x_1 = \frac{97}{2} = 48.5$. We delete the subscript 1 from R and obtain $b_2' = 610$, $b_3' = 395$, $b_4' = 211$.

Iteration 2. We find $a_1 = 0$, $a_2 = \lfloor 100/36 \rfloor = 2$, $a_3 = \lfloor 28/31 \rfloor = 0$, $a_4 = \lfloor 28/14 \rfloor = 2$. Now $x_2 = \min(\frac{610}{2}, \frac{211}{2}) = 105.5$. We delete the subscript 4 from R and obtain $b_2' = 399$, $b_3' = 395$.

Iteration 3. We find $a_1 = 0$, $a_2 = \lfloor 100/36 \rfloor = 2$, $a_3 = \lfloor 28/31 \rfloor = 0$, $a_4 = 0$. Now $x_3 = \frac{399}{2} = 199.5$. We delete the subscript 2 from R and obtain $b_3' = 395$.

Iteration 4. We find $a_1 = 0$, $a_2 = 0$, $a_3 = \lfloor 100/31 \rfloor = 3$, $a_4 = 0$. Now $x_4 = \frac{395}{3} = 131.67$.

Thus we have produced the basic feasible solution with

$$\mathbf{B} = \begin{bmatrix} 2 & 0 & 0 & 0 \\ 0 & 2 & 2 & 0 \\ 0 & 0 & 0 & 3 \\ 0 & 2 & 0 & 0 \end{bmatrix} \quad \text{and} \quad \mathbf{x}_B^* = \begin{bmatrix} 48.5 \\ 105.5 \\ 199.5 \\ 131.67 \end{bmatrix}.$$

Note that this solution is only one simplex iteration away from the optimum.

This procedure is closely related to a method known as *first-fit decreasing*, which produces reasonably good integer-valued solutions for any bin-packing problem. In the jth iteration of first-fit decreasing, we find a way of cutting the jth raw. The iteration begins with certain residual demands $b'_1, b'_2, \ldots, b'_m$; we specify the cutting pattern recursively by

$$a_i = \min \begin{cases} b'_i \\ \left\lfloor \left(r - \sum_{k=1}^{i-1} w_k a_k \right) \Big/ w_i \right\rfloor \end{cases}$$

for $i = 1, 2, \ldots, m$, then replace each b'_i by $b'_i - a_i$, and proceed to the $(j+1)$th iteration. For instance, first-fit decreasing applied to our example yields the following solution:

	a_1	a_2	a_3	a_4
48 times	$a_1 = 2$,	$a_2 = 0$,	$a_3 = 0$,	$a_4 = 0$
once	1	1	0	1
105 times	0	2	0	2
199 times	0	2	0	0
once	0	1	2	0
131 times	0	0	3	0.

A remarkable guarantee of performance for first-fit decreasing has been established by D. S. Johnson (1973): if the integer-valued optimal solution uses n raws, then the integer-valued solution found by first-fit decreasing uses at most $\frac{11}{9}n + 4$ raws. (The proof is very difficult. Nevertheless, it is easy to show that the bound cannot be improved except possibly for the dangling 4; see problem 13.3.) It follows that our initialization procedure always gets within 23% of the fractional optimum, give or take a few raws. □

Additional Complications

The problems arising in the paper industry may be more complicated than the simple examples discussed here. The complications stem from three major sources.

(i) The raws may come in several different widths $r_1, r_2, \ldots, r_N$. Producing a raw of width r_k costs c_k dollars; the problem is to fill the summary of orders at the least cost.

(ii) Raws of different widths are cut on different machines. Limited machine availability may allow no more than v_k raws of width r_k to be cut.

(iii) The number of cutting knives on each machine is limited, usually to six or so. Consequently, a cutting pattern specified by $a_1, a_2, \ldots, a_m$ is admissible only if $\sum a_i$ does not exceed a predetermined bound.

All these complications can be handled by easy modifications of the method just described, and, therefore, we shall comment on them only briefly. First, let us consider the linear programming problem

$$\text{minimize} \quad \mathbf{cx} \qquad \text{subject to} \quad \mathbf{Ax} = \mathbf{b}, \quad \mathbf{x} \geq \mathbf{0}$$

arising from a cutting-stock problem with N raw widths. The columns of $\mathbf{A}$ come in N different groups; columns in group k specify cutting patterns with $\sum_{i=1}^{m} w_i a_i \leq r_k$ and the corresponding components of $\mathbf{c}$ equal c_k. To generate the entering column, we have to find nonnegative integers $a_1, a_2, \ldots, a_m$ and a subscript k such that $\sum_{i=1}^{m} w_i a_i \leq r_k$ and $\sum_{i=1}^{m} y_i a_i > c_k$. This task can be accomplished by solving N knapsack problems, the kth of which reads

$$\begin{aligned} &\text{maximize} && \sum_{i=1}^{m} y_i a_i \\ &\text{subject to} && \sum_{i=1}^{m} w_i a_i \leq r_k \qquad (13.14) \\ & && a_i = \text{nonnegative integer} \end{aligned}$$

and then comparing the optimum values $z_1^*, z_2^*, \ldots, z_N^*$ against $c_1, c_2, \ldots, c_N$. The optimal solution with the largest $z_k^* - c_k$ yields the desired entering column. (In fact, the N enumeration trees can be combined into one. We leave the details for problem 13.9.) Secondly, if at most v_k raws of width w_k can be cut ($k = 1,2, \ldots, N$), then we enlarge our set of constraints by N additional inequalities: the kth of them reads $\sum x_j \leq v_k$, with the summation running through all the variables x_j in the kth group. Now each basic solution includes $m + N$ rather than m basic variables; in step 2 of each simplex iteration, we compare z_k^* against $c_k - y_{m+k}$ rather than c_k. (The product $\mathbf{ya}$ equals $z_k^* + y_{m+k}$ in this case.) Finally, if the number of cutting knives is limited, then the enumeration trees can be easily modified to explore only the admissible patterns.

EXPERIMENTAL RESULTS

The second paper by P. C. Gilmore and R. E. Gomory (1963) on the cutting-stock problem contains a wealth of experimental findings. The results estimate the expected difficulty of a typical cutting-stock problem, provide a basis for comparing different variations on the algorithm, and suggest the likely effect of certain policy rules on the total amount of waste in a paper mill. Our summary will be brief.

The test problems solved by Gilmore and Gomory were made up by amalgamating and selecting from a class of paper industry problems. In the basic set of 20 problems, there were raw rolls of only one size, around 200 in. or less. The number of different final widths ranged from 20 to 40. These widths, generally between 20 in. and 80 in., were specified to $\frac{1}{4}$ in. The number of cutting knives was limited to five, seven, or nine.

The *average* number of simplex iterations for these problems came to about 130. However, the *individual* iteration counts showed a large variance over the sample, from 20 to more than 300. Such variance is quite common: linear programming problems looking quite alike often behave very differently from each other when subjected to the simplex method. As might be expected, problems with smaller numbers of final widths tended to require fewer iterations. Nevertheless, this trend was rather erratic; for example, the average number of iterations required by the 35-final problems was 197, whereas the corresponding figure for the 40-final problems was only 161. (No procedure for finding good initial basic solutions was used; the simplex method seems to find a fairly good fractional solution in the first few iterations.)

Some of the questions answered by Gilmore and Gomory concerned the criterion for selecting a new cutting pattern. For instance, one might either accept the first encountered pattern with $\sum y_i a_i > 1$ or insist on the pattern maximizing $\sum y_i a_i$. Which of the two methods is better? It seems plausible that the first method leads to a larger number of iterations, but it is also clear that the time per iteration is smaller for this method. Without experimental findings, the outcome of the trade-off between the total number of iterations and the time per individual iteration might have been difficult to predict. The Gilmore–Gomory results show that finding the true maximum in the knapsack problem is well worth the effort. In all but one of the 20 problems, the *overall* time required by the second method was smaller. In fact, the average amount of time per problem required by the second method was about half the corresponding figure for the first method.

In terms of Chapter 4, the second method is nothing but the largest-coefficient pivoting rule, whereas the first method is a relative of the smallest-subscript rule. Thus it is only natural to speculate about the largest-improvement rule in the cutting-stock context. A straightforward implementation of this rule would require enumerating an enormous number of different cutting patterns in each iteration, in which case the means would defeat the purpose. However, there is a natural way of mimicking the largest-improvement rule and, at the same time, actually *reducing* the computational effort per iteration. The idea is simple. Some of the final widths are ordered in very small quantities. If our new cutting pattern yields any of these low-demand widths, then the pattern is bound to be used only a few times. Consequently, the resulting improvement won't be very impressive. Led by this observation, Gilmore and Gomory suggested the use of the *median method.* With this method, the final widths are divided into two equally large groups: the low-demand group and the high-demand group. Then, at every second iteration, the cutting pattern is chosen that (i) yields only the high-demand finals and (ii) maximizes the improvement rate among all such patterns. (Note that the corresponding knapsack problem, involving only half of the original variables, is easier to solve.) The median method turned out to be faster in 13 of the 20 basic test problems; furthermore, the problems on which the performance of the median method was worse than that of the original method were mostly the problems that required a small amount of time anyway. The average running time per problem was cut down to about 40% of the original figures.

In each particular instance of the cutting-stock problem, a certain percentage of the paper consumed in the form of raw rolls is directed to the finals, and the rest goes to waste. In the test problems, the waste percentage seems to have been unpredictable, ranging from as little as 0.1% to as much as 10%. An interesting observation is that the high-waste problems were generally solved faster than the low-waste problems. In a typical low-waste problem, the waste dropped sharply in the beginning and then proceeded to creep slowly towards the optimum: although new patterns kept coming up, each led to just a minute improvement. For this reason, Gilmore and Gomory suggested that the calculations be stopped as soon as the waste dropped by less than 0.1% over 10 iterations. The results were encouraging: in certain cases, the running time decreased by 90% or more, while the amount of waste due to premature termination kept always below 0.5%.

Finally, having *several* raw widths available clearly cannot hurt: as new cutting patterns appear, the waste is likely to diminish. To find out how fast the new raw widths drive the waste towards zero, Gilmore and Gomory solved five of their problems (a high-waste group) several times, with varying spectra of raw widths. For a single raw width of 168 in., the average waste was over 7%; this figure dropped to only 1.4% when additional widths of 145 in., 140 in., and 124 in. were also made available. Corresponding running times ranged from 144% to 211% of the original figures. (In this context, it may be interesting to note that human planners usually find the complexity of problems with several raw widths far more bewildering than that of problems with a single raw width.) On the other hand, the limitation of the number of cutting knives did not matter much in the test problems. With that limitation removed, the optimal waste percentage remained unchanged in 19 of the 20 problems (even though the actual solutions did change). □

PROBLEMS

13.1 Suppose that two subscripts i and j in the knapsack problem (13.5) satisfy $ka_i \leq a_j$ and $kc_i \geq c_j$ for some positive integer k. Show that at least one optimal solution has $x_j = 0$.

△ **13.2** (Borrowed and adapted from a report written by M. H. Belz and A. G. Doig for Australian Paper Manufacturers Ltd.) Solve the following cutting stock problems.

a. Raw width 90 in., order summary calling for

3 finals of width 60 in.
21 finals of width 30 in.
94 finals of width $25\frac{1}{2}$ in.
50 finals of width 20 in.
288 finals of width $17\frac{1}{4}$ in.
178 finals of width 15 in.
112 finals of width $12\frac{3}{4}$ in.
144 finals of width 10 in.

b. (Quite difficult.) Raw width 181 in., order summary calling for

90 finals of width $21\frac{5}{8}$ in.
51 finals of width $20\frac{1}{2}$ in.
45 finals of width 20 in.
11 finals of width $17\frac{1}{4}$ in.

13.3 Consider the cutting-stock problem with raw width 100 and the order summary calling for

600 finals of width 52 in.
600 finals of width 29 in.
600 finals of width 27 in.
1,200 finals of width 21 in.

Show that the optimal solution uses 900 raws, whereas the solution produced by first-fit decreasing uses 1,100 raws.

13.4 Let $a_1, a_2, \ldots, a_m$ and $c_1, c_2, \ldots, c_m$ be fixed positive integers. For each nonnegative integer k, let $z(k)$ denote the maximum of

$$\sum_{i=1}^{m} c_i x_i$$

subject to

$$\sum_{i=1}^{m} a_i x_i \leq k$$

$$x_i = \text{nonnegative integer} \qquad (i = 1, 2, \ldots, m).$$

Assuming $k \geq \min a_j$, prove that

$$z(k) = \max(c_i + z(k - a_i)) \tag{13.15}$$

with the maximum taken over all the subscripts i such that $a_i \leq k$.

13.5 Use the result of problem 13.4 to design an algorithm for solving a knapsack problem by successive evaluations of $z(k)$, $k = 0, 1, \ldots, b$. [*Hint*: For each k, keep track of a subscript i, which attains the maximum in (13.15).]

13.6 Let $i(k)$ denote the smallest subscript attaining the maximum in (13.15). Prove: If $i = i(k)$, then $i \leq i(k - a_i)$. How can this fact be used to reduce the number of additions required to evaluate $z(k)$?

13.7 Prove: If $i(k) = 1$ for all $k = n + 1, n + 2, \ldots, n + \max a_i$ and some integer n, then $i(k) = 1$ for all $k > n$. How can this fact be used to streamline the algorithm of problem 13.5?

13.8 Assuming that $c_1/a_1 > c_2/a_2 \geq \cdots \geq c_m/a_m$, prove that $i(k) = 1$ whenever

$$k \geq \frac{c_1}{(c_1/a_1) - (c_2/a_2)}.$$

13.9 Design an algorithm that, given the N knapsack problems (13.14) along with numbers $c_1, c_2, \ldots, c_N$, uses only one (pruned) enumeration tree to find a subscript k and an optimal solution $a_1, a_2, \ldots, a_m$ of the k-th problem (13.14) that maximize $\sum w_i a_i - c_k$.

14

Approximating Data by Linear Functions

The subject of this chapter is finding approximate solutions to possibly unsolvable systems of linear equations. On the intuitive level, an *approximate* solution of

$$\sum_{j=1}^{n} a_{ij}x_j = b_i \qquad (i = 1, 2, \ldots, m) \tag{14.1}$$

is the next best thing to an *exact* solution: it may not make every left-hand side equal to the corresponding right-hand side, but at least it keeps the discrepancies reasonably small. As we shall see, the notion of the *best* approximate solution is more elusive: several criteria for ranking approximate solutions come to mind, and these criteria are not always consistent with each other. Under two criteria that seem very natural, the problem of finding the best approximate solution of (14.1) can be seen as a linear programming problem.

The reader may wonder why anybody should try to "solve" a system of equations that has no solution. One reason might be that, in some applications, the unsolvable system arises from a solvable one whose coefficients have been distorted. One of the many settings in which this pattern occurs will be discussed in some detail.

QUANTITATIVE SPECTROPHOTOMETRY: A MOTIVATION

The amount of light penetrating through layers of various liquids decreases with the thickness of the layers. For example, suppose that a layer of unit thickness absorbs 20% of the light, letting the remaining 80% through. How much light will be absorbed by a layer twice as thick? The layer of double thickness may be thought of as two layers of thickness 1, one on top of the other. In this configuration, the first layer will let through 80% of the light, and the second layer will let through 80% of the 80%. Altogether, the layer of thickness 2 will transmit only 64% of the light. By the same token, a layer of triple thickness will transmit 51.2%, and so on. This is the law formulated in 1729 by P. Bouguer. In addition, increasing the concentration of the light-absorbing substance has the same effect on the proportion of light actually transmitted as increasing the thickness of the sample by the same factor. This is the law formulated in 1852 by A. Beer. In order to make these two laws assume a simple form, we define the *absorbance* of a sample to be the logarithm, to the base 10, of the reciprocal of the proportion of light transmitted by the sample. Quite simply, a sample transmitting 10% of light has absorbance 1, a sample transmitting 1% of light has absorbance 2, and so on. (Of course, typical everyday samples transmit more light than that.) The laws of Bouguer and Beer, then, state that the absorbance is directly proportional to the thickness of a sample and to its concentration. In order to normalize the absorbance figures, we divide them by the product of the two factors. The resulting figure, called *absorptivity*, is independent of both the thickness and the concentration of the sample.

What makes absorptivity useful is its dependence on the wavelength of the light. A graph of absorptivity plotted against the wavelength is referred to as an *absorption spectrum*. Different substances have different absorption spectra, with characteristic peaks at varying wavelengths. Hence, an absorption spectrum is something of a signature for a substance. The absorptivities of five organic compounds at 12 different wavelengths are shown in Table 14.1. They are normalized with respect to molecular weights, and are therefore called *molecular absorptivities*.

Table 14.1 Absorptivities of Five Organic Compounds

Wavelength (nanometers)*	Adenine	Cytosine	Guanine	Thymine	Uracil
220.5	4,243	8,673	5,670	5,866	3,356
230.5	3,301	6,774	6,189	2,284	2,048
240.5	6,049	5,087	9,970	2,857	4,097
250.5	10,510	4,655	10,220	5,401	7,106
260.5	13,410	6,024	7,198	7,692	8,243
270.5	8,561	6,150	7,905	7,302	5,377
280.5	1,569	3,178	7,432	3,902	1,267
290.5	125	361	3,512	741	103
300.5	77	64	802	72	54
310.5	47	27	95	33	38
320.5	33	25	38	14	15
330.5	38	11	16	33	5

* 1 nanometer = 10^{-9} m = 1 millimicron.

Note: These data come from a list of absorbancies measured at more than 130 different wavelengths. They were kindly supplied by G. Barth, E. Bunnenberg, and R. E. Linder of Stanford University.

A law of fundamental importance for quantitative analyses is the *combined Beer–Bouguer's law* for mixtures of independently absorbing components. Let us consider n components such that the absorptivity of component j at wavelength i equals a_{ij}. If the jth component is present with concentration x_j, then, according to the combined Beer–Bouguer's law, the absorbance of the mixture per unit thickness and at wavelength i equals $\sum a_{ij}x_j$. In short, the absorbancies are *additive*. The use of this fact is evident: having measured the absorbancies b_i of the mixture at different wavelengths ($i = 1, 2, \ldots, m$), and having looked up the absorptivities a_{ij} of various components in a catalogue, we may now find the unknown concentrations x_j by solving the system

$$\sum_{j=1}^{n} a_{ij}x_j = b_i \qquad (i = 1, 2, \ldots, m). \tag{14.2}$$

Unfortunately, things are not always that smooth. To begin, there are experimental errors: for example, it may be difficult to set the instrument precisely at the required wavelength or it may be difficult to measure the absorbancies precisely, especially if they are very small. Also, in addition to the substances whose presence in the mixture is suspected, a number of unknown impurities may affect the absorption spectrum of the mixture. Finally, the absorbancies may be not quite additive; like many other laws of the natural sciences, the combined Beer–Bouguer's law is only a first approximation to the unwieldy behavior of the world. All of these distortions conspire to make the system (14.2) unsolvable, especially when m is much larger than n.

The traditional way of getting around this difficulty brings to mind an ostrich with its head stuck in the sand: one takes into account only n of the m available wavelengths. Of course, the accuracy of the result obtained by solving the corresponding n equations in n variables depends very much on the choice of the n wavelengths. To minimize the relative importance of experimental errors, one should choose those wavelengths for which absorbancies are large. In addition, it is important to choose the wavelengths in a range where the spectra of the components have as different shapes as possible. Finally, if the mixture is known not to adhere to the Beer–Bouguer's law in a certain range of wavelengths, then these wavelengths should be avoided.

For illustration, we shall consider a mixture of adenine, cytosine, guanine, thymine, and uracil present in unknown concentrations. Its absorbance per unit thickness is tabulated in the following chart.

Wavelength	Absorbance
220.5	644.4
230.5	551.7
240.5	921.3
250.5	1,153
260.5	1,044
270.5	882.8
280.5	565.5
290.5	238.1
300.5	63.2
310.5	21.3
320.5	15.4
330.5	11.2

Restricting ourselves to the first five wavelengths, we arrive at the system

$$\begin{aligned}
4{,}243x_1 + 8{,}673x_2 + 5{,}670x_3 + 5{,}866x_4 + 3{,}356x_5 &= 644.4\\
3{,}301x_1 + 6{,}774x_2 + 6{,}189x_3 + 2{,}284x_4 + 2{,}048x_5 &= 551.7\\
6{,}049x_1 + 5{,}087x_2 + 9{,}970x_3 + 2{,}857x_4 + 4{,}097x_5 &= 921.3\\
10{,}510x_1 + 4{,}655x_2 + 10{,}220x_3 + 5{,}401x_4 + 7{,}106x_5 &= 1{,}153\\
13{,}410x_1 + 6{,}024x_2 + 7{,}198x_3 + 7{,}692x_4 + 8{,}243x_5 &= 1{,}044
\end{aligned}$$

whose solution is

$$x_1 \doteq -0.016, \quad x_2 \doteq 0.005, \quad x_3 \doteq 0.060, \quad x_4 \doteq 0.001, \quad x_5 \doteq 0.096.$$

Since negative concentrations don't make sense, the value of x_1 seems to suggest that adenine is absent from the mixture. Of course, our choice of wavelengths was quite arbitrary. Had we chosen 230.5, 240.5, 250.5, 260.5, and 270.5 instead, we would have ended up with

$$x_1 \doteq -0.016, \quad x_2 \doteq 0.006, \quad x_3 \doteq 0.059, \quad x_4 \doteq -0.002, \quad x_5 \doteq 0.098.$$

Similarly, the wavelengths of 240.5, 250.5, 260.5, 270.5, and 280.5 yield

$$x_1 \doteq 0.014, \quad x_2 \doteq -0.006, \quad x_3 \doteq 0.066, \quad x_4 \doteq 0.002, \quad x_5 \doteq 0.049$$

and the wavelengths of 250.5, 260.5, 270.5, 280.5, and 290.5 lead to

$$x_1 \doteq 0.010, \quad x_2 \doteq -0.007, \quad x_3 \doteq 0.066, \quad x_4 \doteq 0.003, \quad x_5 \doteq 0.054.$$

Even though we have restricted our choice of wavelengths to the range where the absorbancies are reasonably large, the results are not terribly consistent. It seems more logical to consider the system of 12 equations in five variables as a whole, and then declare the best of its approximate solutions to be the result of the spectrophotometric analysis. That line of attack brings us back to the fundamental question: Which of the many different approximate solutions is the best? □

JUDGING THE GOODNESS OF FIT

Given numbers $x_1^*, x_2^*, \ldots, x_n^*$ and a system

$$\sum_{j=1}^{n} a_{ij}x_j = b_i \qquad (i = 1, 2, \ldots, m)$$

consider the numbers

$$e_i = b_i - \sum_{j=1}^{n} a_{ij}x_j^* \qquad (i = 1, 2, \ldots, m).$$

Clearly, each e_i represents the absolute error with which $x_1^*, x_2^*, \ldots, x_n^*$ fail to satisfy the ith equation: an exact solution would make each $e_i = 0$. If the system is unsolvable, then for every choice of $x_1^*, x_2^*, \ldots, x_n^*$ some of the numbers e_i will be nonzero; good approximate solutions will make these errors small. We shall agree that only the *magnitudes* $|e_i|$ of the errors matter, not their *signs*. For instance, the approximate solution $x_1^* = 2$, $x_2^* = 3$ fits the system

$$\begin{aligned} 3x_1 + x_2 &= 11 \\ x_1 - 2x_2 &= -1 \\ x_1 + x_2 &= 6 \end{aligned}$$

just as well as $x_1^* = 4$, $x_2^* = 1$: whereas the errors of the former are $e_1 = 2$, $e_2 = 3$, $e_3 = 1$, the errors of the latter are $e_1 = -2$, $e_2 = -3$, $e_3 = 1$.

To illustrate the difficulty of selecting the "best" approximate solution, we shall consider the system

$$\begin{aligned} x_1 + x_2 &= 83 \\ 3x_1 - x_2 &= 125 \\ 2x_1 + 7x_2 &= 310 \\ x_1 - 2x_2 &= -7 \\ 3x_1 + 2x_2 &= 215 \end{aligned} \tag{14.3}$$

with three approximate solutions:

(i) $x_1^* = 51,\quad x_2^* = 30$

(ii) $x_1^* = 52,\quad x_2^* = 30$

(iii) $x_1^* = 52,\quad x_2^* = 29.$

The corresponding errors are shown in Table 14.2.

Table 14.2 Errors of Three Solutions

	(i)	(ii)	(iii)
e_1	2	1	2
e_2	2	−1	−2
e_3	−2	−4	3
e_4	2	1	−1
e_5	2	−1	1

In a way, solution (i) seems to be the best of the three: the worst error it makes has magnitude 2, whereas the corresponding figures for solutions (ii) and (iii) have magnitudes 4 and 3. This judgment relies implicitly on the following criteria.

Criterion 1. The goodness of fit is measured by the worst error.
Note that this criterion discriminates against solutions such as solution (ii), with a very good fit for most of the equations spoiled by only one large error. Perhaps the *average* error would be a better measure.

Criterion 2. The goodness of fit is measured by the average error.
Whereas the first criterion ranked the three solutions from best to worst as (i), (iii), (ii), the second criterion ranks them in the opposite order: the average errors are 2 for (i), 1.6 for (ii), and 1.8 for (iii). The criterion most popular in practice is neither of these two; it is a criterion which may be found to be rather artificial.

Criterion 3. The goodness of a fit is measured by the average *squared* error.
In our example, the average squared errors are

$$\frac{1}{5}(4 + 4 + 4 + 4 + 4) = 4 \qquad \text{for (i)}$$

$$\frac{1}{5}(1 + 1 + 16 + 1 + 1) = 4 \qquad \text{for (ii)}$$

$$\frac{1}{5}(4 + 4 + 9 + 1 + 1) = 3.8 \qquad \text{for (iii)}$$

and so solution (iii) ranks best under the third criterion.

Let us put our three criteria in a broader mathematical context. In order to declare one of the approximate solutions $\mathbf{x}^*$ of a system $\mathbf{Ax} = \mathbf{b}$ the best fitting, we have to declare the vector $\mathbf{e} = \mathbf{b} - \mathbf{Ax}^*$ of its errors the smallest. For this purpose, we assign to every vector $\mathbf{e}$ a nonnegative number that is a measure of its magnitude. Our three criteria are based on the measures

$$\max|e_i|, \quad \frac{1}{m}\sum_{i=1}^{m}|e_i| \quad \text{and} \quad \frac{1}{m}\sum_{i=1}^{m}|e_i|^2$$

respectively. When $p \geq 1$, the number

$$\left(\sum_{i=1}^{m}|e_i|^p\right)^{1/p}$$

is called the L_p*-norm* of the vector $\mathbf{e}$ and is denoted by $\|\mathbf{e}\|_p$. Since

$$\lim_{p\to\infty}\|\mathbf{e}\|_p = \max|e_i| \tag{14.4}$$

for every vector $\mathbf{e}$, the right-hand side of (14.4) is called the L_∞*-norm* of $\mathbf{e}$ and is denoted by $\|\mathbf{e}\|_\infty$. [A proof of (14.4) is left for problem 14.4.] Note that criterion 1 directs us to judge the goodness of fit by the L_∞-norm of the error vector. Similarly, criterion 2 is based on the L_1-norm, and criterion 3 is based on the L_2-norm. More generally, every L_p-norm may serve as a measure of vector magnitude. Accordingly, the best L_p*-approximation* of a solution of $\mathbf{Ax} = \mathbf{b}$ is a vector $\mathbf{x}^*$ minimizing $\|\mathbf{b} - \mathbf{Ax}^*\|_p$. We shall, however, restrict ourselves only to the cases $p = 1$, $p = 2$, and $p = \infty$.

Before speculating on the comparative virtues of these three criteria, let us get better acquainted with the nature of the approximate solutions to which they lead.

The differences between the three reveal themselves clearly even for simple systems such as

$$
\begin{aligned}
x &= 24.5\\
x &= 26.25\\
x &= 24.25\\
x &= 24.75\\
x &= 24
\end{aligned}
\tag{14.5}
$$

It seems unnecessarily formal to speak of "best approximate solutions" of (14.5). Quite simply, our task is to replace the five numbers 24.5, 26.25, 24.25, 24.75, and 24 by a single representative value. (The five numbers could be the results of a repeated experiment, the selling prices of the same stock on five consecutive days, or the amounts of a protein found in five different samples of canned tuna.) More generally, we shall consider systems

$$x = b_i \qquad (i = 1, 2, \ldots, m). \tag{14.6}$$

It will do no harm to assume that the right-hand sides are ordered from the smallest to the largest,

$$b_1 \leq b_2 \leq \cdots \leq b_m.$$

Three kinds of values often used to represent the b_i's are the *median*, the *mean*, and the *midrange*. The median is the b_i that appears in the middle of the list: if $m = 2k + 1$, then the median is unequivocally b_{k+1}. (If $m = 2k$, then the definition becomes a little ambiguous: either b_k or b_{k+1}, or any value between the two, might be called a median.) The mean is the ordinary average, $(\sum b_i)/m$, and the midrange is the midpoint $(b_1 + b_m)/2$ between the two extreme values b_1 and b_m.

We propose to show that the median is the best L_1 approximation, the mean is the best L_2 approximation, and the midrange is the best L_∞ approximation to a solution of (14.6). The proof is simple. Its first part is based on the fact that $|t| \geq t$, $|t| \geq 0$, and $|t| \geq -t$ for every number t; in particular, $|b_i - x| \geq b_i - x$, $|b_i - x| \geq 0$, and $|b_i - x| \geq x - b_i$ for each $i = 1, 2, \ldots, m$. Now, in case $m = 2k + 1$, we have

$$\sum_{i=1}^{m} |b_i - x| \geq \sum_{i=1}^{k} (x - b_i) + \sum_{i=k+2}^{2k+1} (b_i - x) = -\sum_{i=1}^{k} b_i + \sum_{i=k+2}^{2k+1} b_i$$

for every x, with equality when x is the median. Similarly, if $m = 2k$, then

$$\sum_{i=1}^{m} |b_i - x| \geq \sum_{i=1}^{k} (x - b_i) + \sum_{i=k+1}^{2k} (b_i - x) = -\sum_{i=1}^{k} b_i + \sum_{i=k+1}^{2k} b_i$$

for every x, with equality when x is a median. To prove the second assertion, note that for any two numbers x and x^* we have

$$\sum_{i=1}^{m} (b_i - x)^2 = \sum_{i=1}^{m} ((b_i - x^*)^2 + 2(b_i - x^*)(x^* - x) + (x^* - x)^2)$$

$$\geq \sum_{i=1}^{m} (b_i - x^*)^2 + 2(x^* - x) \sum_{i=1}^{m} (b_i - x^*).$$

If x^* is the mean, then

$$\sum_{i=1}^{m} (b_i - x^*) = \left(\sum_{i=1}^{m} b_i\right) - mx^* = 0$$

and so

$$\sum_{i=1}^{m} (b_i - x)^2 \geq \sum_{i=1}^{m} (b_i - x^*)^2$$

for every x. The third assertion is nearly trivial. If x is at most the midrange, then

$$\max|b_i - x| = b_m - x \geq (b_m - b_1)/2.$$

If x is at least the midrange, then

$$\max|b_i - x| = x - b_1 \geq (b_m - b_1)/2.$$

The proof is completed.

Note that in (14.5) the second of the five right-hand sides lies conspicuously far from the cluster formed by the remaining four. The midrange 25.125 is affected by this outlier much more than the mean 24.75, and the mean itself is affected by the outlier more than the median 24.5. In fact, the median would remain unchanged even if 26.25 were replaced by an outrageously large number, say, 300. The general insensitivity of the median to the outliers is referred to as its *robustness*. In this terminology, the mean is less robust than the median, and the midrange is by far the least robust of the three.

The question as to which criterion should be adopted in looking for the "best" approximate solution is tricky and often ill-posed. Of our three criteria, minimizing the L_1-norm leads to the most robust answer, whereas minimizing the L_∞-norm avoids gross discrepancies with the data as much as possible. Which of the two is preferable depends on the source of the data, the nature of the distortions that make the system unsolvable, and the intended use of the approximate solution. In many practical applications, the distortions arise from superpositions of numerous small errors and other unaccounted-for factors. In such cases, the statistical distribution of the distortions tends to follow a shape known as the *normal* or *Gaussian* distribution. It can be proved that, as long as the distortions are distributed in this fashion, the L_2-approximations are best in a sense. To put it crudely, they have the best chance of getting close to the "true answer." This fact accounts for their widespread popularity.

In different contexts, however, the distribution of errors may be far from Gaussian. For example, the fluctuations of stock market prices do not adhere well to the Gaussian shape: large fluctuations tend to come up far too frequently for that. B. Mandelbrot (1963, 1966, 1967) suggested that "fatter-tailed" distributions, with a greater likelihood of outliers, may provide a better fitting description of these fluctuations (or, rather, their logarithms). Mandelbrot's hypothesis was supported by an experimental work of E. P. Fama (1965). Hence, for purposes such as estimating the rates of return of stocks, robust estimates such as L_1-approximations seem to be safer than L_2-approximations. [Nevertheless, the differences between the two may not be all that significant; for the results of an empirical study, see W. F. Sharpe (1971).] For additional information on this subject, the reader is referred to R. Blattberg and T. Sargent (1971). Applications of L_1 approximations in geophysical data analysis have been suggested by J. F. Claerbout and F. Muir (1973). The use of L_1 approximations in absorption spectroscopy has been advocated by W. C. White et al. (1963); see also A. W. Pratt et al. (1964).

COMPUTING THE BEST APPROXIMATE SOLUTIONS

Foregoing the argument as to *why* L_1 and L_∞ approximations should (or should not) be computed, we now restrict ourselves to showing *how* they can be computed. Our interest stems from the fact that both of these problems can be seen as linear programming problems.

Computations of best L_2-approximations to solutions of

$$\sum_{j=1}^{n} a_{ij}x_j = b_i \qquad (i = 1, 2, \ldots, m) \tag{14.7}$$

fall outside this framework, and we comment on them only briefly. The L_2 problem is to

$$\text{minimize} \quad \sum_{i=1}^{m} \left(b_i - \sum_{j=1}^{n} a_{ij}x_j \right)^2$$

subject to no constraints whatsoever. As a reader with a minimal background in calculus will recall, the minimizing solution $x_1, x_2, \ldots, x_n$ must satisfy

$$\frac{\partial}{\partial x_k} \sum_{i=1}^{m} \left(b_i - \sum_{j=1}^{n} a_{ij}x_j \right)^2 = 0 \qquad \text{for all} \quad k = 1, 2, \ldots, n,$$

which may be written as

$$\sum_{j=1}^{n} \left(\sum_{i=1}^{m} a_{ik}a_{ij} \right) x_j = \sum_{i=1}^{m} b_i a_{ik} \qquad (k = 1, 2, \ldots, n). \tag{14.8}$$

For example, if (14.3) replaces (14.7) then (14.8) becomes

$$\begin{aligned} 24x_1 + 16x_2 &= 1716 \\ 16x_1 + 59x_2 &= 2572 \end{aligned}$$

and so $x_1 \doteq 51.80$, $x_2 \doteq 29.54$ is the best L_2-approximation to a solution of (14.3). [For "most" systems (14.7), the system (14.8) has a unique solution: see problem 14.5.]

For the L_1-approximations, the problem is to

$$\text{minimize} \quad \sum_{i=1}^{m} \left| b_i - \sum_{j=1}^{n} a_{ij}x_j \right|$$

subject to no restrictions whatsoever. Even though this problem does not look like an LP problem, it may be converted into one quite easily. Consider the problem

$$\begin{aligned} &\text{minimize} && \sum_{i=1}^{m} e_i \\ &\text{subject to} && e_i + \sum_{j=1}^{n} a_{ij}x_j \geq b_i \quad (i = 1, 2, \ldots, m) \\ &&& e_i - \sum_{j=1}^{n} a_{ij}x_j \geq -b_i \quad (i = 1, 2, \ldots, m). \end{aligned} \tag{14.9}$$

It is not difficult to see that in an optimal solution $e_1^*, e_2^*, \ldots, e_m^*, x_1^*, x_2^*, \ldots, x_n^*$ of (14.9), each e_i^* is slammed down against the larger of the two lower bounds $b_i - \sum a_{ij}x_j^*$ and $-b_i + \sum a_{ij}x_j^*$. Hence

$$e_i^* = \max \left\{ \begin{array}{l} b_i - \sum_{j=1}^{n} a_{ij}x_j^* \\ -b_i + \sum_{j=1}^{n} a_{ij}x_j^* \end{array} \right\} = \left| b_i - \sum_{j=1}^{n} a_{ij}x_j^* \right|$$

and the objective function $\sum e_i$ takes the value of

$$\sum_{i=1}^{m} \left| b_i - \sum_{j=1}^{n} a_{ij}x_j^* \right|$$

as desired. We conclude that $x_1^*, x_2^*, \ldots, x_n^*$ is a best L_1 approximation to a solution of (14.7).

A similar trick applies to L_∞ approximations: the problem

$$\text{minimize} \quad \max_i \left| b_i - \sum_{j=1}^{n} a_{ij}x_j \right|$$

gets converted into

$$\text{minimize} \quad z$$

$$\text{subject to} \quad z + \sum_{j=1}^{n} a_{ij}x_j \geq b_i \quad (i = 1, 2, \ldots, m)$$
$$z - \sum_{j=1}^{n} a_{ij}x_j \geq -b_i \quad (i = 1, 2, \ldots, m). \tag{14.10}$$

In every optimal solution $z^*, x_1^*, x_2^*, \ldots, x_n^*$ of (14.10), the number z^* is slammed down against the largest of the $2m$ lower bounds. Hence

$$z^* = \max_i \left| b_i - \sum_{j=1}^{n} a_{ij}x_j^* \right|$$

and $x_1^*, x_2^*, \ldots, x_n^*$ is a best L_∞ approximation to a solution of (14.7).

In typical applications, m is much larger than n: one tries to make up for inaccuracies by collecting as much data as possible. Consequently, if the revised simplex method is applied directly to (14.9) or (14.10), then the size $2m \times 2m$ of the two systems solved in each iteration may be considerable. Fortunately, this size can be drastically reduced. The key to the reduction is the observation that every LP problem may be solved by applying the simplex method to its dual. To begin, note that the dual of (14.9) reads

$$\begin{aligned} \text{maximize} \quad & \sum_{i=1}^{m} b_i v_i - \sum_{i=1}^{m} b_i w_i \\ \text{subject to} \quad & v_i + w_i = 1 \quad (i = 1, 2, \ldots, m) \\ & \sum_{i=1}^{m} a_{ij}v_i - \sum_{i=1}^{m} a_{ij}w_i = 0 \quad (j = 1, 2, \ldots, n) \\ & v_i, w_i \geq 0 \quad (i = 1, 2, \ldots, m). \end{aligned}$$

Substituting $1 - v_i$ for each w_i, we reduce the dual into the form

$$\begin{aligned} \text{maximize} \quad & -\sum_{i=1}^{m} b_i + 2\sum_{i=1}^{m} b_i v_i \\ \text{subject to} \quad & -\sum_{i=1}^{m} a_{ij} + 2\sum_{i=1}^{m} a_{ij}v_i = 0 \quad (j = 1, 2, \ldots, n) \\ & 0 \leq v_i \leq 1 \quad (i = 1, 2, \ldots, m) \end{aligned}$$

or, after simplification,

$$\begin{aligned} \text{maximize} \quad & \sum_{i=1}^{m} b_i v_i \\ \text{subject to} \quad & \sum_{i=1}^{m} a_{ij}v_i = \frac{1}{2}\sum_{i=1}^{m} a_{ij} \quad (j = 1, 2, \ldots, n) \\ & 0 \leq v_i \leq 1 \quad (i = 1, 2, \ldots, m). \end{aligned} \tag{14.11}$$

Solving (14.11) by the revised simplex method is relatively easy; the two systems solved in each iteration have size only $n \times n$. The revised simplex method finds not only an optimal solution $v_1^*, v_2^*, \ldots, v_m^*$ of (14.11) but also numbers $y_1, y_2, \ldots, y_n$ such that

$$v_i^* = 0 \qquad \text{whenever} \quad \sum_{j=1}^{n} a_{ij} y_j > b_i \tag{14.12}$$

and

$$v_i^* = 1 \qquad \text{whenever} \quad \sum_{j=1}^{n} a_{ij} y_j < b_i. \tag{14.13}$$

We claim that the numbers $y_1, y_2, \ldots, y_n$ constitute the best L_1 approximation to a solution of (14.7). To put it differently, we claim that

$$\sum_{i=1}^{m} \left| b_i - \sum_{j=1}^{n} a_{ij} x_j \right| \geq \sum_{i=1}^{m} \left| b_i - \sum_{j=1}^{n} a_{ij} y_j \right| \tag{14.14}$$

for every choice of $x_1, x_2, \ldots, x_n$. Even though this claim follows easily from considerations involving duality, we shall provide a direct proof from scratch. Since $|t| \geq t$ and $|t| \geq -t$ for every t, we have

$$|t| \geq vt + (1 - v)(-t) = (2v - 1)t$$

whenever $0 \leq v \leq 1$. In particular,

$$\left| b_i - \sum_{j=1}^{n} a_{ij} x_j \right| \geq (2v_i^* - 1)\left(b_i - \sum_{j=1}^{n} a_{ij} x_j \right)$$

for all $i = 1, 2, \ldots, m$; thus the left-hand side of (14.14) is bounded from below by

$$\sum_{i=1}^{m} (2v_i^* - 1)\left(b_i - \sum_{j=1}^{n} a_{ij} x_j \right) = \sum_{i=1}^{m} (2v_i^* - 1) b_i.$$

[The last equality follows from the fact that the numbers $v_1^*, v_2^*, \ldots, v_m^*$ constitute a feasible solution of (14.11).] On the other hand, (14.12) and (14.13) guarantee that

$$\left| b_i - \sum_{j=1}^{n} a_{ij} y_j \right| = (2v_i^* - 1)\left(b_i - \sum_{j=1}^{n} a_{ij} y_j \right)$$

for all $i = 1, 2, \ldots, m$. Hence the right-hand side of (14.14) equals

$$\sum_{i=1}^{m} (2v_i^* - 1)\left(b_i - \sum_{j=1}^{n} a_{ij} y_j \right) = \sum_{i=1}^{m} (2v_i^* - 1) b_i.$$

For instance, to find a best L_1 approximation to a solution of (14.3), we consider the LP problem

$$\begin{array}{ll} \text{maximize} & 83v_1 + 125v_2 + 310v_3 - 7v_4 + 215v_5 \\ \text{subject to} & v_1 + 3v_2 + 2v_3 + v_4 + 3v_5 = 5 \\ & v_1 - v_2 + 7v_3 - 2v_4 + 2v_5 = 3.5 \\ & 0 \le v_1, v_2, v_3, v_4, v_5 \le 1. \end{array}$$

Along with the optimal solution

$$v_1^* = 1, \quad v_2^* = 0, \quad v_3^* = \frac{5}{22}, \quad v_4^* = \frac{6}{11}, \quad v_5^* = 1,$$

the revised simplex method delivers the numbers

$$y_1 = \frac{571}{11} \doteq 51.91 \quad \text{and} \quad y_2 = \frac{324}{11} \doteq 29.45$$

which constitute the desired approximation.

Similarly, (14.10) can be solved easily by applying the revised simplex method to to its dual,

$$\begin{array}{lll} \text{maximize} & \sum_{i=1}^{m} b_i v_i - \sum_{i=1}^{m} b_i w_i & \\ \text{subject to} & \sum_{i=1}^{m} a_{ij} v_i - \sum_{i=1}^{m} a_{ij} w_i = 0 & (j = 1, 2, \ldots, n) \\ & \sum_{i=1}^{m} v_i + \sum_{i=1}^{m} w_i = 1 & \\ & v_i, w_i \ge 0 & (i = 1, 2, \ldots, m). \end{array}$$

The two systems solved in each iteration have size only $(n + 1) \times (n + 1)$; upon termination, the method delivers not only an optimal solution of the dual but also numbers $y_1, y_2, \ldots, y_{n+1}$ such that

$$\sum_{j=1}^{n} a_{ij} y_j + y_{n+1} \ge b_i \qquad (i = 1, 2, \ldots, m) \tag{14.15}$$

and

$$-\sum_{j=1}^{n} a_{ij} y_j + y_{n+1} \ge -b_i \qquad (i = 1, 2, \ldots, m) \tag{14.16}$$

with equality in (14.15) whenever $v_i^* > 0$, and with equality in (14.16) whenever $w_i^* > 0$. We claim that the numbers $y_1, y_2, \ldots, y_n$ constitute a best L_∞ approximation to a solution of (14.7). To put it differently, we claim that

$$\max_i \left| b_i - \sum_{j=1}^{n} a_{ij} x_j \right| \ge \max_i \left| b_i - \sum_{j=1}^{n} a_{ij} y_j \right|$$

for every choice of $x_1, x_2, \ldots, x_n$. Again, this claim follows easily from considerations involving duality; again, we shall provide a direct proof from scratch. To begin, we have

$$\begin{aligned}\max_i \left| b_i - \sum_{j=1}^{n} a_{ij}x_j \right| &\geq \sum_{i=1}^{m} (v_i^* + w_i^*) \left| b_i - \sum_{j=1}^{n} a_{ij}x_j \right| \\ &\geq \sum_{i=1}^{m} v_i^* \left(b_i - \sum_{j=1}^{n} a_{ij}x_j \right) + \sum_{i=1}^{m} w_i^* \left(-b_i + \sum_{j=1}^{n} a_{ij}x_j \right) \\ &= \sum_{i=1}^{m} (v_i^* - w_i^*)b_i.\end{aligned}$$

On the other hand, (14.15) and (14.16) imply

$$\max_i \left| b_i - \sum_{j=1}^{n} a_{ij}y_j \right| \leq y_{n+1}$$

and we have

$$\begin{aligned}y_{n+1} &= \sum_{i=1}^{m} v_i^* \left(y_{n+1} + \sum_{j=1}^{m} a_{ij}y_j \right) + \sum_{i=1}^{m} w_i^* \left(y_{n+1} - \sum_{j=1}^{n} a_{ij}y_j \right) \\ &= \sum_{i=1}^{m} (v_i^* - w_i^*)b_i.\end{aligned}$$

For instance, to find a best L_∞ approximation to a solution of (14.3), we consider the LP problem

$$\begin{array}{ll}\text{maximize} & 83v_1 + 125v_2 + 310v_3 - 7v_4 + 215v_5 - 83w_1 - 125w_2 - 310w_3 + 7w_4 - 215w_5 \\ \text{subject to} & v_1 + 3v_2 + 2v_3 + v_4 + 3v_5 - w_1 - 3w_2 - 2w_3 - w_4 - 3w_5 = 0 \\ & v_1 - v_2 + 7v_3 - 2v_4 + 2v_5 - w_1 + w_2 - 7w_3 + 2w_4 - 2w_5 = 0 \\ & v_1 + v_2 + v_3 + v_4 + v_5 + w_1 + w_2 + w_3 + w_4 + w_5 = 1 \\ & \qquad v_1, v_2, v_3, v_4, v_5, w_1, w_2, w_3, w_4, w_5 \geq 0.\end{array}$$

Along with the optimal solution of this dual,

$$v_1^* = \frac{23}{32}, \qquad w_2^* = \frac{5}{32}, \qquad w_3^* = \frac{1}{8},$$

the revised simplex method delivers an optimal solution of the original problem,

$$y_1 = 52, \qquad y_2 = \frac{237}{8}, \qquad y_3 = \frac{11}{8}.$$

Hence $y_1 = 52$ and $y_2 = 29.625$ constitute the desired approximation.

HISTORICAL REMARKS

Development of celestial mechanics in the eighteenth century motivated an interest in approximations of data on star movements. As noted by C. Eisenhart (1961), a geometric method for solving a special L_1 approximation problem was proposed by R. J. Boscovitch sometime between 1755 and 1757. An analytic derivation of Boscovitch's method was supplied by P. S. Laplace (1789) and further theoretical work on L_1 approximations was done by C. F. Gauss (1809). [The way of computing the best L_2 approximations by solving equation (14.8) was proposed only at the beginning of the nineteenth century by A. M. Legendre and C. F. Gauss.] In the 1820s, J. B. J. Fourier outlined a linear programming formulation of L_1 and L_∞ approximation problems, and suggested a simplex method for solving them. We shall quote from his work in Chapter 17. I. Grattan-Guiness (1970) points out that Fourier was an important member of Napoleon Bonaparte's Egyptian campaign of 1798–1801, and organized the calculation of the heights of the Memphis pyramids by separate measurements of the heights of their steps. This work stimulated his interest in minimizing the errors in results deduced from a large number of measurements. More on the history of this subject can be found in E. Seneta and W. L. Steiger (1983).

Our discussion of ways to compute the best L_1 and L_∞ approximations followed H. M. Wagner (1959) with only minor modifications. Since 1974, several very good algorithms for computing the best L_1 and L_∞ approximations have been proposed; the one designed by P. Bloomfield and W. L. Steiger (1980) seems to be the best. For an analysis of this algorithm, and its comparison with other methods, see D. Anderson and W. L. Steiger (1983). □

PROBLEMS

△ **14.1** The mixture whose absorbancies are quoted in the text contained only guanine and uracil. Estimate the concentrations of these components as well as you can in the L_1-sense, L_2-sense, and L_∞-sense.

14.2 Prove or disprove: Among all the best L_1-approximations to a solution of

$$\sum_{j=1}^{n} a_{ij}x_j = b_i \qquad (i = 1, 2, \ldots, m) \tag{14.17}$$

there is always one that satisfies at least n of the m equations.

14.3 Prove or disprove: Among all the best L_∞-approximations to a solution of (14.17), there is always one for which the maximum of

$$\left| b_i - \sum_{j=1}^{n} a_{ij}x_j \right|$$

occurs at at least $n + 1$ distinct subscripts i.

14.4 Prove (14.4).

14.5 Prove that $\mathbf{A}^T\mathbf{A}$ is nonsingular if and only if $\mathbf{A}^T$ has full row rank. (*Hint*: Observe that every solution $\mathbf{x}$ of $(\mathbf{A}^T\mathbf{A})\mathbf{x} = \mathbf{0}$ satisfies $(\mathbf{Ax})^T(\mathbf{Ax}) = 0$. Then observe that $\mathbf{y}^T\mathbf{y} = 0$ if and only if $\mathbf{y} = \mathbf{0}$.) Interpret this result in terms of (14.7) and (14.8).

15

Matrix Games

In this chapter, we shall discuss "finite two-person zero-sum games," also called "matrix games" for short. The first attempts to formalize a theory of such games were made by E. Borel (1921, 1924, 1927); a solid foundation of the theory was laid down by J. von Neumann (1928) who proved the celebrated "Minimax Theorem." His original proof, involving Brouwer's fixed-point theorem, was rather complicated; some twenty years later, von Neumann also pointed out that solving matrix games may be reduced to solving certain linear programming problems. Eventually, through the work of G. B. Dantzig, D. Gale, H. W. Kuhn, A. W. Tucker, and others, the study of matrix games became a part of linear programming. In particular, it turned out that the Minimax Theorem follows easily from the Duality Theorem.

AN INTRODUCTORY EXAMPLE: THE GAME OF MORRA

We shall begin our presentation with the game of *Morra*, which is played by two. Its rules are simple: each player hides one or two francs and tries to guess (aloud) how many francs the other player has hidden. If *only one* player makes the correct guess, then this player wins from the other player an amount of money equal to the *total* amount that has been hidden; in all the other cases, the result is a draw and no money changes hands. (For example, suppose that Trucula hides two francs and guesses two, whereas Claude hides two and guesses one. In that case, Claude has to give Trucula four francs.) Trivially, each player has the choice of four courses of action:

- Hide one, guess one.
- Hide one, guess two.
- Hide two, guess one.
- Hide two, guess two.

These courses of action are called pure *strategies*. We shall denote them, in the above order, by [1, 1], [1, 2], [2, 1], and [2, 2]; thus [x, y] denotes "hide x, guess y."

Now suppose that the two players played a very long match. Claude either stuck to one of his pure strategies in every round or used different pure strategies in different rounds, with or without a discernible pattern in his choices; all we know is that he played [1, 1] in c_1 rounds, [1, 2] in c_2 rounds, [2, 1] in c_3 rounds, and [2, 2] in c_4 rounds. Trucula, however, secretly flipped a coin in each round; then she played either [1, 2] if the coin showed heads or [2, 1] if the coin showed tails. If her coin behaved as an unbiased coin should, then she countered Claude's [1, 1] by her own [1, 2] in $c_1/2$ rounds, countering by [2, 1] in the remaining $c_1/2$ rounds. In fact, she countered each of Claude's pure strategies by [1, 2] half the time and by [2, 1] in the remaining half. Thus a detailed record of the match goes as follows.

In $c_1/2$ rounds, Claude played [1, 1] and Trucula played [1, 2], losing 2 francs.
In $c_1/2$ rounds, Claude played [1, 1] and Trucula played [2, 1], winning 3 francs.
In $c_2/2$ rounds, Claude played [1, 2] and Trucula played [1, 2]: a draw.
In $c_2/2$ rounds, Claude played [1, 2] and Trucula played [2, 1]: a draw.
In $c_3/2$ rounds, Claude played [2, 1] and Trucula played [1, 2]: a draw.
In $c_3/2$ rounds, Claude played [2, 1] and Trucula played [2, 1]: a draw.
In $c_4/2$ rounds, Claude played [2, 2] and Trucula played [1, 2], winning 3 francs.
In $c_4/2$ rounds, Claude played [2, 2] and Trucula played [2, 1], losing 4 francs.

Trucula's total winnings come to $(c_1 - c_4)/2$ francs. This number may be negative: if Claude played [2, 2] more often than [1, 1], then Trucula actually lost. Nevertheless, her average loss per round does not exceed half a franc. We conclude that Trucula can protect herself from expected losses greater than half a franc per round by mixing her pure strategies [1, 2] and [2, 1] in the proportion 1:1. (Of course, she must do so in an unpredictable way exhibiting no regularity; otherwise the consequences would be disastrous. For instance, if she played [1, 2] every odd round and [2, 1] every even round, then Claude would catch on, countering every [1, 2] by [1, 1] and every [2, 1] by [2, 2]. Flipping the coin helps to mix the two pure strategies in the desired proportion and yet creates no discernible pattern.) Could she protect herself even better by using a different mixture of her pure strategies? We shall answer this question in a general setting.

MATRIX GAMES

Every matrix $\mathbf{A} = (a_{ij})$ defines a game for two. In each round, the *row player* selects one of the rows $i = 1, 2, \ldots, m$ and the *column player* selects one of the columns $j = 1, 2, \ldots, n$; the resulting *payoff* to the row player is a_{ij}. (That is to say, the row player receives a_{ij} monetary units from the column player. Of course, if a_{ij} is negative, then it is the row player who pays: receiving a negative amount means paying.) Each player makes a choice unaware of the opponent's choice; however, the *payoff matrix* A is known to both players. Clearly, Morra fits into this format; its payoff matrix is as follows:

		Claude's pure strategies			
		[1, 1]	[1, 2]	[2, 1]	[2, 2]
Trucula's pure strategies	[1, 1]	0	2	−3	0
	[1, 2]	−2	0	0	3
	[2, 1]	3	0	0	−4
	[2, 2]	0	−3	4	0

In a long match, the row player may decide to mix her m pure strategies so that each row i will be selected with a probability x_i in every round. The column player may respond in a regular or random manner; over a long period of time, he will choose the jth column with some relative frequency y_j. (In our example, we considered $x_1 = 0$, $x_2 = \frac{1}{2}$, $x_3 = \frac{1}{2}$, $x_4 = 0$, and $y_j = c_j/N$ with $N = c_1 + c_2 + c_3 + c_4$.) Thus, the row i and the column j will be selected in $x_i y_j N$ of the total N rounds. The resulting average payoff (to the row player) per round equals

$$\sum_{i=1}^{m} \sum_{j=1}^{n} a_{ij} x_i y_j$$

or, in matrix notation, $\mathbf{xAy}$. Here $\mathbf{x}$ stands for the row vector with components $x_1, x_2, \ldots, x_m$ and $\mathbf{y}$ stands for the column vector with components $y_1, y_2, \ldots, y_n$. These two vectors share a characteristic feature: their components are nonnegative, with the sum equal to one. Such vectors are called *stochastic*.

Whenever the row player adopts a *mixed strategy* described by a stochastic row vector **x**, he assures himself of winning at least

$$\min_{\mathbf{y}} \mathbf{xAy}$$

per round on the average, with the minimum taken over all stochastic column vectors **y**. For instance, by adopting the mixed strategy described by $\mathbf{x} = [0, \frac{1}{2}, \frac{1}{2}, 0]$, Trucula assures herself of winning at least -0.5 francs (that is, losing at most 0.5 francs) per round on the average. Thus, a row player desiring the best possible *guarantee* that her expected losses will be curbed and/or her expected winnings kept high should look for a mixed strategy **x** that maximizes the quantity min **xAy**; such a strategy is called *optimal*. Let us note at once that

$$\min_{\mathbf{y}} \mathbf{xAy} = \min_{j} \sum_{i=1}^{m} a_{ij}x_i. \tag{15.1}$$

In words, identity (15.1) asserts that among the most effective replies **y** to the row player's mixed strategy **x**, there is always at least one pure strategy. This claim is not difficult to justify in intuitive terms; a formal proof is as follows. If t stands for the right-hand side in (15.1) and if **y** is an arbitrary stochastic column vector of length n, then

$$\mathbf{xAy} = \sum_{j=1}^{n} y_j\left(\sum_{i=1}^{m} a_{ij}x_i\right) \geq \sum_{j=1}^{n} y_j t = t$$

and so the left-hand side of (15.1) is at least the right-hand side. On the other hand, since each **y** with one component equal to one and the remaining components equal to zero is a candidate for minimizing **xAy**, we have

$$\min_{\mathbf{y}} \mathbf{xAy} \leq \sum_{i=1}^{m} a_{ij}x_i$$

for each $j = 1, 2, \ldots, n$. Hence the left-hand side of (15.1) is at most the right-hand side.

By virtue of (15.1), the problem of finding the row player's optimal strategy reduces to the form

$$\begin{aligned} \text{maximize} \quad & \min_{j} \sum_{i=1}^{m} a_{ij}x_i \\ \text{subject to} \quad & \sum_{i=1}^{m} x_i = 1 \\ & x_i \geq 0 \qquad (i = 1, 2, \ldots, m). \end{aligned} \tag{15.2}$$

The key observation of this chapter is that (15.2) is equivalent to the linear programming problem

$$
\begin{aligned}
\text{maximize} \quad & z \\
\text{subject to} \quad & z - \sum_{i=1}^{m} a_{ij}x_i \le 0 \qquad (j = 1, 2, \ldots, n) \\
& \sum_{i=1}^{m} x_i = 1 \\
& x_i \ge 0 \qquad (i = 1, 2, \ldots, m).
\end{aligned} \tag{15.3}
$$

[To see the equivalence, note that every optimal solution $z^*, x_1^*, \ldots, x_m^*$ of (15.3) satisfies at least one of the constraints $z - \sum a_{ij}x_i \le 0$ with the sign of equality, and so $z^* = \min \sum a_{ij}x_j^*$. A similar trick was used in Chapter 12 and Chapter 14.] Thus the row player can find his optimal strategy by applying the simplex method to (15.3). For instance, since $z^* = 0$, $x_1^* = 0$, $x_2^* = \frac{3}{5}$, $x_3^* = \frac{2}{5}$, $x_4^* = 0$ is one of the optimal solutions of the problem

$$
\begin{aligned}
\text{maximize} \quad & z \\
\text{subject to} \quad & z \phantom{{}- 2x_1} + 2x_2 - 3x_3 \phantom{{}+ 3x_4} \le 0 \\
& z - 2x_1 \phantom{{}+ 2x_2 - 3x_3} + 3x_4 \le 0 \\
& z + 3x_1 \phantom{{}+ 2x_2 - 3x_3} - 4x_4 \le 0 \\
& z \phantom{{}- 2x_1} - 3x_2 + 4x_3 \phantom{{}+ 3x_4} \le 0 \\
& \phantom{z - {}} x_1 + x_2 + x_3 + x_4 = 1 \\
& x_1, x_2, x_3, x_4 \ge 0
\end{aligned}
$$

one of Trucula's optimal strategies is $[0, \frac{3}{5}, \frac{2}{5}, 0]$. Note that by adopting this mixed strategy, Trucula protects herself from positive expected losses.

Similarly, whenever the column player adopts a mixed strategy described by a stochastic column vector $\mathbf{y}$, he assures himself of losing no more than

$$\max_{\mathbf{x}} \mathbf{xAy}$$

per round on the average, with the maximum taken over all stochastic row vectors $\mathbf{x}$; a mixed strategy $\mathbf{y}$ that minimizes the quantity is called *optimal.* Since

$$\max_{\mathbf{x}} \mathbf{xAy} = \max_i \sum_{j=1}^{n} a_{ij}y_j$$

[which can be proved analogously to (15.1)], the problem of finding the column player's optimal strategy reads

$$\text{minimize} \quad \max_i \sum_{j=1}^{n} a_{ij}y_j$$

subject to $$\sum_{j=1}^{n} y_j = 1$$

$$y_j \geq 0 \qquad (j = 1, 2, \ldots, n)$$

or, in the linear programming form,

$$\begin{aligned} &\text{minimize} \quad w \\ &\text{subject to} \quad w - \sum_{j=1}^{n} a_{ij}y_j \geq 0 \qquad (i = 1, 2, \ldots, m) \\ &\qquad \sum_{j=1}^{n} y_j = 1 \\ &\qquad y_j \geq 0 \qquad (j = 1, 2, \ldots, n). \end{aligned} \tag{15.4}$$

Now the main theorem of this chapter can be proved instantaneously.

THE MINIMAX THEOREM

THEOREM 15.1 (The Minimax Theorem). For every $m \times n$ matrix $\mathbf{A}$ there is a stochastic row vector $\mathbf{x}^*$ of length m and a stochastic column vector $\mathbf{y}^*$ length n such that

$$\min_{\mathbf{y}} \mathbf{x}^*\mathbf{A}\mathbf{y} = \max_{\mathbf{x}} \mathbf{x}\mathbf{A}\mathbf{y}^* \tag{15.5}$$

with the minimum taken over all stochastic column vectors $\mathbf{y}$ of length n and the maximum taken over all stochastic row vectors $\mathbf{x}$ of length m.

PROOF. Note that (15.3) and (15.4) are duals of each other and that each of them has feasible solutions. Hence the Duality Theorem guarantees that (15.3) has an optimal solution $z^*, x_1^*, \ldots, x_m^*$ and (15.4) has an optimal solution $w^*, y_1^*, \ldots, y_n^*$ such that $z^* = w^*$. Since z^* equals the left-hand side in (15.5) and w^* equals the right-hand side in (15.5), the desired conclusion follows. ■

When $\mathbf{A}$ is thought of as defining a game, the common value v of the two sides in (15.5) is referred to as the *value* of that game. By adopting the mixed strategy $\mathbf{x}^*$, the row player assures himself of winning at least v units per round on the average. On the other hand, the column player can assure himself of losing no more than v units per round on the average by adopting the mixed strategy $\mathbf{y}^*$. Thus *fair games* have value zero. Games such as Morra, where the roles of the two players are interchangeable, are clearly fair. Such games are called *symmetric*; their payoff matrices satisfy $a_{ij} = -a_{ji}$ for all i and j.

FURTHER REMARKS AND EXAMPLES

The Minimax Theorem has an interesting corollary: as long as your mixed strategy is optimal, you can reveal it to your opponent without hurting your future prospects. This conclusion may seem inconsistent with the mystique of gambling. Apparently Borel found it hard to accept: even though he had proved the theorem for symmetric games of size 3×3 and 5×5, he was led to believe that it may be false for large games. On the subject of symmetric games, he speculated that "Whatever the manner of playing of the second player may be . . . once that manner of playing is determined, the first player can arrange to win for sure; if he knows the manner of playing of the second player, i.e., the probability that the second player plays in such and such a manner." And again, "The player who does not observe the psychology of his partner, and does not modify his manner of playing must necessarily lose against an adversary whose mind is sufficiently flexible to vary his play while taking account of that of the adversary." [From an English translation by J. L. Savage.]

For a further illustration of the power of the theory, let us return to Morra. Unless the two players write their guesses down, they may find it awkward to announce them simultaneously. Eventually, they may agree that Claude will always announce his guess first. That may give Trucula the edge: having heard Claude's guess, she can still adjust her own. However, by simply announcing his *guess*, Claude gives away no information as to the number of coins he has *hidden*. Thus, one may be led to believe that the game remains fair. To find out which is right, let us first construct the payoff matrix for the new version. In addition to the original four pure strategies, Trucula now has four pure strategies that take Claude's guess into account:

- Hide one, make the same guess as Claude.
- Hide one, make a guess different from Claude's.
- Hide two, make the same guess as Claude.
- Hide two, make a guess different from Claude's.

We shall denote these pure strategies by $[1, S]$, $[1, D]$, $[2, S]$, $[2, D]$. The resulting payoff matrix is as follows:

Trucula's pure strategies	Claude's pure strategies: [1, 1]	[1, 2]	[2, 1]	[2, 2]
[1, 1]	0	2	−3	0
[1, 2]	−2	0	0	3
[2, 1]	3	0	0	−4
[2, 2]	0	−3	4	0
[1, S]	0	0	−3	3
[1, D]	−2	2	0	0
[2, S]	3	−3	0	0
[2, D]	0	0	4	−4

By adopting the mixed strategy $[0, 56/99, 40/99, 0, 0, 2/99, 0, 1/99]$, Trucula assures herself of winning at least 4/99 francs per round on the average. On the other hand, by adopting the mixed strategy $[28/99, 30/99, 21/99, 20/99]^T$, Claude assures himself of losing no more than 4/99 francs per round on the average. Thus the value of this game is 4/99.

Bluffing and Underbidding

In card games such as poker, the players sometimes *bluff* by challenging their opponents to a bet even though they are bound to lose if the challenge is accepted. On the other hand, they may also *underbid* by refraining from making such a challenge even though they are sure to win in an open confrontation. In this section, we present an example in which these stratagems are justified as perfectly rational.

The example is a game invented and analyzed by H. W. Kuhn (1950). It is played with a deck of three cards numbered 1, 2, 3. At the beginning of a play, each of the two players bets an *ante* of one unit and receives a card. Then the players take turns either *betting* one additional unit or *passing* without further betting. The play terminates as soon as a bet is answered by a bet, or a pass by a pass, or bet by a pass. The first two eventualities lead to a confrontation in which the player holding the higher card wins the total amount bet by his opponent; a player answering a bet by a pass chooses to lose his ante. Each play takes one of the following five courses:

A passes, B passes . . . payoff 1 to holder of higher card.
A passes, B bets, A passes . . . payoff 1 to B.
A passes, B bets, A bets . . . payoff 2 to holder of higher card.
A bets, B passes . . . payoff 1 to A.
A bets, B bets . . . payoff 2 to holder of higher card.

Once the cards have been dealt, A may proceed along one of three lines:

1. Pass; if B bets, pass again.
2. Pass; if B bets, bet.
3. Bet.

Each complete set of instructions telling A unequivocally what to do in each situation may be described by a triple $x_1x_2x_3$ such that x_j is the line to be used when holding j. For example, 3 1 2 directs A to bet on a 1 in the first round, always pass with a 2, and wait till the second round to bet on a 3. These triples $x_1x_2x_3$ are A's pure strategies. Similarly, B has four different lines:

1. Pass no matter what A did.
2. If A passes, pass; if A bets, bet.
3. If A passes, bet; if A bets, pass.
4. Bet no matter what A did.

Each of B's pure strategies will be denoted by a triple $y_1y_2y_3$ such that y_j is the line to be used when holding j. To evaluate the payoffs for each pair of pure strategies, we have to assume that each of the six possible deals (A holding 1 and B holding 2, A holding 1 and B holding 3, and so on) is equally likely. For example, if A uses 3 1 2 and B uses 1 2 4 then there are six possible outcomes:

A holds 1, B holds 2 . . . A bets, B bets . . . payoff to A $= -2$.
A holds 1, B holds 3 . . . A bets, B bets . . . payoff to A $= -2$.
A holds 2, B holds 1 . . . A passes, B passes . . . payoff to A $= +1$.
A holds 2, B holds 3 . . . A passes, B bets, A passes . . . payoff to A $= -1$.
A holds 3, B holds 1 . . . A passes, B passes . . . payoff to A $= +1$.
A holds 3, B holds 2 . . . A passes, B passes . . . payoff to A $= +1$.

$$\text{Average payoff to A} = \frac{1}{6}(-2-2+1-1+1+1) = -\frac{1}{3}.$$

Obviously, A has $3 \times 3 \times 3 = 27$ pure strategies whereas B has $4 \times 4 \times 4 = 64$ pure strategies. A straightforward analysis of the 27×64 payoff matrix would be quite tedious. Fortunately, we can use common sense to reduce the payoff matrix down to the size of only 8×4. To begin with, note that a player holding a 1 would lose an extra unit if he answered a bet by a bet, rather than a pass. Similarly, a player holding a 3 would lose for no good reason if he answered a bet by a pass; in addition, he cannot go wrong if he answers a pass by a bet. Hence A has at least one optimal mixed strategy in which:

Holding 1, he refrains from line 2.
Holding 3, he refrains from line 1.

Similarly, B has at least one optimal mixed strategy in which:

Holding 1, he refrains from lines 2 and 4.
Holding 3, he refrains from lines 1, 2, 3.

Now we may pretend that the pure strategies $2x_2x_3$ and x_1x_21 are simply unavailable to A and that the pure strategies $2y_2y_3$, $4y_2y_3$, y_1y_21, y_1y_22, y_1y_23 are unavailable to B. Even though some optimal strategies may become lost, at least one optimal strategy for A and at least one optimal strategy for B will remain preserved. In particular, the value of the game will not change.

These eliminations reduce the number of A's pure strategies to 12 and the number of B's pure strategies to 8; in addition, they create possibilities for further simplification. If A holds 2, then he might as well pass in the first round; since B refrains from line 2 when holding 1, and from lines 1, 3 when holding 3, A's line 2 is now as good as 3. Thus we may eliminate A's pure strategies x_13x_3. Similarly, if B holds 2, then his line 1 is as good as 3 and his line 2 is as good as 4. This observation eliminates B's pure strategies y_13y_3 and y_14y_3. The resulting matrix of payoffs to A is as follows:

$$
\begin{array}{c} \\ 1\,1\,2 \\ 1\,1\,3 \\ 1\,2\,2 \\ 1\,2\,3 \\ 3\,1\,2 \\ 3\,1\,3 \\ 3\,2\,2 \\ 3\,2\,3 \end{array}
\begin{array}{c} \begin{array}{cccc} 1\,1\,4 & 1\,2\,4 & 3\,1\,4 & 3\,2\,4 \end{array} \\
\begin{bmatrix}
0 & 0 & -\frac{1}{6} & -\frac{1}{6} \\
0 & \frac{1}{6} & -\frac{1}{3} & -\frac{1}{6} \\
-\frac{1}{6} & -\frac{1}{6} & \frac{1}{6} & \frac{1}{6} \\
-\frac{1}{6} & 0 & 0 & \frac{1}{6} \\
\frac{1}{6} & -\frac{1}{3} & 0 & -\frac{1}{2} \\
\frac{1}{6} & -\frac{1}{6} & -\frac{1}{6} & -\frac{1}{2} \\
0 & -\frac{1}{2} & \frac{1}{3} & -\frac{1}{6} \\
0 & -\frac{1}{3} & \frac{1}{6} & -\frac{1}{6}
\end{bmatrix} \end{array}.
$$

Considering A's mixed strategy $[\frac{1}{3}, 0, 0, \frac{1}{2}, \frac{1}{6}, 0, 0, 0]$ and B's mixed strategy $[\frac{2}{3}, 0, 0, \frac{1}{3}]^T$, we conclude that both of the strategies are optimal and the value of the game is $-\frac{1}{18}$. Our optimal strategy for A may be broken down into the following simple instructions:

- Holding 1, mix lines 1 and 3 in the proportion 5:1.
- Holding 2, mix lines 1 and 2 in the proportion 1:1.
- Holding 3, mix lines 2 and 3 in the proportion 1:1.

Note that these instructions call for bluffing (that is, betting on a 1 in the first round) once out of every six available times, and for underbidding (that is, passing on a 3 in the first round) half the available time. Similarly, our optimal strategy for B breaks down as follows:

- Holding 1, mix lines 1 and 3 in the proportion 2:1.
- Holding 2, mix lines 1 and 2 in the proportion 2:1. (15.6)
- Holding 3, always use line 4.

Hence B should use one-third of his opportunities to bluff; underbidding is not available to him.
A further discussion of this game is deferred to problems 15.9 and 15.10.

PROBLEMS

△ **15.1** Each of two players hides either a nickel or a dime. If the two coins match, A gets both; if they don't match, B gets both. What are the optimal strategies? Is this game fair? What about games with coins of arbitrary but fixed denominations x and y?

15.2 A domino piece can be placed on a 2×3 checkerboard in seven different ways:

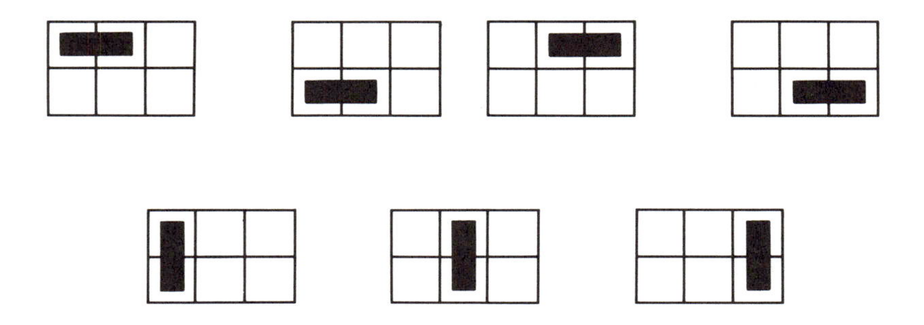

The first player places the domino and the second player selects one of the six squares. If the selected square is covered by the domino, then the second player wins; otherwise the first player wins. Is this game fair? What are the optimal strategies? Can you exploit the symmetries? How do your findings generalize to larger boards?

△ **15.3** Both you and your opponent choose an integer between 1 and 1,000 inclusive. If your number x is smaller than your opponent's number y, then you win, except for $x = y - 1$ in which case you lose. If your number x is larger than your opponent's number, then you lose, except for $x = y + 1$ in which case you win. If $x = y$ then the play is a draw.

15.4 Consider the variant of Morra in which each player can hide one, two, or three coins; for simplicity, assume that the players announce their guesses simultaneously. What are the optimal strategies?

15.5 A row r of the payoff matrix is said to *dominate* a row s if $a_{rj} \geq a_{sj}$ for all $j = 1, 2, \ldots, n$. Similarly, a column r of the payoff matrix is said to *dominate* a column s if $a_{ir} \geq a_{is}$ for all $i = 1, 2, \ldots, m$. Prove:

(i) If a row r is dominated by another row, then the row player has at least one optimal strategy $\mathbf{x}^*$ in which $x_r^* = 0$. In particular, if row r is deleted from the payoff matrix, then the value of the game does not change.

(ii) If a column s dominates another column, then the column player has at least one optimal strategy $\mathbf{y}^*$ in which $y_s^* = 0$. In particular, if column s is deleted from the payoff matrix, then the value of the game does not change.

Use these facts to reduce the following payoff matrix to size 2×2:

$$\begin{bmatrix} -2 & 3 & 0 & -6 & -3 \\ 0 & -4 & 9 & 2 & 1 \\ 6 & -2 & 7 & 4 & 5 \\ 7 & -3 & 8 & 3 & 2 \end{bmatrix}.$$

15.6 Prove that the row player's mixed strategy $\mathbf{x}$ and the column player's mixed strategy $\mathbf{y}$ are simultaneously optimal if and only if

$$x_i = 0 \qquad \text{whenever} \quad \sum_{j=1}^{n} a_{ij}y_j < \max_k \sum_{j=1}^{n} a_{kj}y_j$$

and

$$y_j = 0 \qquad \text{whenever} \quad \sum_{i=1}^{m} a_{ij}x_i > \min_k \sum_{i=1}^{m} a_{ik}x_i.$$

15.7 Use the result of problem 15.6 to describe all optimal strategies for Morra.

15.8 In the game with the payoff matrix

$$\begin{bmatrix} 3 & 2 & 0 & -1 & 5 & -2 \\ -2 & -3 & 2 & 4 & 0 & 4 \\ 5 & -3 & 4 & 0 & 4 & 7 \\ 1 & 3 & 3 & 2 & -6 & 5 \end{bmatrix}$$

the row player's mixed strategy $[\frac{1}{2}, \frac{1}{4}, 0, \frac{1}{4}]$ is optimal. Describe all the optimal strategies of the column player.

15.9 In Kuhn's simplified poker, different mixed strategies may lead to the same explicit instructions. For example, note that B's optimal strategies

$$\begin{bmatrix} \frac{2}{3} \\ 0 \\ 0 \\ \frac{1}{3} \end{bmatrix} \quad \text{and} \quad \begin{bmatrix} \frac{1}{3} \\ \frac{1}{3} \\ \frac{1}{3} \\ 0 \end{bmatrix}$$

lead to the same instructions (15.6). Prove that every optimal strategy for B is described by (15.6). Furthermore, prove that every optimal strategy for A breaks down as follows:

Holding 1, mix lines 1 and 3 in the proportion $(3 - t):t$.

Holding 2, mix lines 1 and 2 in the proportion $(2 - t):(t + 1)$.

Holding 3, mix lines 2 and 3 in the proportion $(1 - t):t$.

Here t is an arbitrary but fixed number such that $0 \le t \le 1$. (Note that the instructions in the text correspond to $t = \frac{1}{2}$.)

15.10 Solve the following variants of simplified poker:

(i) Ante = 2, bet = 1.

(ii) Ante = 2, bet = 3.

(iii) Ante = 2, bet = 4.

15.11 Find necessary and sufficient conditions for the rth pure strategy of the row player and the sth strategy of the column player to be simultaneously optimal.

15.12 The Minimax Theorem is sometimes stated as

$$\max_{\mathbf{x}} \min_{\mathbf{y}} \mathbf{xAy} = \min_{\mathbf{y}} \max_{\mathbf{x}} \mathbf{xAy}.$$

Prove this identity.

15.13 [G. B. Dantzig (1951a).] Describe the relationship between the linear programming problem

$$\begin{array}{ll} \text{maximize} & c_1x_1 + c_2x_2 + c_3x_3 \\ \text{subject to} & a_{11}x_1 + a_{12}x_2 + a_{13}x_3 \le b_1 \\ & a_{21}x_1 + a_{22}x_2 + a_{23}x_3 \le b_2 \\ & x_1, x_2, x_3 \ge 0 \end{array}$$

and the game with the payoff matrix

$$\begin{bmatrix} 0 & -c_1 & -c_2 & -c_3 & b_1 & b_2 \\ c_1 & 0 & 0 & 0 & -a_{11} & -a_{21} \\ c_2 & 0 & 0 & 0 & -a_{12} & -a_{22} \\ c_3 & 0 & 0 & 0 & -a_{13} & -a_{23} \\ -b_1 & a_{11} & a_{12} & a_{13} & 0 & 0 \\ -b_2 & a_{21} & a_{22} & a_{23} & 0 & 0 \end{bmatrix}$$

15.14 On page 128 of R. C. MacLagan (1901), there is the following description of an old Scottish game.

Cheap, Middling, or Dear

> This also is played by two. The letters C, M, D, representing respectively the words from which the game is named, are written on a slate, with some interval between them. Under C the figures 1, 2, 3 are placed, under M 4, 5, 6, and under D 7, 8, 9, thus:
>
> | C | M | D |
> |---|---|---|
> | 1, 2, 3 | 4, 5, 6 | 7, 8, 9 |
>
> Player A, who is to play first, marks one of the figures from any of the groups, concealing it from player B, whom he challenges to guess to which group it belongs, saying "My father bought a horse at a fair." B asks, "Cheap, middling, or dear?" A answers him, naming the group from which he has selected his figure. Thus if his figure were 5, the answer would be "middling." B then guesses one of the three numbers, and if he hits upon 5, that is a gain to him of 5, but if he says 4 or 6, then the 5 is scored to A. In any case the 5 is blotted out. B then leads, each playing in turn, till all the figures have been expunged. The total marks credited to each are then ascertained, and he who has the highest number is the winner.

What are the optimal strategies in the next-to-last round? [For results of a complete analysis, see V. Chvátal (1981).]

16

Systems of Linear Inequalities

In this chapter, we shall study systems of linear inequalities

$$\sum_{j=1}^{n} a_{ij}x_j \leq b_i \qquad (i = 1, 2, \ldots, m). \tag{16.1}$$

Solving such systems is no more difficult than solving linear programming problems: to find a solution of (16.1), or to establish its nonexistence, we need only

$$\begin{aligned} \text{minimize} \quad & x_0 \\ \text{subject to} \quad & -x_0 + \sum_{j=1}^{n} a_{ij}x_j \leq b_i \qquad (i = 1, 2, \ldots, m) \\ & x_0 \geq 0. \end{aligned} \tag{16.2}$$

(Conversely, solving linear programming problems is no more difficult than solving systems of linear inequalities; see problem 16.1.) In particular, every system (16.1) may be solved by applying the simplex method to (16.2). An alternative method for solving (16.1), more intuitive but less efficient, is presented in the next section. The main purpose of this chapter is first to prove a theorem characterizing all solutions of (16.1) and then to prove a converse of this theorem.

THE FOURIER–MOTZKIN METHOD

Here is a quote from an English translation, by D. A. Kohler (1973), of a paper published by J. B. J. Fourier in the 1820s [and accessible in J. B. J. Fourier (1890)].

> We will now sketch a method of solving another more general problem, common to almost all questions involving inequalities. Let $x, y, z, \ldots, u, t$ denote some variables which are required to satisfy a given set of linear inequality constraints. We will proceed to eliminate successively $x, y, z \ldots$. Each of the inequalities is, in relation to x, either of the form
>
> $$x \geq A + By + Cz + \cdots$$
>
> or of the form
>
> $$x \leq \alpha + \beta y + \gamma z + \cdots.$$
>
> We take in turn each of the constraints of the first form with each of the constraints of the second form, and we write
>
> $$\alpha + \beta y + \gamma z + \cdots \geq A + By + Cz + \cdots.$$
>
> By this means we form new inequalities in which x does not appear. It nearly always happens that a rather large number of these new inequalities are redundant and need not be written down. It is usually obvious which inequalities are redundant and their removal simplifies the problem greatly.
>
> When we have replaced the inequalities which contain $x, y, z, \ldots, u, t$ by those which contain only $y, z, \ldots, u, t$ we eliminate y by the same procedure. Continuing the application of this rule, we finally obtain inequalities in which only the unknown t enters. We deduce from this some numerical bounds on this last unknown, some of which are of the form $t \geq a$, and others of the form $t \leq b$. It is only necessary to consider the smallest B of the limits b, and the largest A of the limits a. If it happens that A is greater than B, we conclude that the problem has no solution and from this test we can determine whether the constraints are consistent. When the limit B is not less than the limit A, the problem does not contain any incompatible constraints, and, generally speaking, there are infinitely many solutions. We then assign to t some value between A and B and, substituting this value of t in the constraints which only contain u and t, we find numerical limits for u, and we choose some value for u between these limits. Substituting for u and t their numerical values in the inequalities which contain u, t and another unknown, we determine the bound of this new unknown in the same manner. The application of the same rule yields the value of all the unknowns. It is always possible, as we have said, to find for each unknown a value between its limits except in the

> case of the last unknown t, and that happens only when the problem contains some infeasibility.
>
> The method we have just described is intuitively obvious, but nevertheless a rigorous proof is required. This was given in the earlier paper where we showed that after the elimination of one unknown: (1) the inequalities in $y, z, \ldots, u, t$ must be consistent if the problem does have a solution, and (2) conversely, if these constraints are consistent we can satisfy the original set of constraints. The question of whether the original set of constraints is consistent is thus resolved at the end of the calculations when we have obtained the inequalities for the last variable, t . . . .

For some reason, Fourier refrained from publicizing his method further (for example, he did not include it in his book on equations, published in 1831), and the work was almost forgotten. Later, it was rediscovered by several mathematicians; nowadays, it is known as the *Fourier–Motzkin method*. For an excellent exposition, the reader is directed to H. W. Kuhn (1956). The removal of redundant inequalities, mentioned already by Fourier, deserves a special comment. In general, detecting redundancies in a specified system of linear inequalities may be difficult (see problem 16.4). Nevertheless, the systems encountered in the later stages of the Fourier–Motzkin method arise from the original system in a very special way; when this peculiarity is cleverly exploited, one can detect and remove many of the redundant inequalities with only a minimum of computational effort. For details, see R. J. Duffin (1974). However, even with these refinements, the efficiency of the Fourier–Motzkin method applied to (16.1) is still far worse than that of the simplex method applied to (16.2). □

BASIC FEASIBLE DIRECTIONS

In preparation for our main results, we return to the systems

$$\mathbf{Ax} = \mathbf{b}, \quad \mathbf{l} \leq \mathbf{x} \leq \mathbf{u} \tag{16.3}$$

introduced in Chapter 8. (Each component l_j of the column vector $\mathbf{l}$ is either a number or the symbol $-\infty$ and each component u_j of the column vector $\mathbf{u}$ is either a number or the symbol $+\infty$.) A vector $\mathbf{w}$ will be called a *feasible direction* of (16.3) if $\mathbf{Aw} = \mathbf{0}$ and

$$\begin{aligned} w_j &\geq 0 \quad \text{whenever} \quad l_j \text{ is finite} \\ w_j &\leq 0 \quad \text{whenever} \quad u_j \text{ is finite.} \end{aligned}$$

A feasible direction of (16.3) will be called *basic* if its n components can be partitioned into m "basic" and $n - m$ "nonbasic" ones in such a way that (i) the m columns of $\mathbf{A}$ corresponding to the basic components form a nonsingular matrix, (ii) all but one nonbasic component are zeros, and (iii) the nonzero nonbasic component equals one or minus one. Clearly, every system (16.3) has only finitely many basic feasible directions.

> ***THEOREM 16.1.*** If the problem
>
> $$\text{maximize} \quad \mathbf{cx} \quad \text{subject to} \quad \mathbf{Ax} = \mathbf{b}, \quad \mathbf{l} \le \mathbf{x} \le \mathbf{u} \tag{16.4}$$
>
> is unbounded and if $\mathbf{A}$ has full row rank, then system $\mathbf{Ax} = \mathbf{b}, \mathbf{l} \le \mathbf{x} \le \mathbf{u}$ has a basic feasible direction $\mathbf{w}$ such that $\mathbf{cw} > 0$.

PROOF. If the simplex method is applied to an unbounded problem (16.4) whose matrix $\mathbf{A}$ has full row rank, then it delivers a basic feasible solution $\mathbf{x}^*$ and a vector $\mathbf{w}$ such that $\mathbf{x}^* + t\mathbf{w}$ is a feasible solution of (16.4) for every nonnegative t and such that $\mathbf{c}(\mathbf{x}^* + t\mathbf{w})$ can be made arbitrarily large by making t sufficiently large. Hence $\mathbf{w}$ is a feasible direction of (16.3) and $\mathbf{cw} > 0$. A closer examination of $\mathbf{w}$ shows that this feasible direction is basic. ■

The concepts of normal basic feasible solutions (as defined in Chapter 8) and basic feasible directions can be naturally translated to the context of systems

$$\sum_{j=1}^{n} a_{ij}x_j \le b_i \qquad (i = 1, 2, \ldots, m). \tag{16.5}$$

We shall say that $[v_1, v_2, \ldots, v_n]^T$ is a normal basic feasible solution of (16.5) if the vector $[v_1, v_2, \ldots, v_{n+m}]^T$ with $v_{n+i} = b_i - \sum a_{ij}v_j$ is a normal basic feasible solution of

$$\begin{aligned} \sum_{j=1}^{n} a_{ij}x_j + x_{n+i} &= b_i \qquad (i = 1, 2, \ldots, m) \\ x_{n+i} &\ge 0 \qquad (i = 1, 2, \ldots, m). \end{aligned} \tag{16.6}$$

Similarly, we shall say that $[w_1, w_2, \ldots, w_n]^T$ is a basic feasible direction of (16.5) if the vector $[w_1, w_2, \ldots, w_{n+m}]^T$ with $w_{n+i} = -\sum a_{ij}w_j$ is a basic feasible direction of (16.6). For future reference, note that (i) every system $\mathbf{Ax} \le \mathbf{b}$ has only finitely many normal basic feasible solutions and basic feasible directions; (ii) each basic feasible direction $\mathbf{w}$ of $\mathbf{Ax} \le \mathbf{b}$ satisfies $\mathbf{Aw} \le \mathbf{0}$; (iii) if a problem

$$\text{maximize} \quad \mathbf{cx} \quad \text{subject to} \quad \mathbf{Ax} \le \mathbf{b} \tag{16.7}$$

has an optimal solution, then some basic feasible solution of $\mathbf{Ax} \le \mathbf{b}$ is an optimal solution of this problem; and (iv) if the problem (16.7) is unbounded, then some basic feasible direction $\mathbf{w}$ of $\mathbf{Ax} \le \mathbf{b}$ satisfies $\mathbf{cw} > 0$.

THE FINITE BASIS THEOREM AND ITS COROLLARIES

THEOREM 16.2 (The Finite Basis Theorem). For every system $\mathbf{Ax} \leq \mathbf{b}$ there are column vectors $\mathbf{v}^1, \mathbf{v}^2, \ldots, \mathbf{v}^M$ and $\mathbf{w}^1, \mathbf{w}^2, \ldots, \mathbf{w}^N$ with the following property: a vector $\mathbf{x}^*$ satisfies $\mathbf{Ax}^* \leq \mathbf{b}$ if and only if

$$\mathbf{x}^* = \sum_{r=1}^{M} p_r \mathbf{v}^r + \sum_{s=1}^{N} q_s \mathbf{w}^s$$

for some nonnegative numbers $p_1, p_2, \ldots, p_M, q_1, q_2, \ldots, q_N$ such that $\sum p_r = 1$.

PROOF. We claim that the set of all normal basic feasible solutions $\mathbf{v}^1, \mathbf{v}^2, \ldots, \mathbf{v}^M$ and the set of all basic feasible directions $\mathbf{w}^1, \mathbf{w}^2, \ldots, \mathbf{w}^N$ have the desired property. The "if" part is easy: since $\mathbf{Av}^r \leq \mathbf{b}$ for all r and $\mathbf{Aw}^s \leq \mathbf{0}$ for all s, we have

$$\mathbf{A}\left(\sum_{r=1}^{M} p_r \mathbf{v}^r + \sum_{s=1}^{N} q_s \mathbf{w}^s\right) = \sum_{r=1}^{M} p_r(\mathbf{Av}^r) + \sum_{s=1}^{N} q_s(\mathbf{Aw}^s) \leq \sum_{r=1}^{M} p_r \mathbf{b} = \mathbf{b}$$

whenever $p_1, p_2, \ldots, p_M, q_1, q_2, \ldots, q_N$ are nonnegative numbers such that $\sum p_r = 1$. To prove the "only if" part, consider an arbitrary but fixed vector $\mathbf{x}^*$ such that $\mathbf{Ax}^* \leq \mathbf{b}$. From the assumption that no numbers $p_1, p_2, \ldots, p_M$ and $q_1, q_2, \ldots, q_N$ satisfy

$$\begin{aligned} \sum_{r=1}^{M} p_r \mathbf{v}^r + \sum_{s=1}^{N} q_s \mathbf{w}^s &= \mathbf{x}^* \\ \sum_{r=1}^{M} p_r &= 1 \\ p_1, p_2, \ldots, p_M, q_1, q_2, \ldots, q_N &\geq 0 \end{aligned} \tag{16.8}$$

we shall derive a contradiction. If (16.8) is unsolvable, then Theorem 9.2 guarantees the existence of a row vector $\mathbf{c}$ and a number d such that

$$\begin{aligned} \mathbf{cv}^r + d &\geq 0 \qquad \text{for all} \quad r = 1, 2, \ldots, M \\ \mathbf{cw}^s &\geq 0 \qquad \text{for all} \quad s = 1, 2, \ldots, N \\ \mathbf{cx}^* + d &< 0. \end{aligned}$$

Now consider the linear programming problem

$$\text{minimize} \quad \mathbf{cx} \qquad \text{subject to} \quad \mathbf{Ax} \leq \mathbf{b}.$$

Since $\mathbf{x}^*$ is a feasible solution and since $\mathbf{cw} \geq 0$ for every basic feasible direction, this problem is neither infeasible nor unbounded. Hence, it has an optimal solution; in fact, one of the vectors $\mathbf{v}^r$ must be an optimal solution. But then the optimum value equals the corresponding $\mathbf{cv}^r$, which is at least $-d$. This is the desired contradiction as $\mathbf{x}^*$ is a feasible solution and $\mathbf{cx}^* < -d$. ■

Next, we shall show that we can do without the vectors $\mathbf{v}^1, \mathbf{v}^2, \ldots, \mathbf{v}^M$ as soon as $\mathbf{b} = \mathbf{0}$.

***THEOREM 16.3* [*H. Minkowski* (1911)].** For every system $\mathbf{Ax} \leq \mathbf{0}$ there are column vectors $\mathbf{u}^1, \mathbf{u}^2, \ldots, \mathbf{u}^D$ with the following property: a vector $\mathbf{x}^*$ satisfies $\mathbf{Ax}^* \leq \mathbf{0}$ if and only if

$$\mathbf{x}^* = \sum_{k=1}^{D} t_k \mathbf{u}^k$$

for some nonnegative numbers $t_1, t_2, \ldots, t_D$.

PROOF. Theorem 16.2 applied to the system $\mathbf{Ax} \leq \mathbf{0}$ guarantees the existence of certain vectors $\mathbf{v}^1, \mathbf{v}^2, \ldots, \mathbf{v}^M$ and $\mathbf{w}^1, \mathbf{w}^2, \ldots, \mathbf{w}^N$. We propose to show that these vectors have the property required of $\mathbf{u}^1, \mathbf{u}^2, \ldots, \mathbf{u}^D$. To put it differently, we propose to show that a vector $\mathbf{x}^*$ satisfies $\mathbf{Ax}^* \leq \mathbf{0}$ if and only if

$$\mathbf{x}^* = \sum_{r=1}^{M} p_r \mathbf{v}^r + \sum_{s=1}^{N} q_s \mathbf{v}^s$$

for some nonnegative numbers $p_1, p_2, \ldots, p_M, q_1, q_2, \ldots, q_N$. The "only if" part follows trivially from the "only if" part of Theorem 16.2. To prove the "if" part, note that the "if" part of Theorem 16.2 guarantees $\mathbf{Av}^r \leq \mathbf{b} = \mathbf{0}$ for all r and $\mathbf{Aw}^s \leq \mathbf{0}$ for all s. Hence,

$$\mathbf{A}\left(\sum_{r=1}^{M} p_r \mathbf{v}^r + \sum_{s=1}^{N} q_s \mathbf{w}^s\right) = \sum_{r=1}^{M} p_r(\mathbf{Av}^r) + \sum_{s=1}^{N} q_s(\mathbf{Aw}^s) \leq \mathbf{0}$$

for every choice of nonnegative numbers $p_1, p_2, \ldots, p_M, q_1, q_2, \ldots, q_N$. ■

Finally, we shall prove the converse of Theorem 16.2.

THEOREM 16.4. For every choice of column vectors $\mathbf{v}^1, \mathbf{v}^2, \ldots, \mathbf{v}^M$ and $\mathbf{w}^1, \mathbf{w}^2, \ldots, \mathbf{w}^N$ of the same length there is a system $\mathbf{Ax} \leq \mathbf{b}$ with the following property: a vector $\mathbf{x}^*$ satisfies

$$\mathbf{x}^* = \sum_{r=1}^{M} p_r \mathbf{v}^r + \sum_{s=1}^{N} q_s \mathbf{w}^s$$

for some nonnegative numbers $p_1, p_2, \ldots, p_M, q_1, q_2, \ldots, q_N$ such that $\sum p_r = 1$ if and only if $\mathbf{Ax}^* \leq \mathbf{b}$.

PROOF. Let us say that the column vectors $\mathbf{v}^r$ and $\mathbf{w}^s$ have length n. The sequence $\mathbf{v}^1, \mathbf{v}^2, \ldots, \mathbf{v}^M, \mathbf{w}^1, \mathbf{w}^2, \ldots, \mathbf{w}^N$ may be thought of as defining an $(M + N) \times n$ matrix, with the ith vector in the sequence describing the ith row of the matrix. To the right end of this matrix, we append an $(n + 1)$th column, with the first M components equal to 1 and the remaining N components equal to 0. Let us denote the resulting $(M + N) \times (n + 1)$ matrix by $\mathbf{Q}$. Now Theorem 16.3 guarantees the existence of column vectors $\mathbf{u}^1, \mathbf{u}^2, \ldots, \mathbf{u}^D$ of length $n + 1$ such that a vector $\mathbf{x}$ satisfies $\mathbf{Qx} \geq \mathbf{0}$ if and only if

$$\mathbf{x} = \sum_{k=1}^{D} t_k \mathbf{u}^k \tag{16.9}$$

for some nonnegative numbers $t_1, t_2, \ldots, t_D$. These vectors define the desired $\mathbf{A}$ and $\mathbf{b}$: if the components of $\mathbf{u}^i$ are $z_1, z_2, \ldots, z_{n+1}$, then the ith row of $\mathbf{A}$ is $[-z_1, -z_2, \ldots, -z_n]$ and the ith component of $\mathbf{b}$ is z_{n+1}. To verify that $\mathbf{A}$ and $\mathbf{b}$ have the required property, consider an arbitrary but fixed column vector $\mathbf{x}^*$ of length n. By Theorem 9.2, the system

$$\begin{gathered}\sum_{r=1}^{M} p_r \mathbf{v}^r + \sum_{s=1}^{N} q_s \mathbf{w}^s = \mathbf{x}^* \\ \sum_{r=1}^{M} p_r = 1 \\ p_1, p_2, \ldots, p_M, q_1, q_2, \ldots, q_N \geq 0\end{gathered} \tag{16.10}$$

has no solution if and only if there are a row vector $\mathbf{c} = [c_1, c_2, \ldots, c_n]$ and a number d such that

$$\begin{aligned} \mathbf{cv}^r + d \geq 0 \quad & \text{for all} \quad r = 1, 2, \ldots, M \\ \mathbf{cw}^s \geq 0 \quad & \text{for all} \quad s = 1, 2, \ldots, N \end{aligned} \tag{16.11}$$

and such that

$$\mathbf{cx}^* + d < 0.$$

To put it differently, (16.10) is solvable if and only if every solution of (16.11) satisfies $\mathbf{cx}^* + d \geq 0$. But $\mathbf{c}, d$ is a solution of (16.11) if and only if $\mathbf{c} = -\mathbf{tA}$ and $d = \mathbf{tb}$ for some nonnegative row vector $\mathbf{t}$ of length D: if $\mathbf{x}$ stands for the column vector $[c_1, c_2, \ldots, c_n, d]^T$, then (16.11) reduces into $\mathbf{Qx} \geq \mathbf{0}$ whereas (16.9) reads $\mathbf{t}(-\mathbf{A}) = \mathbf{c}$, $\mathbf{tb} = d$. Thus we conclude that (16.10) is solvable if and only if every nonnegative row vector $\mathbf{t}$ of length D satisfies $\mathbf{t}(-\mathbf{Ax}^* + \mathbf{b}) \geq 0$. Clearly, this is the case if and only if $-\mathbf{Ax}^* + \mathbf{b} \geq \mathbf{0}$. ■

In Chapter 18, we shall present efficient procedures for finding the vectors $\mathbf{v}^1, \mathbf{v}^2, \ldots, \mathbf{v}^M, \mathbf{w}^1, \mathbf{w}^2, \ldots, \mathbf{w}^N$ of Theorem 16.2, the vectors $\mathbf{u}^1, \mathbf{u}^2, \ldots, \mathbf{u}^D$ of Theorem 16.3, and the system $\mathbf{Ax} \leq \mathbf{b}$ of Theorem 16.4.

PROBLEMS

16.1 To show that solving linear programming problems is no more difficult than solving systems of linear inequalities, describe the relationship between the problem

$$\text{maximize } \mathbf{cx} \quad \text{subject to } \mathbf{Ax} \leq \mathbf{b}, \quad \mathbf{x} \geq \mathbf{0}$$

and the system

$$\mathbf{Ax} \leq \mathbf{b}, \quad \mathbf{x} \geq \mathbf{0}, \quad \mathbf{yA} \geq \mathbf{c}, \quad \mathbf{y} \geq \mathbf{0}, \quad \mathbf{cx} \geq \mathbf{yb}.$$

16.2 Prove that in any solvable system

$$\sum_{j=1}^{n} a_{ij}x_j \leq b_i \qquad (i = 1, 2, \ldots, m)$$

there are sets I and J of subscripts such that the system

$$\sum_{j \in J} a_{ij}x_j = b_i \qquad (i \in I)$$

has a unique solution x_j^* $(j \in J)$ and such that

$$\sum_{j \in J} a_{ij}x_j^* \leq b_i$$

for all $i = 1, 2, \ldots, m$. (*Hint*: Consider the outcome of the first phase of the simplex method.)

16.3 Suppose that you have a black box for recognizing solvable systems of linear inequalities. (Given any system $\mathbf{Ax} \leq \mathbf{b}$, the box will answer the question "Is $\mathbf{Ax} \leq \mathbf{b}$ solvable?" without actually returning a solution when the answer is affirmative.) Show that by feeding the box at most $m + n$ systems, each system consisting of at most m linear equations and inequalities in at most n variables, one can find the sets I and J of problem 16.2. In this sense, finding solutions to systems of linear inequalities is no more difficult than merely finding out if a solution exists.

16.4 A particular inequality in a system of linear inequalities is called *redundant* if its removal from the system does not introduce any new solutions. To show that recognizing redundant inequalities is at least as difficult as recognizing solvable systems, prove that $\mathbf{Ax} \leq \mathbf{b}$ is solvable if and only if $\mathbf{yb} \geq 0$ is redundant in $\mathbf{yA} = \mathbf{0}$, $\mathbf{y} \geq \mathbf{0}$, $\mathbf{yb} \geq 0$.

16.5 To establish the converse of the preceding result, show that the first inequality in a solvable system

$$\sum_{j=1}^{n} a_{ij}x_j \leq b_i \qquad (i = 0, 1, \ldots, m)$$

is redundant if and only if the system

$$\sum_{i=1}^{m} a_{ij}y_i = a_{0j} \qquad (j = 1, 2, \ldots, n)$$

$$y_i \geq 0 \qquad (i = 1, 2, \ldots, m)$$

$$\sum_{i=1}^{m} b_i y_i \leq b_0 \text{ is solvable.}$$

△ **16.6** Use the Fourier–Motzkin method to find x, y, z such that

$$\begin{aligned} 3x + y - 2z &\le 0 \\ x - 2y - 4z &\le -14 \\ -x + 3y - 2z &\le -8 \\ -x + y + 4z &\le 14 \\ -2x - 5y + z &\le -6. \end{aligned}$$

16.7 Solve the preceding problem by the simplex method.

△ **16.8** Illustrate Theorem 16.2 on the system

$$\begin{aligned} x_1 + 3x_2 &\le 6 \\ 2x_1 + x_2 &\le -2 \\ x_1 + x_2 &\le 0. \end{aligned}$$

16.9 Prove that the system

$$\sum_{j=1}^{n} a_{ij}x_j \le b_i \qquad (i = 1, 2, \ldots, k)$$

$$\sum_{j=1}^{n} a_{ij}x_j < b_i \qquad (i = k + 1, k + 2, \ldots, m)$$

is unsolvable if and only if the system

$$\sum_{i=1}^{m} a_{ij}y_i = 0 \qquad (j = 1, 2, \ldots, n)$$

$$\sum_{i=1}^{m} b_i y_i \le 0$$

$$y_i \ge 0 \qquad (i = 1, 2, \ldots, m)$$

$$\sum_{i=1}^{k} b_i y_i + \sum_{i=k+1}^{m} (b_i - 1)y_i < 0$$

is solvable.

16.10 Derive the following theorems (with the vector inequality $\mathbf{v} > \mathbf{w}$ meaning, as usual, $v_k > w_k$ for all k) from the result of problem 16.9.

(i) P. Gordan (1873): The system $\mathbf{Ax} < \mathbf{0}$ is unsolvable if and only if the system $\mathbf{yA} = \mathbf{0}$, $\mathbf{y} \ge \mathbf{0}$, $\mathbf{y} \ne \mathbf{0}$ is solvable.

(ii) J. Farkas (1902): The system $\mathbf{Ax} \le \mathbf{0}$, $\mathbf{bx} > 0$ is unsolvable if and only if the system $\mathbf{yA} = \mathbf{b}$, $\mathbf{y} \ge \mathbf{0}$ is solvable.

(iii) E. Stiemke (1915): The system $\mathbf{Ax} = \mathbf{0}$, $\mathbf{x} > \mathbf{0}$ is unsolvable if and only if the system $\mathbf{yA} \ge \mathbf{0}$, $\mathbf{yA} \ne \mathbf{0}$ is solvable.

(iv) J. A. Ville (1938): The system $\mathbf{Ax} < \mathbf{0}$, $\mathbf{x} \ge \mathbf{0}$ is unsolvable if and only if the system $\mathbf{yA} \ge \mathbf{0}$, $\mathbf{y} \ge \mathbf{0}$, $\mathbf{y} \ne \mathbf{0}$ is solvable.

(v) A. W. Tucker (1956): The system $\mathbf{Ax} \ge \mathbf{0}$, $\mathbf{x} \ge \mathbf{0}$ has no solution with $x_k > 0$ if and only if the system $\mathbf{yA} \le \mathbf{0}$, $\mathbf{y} \ge \mathbf{0}$ has a solution with

$$\sum_{i=1}^{m} a_{ik}y_i < 0.$$

16.11 Let

$$\sum_{j=1}^{n} a_{ij}x_j \leq b_i \qquad (i = 1, 2, \ldots, m)$$

be a fixed system of linear inequalities. Call a subset I of $\{1, 2, \ldots, m\}$ *admissible* if the system

$$\sum_{j=1}^{n} a_{ij}x_j \leq b_i \qquad (i \notin I)$$

$$\sum_{j=1}^{n} a_{ij}x_j < b_i \qquad (i \in I)$$

is solvable. Prove that the union of admissible sets is admissible.

16.12 [A. W. Tucker (1956).] Prove that, for every matrix $\mathbf{A}$ and its transpose $\mathbf{A}^T$, the system $\mathbf{Ax} \geq \mathbf{0}$, $\mathbf{x} \geq \mathbf{0}$, $\mathbf{A}^T\mathbf{y}^T \leq \mathbf{0}$, $\mathbf{y}^T \geq \mathbf{0}$, $\mathbf{x} - \mathbf{A}^T\mathbf{y}^T > \mathbf{0}$, $\mathbf{y}^T + \mathbf{Ax} > \mathbf{0}$ is solvable. (*Hint*: Combine the results of problems 16.10(v) and 16.11.)

16.13 Let

$$\sum_{j=1}^{n} a_{ij}x_j \leq b_i \qquad (i = 1, 2, \ldots, m)$$

be a fixed system of inequalities and let $d_1, d_2, \ldots, d_m$ be positive numbers such that, for each k, the system

$$\sum_{j=1}^{n} a_{ij}x_j \leq b_i \qquad (i \neq k)$$

$$\sum_{j=1}^{n} a_{kj}x_j \leq b_k - d_k$$

is solvable.

Prove that the system

$$\sum_{j=1}^{n} a_{ij}x_j \leq b_i - \frac{d_i}{n+1} \qquad (i = 1, 2, \ldots, m)$$

is solvable. [*Hint*: Use Theorem 9.4.]

16.14 Prove: The system $\mathbf{Ax} \leq \mathbf{0}$, $\mathbf{x} \neq \mathbf{0}$, is unsolvable if and only if the system for $\mathbf{yA} = \mathbf{c}$, $\mathbf{y} \geq \mathbf{0}$ is solvable for every right-hand-side $\mathbf{c}$.

17

Connections with Geometry

The concept of *analytic coordinates*, commonly attributed to René Descartes (1596–1650), brought about one of the great breakthroughs in mathematics. Its idea is quite simple: if one draws a horizontal line and a vertical line in a plane, then the position of each point is uniquely determined by its distance x_1 from the vertical line (taken with a positive sign if the point is to the right of the line, and taken with a negative sign if the point is to the left) together with its distance x_2 from the horizontal line (taken with a positive sign if the point is above the line, and with a negative sign if it is below). The resulting one-to-one correspondence between points of the plane and ordered pairs $[x_1, x_2]$ of real numbers creates a correspondence between plane geometry and certain parts of algebra. This correspondence can be exploited in both directions. On the one hand, it elucidates algebraic arguments by providing them with intuitive geometric interpretations; on the other hand, it makes it possible to investigate problems of geometry by applying algebraic techniques.

In this chapter, we shall use the correspondence in both ways. In the beginning, we shall illustrate in intuitive geometric terms the algebraic techniques developed earlier in this book. Later, we shall reverse the direction and see that certain classical results on the geometry of convex sets follow routinely from the algebraic theorems of Chapters 9 and 16.

As a matter of historical interest, let us note that the idea of establishing coordinates in the plane had been around long before Descartes. It occurred to the land surveyors of ancient Egypt; their hieroglyph for a land district was a symbol derived from the square grid. The Greeks, in particular Appolonius of Perga (c. 247–205 B.C.), used concepts that were at least close to the coordinate system. In the Middle Ages, Nicole Oresme (c. 1323–1382), Bishop of Lisieux in Normandy, wrote two tracts on the coordinate system (one of them, "Tractatus de latitudinibus formarum," was published in Padua, 1482). His work was the subject of lectures at the University in Cologne as early as 1398 and influenced both Kepler and Galileo. Pierre Fermat (1601–1665) had the idea of "analytic geometry" before Descartes' work appeared in print (Fermat's paper, "Isagoge ad locos plans et solidos," was published in Toulouse, 1679). The modern term *coordinates*, referring to the real numbers x_1, x_2 associated with each point of the plane, comes from Gottfried Wilhelm von Leibniz (1646–1716).

Descartes' "La Géometrie" appeared in 1637 as one of the three appendices to his "Discourse de la méthode pour bien conduire sa raison et chercher la vérité dans les sciences." Its consistent exploitation of the correspondence between algebra and geometry, and its clear vision of unifying the two, had a profound influence on subsequent mathematical thinking. In fact, the common term *Cartesian geometry* refers to the Latin version of Descartes' name, Renatus Cartesius. More than two centuries after the publication of "La Géometrie," John Stuart Mill praised the invention of analytic geometry as "the greatest single step ever made in the progress of exact science." □

GEOMETRIC INTERPRETATION OF LP PROBLEMS

Let us begin by observing that the line passing through points [0, 4] and [2, 0] in Figure 17.1 represents all the solutions $[x_1, x_2]$ of $2x_1 + x_2 = 4$; the shaded half-plane represents all the solutions $[x_1, x_2]$ of $2x_1 + x_2 \leq 4$. More generally, for every choice of numbers a_1, a_2, b such that at least one a_j is not zero, all the solutions of

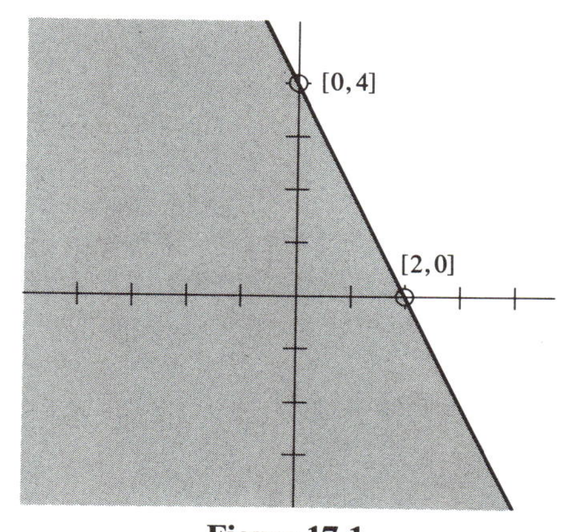

Figure 17.1

$a_1x_1 + a_2x_2 = b$ are represented by a line, whereas all the solutions of $a_1x_1 + a_2x_2 \le b$ are represented by a half-plane bounded by that line. In the space, each point has three coordinates x_1, x_2, x_3. For every choice of numbers a_1, a_2, a_3, b such that at least one a_j is not zero, all the solutions of $a_1x_1 + a_2x_2 + a_3x_3 = b$ are represented by a plane, whereas all the solutions of $a_1x_1 + a_2x_2 + a_3x_3 \le b$ are represented by a half-space bounded by that plane. By analogy, one refers to the set of all the vectors $[x_1, x_2, \ldots, x_n]$ as the *n-dimensional space* R^n, to the set of all the solutions of $a_1x_1 + a_2x_2 + \cdots + a_nx_n \le b$ as a *half-space*, and so on. (According to this definition, the points of R^n are row vectors of length n. Nevertheless, we shall allow ourselves the liberty of viewing these points as column vectors whenever convenient.) Here, however, a word of warning may be in order: do *not* try to visualize n-dimensional objects for $n \ge 4$. Such an effort is not only doomed to failure—it may be dangerous to your mental health. (If you do succeed, then you are in trouble.) To speak of n-dimensional geometry with $n \ge 4$ simply means to speak of a certain part of algebra. Even though it may be convenient to use geometric terms, the corresponding objects are purely algebraic and have no counterpart in our perception of the space.

Now, let us consider the linear programming problem

$$
\begin{array}{lrl}
\text{maximize} & 3x_1 + 2x_2 + 5x_3 & \\
\text{subject to} & 2x_1 + \ \ x_2 \qquad\quad & \le 4 \\
& x_3 & \le 5 \\
& x_1, x_2, x_3 & \ge 0.
\end{array}
$$

Each of the five constraints determines a certain half-space. Hence the *region of feasibility* (that is, the geometric counterpart of the set of all feasible solutions) is the intersection of the five half-spaces. In general, the intersection of a finite number of half-spaces is called a *polyhedron*. In our example, the region of feasibility is the triangular prism (wedge) shown in Figure 17.2. Each of the five facets of the prism corresponds to one of the five constraints in the sense that the points *on* that facet satisfy the corresponding constraint with the sign of *equality*. For example, the triangular bottom facet corresponds to $x_3 \ge 0$, the rectangular facet on the right-hand side corresponds to $2x_1 + x_2 \le 4$, and so on. The correspondence gains elegance when we introduce the slack variables $x_4 = 4 - 2x_1 - x_2$ and $x_5 = 5 - x_3$. Now each facet corresponds to one of the variables $x_1, x_2, \ldots, x_5$ in the sense that each point $[x_1^*, x_2^*, x_3^*]$ on the facet corresponding to x_j satisfies $x_j^* = 0$. Explicitly, the correspondence goes as follows:

bottom	...	x_3
top	...	x_5
front	...	x_2
left	...	x_1
right (shaded)	...	x_4.

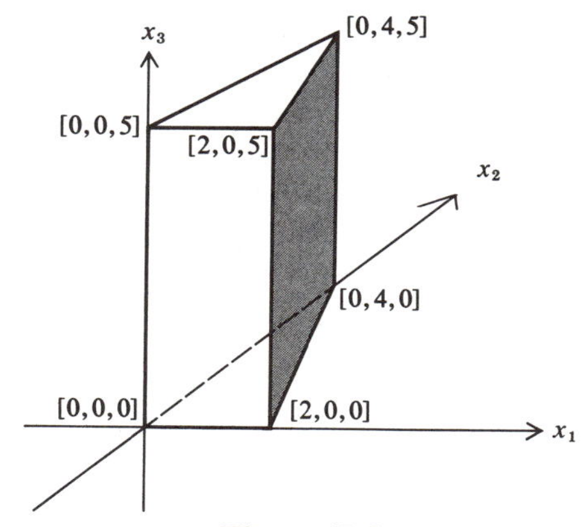

Figure 17.2

Although the problem asks for an optimal solution selected from an *infinite* set of feasible solutions, the simplex method examines only a *finite* number of them: the basic ones. In every basic solution, three of the five variables $x_1, x_2, \ldots, x_5$ are nonbasic. These three variables are set at zero; in fact, setting them at zero determines uniquely the values of the remaining two variables. In geometric terms, this means that every point representing a basic feasible solution is uniquely determined as the intersection of three facets. Such points fit our intuitive notion of *corner-points* or, by another name, *vertices*, of the polyhedron. Hence, at least in this example, the basic feasible solutions (an algebraic concept) correspond to the vertices (a geometric concept) and vice versa.

GEOMETRIC INTERPRETATION OF THE SIMPLEX METHOD

In solving any LP problem, the simplex method constructs a sequence of basic feasible solutions. In our example, this sequence (resulting when the largest-coefficient pivoting rule is used) is as follows:

Initial solution:	$[0, 0, 0, 4, 5]^T$ with x_4, x_5 basic.
After the first iteration:	$[0, 0, 5, 4, 0]^T$ with x_4, x_3 basic.
After the second iteration:	$[2, 0, 5, 0, 0]^T$ with x_1, x_3 basic.
After the third iteration:	$[0, 4, 5, 0, 0]^T$ with x_2, x_3 basic.

These four basic feasible solutions represent four vertices of our polyhedron, namely, [0, 0, 0], [0, 0, 5], [2, 0, 5], and [0, 4, 5]. Note that each iteration corresponds to a continuous motion from one of the vertices in this sequence to the next. For instance, consider the second iteration: having decided to let x_1 enter the basis, we let its value increase while keeping the remaining nonbasic variables x_2 and x_5 at zero. In geometric terms, the condition $x_2 = x_5 = 0$ will confine us to the intersection of the front facet (corresponding to x_2) with the top facet (corresponding to x_5). This intersection is nothing but the edge joining the vertices [0, 0, 5] and [2, 0, 5]. Increasing the value of x_1 means moving along this edge to the right facet, which stops our progress; hence the corresponding variable x_4, whose value drops to zero, leaves the basis.

On nondegenerate problems with no free variables, the simplex method behaves just as it did in our example. It moves along the edges of the polyhedron from one vertex to another, improving the value of the objective function with each iteration, and eventually stopping when no further improvement is possible. Incidentally, it was in the 1820s that J. B. J. Fourier proposed the simplex method in the context of the L_1 and L_∞ approximation problems discussed in Chapter 14. [An accessible reference is J. B. J. Fourier (1890).] In showing how to find the best L_∞ approximation to a solution of a system of linear equations in variables x and y, he used the notion of a point moving along the edges of a polyhedron from one vertex to another. We quote from an English translation by D. A. Kohler (1973):

> We consider a set of linear functions in the variables $x, y, z, \ldots$, the numerical coefficients which enter these functions being given. If the number of functions is not greater than the number of variables, we can find for $x, y, z, \ldots$ a set of numerical values such that the simultaneous substitution of these values in the given functions will set each function to zero. But we cannot in general do this when the number of functions is greater than the number of variables. Let us suppose that we assign to $x, y, z, \ldots$ some numerical values α, β, γ, and that, after substituting these in each function, we compute the value of the functions, which may be positive or negative. We will regard this result as an *error* or *deviation*, and we will take its absolute value as a measure of error.
>
> This done, we ask which numerical values $X, Y, Z, \ldots$ can be assigned to $x, y, z, \ldots$ in order to minimize the maximum error. Alternatively we could attempt to find a set of values $X', Y', Z', \ldots$ of the variables $x, y, z, \ldots$ which minimizes the sum of the absolute errors. Both questions can be resolved by the following general method. It should be stressed that this method can be applied to systems of an arbitrary number of variables and that the basic steps of the algorithm do not change from one problem to the next.
>
> We will use an example in order to describe the method of minimizing the largest absolute error. This example involves only two variables but the reader will be able to see for himself how the method could be generalized to an arbitrary number of variables.
>
> x and y are the coordinates of some point in the horizontal plane. The vertical coordinate z is used to measure the values of the given linear functions of x and y, and each inequality is represented by a plane in three dimensions whose position is given. In the problem which concerns us, the number of these planes is double the number of functions because it is necessary to deal separately with positive and negative

errors. We only consider the parts of the planes which are above the horizontal plane, i.e., the xy-plane. Note that the system of planes forms a *vase* and the shape of this *vase* is that of a convex polyhedron. The lowest point of the *vase* or polyhedron has as coordinates the values X, Y, Z which we wish to find, i.e., Z is the least possible value of the largest deviation and X and Y are the values of x and y which give this minimum.

In order to reach the lowest point of the *vase* as quickly as possible, we start from an arbitrary point (x, y) on the horizontal plane, for example (x, y) could be the origin $(0, 0)$, and select for z the largest absolute error assumed by any function when these values of x and y are used, i.e., among all the points of intersection of the planes with the vertical line through the point (x, y) we choose the one most distant from the xy plane. Let m_1 be this point. It is situated on one of the faces of the *vase*. We descend on this face from the point m_1 to the point m_2 on one edge of the polyhedron and, following this edge, we descend from the point m_2 to the extreme point m_3 at the intersection of three extreme hyperplanes. From the point m_3 we continue to descend following a second edge to a new extreme point m_4; and we continue the application of the same procedure, always following the edge which descends most steeply from the extreme point. In this way we quickly arrive at the lowest point of the polyhedron.

We have described the method in geometrical terms because this is the best way of understanding it. However, there is of course a sequence of numerical operations which corresponds to the geometrical operations we have just described and which can be applied to problems of any size. Unfortunately, there is not the space here to describe them in detail.

THE KLEE–MINTY EXAMPLES REVIEWED

In its quest for an optimal vertex, the simplex method does not always trace the shortest route: in our example, going from $[0, 0, 5]$ to $[0, 4, 5]$, it took a detour via $[2, 0, 5]$. This myopic behavior reaches bizarre proportions in the examples constructed by Klee and Minty. We alluded to these examples in Chapter 4. Now, restricting ourselves to three dimensions, we shall motivate their construction in geometric terms. The idea is

(i) To find a polyhedron P similar to the ordinary cube, together with an objective function that increases steadily along some path through all the eight vertices of P.

(ii) To force the simplex method into following that path by choosing a sufficiently devious algebraic representation of the corresponding LP problem.

To achieve our first objective, we take the ordinary cube represented algebraically by

$$0 \leq x_j \leq 1 \qquad (j = 1, 2, 3). \tag{17.1}$$

It is not difficult to verify that, no matter what objective function $z = c_1x_1 + c_2x_2 + c_3x_3$ we choose, every path along which z increases involves at most four of the eight vertices. Therefore we shall have to squash one cube a little before proceeding. Squashing the cube is a geometric concept; its algebraic counterpart amounts to perturbing the coefficients in (17.1). The reader may easily verify that the perturbed version of (17.1).

$$\begin{aligned} x_1 \qquad\qquad\quad &\leq 1 \\ 0.2x_1 + \quad x_2 \qquad &\leq 1 \\ 0.02x_1 + 0.2x_2 + x_3 &\leq 1 \\ x_1, x_2, x_3 &\geq 0 \end{aligned} \tag{17.2}$$

represents the squashed cube shown in Figure 17.3. The arrows in that figure point out a path along which the value of $z = 100x_1 + 1{,}000x_2 + 10{,}000x_3$ steadily increases. Thus objective (i) is accomplished. Note, however, that objective (ii) is not: it takes only one iteration of the simplex method to maximize z subject to (17.2).

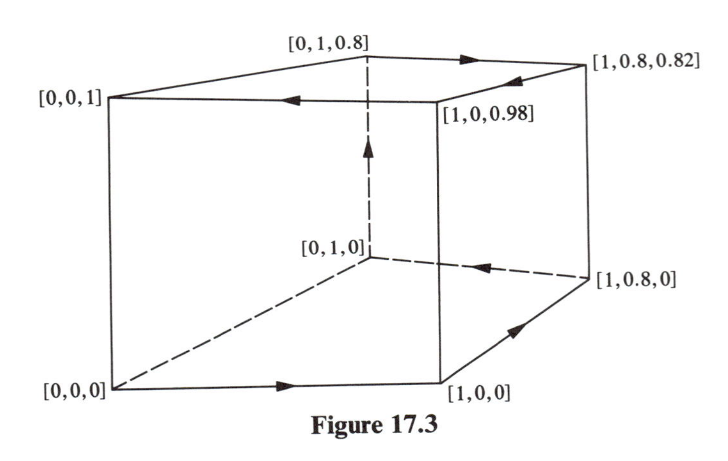

Figure 17.3

To achieve our second objective, let us recall an observation made in Chapter 4: the largest-coefficient pivoting rule is sensitive to the scale on which the variables are measured. Changing the units of this scale amounts to replacing each of the variables x_1, x_2, x_3 and

$$\begin{aligned} x_4 &= 1 - \quad x_1 \\ x_5 &= 1 - \ 0.2x_1 - \quad x_2 \\ x_6 &= 1 - 0.02x_1 - 0.2x_2 - x_3 \end{aligned}$$

by some $s_j\bar{x}_j$ such that $\bar{x}_j$ is a scaled version of x_j, and s_j is the scaling factor: one unit of $\overline{x}_j$ is equivalent to s_j units of x_j. These substitutions (with positive s_j's) convert (17.2) into

$$\begin{aligned} \frac{s_1}{s_4}\overline{x}_1 \qquad\qquad\qquad\qquad\qquad\quad &\leq \frac{1}{s_4} \\ 0.2\frac{s_1}{s_5}\overline{x}_1 + \frac{s_2}{s_5}\overline{x}_2 \qquad\qquad\quad &\leq \frac{1}{s_5} \\ 0.02\frac{s_1}{s_6}\overline{x}_1 + 0.2\frac{s_2}{s_6}\overline{x}_2 + \frac{s_3}{s_6}\overline{x}_3 &\leq \frac{1}{s_6} \\ \overline{x}_1, \overline{x}_2, \overline{x}_3 &\geq 0. \end{aligned}$$

More importantly, they convert each expression

$$z = v + \sum_{j \in N} d_j x_j$$

[with $x_j (j \in N)$ representing the nonbasic variables] into

$$z = v + \sum_{j \in N} (d_j s_j)\overline{x}_j.$$

Now, when it comes to choosing an entering variable, we may stack the odds against $\overline{x}_j$ by making s_j relatively small. Consulting Figure 17.3, we may rank the variables $\overline{x}_j = s x_j$ as to their desirability for entering the basis:

$\overline{x}_1$ and $\overline{x}_4$ ... most desirable

$\overline{x}_2$ and $\overline{x}_5$ ... medium

$\overline{x}_3$ ... least desirable.

Hence we should choose s_2 and s_5 very small and s_3 extremely small with respect to s_1 and s_4; the choice of s_6 affects only the appearance of the resulting problem. A bit of experimenting shows that

$$s_1 = s_4 = 1, \quad s_2 = s_5 = 0.01, \quad s_3 = s_6 = 0.0001$$

will do the job quite nicely. The resulting problem,

$$\begin{aligned} \text{maximize} \quad & 100\overline{x}_1 + 10\overline{x}_2 + \overline{x}_3 \\ \text{subject to} \quad & \overline{x}_1 \qquad\qquad\qquad\quad \leq 1 \\ & 20\overline{x}_1 + \overline{x}_2 \qquad\quad \leq 100 \\ & 200\overline{x}_1 + 20\overline{x}_2 + \overline{x}_3 \leq 10{,}000 \\ & \overline{x}_1, \overline{x}_2, \overline{x}_3 \geq 0 \end{aligned}$$

is precisely the one we worked out in Chapter 4. (The simplex method, misdirected by the largest-coefficient rule, takes seven iterations to solve it.) Since we know better than to visualize in four dimensions, we shall refrain from discussing its n-dimensional counterparts here.

GEOMETRIC INTERPRETATION OF THE PERTURBATION METHOD

Consider the problem

$$\begin{array}{lrl} \text{maximize} & x_1 + x_2 + 4x_3 & \\ \text{subject to} & x_1 \quad\quad + 4x_3 & \leq 4 \\ & x_2 + 4x_3 & \leq 4 \\ & x_1, x_2, x_3 & \geq 0 \end{array} \tag{17.3}$$

whose region of feasibility is shown in Figure 17.4. This polyhedron exhibits a feature unprecedented in the previous examples: *four*, not three, of its facets meet at the vertex [0, 0, 1]. Consequently, if we wish to specify this vertex as an intersection of *three* facets, we can do so in several different ways. The vertex [0, 0, 1] is the intersection of the two vertical facets (front and left) with the light top facet, but it is also the intersection of the two top facets (light and dark) with the left vertical facet, and so on. In algebraic terms, this means that if we set any three of the four variables x_1, x_2, x_4, x_5 at zero, we force the fourth variable to be zero as well. In particular, if three of these four variables are nonbasic, then the fourth basic variable must become zero. Of course, this phenomenon is what we call degeneracy.

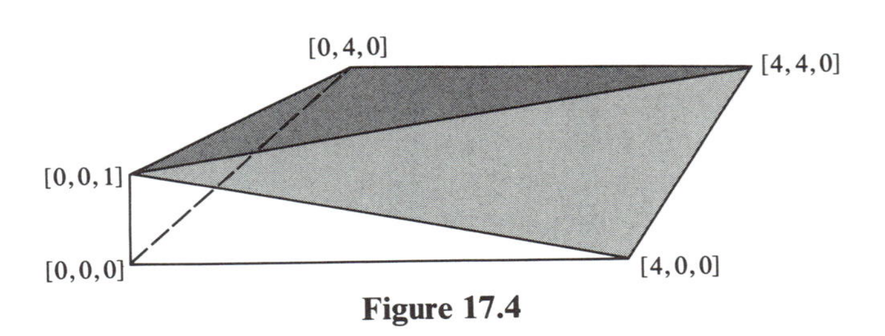

Figure 17.4

The simplex method applied to (17.3) may construct the following sequence of basic feasible solutions.

Initial solution:	$[0, 0, 0, 4, 4]^T$ with x_4, x_5 basic.
After the first iteration:	$[0, 0, 1, 0, 0]^T$ with x_3, x_5 basic.
After the second iteration:	$[0, 0, 1, 0, 0]^T$ with x_3, x_2 basic.
After the third iteration:	$[4, 4, 0, 0, 0]^T$ with x_1, x_2 basic.

In geometric terms, it moves from the vertex [0, 0, 0] to [0, 0, 1] in the first iteration, stays as [0, 0, 1] during the second iteration, and finally moves from [0, 0, 1] to the optimal vertex [4, 4, 0] in the third iteration. The second and the third solutions describe the same degenerate vertex [0, 0, 1]; however, they differ in how they specify that vertex as an intersection of three facets. (Similarly, when the simplex method cycles in certain examples, it does not actually run around their polyhedra in circles. Instead, it remains stationary at a degenerate vertex, changing only the ways of specifying that vertex.)

Applying the perturbation method to (17.3) means replacing the constraints

$$x_1 + 4x_3 \leq 4, \quad x_2 + 4x_3 \leq 4$$

by

$$x_1 + 4x_3 \leq 4 + \varepsilon, \quad x_2 + 4x_3 \leq 4 + \varepsilon^2.$$

Geometrically, the perturbation amounts to almost imperceptible shifts of the two top facets of our polyhedron. Hence, the region of feasibility remains unaffected for all practical purposes. However, watching the vertex [0, 0, 1] through a strong microscope, we might see it split into two nondegenerate vertices $[0, 0, 1 + 0.25\varepsilon^2]$ and $[\varepsilon - \varepsilon^2, 0, 1 + 0.25\varepsilon^2]$ as shown in Figure 17.5. In general, the perturbation method splits every degenerate vertex into a cluster of nondegenerate vertices.

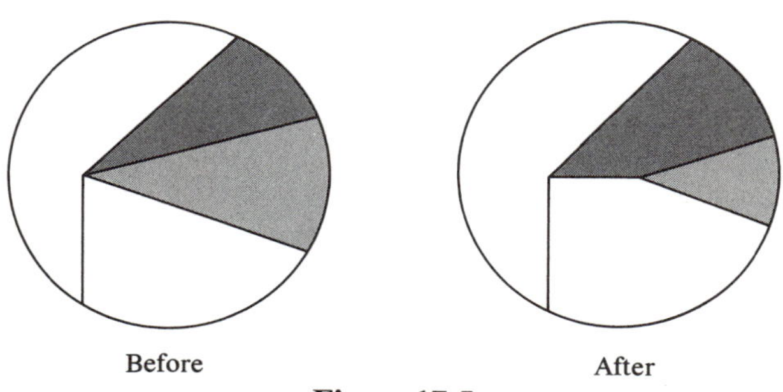

Figure 17.5

Before leaving the subject, let us note that degeneracy does not always manifest itself geometrically. For example, let us consider the problem

$$\begin{array}{ll} \text{maximize} & x_1 \\ \text{subject to} & x_1 - x_2 + x_3 \leq 0 \\ & \quad\;\; x_2 \qquad\; \leq 1 \\ & \quad x_1, x_2, x_3 \geq 0. \end{array}$$

Observe that the initial dictionary is degenerate and yet only three facets meet at the corresponding vertex $[0, 0, 0]$ in Figure 17.6. The reason is that no facet corresponds to $x_2 = 0$. That is simply because the constraint $x_2 \geq 0$ is redundant: it is implied by the conjunction of $x_1 - x_2 + x_3 \leq 0$ with $x_1 \geq 0$ and $x_3 \geq 0$. Since the deletion of $x_2 \geq 0$ does not change the region of feasibility, this constraint cannot manifest itself by a facet.

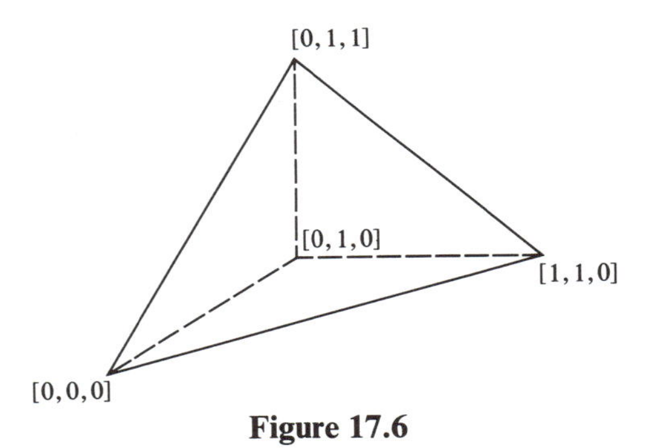

Figure 17.6

THE GRAPHIC METHOD

The geometric interpretation of linear programming leads to a geometric procedure for solving LP problems with two variables. For illustration, let us consider the problem

$$\begin{array}{lrcl} \text{maximize} & x_1 + x_2 & & \\ \text{subject to} & 2x_1 + x_2 & \leq & 14 \\ & -x_1 + 2x_2 & \leq & 8 \\ & 2x_1 - x_2 & \leq & 10 \\ & x_1, x_2 & \geq & 0. \end{array} \tag{17.4}$$

Its region of feasibility is shown in Figure 17.7, together with the line representing all the points

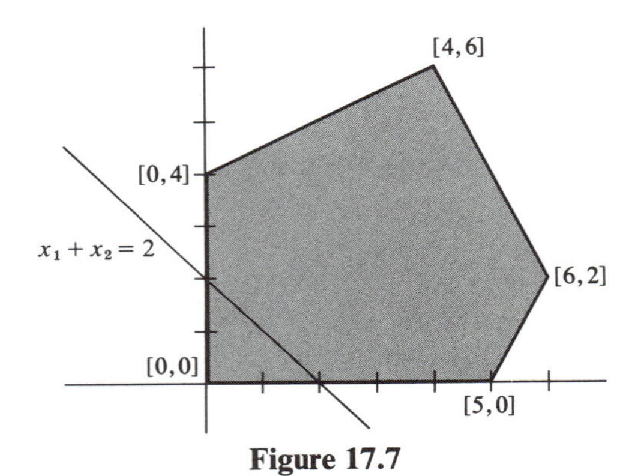

Figure 17.7

$[x_1, x_2]$ such that $x_1 + x_2 = 2$. For every number d, the line $x_1 + x_2 = d$ is parallel to $x_1 + x_2 = 2$. If $d > 2$, then $x_1 + x_2 = d$ lies northeast of $x_1 + x_2 = 2$, if $d < 2$, then $x_1 + x_2 = d$ lies southwest of $x_1 + x_2 = 2$. Now the graphic method for solving problems

$$\begin{array}{lll} \text{maximize} & c_1x_1 + c_2x_2 & \\ \text{subject to} & a_{i1}x_1 + a_{i2}x_2 \le b_i & (i = 1, 2, \ldots, m) \end{array}$$

suggests itself. First plot the region of feasibility and draw some line $c_1x_1 + c_2x_2 = d$ passing through the region. Then slide a straightedge through the plane so that it remains parallel to the line $c_1x_1 + c_2x_2 = d$. Proceeding in the direction of increasing $c_1x_1 + c_2x_2$, stop when the straightedge is about to leave the region of feasibility. The last point of contact represents the optimal solution. (It may also happen that the last point of contact is not unique. In that case, one of the edges of the polygon represents all the optimal solutions.) The final result in our example is shown in Figure 17.8.

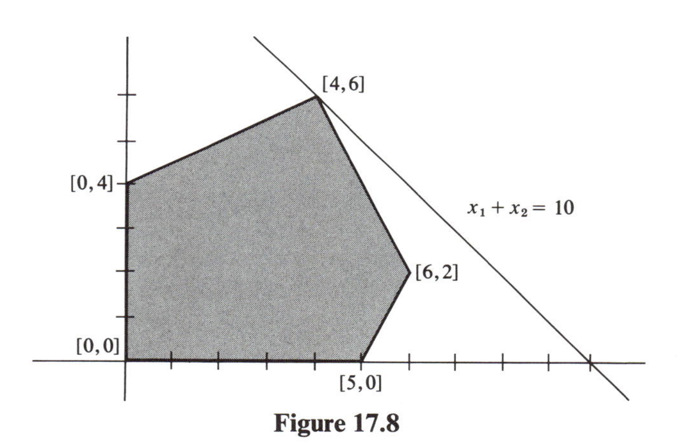

Figure 17.8

GEOMETRIC INTERPRETATION OF DUALITY

The notion of duality was motivated in Chapter 5 by our desire to devise upper bounds on the optimum values of LP problems. In example (17.4), the optimal solution $x_1^* = 4$, $x_2^* = 6$ yields $x_1^* + x_2^* = 10$. To prove that no other feasible solution can do better, we have to verify only that $y_1^* = \frac{3}{5}$, $y_2^* = \frac{1}{5}$, $y_3^* = 0$ is a feasible solution of the dual problem

$$\begin{array}{ll} \text{minimize} & 14y_1 + 8y_2 + 10y_3 \\ \text{subject to} & 2y_1 - y_2 + 2y_3 \ge 1 \\ & y_1 + 2y_2 - y_3 \ge 1 \\ & y_1, y_2, y_3 \ge 0 \end{array}$$

and that $14y_1^* + 8y_2^* + y_3^* = 10$. To put it differently, we multiply the constraint $2x_1 + x_2 \le 14$ by $\frac{3}{5}$ and multiply $-x_1 + 2x_2 \le 8$ by $\frac{1}{5}$. Since the sum of the resulting two inequalities reads $x_1 + x_2 \le 10$, we conclude that every feasible solution x_1, x_2 of (17.4) satisfies $x_1 + x_2 \le 10$.

Interpreted geometrically, this conclusion means that the entire region of feasibility is contained in the half plane $x_1 + x_2 \le 10$; this fact is evident from Figure 17.8. Actually, both Figure 17.8

and the algebraic argument just given support a stronger conclusion: the set of all the points $[x_1, x_2]$ satisfying

$$\begin{aligned} 2x_1 + x_2 &\le 14 \\ -x_1 + 2x_2 &\le 8 \end{aligned} \tag{17.5}$$

is a subset of the half plane $x_1 + x_2 \le 10$. Now, let us consider linear combinations of the two inequalities (17.5). Each of these combinations has the form

$$(2v - w)x_1 + (v + 2w)x_2 \le 14v + 8w$$

for some nonnegative v and w. Geometrically, it represents a half-plane that contains the set represented by (17.5) and whose boundary line passes through the point [4, 6]. Two of these boundary lines, one corresponding to $v = 2, w = 1$ and the other corresponding to $v = 1, w = 2$, are shown in Figure 17.9. The family of all the lines

$$(2v - w)x_1 + (v + 2w)x_2 = 14v + 8w \tag{17.6}$$

may be thought of as a single line that is attached to a hinge at [4, 6] but is free to rotate on this hinge. Continuous changes of the multipliers v and w amount to continuous rotations of the line. In view of this interpretation, it is perhaps not surprising that one can choose nonnegative v and w so as to make (17.6) coincide with the line $x_1 + x_2 = 10$ shown in Figure 17.8. This, of course, is precisely the claim whose validity is guaranteed by the Duality Theorem. □

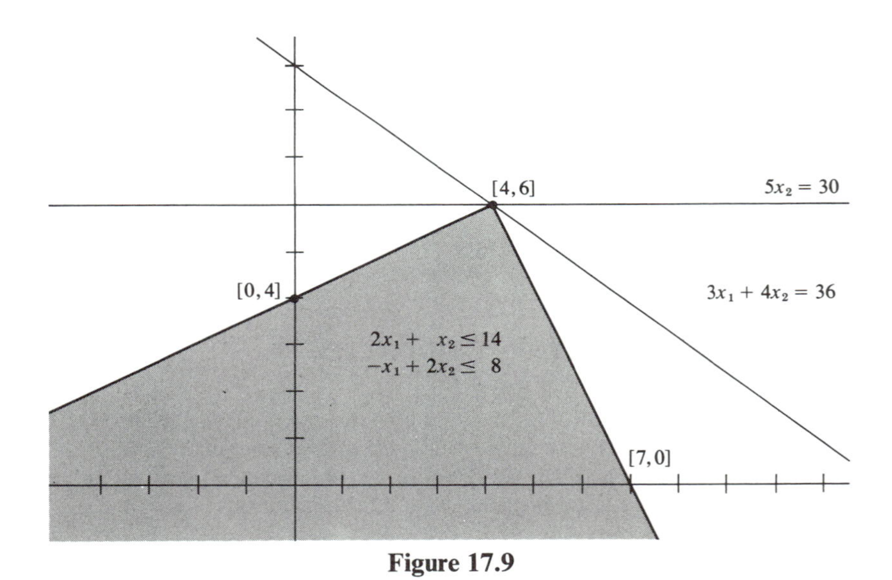

Figure 17.9

CONVEX SETS AND CONVEX HULLS

In the remainder of this chapter, we shall apply the algebraic machinery developed in Chapters 9 and 16 to the geometry of convex sets, a field founded by H. Brunn (1862–1939) and H. Minkowski (1864–1909). To motivate its central notion, let us

Figure 17.10

consider the four sets shown in Figure 17.10. Each of these four sets may be thought of as the floor plan of a room in a somewhat unusual house. A game of hide-and-seek might conceivably hold some attraction for the occupants of S_3 and S_4. However, there is no point in playing hide-and-seek in S_1 or S_2: in each of these two rooms, every point is visible from all the remaining points. Sets with this property are called *convex*.

In order to couch this definition in algebraic terms, we shall first have to make algebraic the notion of visibility. Assuming that light travels along straight lines, we observe that two points **x**, **y** are visible from each other if and only if there is no obstruction on the line segment that joins them. If $\mathbf{x} = [x_1, x_2]$ and $\mathbf{y} = [y_1, y_2]$ are distinct points in the plane, then, as the reader may verify, a point **z** lies on the line segment joining **x** and **y** if and only if

$$\mathbf{z} = t\mathbf{x} + (1 - t)\mathbf{y} = [tx_1 + (1 - t)y_1, tx_2 + (1 - t)y_2]$$

for some t such that $0 \leq t \leq 1$. (Note that $t = 0$ corresponds to $\mathbf{z} = \mathbf{y}$ and that $t = 1$ corresponds to $\mathbf{z} = \mathbf{x}$.) By analogy, the line segment joining distinct points $\mathbf{x}, \mathbf{y} \in R^n$ is defined to consist of all the points $\mathbf{z} = t\mathbf{x} + (1 - t)\mathbf{y}$ such that $0 \leq t \leq 1$. Hence a subset S of R^n is called *convex* if and only if

$$t\mathbf{x} + (1 - t)\mathbf{y} \in S \qquad \text{whenever } \mathbf{x} \in S, \quad \mathbf{y} \in S, \quad \text{and } 0 \leq t \leq 1.$$

One of the immediate corollaries of this definition follows:

If F is a family of convex sets,
then the intersection of all the sets in F is also convex. (17.7)

The proof is trivial: if points **x** and **y** belong to every $C \in F$ and if $\mathbf{z} = t\mathbf{x} + (1 - t)\mathbf{y}$ with $0 \leq t \leq 1$, then **z** belongs to every $C \in F$. In turn, (17.7) implies that for every set $S \subseteq R^n$ there is *the* smallest convex set C^* containing S. More precisely, *for every set $S \subseteq R^n$ there is a convex set C^* containing S and contained in every convex set containing S.* Indeed, let F be the family of all convex sets containing S and let C^* denote the intersection of all $C \in F$. Trivially, $S \subseteq C^*$ and $C^* \subseteq C$ for all $C \in F$. Furthermore, C^* is convex by (17.7). The set C^* is called the *convex hull* of S. Figure 17.11 depicts a planar set and its convex hull.

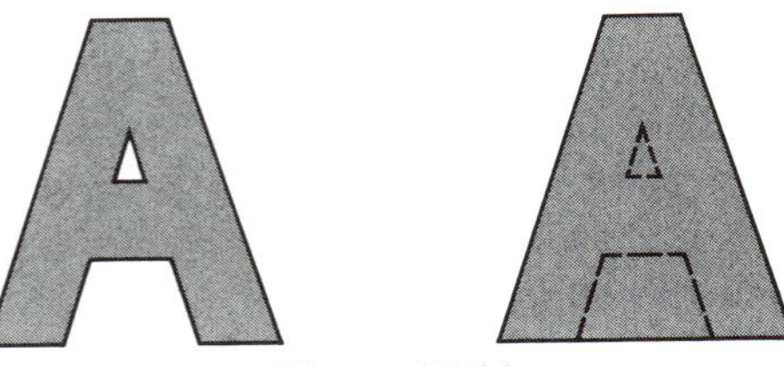

Figure 17.11

It is easy to establish a more explicit characterization of convex hulls. Let S be an arbitrary set in R^n, and let H denote the set of all the points $\mathbf{z} \in R^n$ such that, for some positive integer k, some points $\mathbf{z}_1, \mathbf{z}_2, \ldots, \mathbf{z}_k$ in S, and some *positive* numbers $t_1, t_2, \ldots, t_k$ we have

$$\mathbf{z} = \sum_{i=1}^{k} t_i \mathbf{z}_i \quad \text{and} \quad \sum_{i=1}^{k} t_i = 1. \tag{17.8}$$

We propose to show that

$$H \text{ is the convex hull of } S. \tag{17.9}$$

Our task amounts to routine verifications of the following three facts:

(i) H contains S.

(ii) H is convex.

(iii) Every convex set containing S must contain H.

The first fact is trivial: points of S are precisely the points $\mathbf{z}$ defined by (17.8) with $k = 1$. To establish fact (ii), consider arbitrary points $\mathbf{x} \in H$, $\mathbf{y} \in H$, and $\mathbf{z} = t\mathbf{x} + (1 - t)\mathbf{y}$ such that $0 < t < 1$. Since $\mathbf{x} \in H$ and $\mathbf{y} \in H$, there are points $\mathbf{x}_1, \mathbf{x}_2, \ldots, \mathbf{x}_r, \mathbf{y}_1, \mathbf{y}_2, \ldots, \mathbf{y}_s \in S$ and positive numbers $p_1, p_2, \ldots, p_r, q_1, q_2, \ldots, q_s$ such that $\mathbf{x} = \sum p_i \mathbf{x}_i, \sum p_i = 1, \mathbf{y} = \sum q_i \mathbf{y}_i, \sum q_i = 1$. Writing $t_i = tp_i, \mathbf{z}_i = \mathbf{x}_i$ for $i = 1, 2, \ldots, r$ and $t_{r+i} = (1 - t)q_i, \mathbf{z}_{r+i} = \mathbf{y}_i$ for $i = 1, 2, \ldots, s$ we have

$$\sum_{i=1}^{r+s} t_i \mathbf{z}_i = \sum_{i=1}^{r} tp_i \mathbf{x}_i + \sum_{i=1}^{s} (1 - t) q_i \mathbf{y}_i = t\mathbf{x} + (1 - t)\mathbf{y} = \mathbf{z}$$

$$\sum_{i=1}^{r+s} t_i = \sum_{i=1}^{r} tp_i + \sum_{i=1}^{s} (1 - t) q_i = t + (1 - t) = 1.$$

Hence $\mathbf{z} \in H$. Finally, we shall verify fact (iii). For that purpose, we shall consider an arbitrary convex set C containing S, and use induction on k to show that every point z defined by (17.8) must belong to C. If $k = 1$ then the conclusion is trivial: as we have already observed, point $\mathbf{z}$ defined by (17.8) with $k = 1$ are precisely the points of S. Now consider an arbitrary point $\mathbf{z}$ defined by (17.8) and assume that the conclusion is true for $k - 1$ in place of k. Writing t for $1 - t_k$ and $r_i = t_i/t$ for all $i = 1, 2, \ldots, k - 1$, we have $0 < t < 1$ and

$$\sum_{i=1}^{k-1} r_i = \frac{1}{t}\sum_{i=1}^{k-1} t_i = \frac{1}{t}(1 - t_k) = 1.$$

Hence, by the induction hypothesis, the point $\mathbf{x} = \sum r_i \mathbf{z}_i$ belongs to C. Of course, the point $\mathbf{z}_k \in S$ also belongs to C. But then C, being convex, must contain the point

$$t\mathbf{x} + (1 - t)\mathbf{z}_k = \sum_{i=1}^{k-1} t_i \mathbf{z}_i + t_k \mathbf{z}_k = \mathbf{z}.$$

This is the desired conclusion.

APPLICATIONS OF LINEAR PROGRAMMING TO THE GEOMETRY OF CONVEX SETS

Our characterization of convex hulls shows that a point $\mathbf{z}$ belongs to the convex hull of a set $S \subseteq R^n$ if and only if $\mathbf{z}$ belongs to the convex hull of some set of finitely many points $\mathbf{z}_1\ \mathbf{z}_2, \ldots, \mathbf{z}_k \in S$. For infinite sets S, this assertion may not be entirely obvious. The following theorem, due to C. Carathéodory (1907), strengthens it even further, replacing the phrase "finitely many" by "at most $n + 1$."

THEOREM 17.1. If $S \subseteq R^n$ then a point $\mathbf{z}$ belongs to the convex hull of S if and only if there are points $\mathbf{z}_1, \mathbf{z}_2, \ldots, \mathbf{z}_k$ in S and positive numbers $t_1, t_2, \ldots, t_k$ such that $k \leq n + 1$, $\sum t_i = 1$, and $\mathbf{z} = \sum t_i \mathbf{z}_i$.

PROOF. The "if" part follows from (17.9). To prove the "only if" part, consider an arbitrary point $\mathbf{z} = [z_1, z_2, \ldots, z_n]$ in the convex hull of S. By (17.9) again, we have (17.8) for some positive integer k, some points $\mathbf{z}_1, \mathbf{z}_2, \ldots, \mathbf{z}_k$ in S, and some positive numbers $t_1, t_2, \ldots, t_k$. That is, the system $\sum t_i \mathbf{z}_i = \mathbf{z}$, $\sum t_i = 1$ of $n + 1$ linear equations in variables t_i has a positive solution. Hence, by Theorem 9.3, it has a nonnegative solution with at most $n + 1$ variables positive. ■

In particular, Theorem 17.1 asserts that a point belongs to the convex hull of a set $S \subseteq R^2$ if and only if it belongs to the convex hull of some three-point subset of S. With "three" replaced by "two," the claim may fail; for example, $[1, 1] = \frac{1}{3}[3, 0] + \frac{1}{3}[0, 3] + \frac{1}{3}[0, 0]$, but $[1, 1]$ does *not* belong to the convex hull of any two of the three points $[3, 0]$, $[0, 3]$, $[0, 0]$. More generally, the point $[1, 1, \ldots, 1] \in R^n$ belongs to the convex hull of the $n+1$ points $[n+1, 0, \ldots, 0]$, $[0, n+1, \ldots, 0], \ldots, [0, 0, \ldots, n + 1]$ and $[0, 0, \ldots, 0]$, but it does not belong to the convex hull of any n of these $n + 1$ points.

Next, let us note that

every half-space is convex. (17.10)

The proof is straightforward: if $\sum a_j x_j \leq b$, $\sum a_j y_j \leq b$, and $\mathbf{z} = t\mathbf{x} + (1 - t)\mathbf{y}$ with $0 \leq t \leq 1$, then $\sum a_j z_j = \sum a_j(tx_j + (1 - t)y_j) = t\sum a_j x_j + (1 - t)\sum a_j y_j \leq tb + (1 - t)b = b$. Combining (17.10) with (17.7) we conclude that *every polyhedron is convex*. Polyhedra play a special role in the theory of convex sets: Theorem 16.4 with $N = 0$ asserts that

the convex hull of a finite set is always a polyhedron. (17.11)

Our next theorem is an important result discovered by E. Helly (1923) in 1913 but not published by him until ten years later.

THEOREM 17.2. Let F be a *finite* family of at least $n + 1$ convex sets in R^n such that every $n + 1$ sets in F have a point in common. Then all the sets in F have a point in common.

PROOF. For definiteness, let us enumerate the members of F as $C_1, C_2, \ldots, C_k$. Our first task is finding polyhedra $P_1, P_2, \ldots, P_k$ such that

$$P_r \subseteq C_r \qquad \text{for all } r = 1, 2, \ldots, k \tag{17.12}$$

and such that

every $n + 1$ of the polyhedra P_r have a point in common. (17.13)

For this purpose, let us observe that there is a *finite* set $S \subseteq R^n$ that, for every choice of $n + 1$ sets in F, contains at least one point common to these sets. (Such a set may be constructed by examining each of the finitely many choices of $n + 1$ sets in F. For each choice of $n + 1$ sets in F, there is at least one point common to these sets, and this point may be included in S.) For each $r = 1, 2, \ldots, k$, let P_r denote the convex hull of the *finite* set $S \cap C_r$. By (17.11), P_r is a polyhedron; since $S \cap C_r \subseteq C_r$ and since C_r is convex, we have $P_r \subseteq C_r$. The remaining property (17.13) follows from the property of S stipulated previously; for every choice of $n + 1$ sets C_r, the corresponding sets $S \cap C_r$ have a point in common. Hence the $n + 1$ polyhedra P_r, each containing $S \cap C_r$, have a point in common.

The rest is straightforward. By the definition of a polyhedron, each P_r consists of all the solutions to some system of linear inequalities. By (17.13), each system of $n + 1$ linear inequalities chosen from the aggregate of our k systems is solvable. Hence, by Theorem 9.4, the entire aggregate is solvable. To put it differently, the polyhedra $P_1, P_2, \ldots, P_k$ have at least one point in common. By (17.12), this point belongs to all the sets $C_1, C_2, \ldots, C_k$. ■

Without the assumption that F is *finite*, the conclusion of Theorem 17.2 may fail. For example, let H_k denote the half-space

$$x_1 + x_2 + \cdots + x_n \geq k$$

and let F be the infinite family consisting of H_1, H_2, H_3, and so on. There is no point common to all of these half-spaces, even though every finite number of them do have a common point.

In case $n = 2$, Theorem 17.2 asserts that convex sets $C_1, C_2, \ldots, C_k$ $(k \geq 3)$ in the plane have a point in common if and only if every three of them have a point in common. With "three" replaced by "two," the claim may fail. For example, each of the three sides of a triangle is a line segment and therefore a convex set; every two of the sides meet in a vertex but there is no point common to all three of them. More generally, let P_i denote the polyhedron defined by the system

$$x_j \geq 0 \qquad (j = 1, 2, \ldots, n)$$

$$\sum_{j=1}^{n} x_j \leq 1$$

in which the ith inequality has been replaced by an equation. It is not difficult to verify that every n of the polyhedra $P_1, P_2, \ldots, P_{n+1}$ have a point in common but that there is no point common to all $n + 1$ of them.

Other theorems on convex sets that follow easily from the results of Chapters 9 and 16 include Kirchberger's separation theorem (problem 17.2) and a theorem of Blaschke on the size of balls contained in convex sets (problems 17.3–17.7). For a host of information on Helly's theorem and related results, the reader is referred to L. Danzer, B. Grünbaum, and V. Klee (1963). We close this chapter with a theorem known as "the separation theorem for polyhedra."

THEOREM 17.3. For every pair of nonempty disjoint polyhedra P_1, P_2 in R^n there is a pair of disjoint half-spaces H_1, H_2 such that $P_1 \subseteq H_1$ and $P_2 \subseteq H_2$.

PROOF. By definition, P_1 consists of all the solutions to some system $\mathbf{Ax} \leq \mathbf{b}$ and P_2 consists of all the solutions to some system $\tilde{\mathbf{A}}\mathbf{x} \leq \tilde{\mathbf{b}}$. Since P_1 and P_2 are disjoint, the aggregate system $\mathbf{Ax} \leq \mathbf{b}$, $\tilde{\mathbf{A}}\mathbf{x} \leq \tilde{\mathbf{b}}$ is unsolvable. Hence, by Theorem 9.2, it must be inconsistent in the sense that $\mathbf{yA} + \tilde{\mathbf{y}}\tilde{\mathbf{A}} = \mathbf{0}$ and $\mathbf{yb} + \tilde{\mathbf{y}}\tilde{\mathbf{b}} < 0$ for some non-negative row vectors $\mathbf{y}$ and $\tilde{\mathbf{y}}$. Without loss of generality, we may assume that $\mathbf{yb} < 0$. Now it follows that $\mathbf{yA} \neq \mathbf{0}$, for otherwise the system $\mathbf{Ax} \leq \mathbf{b}$ would be inconsistent, contradicting the assumption that P_1 is nonempty. We conclude that the sets H_1 and H_2 defined by $(\mathbf{yA})\mathbf{x} \leq \mathbf{yb}$ and $(\mathbf{yA})\mathbf{x} \geq -\tilde{\mathbf{y}}\tilde{\mathbf{b}}$, respectively, are halfspaces. Since $\mathbf{yb} < -\tilde{\mathbf{y}}\tilde{\mathbf{b}}$, these two halfspaces are disjoint. Since every solution $\mathbf{x}$ of $\mathbf{Ax} \leq \mathbf{b}$

satisfies $(\mathbf{yA})\mathbf{x} \le \mathbf{yb}$, we have $P_1 \subseteq H_1$. Since $\mathbf{yA} = -\tilde{\mathbf{y}}\tilde{\mathbf{A}}$, every solution $\mathbf{x}$ of $\tilde{\mathbf{A}}\mathbf{x} \le \tilde{\mathbf{b}}$ satisfies $(-\mathbf{yA})\mathbf{x} = (\tilde{\mathbf{y}}\tilde{\mathbf{A}})\mathbf{x} \le \tilde{\mathbf{y}}\tilde{\mathbf{b}}$, and so we have $P_2 \subseteq H_2$. ■

A generalization of Theorem 17.3 to arbitrary convex sets C_1, C_2 in place of the polyhedra P_1, P_2 is no longer valid. For example, consider the sets C_1, C_2 in the plane such that $[x_1, x_2] \in C_1$ if and only if $x_2 \le 0$ and such that $[x_1, x_2] \in C_2$ if and only if $x_1x_2 \ge 1$ and $x_2 > 0$. Trivially, C_1, C_2 are disjoint, and the half-space C_1 is convex. Establishing the convexity of C_2 is a little trickier. To begin with, note that every number r satisfies $r^2 - 2r + 1 = (r - 1)^2 \ge 0$ and so, as long as $r > 0$, we have

$$r + (1/r) \ge 2.$$

Hence every $\mathbf{x} = [x_1, x_2] \in C_2$ and every $\mathbf{y} = [y_1, y_2] \in C_2$ satisfy

$$x_1y_2 + y_1x_2 \ge (x_1/y_1) + (y_1/x_1) \ge 2.$$

If now $\mathbf{z} = t\mathbf{x} + (1 - t)\mathbf{y}$ with $0 < t < 1$, then

$$\begin{aligned} z_1z_2 &= t^2x_1x_2 + t(1 - t)(x_1y_2 + y_1x_2) + (1 - t)^2y_1y_2 \\ &\ge t^2 + 2t(1 - t) + (1 - t)^2 = 1 \end{aligned}$$

and so $\mathbf{z} \in C_2$. The sets C_1 and C_2 are shown in Figure 17.12. It is easy to see that every half-space disjoint from C_1 must have the form $x_2 \ge b$ for some *positive* constant b. However, no such half-space contains C_2.

A counterexample of a different nature is provided by sets C_1 such that $[x_1, x_2] \in C_1$ if and only if $-1 \le x_1 \le 1$ and $x_2 = 0$, and such that $[x_1, x_2] \in C_2$ if and only if $x_1 = 0$ and $0 < x_2 \le 1$.

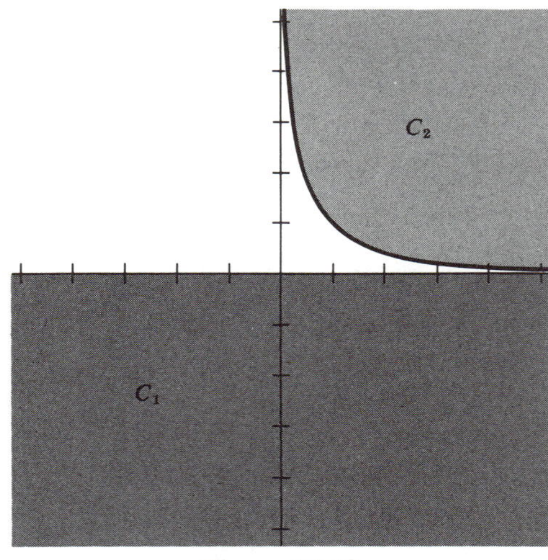

Figure 17.12

Note that $[0, 0] \notin C_2$ and so C_1, C_2 are disjoint; the convexity of both C_1, C_2 is trivial as well. It is not difficult to show that the only half-space H_1 containing C_1 and disjoint from C_2 is the half-space $x_2 \leq 0$. Again, every half-space disjoint from H_1 must have the form $x_2 \geq b$ such that b is positive; trivially, no such half-space contains C_2.

The theorem known as "the separation theorem for convex sets" asserts the following: For every pair of disjoint convex sets C_1, C_2 there are half-spaces H_1 and H_2 defined by $\sum c_j x_j \leq d$ and $\sum c_j x_j \geq d$, respectively, such that $C_1 \subseteq H_1$ and $C_2 \subseteq H_2$. Note that the half-spaces H_1 and H_2 are *not* disjoint; they share all the points $\mathbf{x}$ such that $\sum c_j x_j = d$. Proofs of this theorem require notions of elementary topology, which exceed the scope of this book. Those interested may find two relatively simple proofs in T. Botts (1942).

A necessary and sufficient condition for the existence of *disjoint* half-spaces H_1, H_2 such that $C_1 \subseteq H_1$ and $C_2 \subseteq H_2$ is the existence of a positive d such that, in terms not yet defined here, the distance between every $\mathbf{x} \in C_1$ and every $\mathbf{y} \in C_2$ is at least d. [See, for example, Theorem 11.4 in R. T. Rockafellar (1970).] We shall present a slightly weaker theorem in problem 17.8. □

PROBLEMS

17.1 Solve each of the three LP problems in problem 4.1 by the graphic method. Describe in geometric terms the effect of the three different pivoting rules on the performance of the simplex method.

17.2 [P. Kirchberger (1903)] Let A and B be finite sets in R^n. Prove that the following assertions are equivalent:

(i) There are disjoint half-spaces H_A and H_B such that $A \subseteq H_A$ and $B \subseteq H_B$.

(ii) For every set S of at most $n + 2$ points in $A \cup B$, there are disjoint half-spaces H_A and H_B such that $S \cap A \subseteq H_A$ and $S \cap B \subseteq H_B$.

[*Hint*: If 17.2(i) fails, then a certain system of linear equalities in $n + 1$ variables is unsolvable. Apply Theorem 9.4 to this system.]

17.3 If $\mathbf{x}$ is a vector with components $x_1, x_2, \ldots, x_n$ and $\mathbf{y}$ is a vector with components $y_1, y_2, \ldots, y_n$, then the *scalar product* $\mathbf{xy}$ is defined by

$$\mathbf{xy} = x_1 y_1 + x_2 y_2 + \cdots + x_n y_n$$

the *norm* $\|x\|$ is defined by

$$\|\mathbf{x}\| = (\mathbf{xx})^{1/2}$$

and the *distance* dist($\mathbf{x}$, $\mathbf{y}$) is defined by

$$\text{dist}(\mathbf{x}, \mathbf{y}) = \|\mathbf{x} - \mathbf{y}\|.$$

Prove that for every choice of a vector $\mathbf{a}$ with components $a_1, a_2, \ldots, a_n$, a number b, and a vector $\mathbf{x}$ with components $x_1, x_2, \ldots, x_n$, there is a vector $\mathbf{y}^*$ with components $y_1^*, y_2^*, \ldots, y_n^*$ such that $\sum a_j y_j^* = b$ and $\text{dist}(\mathbf{x}, \mathbf{y}^*) = |b - \sum a_j x_j| / \|\mathbf{a}\|$ and $\text{dist}(\mathbf{x}, \mathbf{y}) \geq \text{dist}(\mathbf{x}, \mathbf{y}^*)$ whenever $\sum a_j y_j = b$.

[*Hint*: Consider $\mathbf{y}^* = t\mathbf{a} + \mathbf{x}$ with $t = (b - \sum a_j x_j)/\|\mathbf{a}\|^2$. Observe that $(\mathbf{x} - \mathbf{y}^*)(\mathbf{y}^* - \mathbf{y}) = 0$ and so $\|\mathbf{x} - \mathbf{y}\|^2 = \|(\mathbf{x} - \mathbf{y}^*) + (\mathbf{y}^* - \mathbf{y})\|^2 = \|\mathbf{x} - \mathbf{y}^*\|^2 + \|\mathbf{y}^* - \mathbf{y}\|^2$.]

17.4 The intersection of half-spaces $\sum a_j x_j \le d$ and $\sum a_j x_j \ge c$ such that $d > c$ is called a *slab of thickness* $(d - c)/\|\mathbf{a}\|$. [By the result of problem 17.3, $(d - c)/\|\mathbf{a}\|$ is the distance between the two parallel planes bounding the slab.] Now let $\mathbf{a}_1, \mathbf{a}_2, \ldots, \mathbf{a}_m$ be row vectors of length n, and let $\mathbf{b}$ be a column vector of length m. Prove that the polyhedron

$$\mathbf{a}_i\mathbf{x} \le b_i \qquad (i = 1, 2, \ldots, m)$$

is contained in no slab of thickness less than t only if, for each $k = 1, 2, \ldots, m$, the system

$$\mathbf{a}_i\mathbf{x} \le b_i \qquad (i \ne k)$$
$$\mathbf{a}_k\mathbf{x} \le b_k - t\|\mathbf{a}_k\|$$

is solvable.

17.5 A *ball* with a center $\mathbf{x}$ and a radius r is the set of all points $\mathbf{y}$ such that $\text{dist}(\mathbf{x}, \mathbf{y}) \le r$. Prove: If a polyhedron in R^n is contained in no slab of thickness less than t, then this polyhedron contains a ball of radius $t/(n + 1)$.

[*Hint*: Combine the results of problems 17.4 and 16.13 to establish solvability of a certain system of linear inequalities. Then use the result of problem 17.3 to show that every solution of this system is the center of a ball with desired properties.]

17.6 A subset C of R^n is called *bounded* if there is a constant M such that every $\mathbf{x} = [x_1, x_2, \ldots, x_n]$ in C satisfies $|x_j| \le M$ for all j. Prove that for every bounded convex set C and for every positive δ there is a polyhedron $P \subseteq C$ with the following property: for every $\mathbf{x} \in C$ there is a $\mathbf{y} \in P$ such that $\text{dist}(\mathbf{x}, \mathbf{y}) < \delta$.

[*Hint*: First find a finite set S with the following property: for every $x \in R^n$ such that $|x_j| \le M$ for all j there is a $\mathbf{y} \in S$ such that $\text{dist}(\mathbf{x}, \mathbf{y}) < \delta/2$.]

17.7 [W. Blaschke (1916).] Prove: If a bounded convex set in R^n is contained in no slab of thickness less than t, then, for every positive ε, this set contains a ball of radius $(t - \varepsilon)/(n + 1)$.

(*Hint*: Combine the results of problem 17.6 with $\delta = \varepsilon/2$ and of problem 17.5.)

17.8 Let A, B be disjoint convex sets with A containing a point $\mathbf{a}$ and B containing a point $\mathbf{b}$ such that $\text{dist}(\mathbf{x}, \mathbf{y}) \ge \text{dist}(\mathbf{a}, \mathbf{b})$ for all $\mathbf{x}$ in A and $\mathbf{y}$ in B. Prove that there are disjoint half-spaces H_A, H_B such that $A \subseteq H_A$ and $B \subseteq H_B$.

[*Hint*: Define H_A and H_B as the sets of all points $\mathbf{z}$ such that $(\mathbf{b} - \mathbf{a})\mathbf{z} \le (\mathbf{b} - \mathbf{a})\mathbf{a}$ and $(\mathbf{b} - \mathbf{a})\mathbf{z} \ge (\mathbf{b} - \mathbf{a})\mathbf{b}$, respectively. To prove that $A \subseteq H_A$, derive a contradiction from the hypothesis that $(\mathbf{b} - \mathbf{a})\mathbf{x} > (\mathbf{b} - \mathbf{a})\mathbf{a}$ for some $\mathbf{x}$ in A. More precisely, prove that $\text{dist}(t\mathbf{x} + (1 - t)\mathbf{a}, \mathbf{b}) < \text{dist}(\mathbf{a}, \mathbf{b})$ whenever t is a positive number smaller than $(\mathbf{b} - \mathbf{a})(\mathbf{x} - \mathbf{a})/\|\mathbf{x} - \mathbf{a}\|^2$.]

18

Finding All Vertices of a Polyhedron

Even though our proof of Theorem 16.2 is constructive, it does not suggest a very efficient procedure for finding the vectors $\mathbf{v}^1, \mathbf{v}^2, \ldots, \mathbf{v}^M$ and $\mathbf{w}^1, \mathbf{w}^2, \ldots, \mathbf{w}^N$. The main purpose of this chapter is to present such a procedure. As we shall see, the task of finding the vectors $\mathbf{v}^r$ and $\mathbf{w}^s$ is closely related to the task of finding all the vertices of a polyhedron. Toward the end of the chapter, we shall consider other related problems, such as the problem of describing the convex hull of a specified finite set as an intersection of half-spaces.

VERTICES OF POLYHEDRA

In the preceding chapter, we defined a polyhedron as the intersection of a finite number of half-spaces. Thus, a polyhedron is the set of all the points $\mathbf{x} = [x_1, x_2, \ldots, x_n]^T$ satisfying a prescribed system of linear inequalities

$$\sum_{j=1}^{n} a_{ij}x_j \leq b_i \qquad (i = 1, 2, \ldots, m). \tag{18.1}$$

A *vertex* of this polyhedron is any of its points that can be specified as an intersection of facets. To put it algebraically, a point $\mathbf{x}$ is a vertex of the polyhedron defined by (18.1) if and only if it satisfies (18.1) and is the unique solution of some system

$$\sum_{j=1}^{n} a_{ij}x_j = b_i \qquad (i \in I). \tag{18.2}$$

(For two alternative definitions of a vertex, equivalent to this one, see problems 18.3 and 18.4.)

Our definition of a vertex imposes no restriction on the size of the set I in (18.2). As we are about to show, the definition does not change its meaning if $|I| = n$ is required.

> ***THEOREM 18.1.*** A point $\mathbf{x}$ is a vertex of the polyhedron defined by (18.1) if and only if it satisfies (18.1) and is the unique solution of some system (18.2) with $|I| = n$.

PROOF. Since the "if" part is trivial, we need only prove the "only if" part. In fact, we need only prove that every system (18.2) with a unique solution contains a subsystem

$$\sum_{j=1}^{n} a_{ij}x_j = b_i \qquad (i \in I^*) \tag{18.3}$$

with a unique solution and with $|I^*| = n$. For this purpose, note that (18.2) can be solved by Gaussian elimination. [The fact that the number of equations in (18.2) may not match the number of variables presents no difficulty.] In particular, if (18.2) has a unique solution, then Gaussian elimination transforms (18.2) into a system of n substitution formulas for $x_1, x_2, \ldots, x_n$. Actually, this system of substitution formulas arises from some subsystem (18.3) with $|I^*| = n$, and therefore (18.3) has a unique solution. ■

The Number of Vertices

Theorem 18.1 implies that the number of vertices of (18.1) never exceeds the number of ways in which n objects (namely, the subscripts forming I) can be chosen from a set of m objects (namely, the subscripts $1, 2, \ldots, m$). This number, denoted by $\binom{m}{n}$ and equal to

$$\frac{m(m-1)\cdots(m-n+1)}{n(n-1)\cdots 3\cdot 2\cdot 1}$$

turns out to be a gross overestimate: as proved by P. McMullen (1970), the polyhedron (18.1) has at most

$$f(m, n) = \binom{m - \left\lfloor \frac{n+1}{2} \right\rfloor}{m - n} + \binom{m - \left\lfloor \frac{n+2}{2} \right\rfloor}{m - n} \tag{18.4}$$

vertices. (Here, as usual, $\lfloor x \rfloor$ stands for x rounded down to the nearest integer.) The values of $f(m, n)$ are consistently smaller than those of $\binom{m}{n}$; in fact, $f(m, n)$ can be only a fraction of $\binom{m}{n}$ even when m and n are quite small. For instance, $\binom{16}{8} = 12{,}870$ but $f(16, 8) = 660$. [McMullen's bound cannot be further improved: D. Gale (1963) constructed examples of polyhedra defined by (18.1) and having precisely $f(m, n)$ vertices. We shall present these examples in problems 18.6 – 18.11.]

Even though "typical" polyhedra defined by (18.1) will not have as many as $f(m, n)$ vertices, they will still have quite a few. For instance, suppose that the numbers a_{ij} and b_i are chosen at random and independently of each other in such a way that (i) for every constant c, the probability that $a_{ij} = c$ is zero and (ii) for every pair of constants c and d, the probability that $c < a_{ij} < d$ equals the probability that $-d < a_{ij} < -c$, and the probability that $c < b < d$ equals the probability that $-d < b < -c$. Now an easy argument, suggested by A. Prékopa (1972), shows that the expected number of vertices of (18.1) equals

$$g(m, n) = \binom{m}{n} 2^{n-m}.$$

[Let us sketch the proof only briefly. Assumption (i) guarantees that, with probability 1, each of the $\binom{m}{n}$ systems (18.2) with $|I| = n$ has a unique solution $\mathbf{x}^*$, and that $\sum a_{ij}x_j^* \neq b_i$ whenever $i \in I$. Now assumption (ii) guarantees that each of the $m - n$ inequalities $\sum a_{ij}x_j \leq b_i$ with $i \notin I$ is as likely to get satisfied by $\mathbf{x}^*$ as it is to get violated. Hence $\mathbf{x}^*$ satisfies (18.1) with probability $(\frac{1}{2})^{m-n}$.] The values of $g(m, n)$ can get quite large even when m and n are moderately small: for instance, $g(30, 15) \doteq 4{,}734$. When n is large, $g(2n, n)$ is well approximated by $2^n/\sqrt{\pi n}$.

Vertices and Basic Feasible Solutions

Some of the inequalities in (18.1) may amount to explicit lower and/or upper bounds on individual variables. We shall find it convenient to set such bounds apart from the remaining inequalities, just as we did in Chapter 8. Thus, we shall record every system of linear inequalities as

$$\begin{aligned} \sum_{j=1}^{n} a_{ij}x_j &\leq b_i \qquad (i = 1, 2, \ldots, m) \\ l_j \leq x_j &\leq u_j \qquad (j = 1, 2, \ldots, n). \end{aligned} \tag{18.5}$$

As usual, each l_j in (18.5) is either a number or the symbol $-\infty$ (meaning that no lower bound is imposed on x_j) and each u_j is either a number or the symbol $+\infty$

(meaning that no upper bound is imposed on x_j). By Theorem 18.1, a point $\mathbf{x}$ is a vertex of the polyhedron defined by (18.5) if and only if it satisfies (18.5) and is the unique solution of some system

$$\begin{aligned} \sum_{j=1}^{n} a_{ij}x_j &= b_i && (i \in I) \\ x_j &= l_j && (j \in L) \\ x_j &= u_j && (j \in U) \end{aligned} \tag{18.6}$$

with $|I| + |L| + |U| = n$. Now the relationship between vertices of polyhedra and basic feasible solutions of LP problems, alluded to in the preceding chapter, can be described precisely.

THEOREM 18.2. A vector $[x_1^*, x_2^*, \ldots, x_n^*]^T$ is a vertex of the polyhedron defined by (18.5) if and only if the system

$$\begin{aligned} \sum_{j=1}^{n} a_{ij}x_j + x_{n+i} &= b_i && (i = 1, 2, \ldots, m) \\ l_j \le x_j &\le u_j && (j = 1, 2, \ldots, n) \\ 0 \le x_{n+i} &\le +\infty && (i = 1, 2, \ldots, m) \end{aligned} \tag{18.7}$$

has a basic feasible solution in which (i) each x_j with $j = 1, 2, \ldots, n$ has value x_j^*, and (ii) each free variable is basic.

PROOF. To establish the "if" part, consider a basic feasible solution $[x_1^*, x_2^*, \ldots, x_{n+m}^*]^T$ of (18.7) with all the free variables in the basis. Write

$j \in J$ if $1 \le j \le n$ and x_j is basic

$j \in L$ if $1 \le j \le n$, x_j is nonbasic, and $x_j^* = l_j < u_j$

$j \in U$ if $1 \le j \le n$, x_j is nonbasic, and $x_j^* = u_j$

$i \in I$ if $1 \le i \le m$ and x_{n+i} is nonbasic.

Since the numbers x_j^* $(j \in J)$ and x_{n+i}^* $(i \notin I)$ constitute the unique solution of the system

$$\begin{aligned} \sum_{j \in J} a_{ij}x_j &= b_i - \sum_{j \in L} a_{ij}l_j - \sum_{j \in U} a_{ij}u_j && (i \in I) \\ \sum_{j \in J} a_{ij}x_j + x_{n+i} &= b_i - \sum_{j \in L} a_{ij}l_j - \sum_{j \in U} a_{ij}u_j && (i \notin I) \end{aligned} \tag{18.8}$$

the numbers $x_1^*, x_2^*, \ldots, x_n^*$ constitute the unique solution of (18.6). Since the numbers $x_1^*, x_2^*, \ldots, x_{n+m}^*$ satisfy (18.7), the numbers $x_1^*, x_2^*, \ldots, x_n^*$ satisfy (18.5). Thus, $[x_1^*, x_2^*, \ldots, x_n^*]^T$ is a vertex of the polyhedron defined by (18.5).

To establish the "only if" part, consider a vertex $[x_1^*, x_2^*, \ldots, x_n^*]^T$ of the polyhedron defined by (18.5), and write

$$x_{n+i}^* = b_i - \sum_{j=1}^{n} a_{ij}x_j^* \qquad (i = 1, 2, \ldots, m).$$

Since $[x_1^*, x_2^*, \ldots, x_n^*]^T$ satisfies (18.5), the vector $[x_1^*, x_2^*, \ldots, x_{n+m}^*]^T$ is a feasible solution of (18.7). Since $[x_1^*, x_2^*, \ldots, x_n^*]^T$ is a vertex, it constitutes the unique solution of some system (18.6) with $|I| + |L| + |U| = n$. No subscript j belongs to L and U at the same time: otherwise the matrix of the system (18.6), having two identical rows, would be singular, and so (18.6) could not have a unique solution. [We cannot argue that (18.6) would be unsolvable, for we may have $l_j = u_j$.] Now, writing $j \in J$ if $1 \le j \le n$ and $j \notin L, j \notin U$, we conclude that the numbers x_j^* $(j \in J)$ and x_{n+i}^* $(i \notin I)$ constitute the unique solution of (18.8). Since $|I| = n - |L| - |U| = |J|$, the number $|J| + (m - |I|)$ of variables in (18.8) equals m. Hence $[x_1^*, x_2^*, \ldots, x_{n+m}^*]^T$ is a basic feasible solution; its basic variables are x_j with $j \in J$ and x_{n+i} with $i \notin I$. Since $j \in J$ whenever x_j is free, each free variable is basic. ■

Polyhedra with No Vertices

Not every polyhedron has a vertex. A trivial reason why a polyhedron may have no vertices is that the polyhedron may be empty. However, there are also nonempty polyhedra with no vertices. For instance, a half-space in R^n is a polyhedron, and yet it has no vertices (unless $n = 1$). To take a more sophisticated example, consider the polyhedron defined in R^3 by

$$\begin{aligned} x_1 + x_2 - x_3 &\le 1 \\ 3x_1 - x_2 - 2x_3 &\le 2 \\ -4x_1 \phantom{{}+x_2} + 3x_3 &\le 3 \\ -2x_1 + 2x_2 + x_3 &\le 1 \\ x_1 - 3x_2 \phantom{{}+x_3} &\le 2. \end{aligned} \tag{18.9}$$

This polyhedron is nonempty (it contains, for instance, the point $[0, 0, 0]^T$), but it has no vertices. To see that this is the case, observe that $w_1 = 3, w_2 = 1, w_3 = 4$ is a solution of the system

$$\begin{aligned} w_1 + w_2 - w_3 &= 0 \\ 3w_1 - w_2 - 2w_3 &= 0 \\ -4w_1 \phantom{{}+w_2} + 3w_3 &= 0 \\ -2w_1 + 2w_2 + w_3 &= 0 \\ w_1 - 3w_2 \phantom{{}+w_3} &= 0. \end{aligned}$$

It follows that no system arising from (18.9) in the way (18.2) arises from (18.1) can have a unique solution: if $[x_1, x_2, x_3]^T$ is one solution then $[x_1 + w_1, x_2 + w_2, x_3 + w_3]^T$ is another. More generally, consider an arbitrary polyhedron defined by (18.5) and assume that the system

$$\begin{aligned} \sum_{j=1}^{n} a_{ij}w_j &= 0 \qquad (i = 1, 2, \ldots, m) \\ w_j &= 0 \qquad \text{whenever } l_j \text{ or } u_j \text{ is finite} \end{aligned} \tag{18.10}$$

has a nonzero solution. Under this assumption, no system (18.6) can have a unique solution: if $[x_1, x_2, \ldots, x_n]^T$ is one solution then $[x_1 + w_1, x_2 + w_2, \ldots, x_n + w_n]^T$ is another. Hence the polyhedron (18.5) has no vertices in this case.

Existence of a nonzero solution $\mathbf{w} = [w_1, w_2, \ldots, w_n]^T$ of the system (18.10) has a natural geometric interpretation. First, take any solution $\mathbf{x}^* = [x_1^*, x_2^*, \ldots, x_n^*]$ of (18.5) and observe that $\mathbf{x}^* + \mathbf{w}$ is another solution of (18.5); in fact, $\mathbf{x}^* + t\mathbf{w}$ is a solution of (18.5) for all real numbers t. For instance, taking $\mathbf{x}^* = [1, 0, 2]^T$ in our illustrative example, we observe that

$$\mathbf{x}^* + t\mathbf{w} = [1 + 3t, t, 2 + 4t]^T \tag{18.11}$$

is a solution of (18.9) for all real numbers t. Geometrically, the set of all the points (18.11) is represented by a line passing through the point $[1, 0, 2]^T$. More generally, a *line* in R^n is the set of all the points $\mathbf{x}^* + t\mathbf{w}$ such that $\mathbf{x}^*$ is some fixed point in R^n and $\mathbf{w}$ is some fixed nonzero vector of length n. It is easy to show that, for every nonempty polyhedron P, the following two conditions are equivalent: (i) the system (18.10) has a nonzero solution, (ii) P contains a line. [Since P is nonempty, it contains some point $\mathbf{x}^*$. If (18.10) has a nonzero solution $\mathbf{w}$ then P contains the line $\mathbf{x}^* + t\mathbf{w}$. Conversely, if P contains a line $\tilde{\mathbf{x}} + t\mathbf{w}$ then $\mathbf{w}$ must be a solution of (18.10).]

To summarize, one reason why a nonempty polyhedron may have no vertices is that the polyhedron may contain a line. The following theorem shows that this is the only reason.

THEOREM 18.3. A nonempty polyhedron has no vertex if and only if it contains a line.

PROOF. Since the "if" part has just been established, we need only establish the "only if" part. For this purpose, we shall use a constructive procedure that, given any nonempty polyhedron P defined by (18.5), delivers either a nonzero solution of (18.10) or a vertex of P. To begin, write $j \in F$ if and only if x_j is free in (18.5); then observe that (18.10) may be presented as

$$w_j = 0 \qquad (j \notin F)$$

along with

$$\sum_{j \in F} a_{ij} w_j = 0 \qquad (i = 1, 2, \ldots, m). \tag{18.12}$$

Finding a nonzero solution of (18.12), or showing that there is none, is an easy matter, which can be handled routinely by Gaussian elimination. If (18.12) turns out to have no nonzero solutions then, as in the proof of Theorem 18.1, Gaussian elimination delivers a subsystem

$$\sum_{j \in F} a_{ij} x_j = 0 \qquad (i \in I)$$

of (18.12) with no nonzero solutions and with $|I| = |F|$. But then a basic (although not necessarily feasible) solution $\mathbf{x}^*$ of (18.7) is readily available: its basic variables are all the free variables along with all the slack variables x_{n+i} such that $i \notin I$. To compute $\mathbf{x}^*$, we need only set $x_j^* = l_j$ or $x_j^* = u_j$ for each bounded variable x_j with $1 \leq j \leq n$, set $x_{n+i}^* = 0$ whenever $i \in I$, and then compute $\mathbf{x}_B^*$ by solving the system

$$\sum_{j \in F} a_{ij} x_j^* \qquad\qquad = b_i - \sum_{j \notin F} a_{ij} x_j^* \qquad (i \in I)$$

$$\sum_{j \in F} a_{ij} x_j^* + x_{n+i}^* = b_i - \sum_{j \notin F} a_{ij} x_j^* \qquad (i \notin I).$$

This basic solution may be used to initialize the first phase of the two-phase simplex method as described in Chapter 8. Since P is nonempty, the first phase delivers a basic feasible solution $\tilde{\mathbf{x}}$ of (18.7). Since all the free variables are basic in $\mathbf{x}^*$, and since free variables never leave the basis, all of them are basic in $\tilde{\mathbf{x}}$. By Theorem 18.2, $[\tilde{x}_1, \tilde{x}_2, \ldots, \tilde{x}_n]^T$ is a vertex of P. ■

Finding a nonzero solution $\mathbf{w}$ of (18.10) helps in characterizing all solutions of (18.5). More precisely, let w_k be any nonzero component of $\mathbf{w}$; it is easy to show that $\mathbf{x}$ is a solution of (18.5) if and only if $\mathbf{x} = \tilde{\mathbf{x}} + t\mathbf{w}$ for some solution $\tilde{\mathbf{x}}$ of (18.5) with $\tilde{x}_k = 0$, and for some number t. (The "if" part is trivial; to establish the "only if" part, it suffices to set $t = x_k/w_k$.) Thus characterizing all solutions of (18.5) reduces to characterizing all solutions of the system

$$\sum_{j \neq k} a_{ij} x_j \leq b_i \qquad (i = 1, 2, \ldots, m)$$

$$l_j \leq x_j \leq u_j \qquad (j \neq k)$$

which involves only $n - 1$ variables. This trick may be repeated over and over, until a polyhedron containing no line is obtained. Thus, where characterizing all solutions of (18.5) is concerned, no generality is lost by assuming that the polyhedron defined by (18.5) contains a vertex.

Basic Feasible Partitions

Let us investigate basic feasible solutions of systems $\mathbf{Ax} = \mathbf{b}, \mathbf{l} \le \mathbf{x} \le \mathbf{u}$ such that all the free variables are basic. Every such solution $\mathbf{x}^*$ is fully determined by specifying which of the variables are basic, which nonbasic variables are at their lower bounds, and which nonbasic variables are at their upper bounds. The resulting partition of $\mathbf{x}$ into three parts will be referred to as a *basic feasible partition*; if $\mathbf{A}$ has size $m \times n$, then the basic feasible partition may be represented by the vector $\mathbf{e} = [e_1, e_2, \ldots, e_n]$ such that

$$
\begin{aligned}
e_k &= 1 \quad \text{if} \quad x_k \text{ is basic} \\
e_k &= 0 \quad \text{if} \quad x_k \text{ is nonbasic and } x_k^* = l_k < u_k \\
e_k &= 2 \quad \text{if} \quad x_k \text{ is nonbasic and } x_k^* = u_k.
\end{aligned}
$$

Two basic feasible partitions will be said to be *neighbors* of each other if they can be obtained from each other by a single pivot. That is, two partitions $\mathbf{e} = [e_1, e_2, \ldots, e_n]$ and $\tilde{\mathbf{e}} = [\tilde{e}_1, \tilde{e}_2, \ldots, \tilde{e}_n]$ are neighbors of each other if and only if either (i) they differ only in two components such that $e_i \ne 1, \tilde{e}_i = 1$, and $e_j = 1, \tilde{e}_j \ne 1$ or (ii) they differ only in one component such that $e_i \ne \tilde{e}_i$ and $e_i \ne 1$, $\tilde{e}_i \ne 1$. (In the first case, $\tilde{\mathbf{e}}$ arises from $\mathbf{e}$ by a single pivot with x_i entering and x_j leaving the basis; in the second case, $\tilde{\mathbf{e}}$ arises from $\mathbf{e}$ by a single pivot with the value of x_i switching from one bound to the other and the basis remaining unchanged.) In addition, we shall say that a basic feasible direction is a *neighbor* of a basic feasible partition if the two have the same set of basic variables. For illustration, consider $\mathbf{Ax} = \mathbf{b}, \mathbf{l} \le \mathbf{x} \le \mathbf{u}$ with

$$
\mathbf{A} = \begin{bmatrix} 3 & 1 & 4 & 1 & 0 \\ 4 & 1 & 5 & 0 & 1 \end{bmatrix}, \qquad \mathbf{b} = \begin{bmatrix} 23 \\ 25 \end{bmatrix}
$$

and

$$
\mathbf{l} = [-2, 4, -\infty, 0, 0]^T, \qquad \mathbf{u} = [3, +\infty, 5, +\infty, +\infty]^T.
$$

The basic feasible partition $[0, 0, 2, 1, 1]$ determines the basic feasible solution $[-2, 4, 5, 5, 4]^T$, and its three neighbors are

basic feasible partition $[1, 0, 2, 1, 0]$
basic feasible partition $[0, 1, 2, 1, 0]$
basic feasible direction $[0, 0, -1, 4, 5]^T$.

More generally, to produce all the neighbors of a basic feasible partition $\mathbf{e}$, we have to consider only each nonbasic variable x_i with $l_i < u_i$ in turn and to replace its value x_i^* by $x_i^* + t$ (if $x_i^* = l_i$) or by $x_i^* - t$ (if $x_i^* = u_i$) with t as large as feasibility allows. If t can be arbitrarily large, then we discover a basic feasible direction that is a neighbor of $\mathbf{e}$; otherwise, we discover one or more basic feasible partitions that are neighbors of $\mathbf{e}$.

A list $\mathbf{e}^1, \mathbf{e}^2, \ldots, \mathbf{e}^M, \mathbf{w}^1, \mathbf{w}^2, \ldots, \mathbf{w}^N$ of basic feasible partitions $\mathbf{e}^r$ and basic feasible directions $\mathbf{w}^s$ will be called *closed* if it includes all the neighbors of each $\mathbf{e}^r$. This concept is featured in the following easy variation on Theorem 16.2.

THEOREM 18.4. Let $\mathbf{e}^1, \mathbf{e}^2, \ldots, \mathbf{e}^M, \mathbf{w}^1, \mathbf{w}^2, \ldots, \mathbf{w}^N$ be a closed list of basic feasible partitions $\mathbf{e}^r$ and basic feasible directions $\mathbf{w}^s$ of

$$\mathbf{Ax} = \mathbf{b}, \qquad \mathbf{l} \le \mathbf{x} \le \mathbf{u}. \tag{18.13}$$

Let $\mathbf{v}^1, \mathbf{v}^2, \ldots, \mathbf{v}^M$ be the basic feasible solutions determined by $\mathbf{e}^1, \mathbf{e}^2, \ldots, \mathbf{e}^M$. If $M > 0$ then:

(i) A vector $\mathbf{x}^*$ satisfies (18.13) in place of $\mathbf{x}$ if and only if

$$\mathbf{x}^* = \sum_{r=1}^{M} p_r \mathbf{v}^r + \sum_{s=1}^{N} q_s \mathbf{w}^s$$

for some nonnegative numbers $p_1, p_2, \ldots, p_M, q_1, q_2, \ldots, q_N$ such that

$$\sum_{r=1}^{M} p_r = 1.$$

(ii) Each basic feasible solution of (18.13) with all free variables basic is one of the vectors $\mathbf{v}^r$.

PROOF. The "if" part of (i) is easy: we have

$$\mathbf{Ax}^* = \sum_{r=1}^{M} p_r(\mathbf{Av}^r) + \sum_{s=1}^{N} q_s(\mathbf{Aw}^s) = \mathbf{b} \sum_{r=1}^{M} p_r = \mathbf{b}$$

and, similarly, $\mathbf{l} \le \mathbf{x}^* \le \mathbf{u}$. To prove the "only if" part, consider an arbitrary but fixed vector $\mathbf{x}^*$ satisfying (18.13) in place of $\mathbf{x}$. It will suffice to derive a contradiction from the assumption that no numbers $p_1, p_2, \ldots, p_M, q_1, q_2, \ldots, q_N$ satisfy

$$\begin{aligned} \sum_{r=1}^{M} p_r \mathbf{v}^r + \sum_{s=1}^{N} q_s \mathbf{w}^s &= \mathbf{x}^* \\ \sum_{r=1}^{M} p_r &= 1 \\ p_1, p_2, \ldots, p_M, q_1, q_2, \ldots, q_N &\ge 0. \end{aligned} \tag{18.14}$$

Theorem 9.2 guarantees that the unsolvable system (18.14) is inconsistent in the sense that some row vector $\mathbf{c}$ and a number d satisfy

$$\begin{aligned} \mathbf{cv}^r + d \ge 0 \quad & \text{for all} \quad r = 1, 2, \ldots, M \\ \mathbf{cw}^s \ge 0 \quad & \text{for all} \quad s = 1, 2, \ldots, N \\ \mathbf{cx}^* + d < 0. & \end{aligned} \tag{18.15}$$

Now consider the LP problem

$$\text{minimize} \quad \mathbf{cx} \qquad \text{subject to} \quad \mathbf{Ax} = \mathbf{b}, \quad \mathbf{l} \le \mathbf{x} \le \mathbf{u}. \tag{18.16}$$

The simplex method, initialized by the basic feasible partition $\mathbf{e}^1$, constructs a sequence of basic feasible partitions, each of them appearing on the list $\mathbf{e}^1, \mathbf{e}^2, \ldots, \mathbf{e}^M$, until it either discovers that (18.16) is unbounded or delivers an optimal solution. In the former case, the method actually discovers a basic feasible direction $\mathbf{w}$ such that $\mathbf{cw} < 0$. Since this basic feasible direction is a neighbor of one of the partitions on the list, we must have $\mathbf{w} = \mathbf{w}^s$ for some s. By the second line of (18.15), this is impossible, and so the latter case must take place. Now the optimal solution of (18.16) delivered by the simplex method is one of the vectors $\mathbf{v}^r$ and so, by the first line of (18.15), the optimal value of (18.16) is at least $-d$. This is the desired contradiction as $\mathbf{x}^*$ is a feasible solution of (18.16) and $\mathbf{cx}^* < -d$ by the third line of (18.15).

Finally, to prove (ii), consider an arbitrary basic feasible solution $\mathbf{x}^*$ of (18.13) with all free variables basic. Define a vector $\mathbf{c} = [c_1, c_2, \ldots, c_n]$ by

$$\begin{aligned} c_j &= 0 \quad \text{if} \quad x_j \text{ is basic} \\ c_j &= -1 \quad \text{if} \quad x_j \text{ is nonbasic and } x_j^* = l_j < u_j \\ c_j &= +1 \quad \text{if} \quad x_j \text{ is nonbasic and } x_j^* = u_j \end{aligned}$$

and observe that $\mathbf{x}^*$ is the unique optimal solution of the problem

$$\text{maximize} \quad \mathbf{cx} \qquad \text{subject to} \quad \mathbf{Ax} = \mathbf{b}, \quad \mathbf{l} \le \mathbf{x} \le \mathbf{u}.$$

Hence the simplex method, applied to this problem and initialized by $\mathbf{e}^1$, will deliver one of the basic feasible partitions $\mathbf{e}^r$, and this $\mathbf{e}^r$ will determine $\mathbf{x}^*$. ■

Breadth-First Search

The results presented so far reduce the task of characterizing all solutions of a system of linear equations, as well as the task of finding all the vertices of a polyhedron, into the task of producing a closed list of basic feasible partitions and basic feasible directions of a system $\mathbf{Ax} = \mathbf{b}, \mathbf{l} \le \mathbf{x} \le \mathbf{u}$. In fact, we may even assume that a basic feasible partition $\mathbf{e}^*$ of this system is readily available. A straightforward procedure for creating the closed list is known as *breadth-first search*; it is described in Box 18.1.

BOX 18.1 Breadth-First Search

Step 0. Set $\mathbf{e}^1 = \mathbf{e}^*$, $M = 1$, $N = 0$, $k = 1$, and let $\mathbf{v}^1$ be the basic feasible solution determined by $\mathbf{e}^1$.

Step 1. Produce all the neighbors of $\mathbf{e}^k$. Whenever you produce a basic feasible partition $\mathbf{e}$ that is not on the current list $\mathbf{e}^1, \mathbf{e}^2, \ldots, \mathbf{e}^M$, set $\mathbf{e}^{M+1}$ equal to $\mathbf{e}$, let $\mathbf{v}^{M+1}$ be the basic feasible solution determined by $\mathbf{e}^{M+1}$, replace M by $M + 1$, and continue. Whenever you produce a basic feasible direction $\mathbf{w}$ that is not on the current list $\mathbf{w}^1, \mathbf{w}^2, \ldots, \mathbf{w}^N$, set $\mathbf{w}^{N+1}$ equal to $\mathbf{w}$, replace N by $N + 1$, and continue.

Step 2. If $k = M$ then stop; otherwise, replace k by $k + 1$ and return to step 1.

For illustration, let us return to the system $\mathbf{Ax} = \mathbf{b}, \mathbf{l} \le \mathbf{x} \le \mathbf{u}$ with

$$\mathbf{A} = \begin{bmatrix} 3 & 1 & 4 & 1 & 0 \\ 4 & 1 & 5 & 0 & 1 \end{bmatrix}, \quad \mathbf{b} = \begin{bmatrix} 23 \\ 25 \end{bmatrix},$$

$$\begin{aligned} \mathbf{l} &= [-2, \quad 4, \quad -\infty, \quad 0, \quad 0]^T, \\ \mathbf{u} &= [\ \ 3, \quad +\infty, \quad 5, \quad +\infty, \quad +\infty]^T \end{aligned}$$

and initialize breadth-first search by

$$\begin{aligned} \mathbf{e}^1 &= [\ \ 0, \quad 0, \quad 2, \quad 1, \quad 1] \\ \mathbf{v}^1 &= [-2, \quad 4, \quad 5, \quad 5, \quad 4]^T. \end{aligned}$$

In the first iteration, we produce three neighbors of $\mathbf{e}^1$ and add

$$\begin{aligned} \mathbf{e}^2 &= [\ \ 1, \quad 0, \quad 2, \quad 1, \quad 0] \\ \mathbf{v}^2 &= [-1, \quad 4, \quad 5, \quad 2, \quad 0]^T \\ \mathbf{e}^3 &= [\ \ 0, \quad 1, \quad 2, \quad 1, \quad 0] \\ \mathbf{v}^3 &= [-2, \quad 8, \quad 5, \quad 1, \quad 0]^T \\ \mathbf{w}^1 &= [\ \ 0, \quad 0, \quad -1, \quad 4, \quad 5]^T \end{aligned}$$

to the list. In the second iteration, we produce three neighbors of $\mathbf{e}^2$, add

$$\begin{aligned} \mathbf{e}^4 &= [\ \ 2, \quad 0, \quad 1, \quad 1, \quad 0] \\ \mathbf{v}^4 &= [\ \ 3, \quad 4, \quad 1.8, \quad 2.8, \quad 0]^T \end{aligned}$$

to the list and find the remaining two neighbors on the list as $\mathbf{e}^1, \mathbf{e}^3$. In the third iteration, we produce three neighbors of $\mathbf{e}^3$, add

$$\begin{aligned} \mathbf{e}^5 &= [\ \ 0, \quad 1, \quad 1, \quad 0, \quad 0] \\ \mathbf{v}^5 &= [-2, \quad 13, \quad 4, \quad 0, \quad 0]^T \end{aligned}$$

to the list and find the remaining two neighbors on the list as $\mathbf{e}^1, \mathbf{e}^2$. In the fourth iteration, we produce three neighbors of $\mathbf{e}^4$, add

$$\begin{aligned} \mathbf{e}^6 &= [\ \ 2, \quad 1, \quad 1, \quad 0, \quad 0] \\ \mathbf{v}^6 &= [\ \ 3, \quad 18, \quad -1, \quad 0, \quad 0]^T \\ \mathbf{w}^2 &= [\ \ 0, \quad 0, \quad -0.2, \quad 0.8, \quad 1]^T \end{aligned}$$

to the list and find the remaining neighbor on the list as $\mathbf{e}^2$. In the fifth iteration, we produce three neighbors of $\mathbf{e}^5$, add

$$\mathbf{w}^3 = [\ \ 0, \quad 4, \quad -1, \quad 0, \quad 1]$$

to the list and find the remaining two neighbors on the list as $\mathbf{e}^3$, $\mathbf{e}^6$. In the sixth iteration, we produce three neighbors of $\mathbf{e}^6$ and find them on the list as $\mathbf{e}^4$, $\mathbf{e}^5$, $\mathbf{w}^3$. Since $\mathbf{e}^6$ is the last partition on the current list, the procedure terminates here; the list $\mathbf{e}^1, \mathbf{e}^2, \ldots, \mathbf{e}^6, \mathbf{w}^1, \mathbf{w}^2, \mathbf{w}^3$ is closed. Now we may conclude that $\mathbf{x} = [x_1, x_2, x_3]^T$ is a solution of

$$\begin{aligned}
3x_1 + x_2 + 4x_3 &\le 23 \\
4x_1 + x_2 + 5x_3 &\le 25 \qquad (18.17) \\
-2 \le x_1 \le 3, \quad x_2 \ge 4, &\quad x_3 \le 5
\end{aligned}$$

if and only if

$$\begin{bmatrix} x_1 \\ x_2 \\ x_3 \end{bmatrix} = p_1 \begin{bmatrix} -2 \\ 4 \\ 5 \end{bmatrix} + p_2 \begin{bmatrix} -1 \\ 4 \\ 5 \end{bmatrix} + p_3 \begin{bmatrix} -2 \\ 8 \\ 5 \end{bmatrix} + p_4 \begin{bmatrix} 3 \\ 4 \\ 1.8 \end{bmatrix} + p_5 \begin{bmatrix} -2 \\ 13 \\ 4 \end{bmatrix} + p_6 \begin{bmatrix} 3 \\ 18 \\ -1 \end{bmatrix} +$$

$$q_1 \begin{bmatrix} 0 \\ 0 \\ -1 \end{bmatrix} + q_2 \begin{bmatrix} 0 \\ 0 \\ -0.2 \end{bmatrix} + q_3 \begin{bmatrix} 0 \\ 4 \\ -1 \end{bmatrix}$$

for some nonnegative numbers $p_1, p_2, \ldots, p_6, q_1, q_2, q_3$ such that $p_1 + p_2 + \cdots + p_6 = 1$. Furthermore, we may conclude that

$$\begin{bmatrix} -2 \\ 4 \\ 5 \end{bmatrix}, \begin{bmatrix} -1 \\ 4 \\ 5 \end{bmatrix}, \begin{bmatrix} -2 \\ 8 \\ 5 \end{bmatrix}, \begin{bmatrix} 3 \\ 4 \\ 1.8 \end{bmatrix}, \begin{bmatrix} -2 \\ 13 \\ 4 \end{bmatrix}, \begin{bmatrix} 3 \\ 18 \\ -1 \end{bmatrix}$$

are all the vertices of the polyhedron defined by (18.17).

The main computational burden involved in each execution of step 1 comes from solving the $n - m$ systems $\mathbf{Bd} = \mathbf{a}$ with $\mathbf{B}$ standing for the basis matrix and $\mathbf{a}$ running through all the nonbasic columns. This task requires no more than $m^2(n - 2m/3)$ multiplications (up to $m^3/3$ to find a triangular factorization of $\mathbf{B}$, and up to m^2 to solve each of the $n - m$ systems afterwards). Finding out if the most recently produced neighbor of $\mathbf{e}^k$ is already on the list is a different matter: frequent sequential scans of the list might become prohibitively time-consuming when M and N get very large. Fortunately, data structures known as *balanced trees* implement each look-up in time proportional only to n; another alternative, satisfactory at least from the practical point of view, is the use of so-called *hashing tables*. For more details, those interested are referred to D. E. Knuth (1973) or A. V. Aho, J. E. Hopcroft, and J. D. Ullman (1983). To summarize, each execution of step 1 takes time proportional to at most $m^2 n$, and so the total running time of breadth-first search is at most proportional to $m^2 nM$. This guarantee is quite satisfactory: any algorithm for producing the list $\mathbf{v}^1, \mathbf{v}^2, \ldots, \mathbf{v}^M$ has to take at least n units of time to record each $\mathbf{v}^r$, and so its running time is at least proportional to nM.

For further discussion and an extensive bibliography of related algorithms, those interested are referred to M. E. Dyer and L. G. Proll (1977), T. H. Mattheiss and D. S. Rubin (1980), and to T. H. Mattheiss and B. K. Schmidt (1980).

RELATED PROBLEMS

When applied to systems

$$\sum_{j=1}^{n} a_{ij}x_j \leq 0 \qquad (i = 1, 2, \ldots, m) \tag{18.18}$$

the breadth-first search may waste much time by examining many different basic feasible partitions $\mathbf{e}^1, \mathbf{e}^2, \ldots, \mathbf{e}^M$ of the system

$$\sum_{j=1}^{n} a_{ij}x_j + x_{n+i} = 0 \qquad (i = 1, 2, \ldots, m)$$

$$x_{n+i} \geq 0 \qquad (i = 1, 2, \ldots, m)$$

all of which define the same basic feasible solution $[0, 0, \ldots, 0]^T$. We are going to describe a way of avoiding such unnecessary work. Without loss of generality, we may assume to have found a set I such that the system

$$\sum_{j=1}^{n} a_{ij}x_j = 0 \qquad (i \in I)$$

has no solution other than $x_j = 0$ for all j. Writing

$$a_j = \sum_{i \in I} a_{ij} \qquad (j = 1, 2, \ldots, n)$$

we observe that each nonzero solution of (18.18) satisfies the inequality $\sum a_j x_j < 0$. In particular, if $a_j = 0$ for all j, then (18.18) has no solution other than $x_j = 0$ for all j. On the other hand, if $a_k \neq 0$ for some k, then we may convert the system

$$\begin{aligned} \sum_{j=1}^{n} a_{ij}x_j &\leq \ \ 0 \qquad (i = 1, 2, \ldots, m) \\ \sum_{j=1}^{n} a_j x_j &= -1 \end{aligned} \tag{18.19}$$

into a system of m inequalities in $n - 1$ variables by converting the equation $\sum a_j x_j = -1$ into a formula for x_k and substituting from this formula throughout (18.19). Then we are all set for finding vectors $\mathbf{v}^1, \mathbf{v}^2, \ldots, \mathbf{v}^M$ and $\mathbf{w}^1, \mathbf{w}^2, \ldots, \mathbf{w}^N$ such that $\mathbf{x}$ satisfies (18.19) if and only if

$$\mathbf{x} = \sum_{r=1}^{M} p_r \mathbf{v}^r + \sum_{s=1}^{N} q_s \mathbf{w}^s \tag{18.20}$$

for some nonnegative numbers $p_1, p_2, \ldots, p_M, q_1, q_2, \ldots, q_N$ with $\sum p_r = 1$. Finally, we claim that $\mathbf{x}$ satisfies (18.18) if and only if

$$\mathbf{x} = \sum_{r=1}^{M} t_r \mathbf{v}^r + \sum_{s=1}^{N} t_{M+s} \mathbf{w}^s \tag{18.21}$$

for some nonnegative numbers $t_1, t_2, \ldots, t_{M+N}$. To justify the "if" part of this claim, it suffices to observe that each of the vectors $\mathbf{v}^1, \mathbf{v}^2, \ldots, \mathbf{v}^M, \mathbf{w}^1, \mathbf{w}^2, \ldots, \mathbf{w}^N$ satisfies (18.18) in place of $\mathbf{x}$. To justify the "only if" part, consider an arbitrary solution $\mathbf{x}$ of (18.18). If $\mathbf{x} = [0, 0, \ldots, 0]^T$, then we may satisfy (18.21) by $t_1 = t_2 = \cdots = t_{M+N} = 0$; otherwise, $\sum a_j x_j = -t$ for some positive t. In this latter case, the vector $\mathbf{x}/t$ satisfies (18.19) in place of $\mathbf{x}$, and so it equals the right-hand side of (18.20) for some nonnegative numbers $p_1, p_2, \ldots, p_M, q_1, q_2, \ldots, q_N$. But then $\mathbf{x}$ satisfies (18.21) with $t_r = tp_r$ and $t_{M+S} = tq_S$.

For illustration, let us consider the system of inequalities

$$\begin{aligned} 4x_1 - 2x_2 - x_3 - x_4 &\le 0 \\ x_1 - x_2 - x_4 &\le 0 \\ 2x_2 + x_3 - x_4 &\le 0 \\ -x_1 + 2x_2 - x_3 - x_4 &\le 0 \\ -4x_1 + 2x_2 - x_3 - x_4 &\le 0. \end{aligned} \tag{18.22}$$

We might establish, for instance, that the system of equations

$$\begin{aligned} 4x_1 - 2x_2 - x_3 - x_4 &= 0 \\ x_1 - x_2 - x_4 &= 0 \\ 2x_2 + x_3 - x_4 &= 0 \\ -x_1 + 2x_2 - x_3 - x_4 &= 0 \end{aligned}$$

has no solution other than $x_1 = x_2 = x_3 = x_4 = 0$; in this case, the equation $\sum a_j x_j = -1$ assumes the form

$$4x_1 + x_2 - x_3 - 4x_4 = -1. \tag{18.23}$$

Substituting for x_3 from (18.23) into (18.22) we obtain the system

$$\begin{aligned} -3x_2 + 3x_4 &\le 1 \\ x_1 - x_2 - x_4 &\le 0 \\ 4x_1 + 3x_2 - 5x_4 &\le -1 \\ -5x_1 + x_2 + 3x_4 &\le 1 \\ -8x_1 + x_2 + 3x_4 &\le 1. \end{aligned} \tag{18.24}$$

Now a routine application of breadth-first search shows that $[x_1, x_2, x_4]^T$ satisfies (18.24) if and only if

$$\begin{bmatrix} x_1 \\ x_2 \\ x_4 \end{bmatrix} = p_1 \begin{bmatrix} \frac{1}{9} \\ -\frac{1}{9} \\ \frac{2}{9} \end{bmatrix} + p_2 \begin{bmatrix} -\frac{1}{7} \\ -\frac{1}{7} \\ 0 \end{bmatrix} + p_3 \begin{bmatrix} \frac{4}{9} \\ \frac{5}{9} \\ \frac{8}{9} \end{bmatrix} + p_4 \begin{bmatrix} -\frac{1}{9} \\ -\frac{2}{9} \\ \frac{1}{9} \end{bmatrix} + p_5 \begin{bmatrix} 0 \\ \frac{1}{7} \\ \frac{2}{7} \end{bmatrix} + p_6 \begin{bmatrix} 0 \\ 0 \\ \frac{1}{3} \end{bmatrix}$$

for some nonnegative numbers $p_1, p_2, \ldots, p_6$ such that $\sum p_r = 1$. Hence, $[x_1, x_2, x_3, x_4]^T$ satisfies (18.22) and (18.23) if and only if

$$\begin{bmatrix} x_1 \\ x_2 \\ x_3 \\ x_4 \end{bmatrix} = p_1 \begin{bmatrix} \frac{1}{9} \\ -\frac{1}{9} \\ \frac{4}{9} \\ \frac{2}{9} \end{bmatrix} + p_2 \begin{bmatrix} -\frac{1}{7} \\ -\frac{1}{7} \\ \frac{2}{7} \\ 0 \end{bmatrix} + p_3 \begin{bmatrix} \frac{4}{9} \\ \frac{5}{9} \\ -\frac{2}{9} \\ \frac{8}{9} \end{bmatrix} + p_4 \begin{bmatrix} -\frac{1}{9} \\ -\frac{2}{9} \\ -\frac{1}{9} \\ \frac{1}{9} \end{bmatrix} + p_5 \begin{bmatrix} 0 \\ \frac{1}{7} \\ 0 \\ \frac{2}{7} \end{bmatrix} + p_6 \begin{bmatrix} 0 \\ 0 \\ -\frac{1}{3} \\ \frac{1}{3} \end{bmatrix}$$

for some nonnegative numbers $p_1, p_2, \ldots, p_6$ such that $\sum p_r = 1$. We conclude that $[x_1, x_2, x_3, x_4]^T$ satisfies (18.22) if and only if

$$\begin{bmatrix} x_1 \\ x_2 \\ x_3 \\ x_4 \end{bmatrix} = t_1 \begin{bmatrix} 1 \\ -1 \\ 4 \\ 2 \end{bmatrix} + t_2 \begin{bmatrix} -1 \\ -1 \\ 2 \\ 0 \end{bmatrix} + t_3 \begin{bmatrix} 4 \\ 5 \\ -2 \\ 8 \end{bmatrix} + t_4 \begin{bmatrix} -1 \\ -2 \\ -1 \\ 1 \end{bmatrix} + t_5 \begin{bmatrix} 0 \\ 1 \\ 0 \\ 2 \end{bmatrix} + t_6 \begin{bmatrix} 0 \\ 0 \\ -1 \\ 1 \end{bmatrix} \tag{18.25}$$

for some nonnegative numbers $t_1, t_2, \ldots, t_6$.

In closing, we shall comment on two problems related to the problem of characterizing all solutions of (18.18). The first of these is the problem of describing the convex hull of a set of points $\mathbf{v}^1, \mathbf{v}^2, \ldots, \mathbf{v}^M$ as an intersection of half-spaces. The proof of Theorem 16.4 reduces this problem into the problem of characterizing all solutions of a certain system $\mathbf{Qx} \geq \mathbf{0}$. For instance, if

$$\mathbf{v}^1 = \begin{bmatrix} -4 \\ 2 \\ 1 \end{bmatrix}, \quad \mathbf{v}^2 = \begin{bmatrix} -1 \\ 1 \\ 0 \end{bmatrix}, \quad \mathbf{v}^3 = \begin{bmatrix} 0 \\ -2 \\ -1 \end{bmatrix}, \quad \mathbf{v}^4 = \begin{bmatrix} 1 \\ -2 \\ 1 \end{bmatrix}, \quad \mathbf{v}^5 = \begin{bmatrix} 4 \\ -2 \\ 1 \end{bmatrix}$$

then

$$\mathbf{Q} = \begin{bmatrix} -4 & 2 & 1 & 1 \\ -1 & 1 & 0 & 1 \\ 0 & -2 & -1 & 1 \\ 1 & -2 & 1 & 1 \\ 4 & -2 & 1 & 1 \end{bmatrix}$$

and the characterization (18.25) of all solutions of $\mathbf{Qx} \geq \mathbf{0}$ points out that $[x_1, x_2, x_3]^T$ belongs to the convex hull of $\mathbf{v}^1, \mathbf{v}^2, \ldots, \mathbf{v}^5$ if and only if

$$\begin{aligned} -x_1 + x_2 - 4x_3 &\leq 2 \\ x_1 + x_2 - 2x_3 &\leq 0 \\ -4x_1 - 5x_2 + 2x_3 &\leq 8 \\ x_1 + 2x_2 + x_3 &\leq 1 \\ -x_2 &\leq 2 \\ x_3 &\leq 1. \end{aligned}$$

The other problem goes as follows: Given a matrix $\mathbf{A}$, find a system of linear inequalities that is satisfied by a vector $\mathbf{b}$ if and only if the system $\mathbf{Ax} \leq \mathbf{b}$ is solvable.

To solve this problem, we need only find row vectors $\mathbf{u}^1, \mathbf{u}^2, \ldots, \mathbf{u}^M$ such that a row vector $\mathbf{y}$ satisfies $\mathbf{yA} = \mathbf{0}$, $\mathbf{y} \geq \mathbf{0}$ if and only if $\mathbf{y} = \sum t_k \mathbf{u}^k$ for some nonnegative numbers $t_1, t_2, \ldots, t_M$: Theorem 9.2 implies that $\mathbf{Ax} \leq \mathbf{b}$ is solvable if and only if $\mathbf{u}^k\mathbf{b} \geq 0$ for all $k = 1, 2, \ldots, M$. In finding the vectors $\mathbf{u}^k$, we may write $\mathbf{yA} = \mathbf{0}$, $\mathbf{y} \geq \mathbf{0}$ as $\mathbf{A}^T\mathbf{y}^T = \mathbf{0}$, $\mathbf{y}^T \geq \mathbf{0}$ and assume that the $m \times n$ matrix $\mathbf{A}^T = (a_{ij})$ has full row rank; then it suffices to find, by breadth-first search, all basic feasible solutions of the system

$$\begin{aligned} \sum_{j=1}^{n} a_{ij} y_j &= 0 \qquad (i = 1, 2, \ldots, m) \\ \sum_{j=1}^{n} y_j &= 1 \\ y_j &\geq 0 \qquad (j = 1, 2, \ldots, n). \end{aligned} \tag{18.26}$$

[It is not difficult to see that (18.26) has no feasible directions.] For illustration, we borrow an example from R. J. Duffin (1974):

$$\mathbf{A} = \begin{bmatrix} 5 & 1 & 3 & -1 \\ -2 & 1 & 1 & -1 \\ 2 & -2 & -4 & 2 \\ -3 & -1 & -2 & 1 \\ -1 & 0 & 1 & 0 \\ 6 & -1 & -5 & -1 \end{bmatrix}$$

Here system (18.26) reads

$$\begin{aligned} 5y_1 - 2y_2 + 2y_3 - 3y_4 - y_5 + 6y_6 &= 0 \\ y_1 + y_2 - 2y_3 - y_4 \phantom{{}- y_5} - y_6 &= 0 \\ 3y_1 + y_2 - 4y_3 - 2y_4 + y_5 - 5y_6 &= 0 \\ -y_1 - y_2 + 2y_3 + y_4 \phantom{{}- y_5} - y_6 &= 0 \\ y_1 + y_2 + y_3 + y_4 + y_5 + y_6 &= 1 \\ y_1, y_2, \ldots, y_6 &\geq 0. \end{aligned}$$

Since its only two basic feasible solutions are

$$\left[\frac{8}{29}, \frac{8}{29}, \frac{3}{29}, \frac{10}{29}, 0, 0\right]^T \quad \text{and} \quad \left[\frac{4}{37}, \frac{14}{37}, \frac{9}{37}, 0, \frac{10}{37}, 0\right]^T$$

a vector $\mathbf{y}$ satisfies $\mathbf{yA} = \mathbf{0}$, $\mathbf{y} \geq \mathbf{0}$ if and only if

$$\mathbf{y} = t_1[8, 8, 3, 10, 0, 0] + t_2[4, 14, 9, 0, 10, 0]$$

for some nonnegative t_1 and t_2. We conclude that $\mathbf{Ax} \leq \mathbf{b}$ is solvable if and only if

$$\begin{aligned} 8b_1 + 8b_2 + 3b_3 + 10b_4 \quad &\geq 0 \\ 4b_1 + 14b_2 + 9b_3 \quad + 10b_5 &\geq 0. \end{aligned}$$

PROBLEMS

18.1 Illustrate Theorem 16.2 on the system

$$\begin{aligned} x_1 + 2x_2 + 3x_3 &\leq -2 \\ x_2 + 2x_3 &\leq 1 \\ x_2 + 3x_3 &\leq -3 \\ x_1 + 2x_2 + 4x_3 &\leq -4 \\ -x_2 - x_3 &\leq -1 \\ -x_1 - 2x_2 - 3x_3 &\leq 4. \end{aligned}$$

△ **18.2** Illustrate Theorem 16.4 on

$$\mathbf{v}^1 = \begin{bmatrix} 3 \\ -1 \\ 4 \\ -1 \end{bmatrix}, \quad \mathbf{v}^2 = \begin{bmatrix} 5 \\ 1 \\ 8 \\ 3 \end{bmatrix}, \quad \mathbf{v}^3 = \begin{bmatrix} 5 \\ 2 \\ 10 \\ 4 \end{bmatrix}, \quad \mathbf{v}^4 = \begin{bmatrix} 3 \\ 0 \\ 6 \\ 0 \end{bmatrix}, \quad \mathbf{v}^5 = \begin{bmatrix} 5 \\ 1 \\ 6 \\ 1 \end{bmatrix}, \quad \mathbf{v}^6 = \begin{bmatrix} 5 \\ -1 \\ 6 \\ 3 \end{bmatrix}.$$

18.3 Prove that a point $\mathbf{x}^*$ in a polyhedron P is a vertex of P if and only if the set $P - \mathbf{x}^*$ obtained from P by deleting $\mathbf{x}^*$ remains convex.

18.4 Prove that a solution $\mathbf{x}^*$ of a system $\mathbf{Ax} \leq \mathbf{b}$ is a vertex of the corresponding polyhedron if and only if $\mathbf{x}^*$ is the unique optimal solution of some linear programming problem

$$\text{maximize } \mathbf{cx} \quad \text{subject to } \mathbf{Ax} \leq \mathbf{b}.$$

18.5 As in Problem 17.6, a subset S of R^n is called *bounded* if there is a constant c such that $|x_j| \leq c$ for each component x_j of every $\mathbf{x}$ in S. A bounded polyhedron is called a *polytope*. Prove that every polytope is the convex hull of its vertices.

18.6 For fixed positive integers s and t, let $h(s, t)$ denote the number of sequences $c_1, c_2, \ldots, c_t$ such that each c_i is a nonnegative integer and $\sum c_i = s$. Prove that

$$h(s, t) = \binom{s + t - 1}{t - 1}.$$

(*Hint*: Consider sequences $d_1, d_2, \ldots, d_{t-1}$ defined by $d_i = i + c_1 + c_2 + \cdots + c_i$.)

18.7 A set I of n integers selected from $1, 2, \ldots, m$ will be called *admissible* if between every two $r, s \notin I$ there are an even number (possibly zero) of elements of I. Prove that the number of admissible sets equals $f(m, n)$, as defined by (18.4).

(*Hint*: Observe that each admissible set is characterized by a sequence of nonnegative integers $d_0, d_1, \ldots, d_{m-n}$ such that $\sum d_i = n$ and each of the terms $d_1, d_2, \ldots, d_{m-n-1}$ is even. Then observe that the number of such sequences equals $h(n/2, m - n + 1) + h((n - 2)/2, m - n + 1)$ if n is even and $2h((n - 1)/2, m - n + 1)$ if n is odd.)

18.8 Let $t_1, t_2, \ldots, t_m$ be numbers such that $t_1 < t_2 < \cdots < t_m$ and let I be a set of n integers selected from $1, 2, \ldots, m$. Prove that the system

$$\sum_{j=1}^{n}\left(\sum_{k=1}^{m}(t_k^j - t_i^j)\right)x_j = 1 \qquad (i \in I) \tag{18.27}$$

(with t_k^j standing for the jth power of t_k) has at most one solution.

(*Hint*: Consider solutions $v_1, v_2, \ldots, v_n$ and $w_1, w_2, \ldots, w_n$. Define

$$f(t) = \left(1 - \sum_{k=1}^{m}\sum_{j=1}^{n} t_k^j v_j\right) + m\sum_{j=1}^{n} v_j t^j$$

and

$$g(t) = \left(1 - \sum_{k=1}^{m}\sum_{j=1}^{n} t_k^j w_j\right) + m\sum_{j=1}^{n} w_j t^j.$$

Observe that $f(t_i) = g(t_i) = 0$ whenever $i \in I$, and so $f(t) = cg(t)$ for all t and some constant c. Conclude that $v_j = cw_j$ for all j and show that $c = 1$.)

18.9 Prove: If I is admissible, then there are numbers $x_1, x_2, \ldots, x_n$ satisfying (18.27) and such that

$$\sum_{j=1}^{n}\sum_{k=1}^{m}(t_k^j - t_i^j)x_j < 1 \qquad \text{for all} \quad i \notin I.$$

(*Hint*: Write

$$f(t) = \prod_{i \in I}(t - t_i)$$

and observe that, for every $r, s, \notin I$, the two values $f(t_r)$, $f(t_s)$ have the same sign. Hence, each $f(t_r)$ with $r \notin I$ has the same sign as

$$d = \sum_{k=1}^{m} f(t_k).$$

Now define $c_0, c_1, \ldots, c_n$ by

$$f(t) = \sum_{j=0}^{n} c_j t^j$$

and verify that the numbers $x_1, x_2, \ldots, x_n$ defined by $x_j = c_j/d$ have the desired property.)

18.10 Prove: If I is not admissible, then there are no numbers $x_1, x_2, \ldots, x_n$ satisfying (18.27) and such that

$$\sum_{j=1}^{n}\sum_{k=1}^{m}(t_k^j - t_i^j)x_j \le 1 \qquad \text{for all} \quad i \notin I.$$

(*Hint*: Assuming that there are numbers $x_1, x_2, \ldots, x_n$ satisfying (18.27), define

$$f(t) = \prod_{i \in I}(t - t_i)$$

and

$$g(t) = m\sum_{j=1}^{n} x_j t^j + \left(1 - \sum_{k=1}^{m}\sum_{j=1}^{n} t_k^j x_j\right).$$

Show that there is a nonzero constant c such that $cf(t) = g(t)$ for all t. Observe that there are $r, s \notin I$ such that $f(t_r)$ and $f(t_s)$ have different signs. Conclude that at least one of $g(t_r)$, $g(t_s)$ is negative.)

18.11 Prove that the polyhedron

$$\sum_{j=1}^{n}\left(\sum_{k=1}^{m}(t_k^j - t_i^j)\right)x_j \le 1 \qquad (i = 1, 2, \ldots, m)$$

has precisely $f(m, n)$ vertices, as defined by (18.4).

III NETWORK FLOW PROBLEMS

19

The Network Simplex Method

The special structure found in certain LP problems can often be exploited in the design of highly efficient solution techniques. This effect of the special structure is at its most impressive for so-called *transshipment problems* or *network flow problems*, the subject of this chapter and the next four. As shown by G. B. Dantzig (1951b) and A. Orden (1956), the special structure of transshipment problems allows radical simplifications in the revised simplex method. The resulting *network simplex method* may be developed either by streamlining the revised simplex method or, just as easily, from scratch. To make our exposition self-contained, we shall adopt the latter option. (The connection between the revised simplex method and the network simplex method is pinpointed in Problem 19.9.)

THE TRANSSHIPMENT PROBLEM

The transshipment problem concerns finding the cheapest way to ship prescribed amounts of a commodity such as oil, oranges or empty railroad cars from specified origins to specified destinations through a concrete transportation network. The network may be represented by an illustration such as Figure 19.1. Here, each of the small circles labeled 1, 2, . . . , 7 may be thought of as one of seven cities and each of the arrows may be thought of as a one-way highway or a railroad line. (No generality is lost by considering one-way rather than two-way communication links—quite

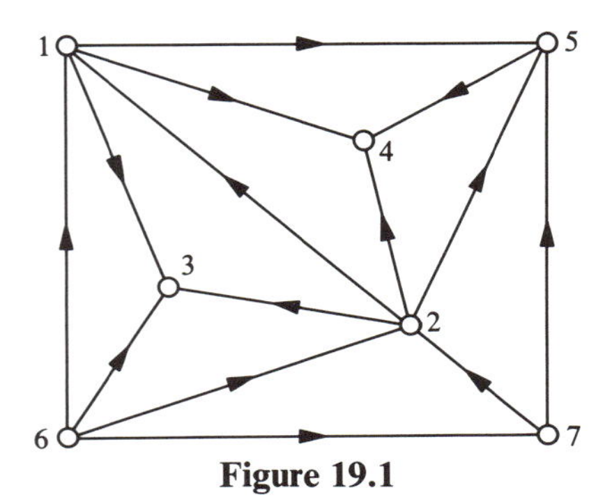

Figure 19.1

the contrary: each two-way highway can be represented by a pair of arrows pointing in opposite directions between the two particular cities.) More rigorously, a *network* is a set of elements called *nodes* and a set of elements called *arcs*, each arc e being an ordered pair (i, j) of distinct nodes i and j. It is convenient to write $e = ij$ rather than $e = (i, j)$; for instance, the network in Figure 19.1 consists of nodes 1, 2, 3, ... and arcs 15, 14, 21, ... If $e = ij$ is an arc, then the node i is called the *tail* of e and the node j is called the *head* of e; the two nodes i and j may be also referred to as the *endpoints* of e.

A network constitutes only a part of the data in a transshipment problem: there is also a demand for some commodity at certain nodes and a supply of the same commodity at other nodes. The nodes with a demand for the commodity are called *sinks* and the nodes with a supply of the commodity are called *sources*. There may also be nodes that, although present in the network, have neither a demand for nor a supply of the commodity. Such nodes are referred to as *intermediate*. (Note that this classification of nodes into three kinds is completely independent of the structure of the network: it is defined only by the numerical data specifying supplies and demands.) For instance, if there is a demand for 6 units at the node 3 a demand for 10 units at the node 4 a demand for 8 units at the node 5 and a supply of 9 units at the node 6 a supply of 15 units at the node 7 then nodes 3, 4, 5 are sinks, nodes 6, 7 are sources and nodes 1, 2 are intermediate. Throughout this chapter, we shall maintain a convenient but unrealistic assumption:

The total supply equals the total demand. (19.1)

The lack of realism in this assumption is obvious: it would be quite surprising to find that, say, the total demand for oranges matches their supply precisely. Nevertheless, the assumption does not limit the applicability of the resulting theory: as we shall point out in the next chapter, problems involving actual shipments of goods can always be converted into a form which does satisfy (19.1). On the other hand, assumption (19.1) makes the theory simpler and more elegant: as soon as it holds, every schedule of shipments that satisfies the demand at each sink must satisfy this demand exactly and it must fully exhaust the supply at each source.

We shall always describe a schedule of shipments by specifying, for each arc ij, the amount x_{ij} of the commodity shipped directly from i to j. For instance, the schedule of shipping

6 units from 6 to 3 directly
3 units from 6 to 5 via 7
10 units from 7 to 4 via 2
5 units from 7 to 5 directly

will be represented by $x_{63} = 6, \quad x_{67} = 3, \quad x_{72} = 10, \quad x_{24} = 10, \quad x_{75} = 8$

and $x_{ij} = 0$ for all the remaining arcs ij. True, this representation is ambiguous: it could also mean shipping

6 units from 6 to 3 directly
7 units from 7 to 4 via 2
3 units from 6 to 4 via 7 and 2
8 units from 7 to 5 directly.

Nevertheless, the ambiguity is harmless: when two crates of oranges are considered interchangeable, the actual trajectory of an individual crate is of little interest. In small examples, it is often convenient to write each positive x_{ij} next to the corresponding arc ij and to completely ignore arcs ij with $x_{ij} = 0$. For instance, a schedule of shipments might be presented in the form shown in Figure 19.2.

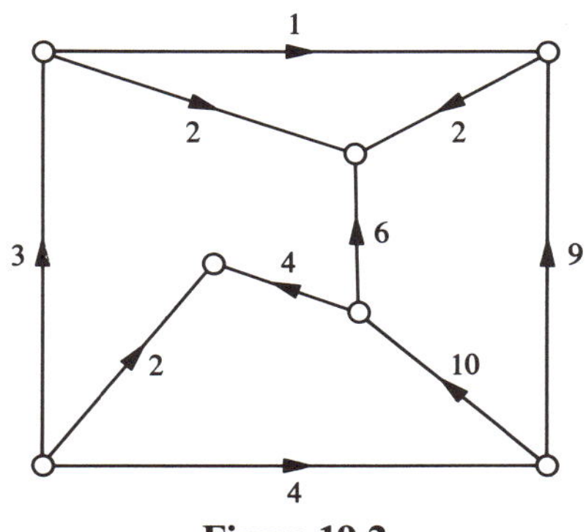

Figure 19.2

A schedule of shipments described by the amounts shipped along individual arcs can be implemented *if and only if*:

(i) The amount entering each intermediate node equals the amount leaving that node (for instance, $10 = 4 + 6$ at intermediate node 2 in our case).

(ii) The amount entering each sink minus the amount leaving that sink equals the demand at that sink (for instance, $9 + 1 - 2 = 8$ at sink 5 in our case).

(iii) The amount leaving each source minus the amount entering that source equals the supply at that source (for instance, $10 + 9 - 4 = 15$ at source 7 in our case).

(iv) Each amount shipped along an individual arc is nonnegative.

(The "only if" part of this claim is easy to establish; a rigorous proof of the "if" part is a little trickier. The details are left to the reader.)

These requirements can be expressed as

$$\mathbf{Ax} = \mathbf{b}, \qquad \mathbf{x} \geq \mathbf{0} \tag{19.2}$$

with

$$\mathbf{A} = \begin{bmatrix} -1 & -1 & -1 & 1 & & & & & 1 & & & & & \\ & & & -1 & -1 & -1 & -1 & & & 1 & & & 1 & \\ 1 & & & & 1 & & & & & & 1 & & & \\ & 1 & & & & 1 & & 1 & & & & & & \\ & & 1 & & & & 1 & -1 & & & & & & 1 \\ & & & & & & & & -1 & -1 & -1 & -1 & & \\ & & & & & & & & & & & 1 & -1 & -1 \end{bmatrix},$$

$$\mathbf{x} = \begin{bmatrix} x_{13} \\ x_{14} \\ x_{15} \\ x_{21} \\ x_{23} \\ x_{24} \\ x_{25} \\ x_{54} \\ x_{61} \\ x_{62} \\ x_{63} \\ x_{67} \\ x_{72} \\ x_{75} \end{bmatrix}, \qquad \mathbf{b} = \begin{bmatrix} 0 \\ 0 \\ 6 \\ 10 \\ 8 \\ -9 \\ -15 \end{bmatrix}.$$

(This notation deviates from the convention that reserves double subscripts ij for entries of a matrix rather than components of a vector.) Matrix $\mathbf{A}$ is called the *incidence matrix* of our network. More generally, the incidence matrix of a network with n nodes and m arcs has n rows and m columns; each arc ij corresponds to a column

$$\begin{bmatrix} a_1 \\ a_2 \\ \vdots \\ a_n \end{bmatrix} \quad \text{defined by} \quad a_k = \begin{cases} -1 & \text{if } k = i \\ 1 & \text{if } k = j \\ 0 & \text{otherwise.} \end{cases}$$

(Although in the past we reserved the letter m for the number of rows in a matrix, and the letter n for the number of columns, the roles of the two letters are now interchanged: $\mathbf{A}$ has size $n \times m$. This switch is forced by a convention of network theory: it is customary to denote the number of nodes by n and the number of arcs by m.) Each component b_i of the *demand vector* $\mathbf{b}$ specifies the demand at node i, with supplies interpreted as negative demands. We shall denote the cost of shipping a unit amount along ij by c_{ij} and refer to the row vector $\mathbf{c}$ with components c_{ij} as the *cost vector*. In our example, we shall consider

$$\begin{aligned} c &= [c_{13}, c_{14}, c_{15}, c_{21}, c_{23}, c_{24}, c_{25}, c_{54}, c_{61}, c_{62}, c_{63}, c_{67}, c_{72}, c_{75}] \\ &= [\ 53,\ 18,\ 29,\ 8,\ 60,\ 28,\ 37,\ 5,\ 44,\ 38,\ 98,\ 14,\ 23,\ 59]. \end{aligned}$$

Now the total cost of a schedule $\mathbf{x}$ equals

$$\mathbf{cx} = \sum c_{ij} x_{ij}$$

and so the natural problem of finding the cheapest schedule amounts to minimizing $\mathbf{cx}$ subject to (19.2). More generally, the *transshipment problem* is any problem

$$\text{minimize} \quad \mathbf{cx} \qquad \text{subject to} \quad \mathbf{Ax} = \mathbf{b}, \quad \mathbf{x} \geq \mathbf{0}$$

such that $\mathbf{A}$ is the $n \times m$ incidence matrix of some network and such that

$$\sum_{i=1}^{n} b_i = 0.$$

[The last requirement stipulates assumption (19.1): The total supply *equals* the total demand.] Note that the sum of the n equations $\mathbf{Ax} = \mathbf{b}$ reads $0 = 0$, and so any one of them amounts to the sum of the remaining $n - 1$. Hence, removing the last of the n equations, we may state our problem in the form

$$\text{minimize} \quad \mathbf{cx} \qquad \text{subject to} \quad \tilde{\mathbf{A}}\mathbf{x} = \tilde{\mathbf{b}}, \quad \mathbf{x} \geq \mathbf{0}$$

such that matrix $\tilde{\mathbf{A}}$ has only $n - 1$ rows and vector $\tilde{\mathbf{b}}$ has only $n - 1$ components. For instance, deleting the last equation in our illustrative problem we obtain

$$\tilde{\mathbf{A}} = \begin{bmatrix} -1 & -1 & -1 & 1 & & & & & 1 & & & & & \\ & & & -1 & -1 & -1 & -1 & & & 1 & & & 1 & \\ 1 & & & & 1 & & & & & & 1 & & & \\ & 1 & & & & 1 & & 1 & & & & & & \\ & & 1 & & & & 1 & -1 & & & & & & 1 \\ & & & & & & & & -1 & -1 & -1 & -1 & & \end{bmatrix} \qquad \tilde{\mathbf{b}} = \begin{bmatrix} 0 \\ 0 \\ 6 \\ 10 \\ 8 \\ -9 \end{bmatrix}.$$

The matrix $\tilde{\mathbf{A}}$ will be referred to as the *truncated incidence matrix* of the network.

TREES AND FEASIBLE TREE SOLUTIONS

A *path* between nodes u and v is a network with nodes $w_1, w_2, \ldots, w_k$ and arcs $e_1, e_2, \ldots, e_{k-1}$ such that $u = w_1$, $v = w_k$ and the two endpoints of each e_i are w_i and w_{i+1}. A *cycle* is a network arising from this path by adding a new arc with endpoints w_1 and w_k. (In particular, two distinct arcs with the same pair of endpoints form a cycle.) A network is called *connected* if between every two of its nodes there is a path; it is called *acyclic* if it contains no cycle. A *tree* is a connected and acyclic network; an example of a tree is shown in Figure 19.3. Note that this tree T and our original network N have the same set of nodes; in addition, each arc of T is an arc of N. Such trees are called *spanning trees* of N. A *feasible tree solution* is a feasible solution $\mathbf{x}$ associated with a spanning tree T in such a way that $x_{ij} = 0$ whenever $ij \notin T$. For instance, the tree in Figure 19.3 determines the feasible tree solution in Figure 19.4.

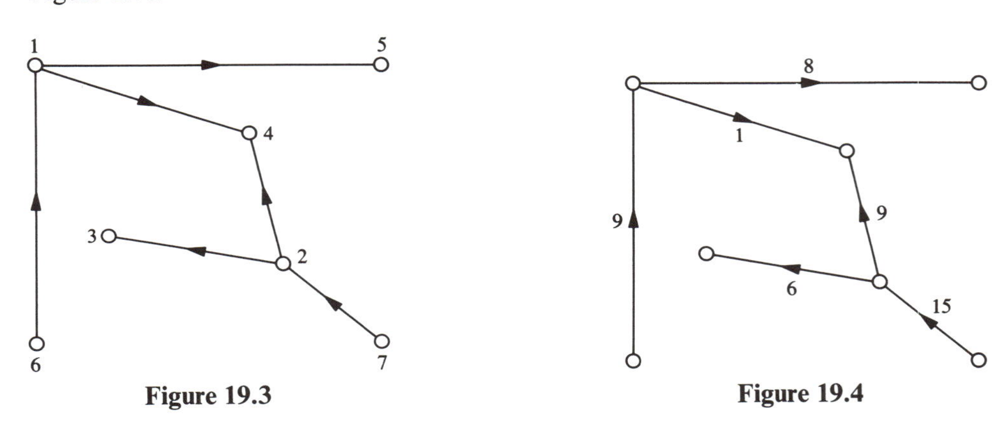

Figure 19.3

Figure 19.4

It is not difficult to prove that every tree has the following property: no matter how a node v_1 is chosen, the remaining nodes can be enumerated as $v_2, v_3, \ldots, v_n$ in such a way that:

$$\text{For every } i \geq 2\text{, there is precisely one arc with one endpoint equal to } v_i \text{ and the other endpoint among the nodes } v_1, v_2, \ldots, v_{i-1}. \tag{19.3}$$

[For instance, if $v_1 = 7$ is chosen in our example, then we may set $v_2 = 2$, $v_3 = 3$, $v_4 = 4$, $v_5 = 1$, $v_6 = 5$, $v_7 = 6$ to satisfy (19.3).] We leave the proof for problem 19.5. In particular, every tree with n nodes has precisely $n - 1$ arcs: if $v_1, v_2, \ldots, v_n$ satisfy (19.3), then the arcs can be enumerated as $e_2, e_3, \ldots, e_n$ in such a way that one endpoint of each e_i equals v_i and the other endpoint is among $v_1, v_2, \ldots, v_{i-1}$. (In our example, $e_2 = 72$, $e_3 = 23$, $e_4 = 24$, $e_5 = 14$, $e_6 = 15$, $e_7 = 61$.)

Now consider the truncated incidence matrix $\mathbf{B}$ of this tree, with a row for each node except $v_1 = n$ and a column for each arc. When the rows and columns of $\mathbf{B}$ are permuted so that the rows correspond to $v_2, v_3, \ldots, v_n$ in this order, and the columns correspond to $e_2, e_3, \ldots, e_n$ in this order, the resulting matrix is upper triangular with nonzero entries on the diagonal. Hence the system $\mathbf{Bx}^* = \tilde{\mathbf{b}}$ has a *unique* solution $\mathbf{x}^*$. We conclude that each feasible tree solution $\mathbf{x}$ is uniquely determined by the associated tree: if $\mathbf{x}^*$ stands for the column vector with $n - 1$ components x_{ij}, one for each arc of the tree, then $\mathbf{Bx}^* = \tilde{\mathbf{b}}$.

THE NETWORK SIMPLEX METHOD: AN ECONOMIC MOTIVATION

The network simplex method works exclusively with feasible tree solutions. In each iteration, it aims to improve the current solution. As a result, either it produces a feasible solution defined by a new tree or it finds out that the current solution is optimal. There are three parts to each iteration. We shall illustrate them on the feasible tree solution shown above and motivate them in intuitive economic terms; rigorous general treatment will follow in the next section.

In the first part of the iteration, we imagine that the current feasible tree solution has been adopted by a transportation company. Because of the shipping costs, the market price of the commodity will vary with location: for example, oranges cost more in Alaska than they do in Florida. If the unit price is y_i at a node i and if the company ships the commodity along an arc ij, then it is fair to expect a unit price of $y_i + c_{ij}$ at the node j: a lower price would make shipping along the arc ij wasteful, whereas a higher price would allow competitors to undersell the company at j and still make a profit for themselves. Our first task is to determine a set of "fair prices" $y_1, y_2, \ldots, y_n$ such that

$$y_i + c_{ij} = y_j \qquad \text{for each} \quad ij \in T. \tag{19.4}$$

Note that (19.4) amounts to only $n - 1$ equations (one for each arc of T), in n variables (one for each node of T). Hence it does not determine the fair prices uniquely: if d is an arbitrary constant then $y_1 + d, y_2 + d, \ldots, y_n + d$ satisfy (19.4) whenever $y_1, y_2, \ldots, y_n$ do. (To put it differently, the fairness of the prices depends only on their relative differences $y_j - y_i$ and remains unaffected when all the prices rise or fall uniformly.) In particular, we may assume that $y_n = 0$; as we shall see in a moment, the resulting system of $n - 1$ equations in $y_1, y_2, \ldots, y_{n-1}$ has a unique solution. In our example, this system reads

$$\begin{array}{rrrrrrl}
 & y_2 & & & & & = 23 \\
 & -y_2 & +y_3 & & & & = 60 \\
 & -y_2 & & +y_4 & & & = 28 \\
-y_1 & & & +y_4 & & & = 18 \\
-y_1 & & & & +y_5 & & = 29 \\
y_1 & & & & & -y_6 & = 44
\end{array}$$

and its unique solution is

$$y_2 = 23, \quad y_3 = 83, \quad y_4 = 51, \quad y_1 = 33, \quad y_5 = 62, \quad y_6 = -11.$$

(Even though one of these "fair prices" is negative, the differences $y_j - y_i$ retain their economic significance.) In general, the left-hand side of the system may be thought of as the row vector $\mathbf{yB}$ such that $\mathbf{y}$ is the row vector with components $y_1, y_2, \ldots, y_{n-1}$ and $\mathbf{B}$ is the truncated incidence matrix of the tree. As observed previously, the rows and columns of $\mathbf{B}$ may be permuted in such a way that the resulting matrix is upper triangular with nonzero entries on the diagonal. Hence the system has a unique solution.

In the second step of each iteration, we put ourselves in the role of a competitor: would it pay to buy the commodity at some node i, ship it along an arc ij and sell at the node j? In our example, the answer is affirmative for several different arcs ij. For instance, we could buy at the node 2 for $y_2 = 23$ and ship along the arc 25 for $c_{25} = 37$: the total expense of $y_2 + c_{25} = 60$ compares favorably with the selling price of $y_5 = 62$. In general, this second step consists of choosing an arc ij such that

$$y_i + c_{ij} < y_j. \tag{19.5}$$

Thereafter, this arc is referred to as the *entering arc*. [If no arc satisfies (19.5), then the current solution is optimal; more about that in the next section.] Of course, (19.4) implies that only the out-of-tree arcs have to be examined in this step. In hand calculations on small problems, it is customary to choose that arc which maximizes the difference $y_j - y_i - c_{ij}$ between the two sides of (19.5). When problems with hundreds of thousands of arcs are handled on a computer, this rule is abandoned in favor of some less time-consuming strategy. In our example, it doesn't take long to find the "most tempting" arc 75 with the net profit of $y_5 - y_7 - c_{75} = 3$ per unit shipped. This will be our entering arc.

In the third step of each iteration, we imagine that the first company has found out about its competitor's plan and hastens to use the information to its own advantage. It will now ship t units through the entering arc and keep ignoring other out-of-tree arcs; the shipments along the arcs of T will have to be adjusted so as to maintain feasibility of the resulting schedule. The adjustments, easy to work out, are shown in Figure 19.5. Our intuitive reasoning suggests that each unit shipped along the arc 75 in the new schedule cuts down the total cost. Hence $x_{75} = t$ should be made as large as possible. The requirements

$$x_{15} = 8 - t \geq 0, \quad x_{24} = 9 - t \geq 0, \quad \text{and} \quad x_{72} = 15 - t \geq 0$$

prevent us from making t too large; the largest value we get away with is $t = 8$. The corresponding solution is shown in Figure 19.6. (Note that the new solution is again a feasible tree solution. As we shall explain in the next section, this is no accident.) In order to illustrate this third step again, we shall continue solving our problem.

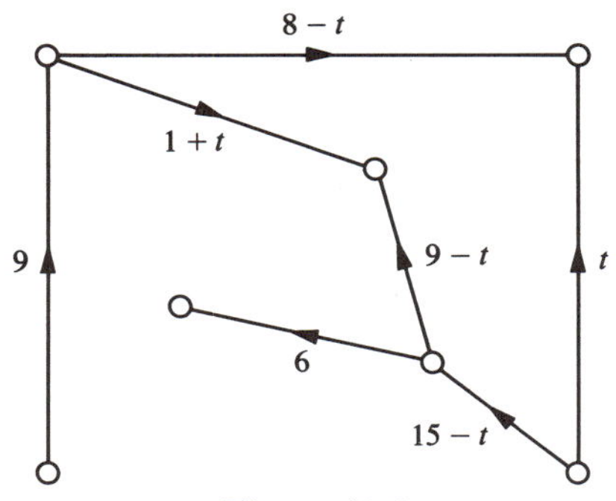

Figure 19.5

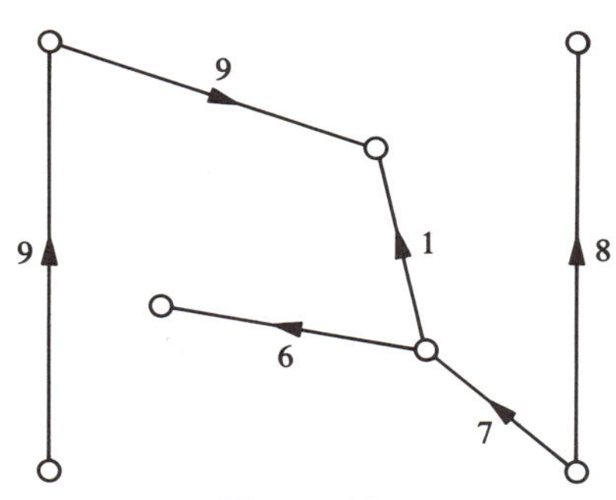

Figure 19.6

The second iteration begins.

Step 1. $y_7 = 0, y_5 = 59, y_2 = 23, y_3 = 83, y_4 = 51, y_1 = 33, y_6 = -11.$

Step 2. We let 21 be the entering arc $(y_1 - y_2 - c_{21} = 2)$.

Step 3. Illustrated in Figure 19.7.

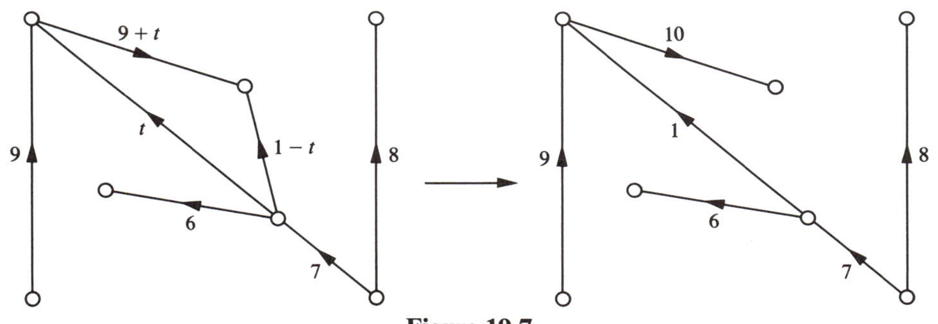

Figure 19.7

The third iteration begins.

Step 1. $y_7 = 0, y_5 = 59, y_2 = 23, y_3 = 83, y_1 = 31, y_4 = 49, y_6 = -13.$

Step 2. Now $y_j - y_i - c_{ij} \leq 0$ for all arcs ij.

Now our current solution is optimal. To see that this is the case, suppose that we approach the company and propose to take over its services for a fee. To make the proposal more persuasive, we do not name the fee explicitly; instead, we exhibit our list of local prices $y_1, y_2, \ldots, y_n$ and offer to trade at these prices. Since $y_j - y_i \leq c_{ij}$ for all arcs ij, the company has no incentive to stay in business: in every single instance, shipping from i to j would cost it at least as much as selling to us at i and buying back at j. Hence its total operating cost must be at least the lump sum $\sum y_i b_i$ we are implicitly asking for. On the other hand, this lump sum equals our total operating cost: on each individual shipment, we break even.

AN ALGEBRAIC DESCRIPTION

Each iteration of the network simplex method begins with a feasible tree solution $\mathbf{x}$ defined by a tree T. In Step 1, we calculate $y_1, y_2, \ldots, y_n$ such that $y_i + c_{ij} = y_j$ whenever $ij \in T$. Let $\mathbf{y}$ stand for the row vector with components $y_1, y_2, \ldots, y_n$ and write $\bar{\mathbf{c}} = \mathbf{c} - \mathbf{yA}$. Since $\bar{c}_{ij} = c_{ij} + y_i - y_j = 0$ whenever $ij \in T$ and $x_{ij} = 0$ whenever $ij \notin T$, we have $\bar{\mathbf{c}}\mathbf{x} = 0$. On the other hand, every vector $\bar{\mathbf{x}}$ satisfying $\mathbf{A}\bar{\mathbf{x}} = \mathbf{b}$ satisfies $\mathbf{c}\bar{\mathbf{x}} = \bar{\mathbf{c}}\bar{\mathbf{x}} + \mathbf{yA}\bar{\mathbf{x}} = \bar{\mathbf{c}}\bar{\mathbf{x}} + \mathbf{yb}$; in particular, $\mathbf{cx} = \bar{\mathbf{c}}\mathbf{x} + \mathbf{yb} = \mathbf{yb}$. We conclude that

$$\mathbf{c}\bar{\mathbf{x}} = \mathbf{cx} + \bar{\mathbf{c}}\bar{\mathbf{x}} \tag{19.6}$$

whenever $\mathbf{A}\bar{\mathbf{x}} = \mathbf{b}$.

In Step 2, we find an arc $e = uv$ such that $y_u + c_{uv} < y_v$. [If there is no such arc then $\mathbf{x}$ is optimal. This claim, previously justified in economic terms, may be justified formally as follows. If $c_{ij} + y_i - y_j \geq 0$ for all arcs ij, then $\bar{\mathbf{c}} \geq \mathbf{0}$ and so $\bar{\mathbf{c}}\bar{\mathbf{x}} \geq 0$ whenever $\bar{\mathbf{x}} \geq \mathbf{0}$. Hence (19.6) implies $\mathbf{c}\bar{\mathbf{x}} \geq \mathbf{cx}$ for every feasible solution $\bar{\mathbf{x}}$.]

Step 3 relies on further properties of trees. Since T is connected, it contains a path between u and v; it is not difficult to show that this path is unique (problem 19.6). Now it follows that the network $T + e$, obtained by adding the *entering* arc $e = uv$ to T, contains a unique cycle. Traversing this cycle in the direction of the entering arc e, we distinguish between *forward* arcs, pointing the same way as e, and *reverse* arcs, pointing the other way. Then we set

$$\bar{x}_{ij} = \begin{cases} x_{ij} + t & \text{if } ij \text{ is a forward arc,} \\ x_{ij} - t & \text{if } ij \text{ is a reverse arc,} \\ x_{ij} & \text{if } ij \text{ is not on the cycle,} \end{cases}$$

for some value of t. Note that $\mathbf{A}\bar{\mathbf{x}} = \mathbf{Ax}$: the two extra contributions $\pm t$ at each node of the cycle cancel each other out. Hence $\bar{\mathbf{x}}$ satisfies (19.6); since e is the only arc ij with $\bar{c}_{ij} \neq 0$ and $\bar{x}_{ij} \neq 0$, we have $\bar{\mathbf{c}}\bar{\mathbf{x}} = \bar{c}_e\bar{x}_e = \bar{c}_e t$ and (19.6) reduces to

$$\mathbf{c}\bar{\mathbf{x}} = \mathbf{cx} + \bar{c}_e t.$$

We wish to choose t in such a way that $\bar{\mathbf{x}}$ is feasible and $\mathbf{c}\bar{\mathbf{x}}$ is as small as possible. Since $\bar{c}_e < 0$, we are led to maximize t subject to $\bar{\mathbf{x}} \geq \mathbf{0}$. To achieve this objective, we find a reverse arc f such that $x_f \leq x_{ij}$ for all reverse arcs ij, and then we set $t = x_f$. (If there are no reverse arcs, then every positive t defines a feasible $\bar{\mathbf{x}}$. In that case, our problem is *unbounded* in the sense that for every positive M there is a feasible solution $\bar{\mathbf{x}}$ with $\mathbf{c}\bar{\mathbf{x}} < -M$. Note that problems with $\mathbf{c} \geq \mathbf{0}$ are never unbounded: all their feasible solutions $\mathbf{x}$ satisfy $\mathbf{cx} \geq 0$.) With $t = x_f$, the new feasible solution $\bar{\mathbf{x}}$ has $\bar{x}_f = 0$ as well as $\bar{x}_{ij} = 0$ whenever $ij \notin T + e$. To put it differently, with $T + e - f$ standing for the network obtained from $T + e$ by deleting the *leaving* arc f, we have $\bar{x}_{ij} = 0$ whenever $ij \notin T + e - f$. Since the only cycle in $T + e$ contained f, the network $T + e - f$ is acyclic. In fact, it is not difficult to show that $T + e - f$ is a

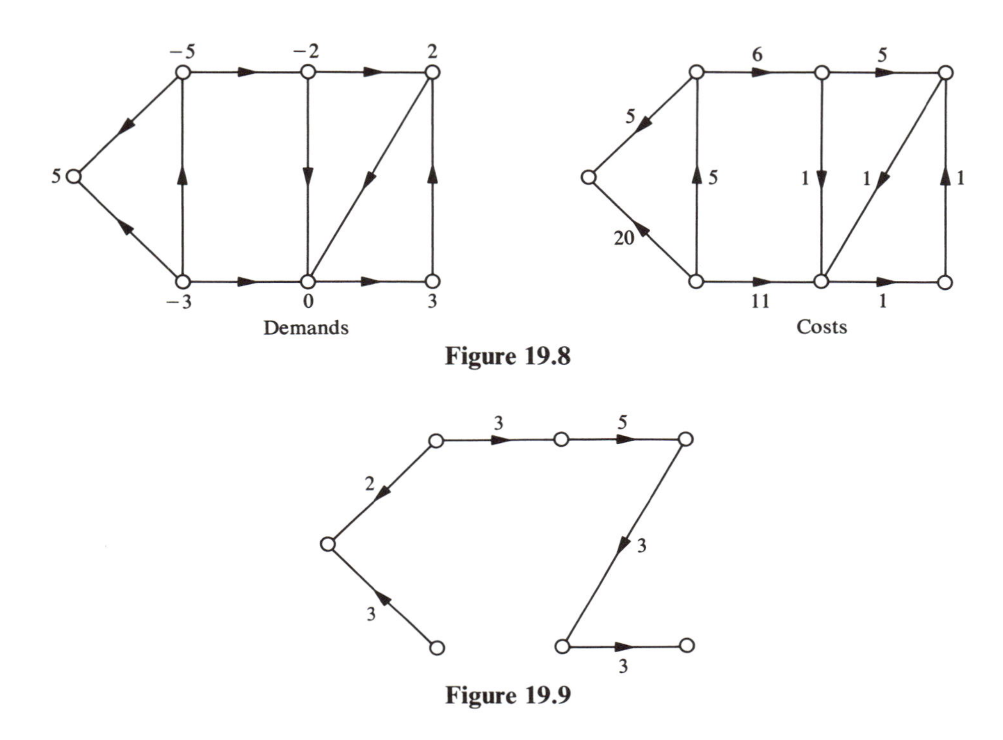

Figure 19.8

Figure 19.9

tree (Problem 19.7), and so $\bar{\mathbf{x}}$ is a feasible tree solution. This third step of each iteration is called a *pivot*.

Degeneracy and Cycling

Even though one may tend to think of each iteration as reducing the value of the objective function, this is not always the case: sometimes $t = 0$ is forced in the new feasible solution $\bar{\mathbf{x}}$. For illustration, consider the data in Figure 19.8 and the initial feasible tree solution in Figure 19.9.

In the first iteration, we are led to consider solutions shown in Figure 19.10. There are three candidates for the leaving arc. We choose one of them arbitrarily. The next solution, shown in Figure 19.11, differs from all the previously encountered feasible

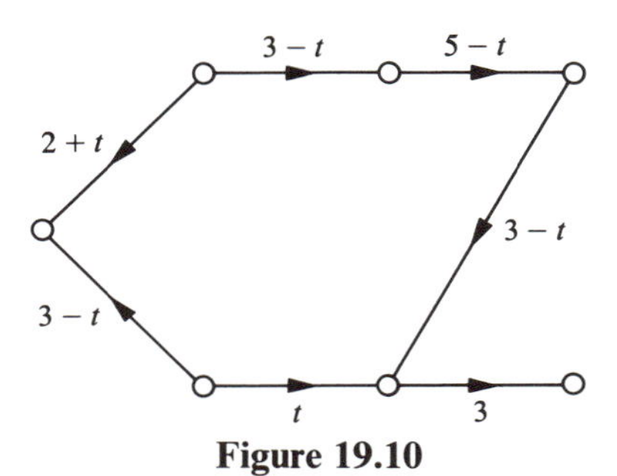

Figure 19.10

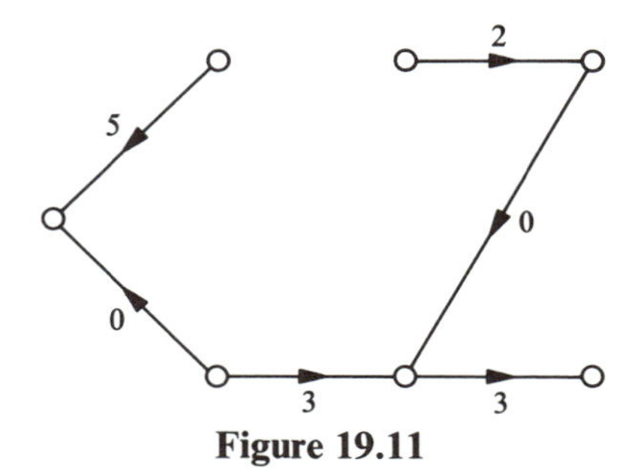

Figure 19.11

tree solutions in an important respect: $x_{ij} = 0$ not only for all the arcs ij outside the tree, but also for certain arcs *ij in the tree*. Such feasible tree solutions are called *degenerate*. Although degeneracy is quite harmless in its own right, it may indicate serious trouble ahead. To illustrate this trouble, we shall continue solving our example.

In the next iteration, we are led to consider solutions shown in Figure 19.12. Now $x_f = 0$ for the leaving arc f, and so the new solution $\bar{\mathbf{x}}$ will not differ from the current solution $\mathbf{x}$. The only change occurs in the underlying tree; it is illustrated in Figure 19.13.

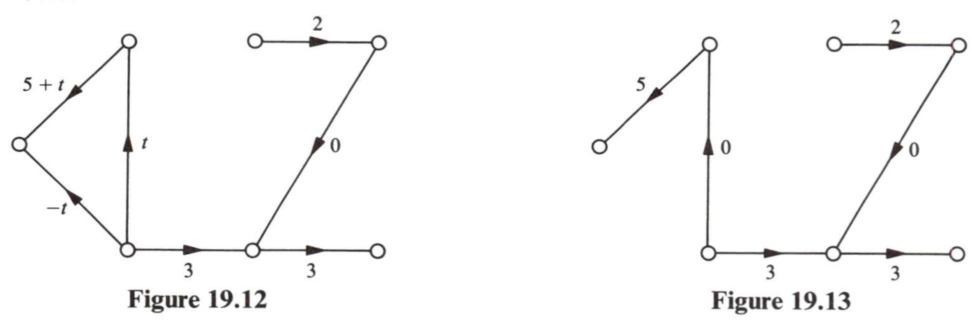

Figure 19.12 Figure 19.13

More generally, pivots with $x_f = 0$ for the leaving arc f are called *degenerate*. Such pivots change only the tree but leave $\mathbf{x}$ unaltered. (In our example, the next pivot happens to be degenerate again; after that, one more pivot brings us directly to the optimal solution.) These degenerate pivots are a nuisance: they cause the network simplex method to "stall" without showing any signs of progress. In fact, they can lead to something far worse than stalling. For example, consider a network with sources 1, 2, 3, 4, sinks 5, 6, 7, 8, and an arc ij for each source i and each sink j. There is a unit supply at each source and a unit demand at each sink; the costs are specified as

$$c_{16} = c_{17} = c_{25} = c_{27} = c_{35} = c_{36} = c_{48} = 1$$

and $c_{ij} = 0$ otherwise. The network simplex method, initialized by the tree with arcs 15, 16, 26, 28, 48, 47, 37 (so that $x_{15} = x_{26} = x_{48} = x_{37} = 1$ and $x_{ij} = 0$ otherwise) might proceed as shown in Table 19.1.

Table 19.1 A Cycle of Twelve Iterations

Iteration number	Entering arc	Leaving arc
1	18	28
2	36	16
3	46	47
4	35	36
5	38	18
6	25	35
7	45	46
8	27	25
9	28	38
10	17	27
11	47	45
12	16	17

After twelve iterations, we find ourselves looking at the initial tree again. Now we might go through these twelve iterations over and over again, never reaching an optimal solution (such as $x_{15} = x_{28} = x_{37} = x_{46} = 1$ and $x_{ij} = 0$ otherwise). This phenomenon, called *cycling*, seems to make the network simplex method unsound. Nevertheless, its threat is only illusory: cycling can be avoided by judicious choices of leaving arcs. The most elegant way to avoid cycling, and also the easiest to implement in practice, was proposed by W. H. Cunningham (1976); we shall return to it later in this chapter. Furthermore, cycling is extremely rare. Its occurrence in practical problems has never been reported; the first example was constructed artificially by B. J. Gassner (1964) and published 13 years after the appearance of the network simplex method. [The example given here is its simplified version. For other examples, see W. H. Cunningham (1979) and W. H. Cunningham and J. G. Klincewicz (1983).] Finally, it is not known whether cycling can occur when each entering arc maximizes $y_j - y_i - c_{ij}$. (However, the interest of this question is only theoretical: efficient computer implementations of the network simplex method do not select the entering arcs in this way.)

TERMINATION AND INITIALIZATION

For the moment, let us simply assume that cycling does not occur. In other words, let us assume that *each tree is looked at in at most one iteration.* Under this assumption, the network simplex method must terminate: since there are only a finite number of trees to be possibly looked at, there can be only a finite number of iterations. Hence the only remaining problem is that of finding a feasible tree solution to begin with. This problem is far from trivial: not every transshipment problem has a feasible solution, and the nonexistence of a feasible solution is not always apparent.

To get around this difficulty, we first observe that a feasible tree solution is readily available whenever our network is sufficiently rich in arcs. More precisely, it suffices to have an arc from a fixed node w to each sink or intermediate node other than w itself, and an arc to w from each source other than w itself. These $n - 1$ arcs, one for each node other than w, constitute a tree T_w. Now we may set

$$\begin{aligned} x_{wj} &= b_j && \text{whenever} \quad b_j \geq 0 \quad \text{and} \quad j \neq w \\ x_{iw} &= -b_i && \text{whenever} \quad b_i < 0 \quad \text{and} \quad i \neq w \\ x_{ij} &= 0 && \text{whenever} \quad ij \notin T_w. \end{aligned}$$

To verify that this is a feasible tree solution, we need only check that

$$\sum_i x_{ik} - \sum_j x_{kj} = b_k \tag{19.7}$$

for every node k. If $k \neq w$ then (19.7) is obvious: only one of the left-hand side variables corresponds to an arc of T_w. On the other hand,

$$\sum_i x_{iw} - \sum_j x_{wj} = - \sum_{k \neq w} b_k = b_w$$

and so (19.7) holds even for $k = w$.

Of course, it may happen that not all of the $n - 1$ arcs constituting T_w are present in our network. In that case, we resort to an intuitive trick: we make the absent arcs available but penalize their use. More precisely, we choose w arbitrarily and enlarge the network by adding the missing arcs of T_w. The added arcs are referred to as *artificial*. As we have just observed, a feasible tree solution in the enlarged network is readily available; what we really want is a feasible solution $\mathbf{x}$ such that $x_{ij} = 0$ for each artificial arc ij. Toward this end, we associate a penalty $p_{ij} = 1$ with each artificial arc ij, set $p_{ij} = 0$ for all the original arcs ij, and seek a feasible solution $\mathbf{x}^*$ minimizing $\sum p_{ij} x_{ij}$. The resulting transshipment problem is called the *auxiliary problem*; to find $\mathbf{x}^*$, we may use the network simplex method initialized by T_w. (Clearly, the auxiliary problem has a feasible solution and is not unbounded.) It will be convenient to distinguish three possible outcomes, depending on the optimal solution $\mathbf{x}^*$ and the corresponding tree T in the auxiliary problem.

(i) T contains an artificial arc uv with $x^*_{uv} > 0$.

(ii) T contains no artificial arc.

(iii) T contains at least one artificial arc, but every artificial arc ij has $x^*_{ij} = 0$.

In case (i), the original problem has no feasible solution: such a solution $\mathbf{x}$ would satisfy $\sum p_{ij} x_{ij} = 0 < \sum p_{ij} x^*_{ij}$, contradicting the optimality of $\mathbf{x}^*$. In case (ii), the network simplex method on the original problem can be initialized instantly by T and

$\mathbf{x}^*$. The somewhat atypical case (iii) is perhaps the most interesting: even though the original problem has a feasible solution (specified by the values of $\mathbf{x}^*$ on the original arcs), it may have no feasible tree solution (for instance, the original network may be not connected). We are going to show that in case (iii) the original network breaks down into smaller subproblems, which may be solved separately.

Decomposition into Subproblems

To begin, let us note a simple but useful identity: for each set S of nodes and for every feasible solution $\mathbf{x}$, we have

$$\sum_{\substack{i \notin S \\ j \in S}} x_{ij} - \sum_{\substack{i \in S \\ j \notin S}} x_{ij} = \sum_{k \in S} b_k. \tag{19.8}$$

In intuitive terms, (19.8) makes perfect sense: its left-hand side represents the import to S minus the export from S, whereas the right-hand side represents the net demand in S. A rigorous proof is equally easy: in the system $\mathbf{Ax} = \mathbf{b}$, take the sum of those equations that correspond to the nodes of S.

Next, suppose that the set of nodes of our network can be partitioned into non-empty subsets R and S in such a way that

$$\sum_{k \in S} b_k = 0 \quad \text{and there is no arc } ij \text{ with } i \in R, j \in S. \tag{19.9}$$

In intuitive terms, S represents a region whose domestic demand matches its domestic supply and which has no means of import. Clearly, such a region cannot afford to export. To put it formally, every feasible solution $\mathbf{x}$ must have

$$x_{ij} = 0 \qquad \text{whenever} \quad i \in S, \quad j \in R.$$

[A rigorous proof of this fact is obtained instantly by combining (19.8) with (19.9).] Hence the arcs ij with $i \in S, j \in R$ are useless and might just as well be deleted. As soon as this is done, the original problem splits into two subproblems with node sets R and S, respectively. For illustration, consider the network and demands in Figure 19.14. If R stands for the two rightmost nodes and S stands for the remaining

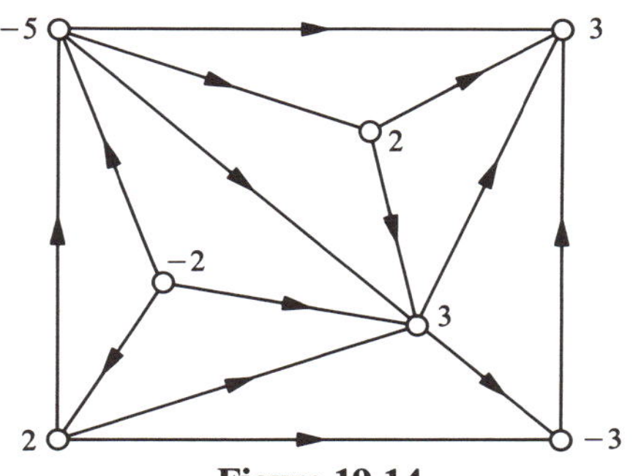

Figure 19.14

five nodes, then (19.9) holds. Hence the problem splits into the two subproblems (with cost coefficients inherited from the original problem) shown in Figure 19.15. Solving the two subproblems separately requires less time and space than solving the original problem as a whole.

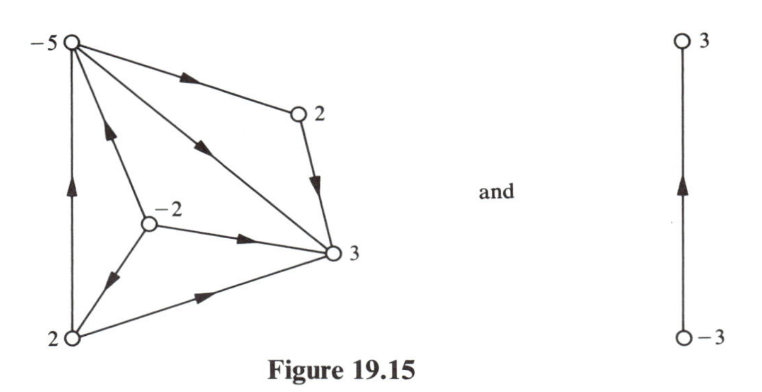

Figure 19.15

We claim that in case (iii), the original problem always decomposes in this way. In fact, the decomposition is pointed out by the node numbers $y_1, y_2, \ldots, y_n$ associated with the optimal solution $\mathbf{x}^*$ of the auxiliary problem. More precisely, taking an arbitrary artificial arc uv in the optimal tree T, we may set $k \in R$ if $y_k \leq y_u$ and $k \in S$ if $y_k > y_u$. Since R and S are nonempty (we have $u \in R$ trivially and $v \in S$ since $y_v = y_u + p_{uv} = y_u + 1$), we need only verify that (19.9) holds. As $y_i + p_{ij} \geq y_j$ for all arcs ij, no original arc ij has $i \in R, j \in S$ (otherwise $y_i + p_{ij} = y_i \leq y_u < y_j$). Thus the second part of (19.9) holds. Furthermore, substituting $\mathbf{x}^*$ for $\mathbf{x}$ in (19.8) and recalling that $x_{ij}^* = 0$ for all artificial arcs ij, we find that

$$\sum_{k \in S} b_k = -\sum x_{ij}^*$$

with the right-hand sum taken over all the original arcs ij such that $i \in S, j \in R$. But none of these arcs can be in T (otherwise $y_j = y_i + p_{ij} = y_i$ and $y_i > y_u \geq y_j$ at the same time) and so each of them has $x_{ij}^* = 0$. Thus the first part of (19.9) holds as well.

Updating the Node Numbers

Now the problems of initialization are out of the way. Before returning to the unfinished business of cycling, we shall make a few general comments on the nature of numbers $y_1, y_2, \ldots, y_n$ satisfying

$$y_i + c_{ij} = y_j \qquad \text{whenever} \quad ij \in T. \tag{19.10}$$

As previously observed, there are unique numbers $y_1^*, y_2^*, \ldots, y_n^*$ such that

$$y_n^* = 0 \quad \text{and} \quad y_i^* + c_{ij} = y_j^* \quad \text{whenever} \quad ij \in T. \tag{19.11}$$

It follows easily that $y_1, y_2, \ldots, y_n$ satisfy (19.10) if and only if

$$y_k = y_k^* + d \qquad (k = 1, 2, \ldots, n)$$

for some constant d. [The "if" part is trivial; to prove the "only if" part, observe that $y_k - y_n$ satisfy (19.11) in place of y_k^*, and so $d = y_n$.] In particular, all choices of numbers $y_1, y_2, \ldots, y_n$ satisfying (19.10) determine the same vector $\bar{\mathbf{c}} = \mathbf{c} - \mathbf{yA}$ with components $\bar{c}_{ij} = c_{ij} + y_i - y_j$. (Any side conditions such as $y_n = 0$, imposed in our example, are irrelevant.)

We shall show that the numbers $y_1, y_2, \ldots, y_n$ can be updated by a simple formula rather than computed from scratch in each iteration. The update relies on the fact that every tree with an arc e deleted splits into two disjoint trees, each of which contains one endpoint of e. (A proof is left for problem 19.8.) In particular, if $e = uv$ is the entering arc in an iteration leading from T to $T + e - f$, then $(T + e - f) - e$ consists of two disjoint trees T_u and T_v such that $u \in T_u$ and $v \in T_v$. If $y_1, y_2, \ldots, y_n$ satisfy (19.10), then the numbers $\bar{y}_1, \bar{y}_2, \ldots, \bar{y}_n$ defined by

$$\bar{y}_k = \begin{cases} y_k & (k \in T_u) \\ y_k + \bar{c}_e & (k \in T_v) \end{cases} \tag{19.12}$$

with $\bar{c}_e = c_e + y_u - y_v$ satisfy

$$\bar{y}_i + c_{ij} = \bar{y}_j \qquad \text{whenever} \quad ij \in T + e - f.$$

[The last claim is easy to justify. We have $\bar{y}_j - \bar{y}_i = y_j - y_i$ whenever $ij \in (T + e - f) - e$; in addition, $\bar{y}_u + c_e = y_u + c_e = y_v + \bar{c}_e = \bar{y}_v$.] Practical implementations of the network simplex method make use of the updating formula (19.12) or its relatives such as

$$\bar{y}_k = \begin{cases} y_k - \bar{c}_e & (k \in T_u) \\ y_k & (k \in T_v) \end{cases} \tag{19.13}$$

whose right-hand side is obtained by subtracting $\bar{c}_e$ from the right-hand side of (19.12).

An Easy Way to Avoid Cycling

To state a certain theoretical result on cycling, we need one more definition. Let T^* be a tree and let w be a node of T^*. In an intuitive sense, each arc of T^* is directed either towards w or away from it. For instance, the arcs 14, 29, 49, 62, and 75 in the tree of Figure 19.16 are directed towards the node 9, whereas 28, 43, and 95 are directed away from it. This notion can be made precise by referring to the partition of $T^* - uv$ into trees T_u and T_v such that $u \in T_u$ and $v \in T_v$. If $w \in T_v$ then uv is directed *toward* w; if $w \in T_u$, then uv is directed *away from* w. In the theorem below, we choose an arbitrary node w, call it the *root* and keep it fixed through all iterations of the network simplex method.

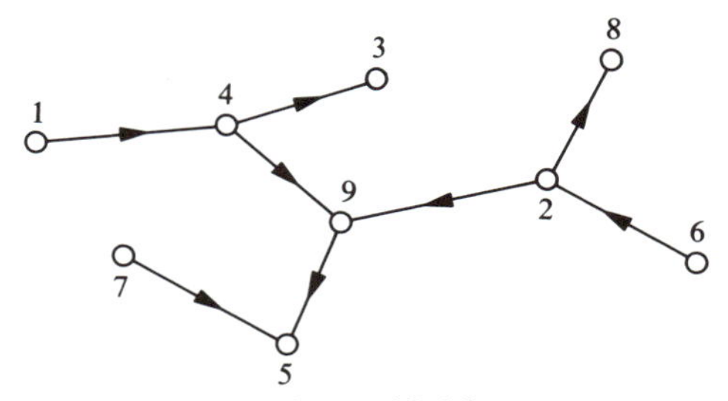

Figure 19.16

THEOREM 19.1. If, in each degenerate pivot leading from T to $T + e - f$, the entering arc e is directed away from the root in $T + e - f$ then the network simplex method does not cycle.

PROOF. We shall associate two numbers $g(T)$ and $h(T)$ with each tree T examined by the network simplex method. To define $g(T)$, we recall that T uniquely determines its feasible tree solution $\mathbf{x}$; then we set $g(T) = \mathbf{cx}$. The other number $h(T)$ arises from numbers $y_1, y_2, \ldots, y_n$ satisfying (19.10): even though these numbers are not uniquely determined by T, the quantity

$$h(T) = \sum_{k=1}^{n} (y_k - y_w)$$

is.

Now consider two consecutive trees, $T_i = T$ and $T_{i+1} = T + e - f$, in the sequence $T_1, T_2, T_3, \ldots$ constructed by the network simplex method. Clearly, we have

$$g(T_{i+1}) \leq g(T_i). \tag{19.14}$$

If $g(T_{i+1}) = g(T_i)$, then the pivot leading from T_i to T_{i+1} is degenerate and, by our hypothesis, e is directed away from the root w in T_{i+1}. To put it differently, $w \in T_u$ in (19.12) or (19.13). But then

$$\sum_{k=1}^{n} (\bar{y}_k - \bar{y}_w) = \sum_{k=1}^{n} (y_k - y_w) + \bar{c}_e|T_v|$$

and $\bar{c}_e < 0$ as e is the entering arc. In short,

$$h(T_{i+1}) < h(T_i) \qquad \text{whenever} \quad g(T_{i+1}) = g(T_i). \tag{19.15}$$

If the conclusion of the theorem fails, then $T_i = T_j$ but $i < j$ for some subscripts i and j. In that case, (19.14) and $g(T_i) = g(T_j)$ imply $g(T_i) = g(T_{i+1}) = \cdots = g(T_j)$. But then (19.15) implies $h(T_i) > h(T_{i+1}) > \cdots > h(T_j)$, contradicting $h(T_i) = h(T_j)$. ■

The anticycling strategy devised by W. H. Cunningham (1976) relies on Theorem 19.1. Let us say that a tree T is *strongly feasible* if the feasible tree solution $\mathbf{x}$ defined by T has the following property: *Every arc* $ij \in T$ *with* $x_{ij} = 0$ *is directed away from the root.* If all the trees in the sequence $T_1, T_2, T_3, \ldots$ constructed by the network simplex method are strongly feasible, then the hypothesis of Theorem 19.1 is satisfied: in each degenerate pivot leading from T to $T + e - f$, the entering arc e maintains $x_e = 0$, and so it must be directed away from the root in the strongly feasible $T + e - f$. Now it follows that cycling in the network simplex method never occurs as long as (i) the initial tree is strongly feasible and (ii) each pivot, departing from a strongly feasible tree T, arrives at a strongly feasible tree $T + e - f$.

Getting hold of an initial strongly feasible tree presents no difficulty: we need note only that the initial tree in the auxiliary problem is always strongly feasible. Furthermore, given any feasible tree solution $\mathbf{x}$, we can easily find either a strongly feasible tree defining $\mathbf{x}$ or a decomposition of the problem into two subproblems. More precisely, consider a tree T defining $\mathbf{x}$. An arc $uv \in T$ will be called *bad* if it prevents T from being strongly feasible: that is, uv is bad if it is directed towards the root and yet $x_{uv} = 0$. As above, $T - uv$ consists of disjoint trees T_u, T_v such that $u \in T_u$, $v \in T_v$; let R and S stand for the node sets of T_v and T_u, respectively. Trivially, $T - uv$ contains no arc with one endpoint in R and the other endpoint in S. Hence $x_{ij} = 0$ whenever $i \in R, j \in S$ or $i \in S, j \in R$. Substituting into (19.8), we see that S satisfies the first part of (19.9). Now there are two possibilities. If there is no arc ij with $i \in R$, $j \in S$ then (19.9) holds and the problem decomposes. If there is an arc ij with $i \in R, j \in S$, then $\mathbf{x}$ may be defined by the tree $T + ij - uv$ which contains fewer bad arcs than T. This procedure may be iterated as many times as necessary; the result is either the decomposition or the strongly feasible tree.

Now we need ensure only that each pivot, departing from a strongly feasible tree T, arrives at a strongly feasible tree $T + e - f$. This objective may be achieved with a remarkable elegance and simplicity by an appropriate choice of the leaving arc f, allowing a total freedom in the choice of the entering arc e. (Readers acquainted with Chapter 3 will note the contrast with the smallest-subscript pivoting rule, which prescribes the entering as well as the leaving variable.) On the cycle C in $T + e$, we locate a special node called the *join*: this is the first node at which the path from one endpoint of e to the root in T meeets the path from the other endpoint of e to the root in T. (See Figure 19.17.) If there are two or more candidates for the leaving arc, then break the tie by the following rule:

Choose the first candidate encountered when C is traversed in the direction of e, starting at the join. (19.16)

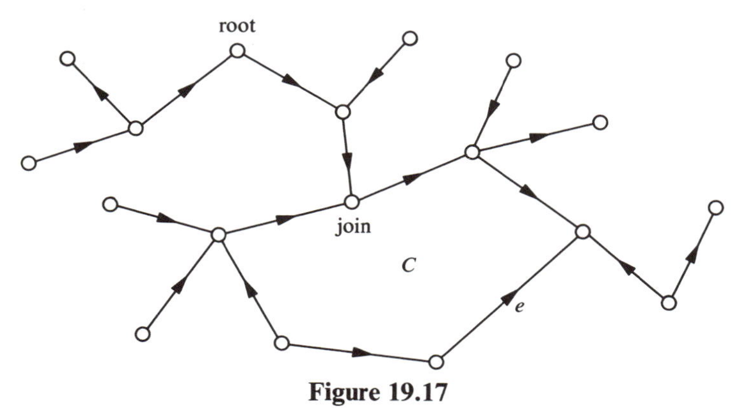

Figure 19.17

To verify that (19.16) preserves strong feasibility, consider an arbitrary pivot leading from T to $T + e - f$. Assuming that T is strongly feasible, we wish to establish strong feasibility of $T + e - f$. For that purpose, we may restrict our attention to the arcs of C: the remaining arcs ij change neither their direction with respect to the root nor the corresponding values of x_{ij}. Now we distinguish between two cases.

Case 1. The pivot is nondegenerate. In this case, the values of x_{ij} increase on all forward arcs and decrease on all reverse arcs. Hence the arcs $ij \in C$ with $\bar{x}_{ij} = 0$ in the new solution $\bar{\mathbf{x}}$ are precisely the candidates for the leaving arc. Now (19.16) guarantees that all of these arcs are directed away from the root in $T + e - f$. (See Figure 19.18.)

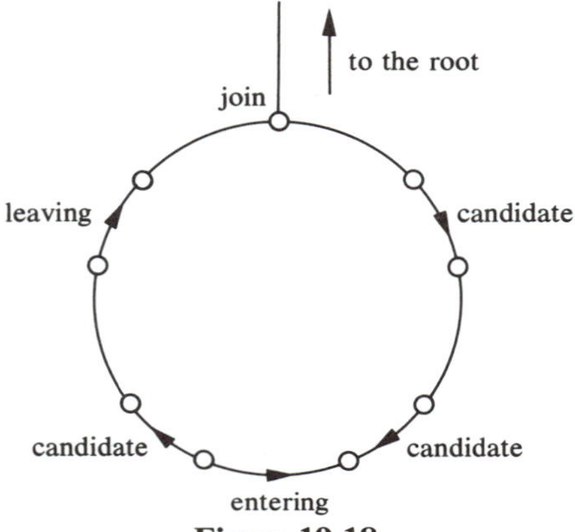

Figure 19.18

Case 2. The pivot is degenerate. In this case, the solution does not change; we shall refer to the arcs $ij \in C$ with $x_{ij} = 0$ as *zero arcs*. The entering arc $e = uv$ splits C into three parts: first the path from the join to u, then the arc uv itself, and finally the path from v back to the join. Strong feasibility of T guarantees that every zero arc on the first path is forward and every zero arc on the second path is reverse. (In particular, the leaving arc is on the second path.) Now the desired conclusion follows again from (19.16). (See Figure 19.19.)

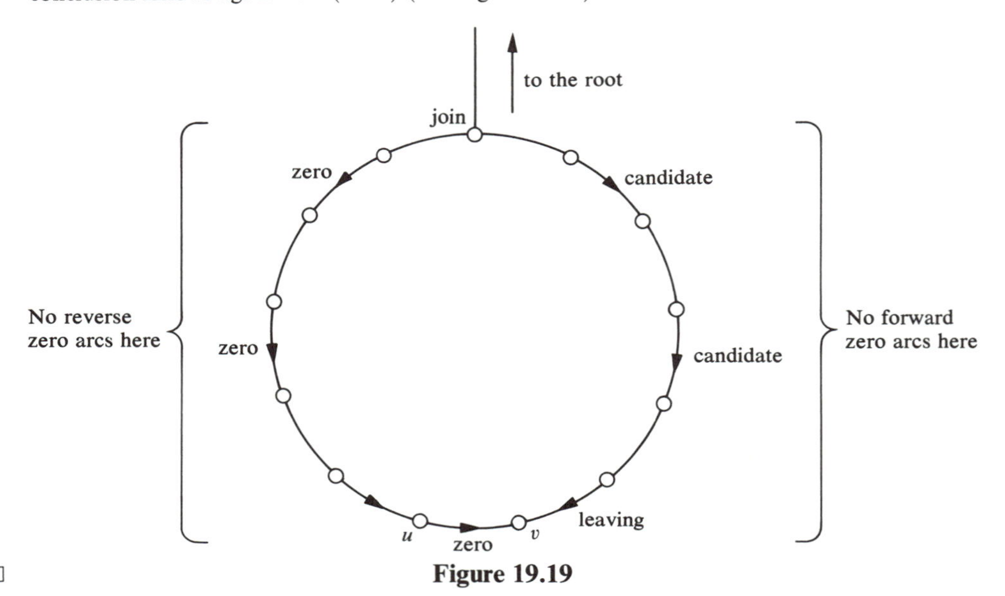

Figure 19.19

□

THE NUMBER OF ITERATIONS

With cycling effectively prevented, the network simplex method will always terminate after a finite number of iterations. It is natural to ask just how large the number of iterations can get. The answer depends not only on the size of the problem, but also on the way in which the entering arcs are selected.

Specific rules for the selection of entering arcs are called *pivoting rules*. For example, the rule of always choosing that arc which maximizes $y_j - y_i - c_{ij}$ is a pivoting rule; we shall refer to it as the rule of the *largest merit*. Another example is the rule of the *first candidate*: Always choose the first encountered arc ij with $y_j - y_i - c_{ij} > 0$. Efficient implementations of the network simplex method use pivoting rules (sketched in the next section) that are more sophisticated than first candidate and yet less time-consuming than largest merit. In practice, such rules have been found quite satisfactory: the typical number of iterations that they require is comparable to (and often smaller than) the number of nodes in the network. [When it comes to *guarantees* of a reasonably quick termination, the situation is far less encouraging. The largest-merit rule, in spite of its intuitive appeal, may lead to a disaster: N. Zadeh (1973) constructed a sequence of transshipment problems such that the kth problem has only $2k + 2$ nodes but the largest-merit rule leads to $2^k + 2^{k-2} - 2$ iterations. For instance, in a network with 100 nodes, the method goes through nearly 10^{15} iterations, which is not "reasonably quick" by any standards. Similar examples may exist for other popular pivoting rules, but they must be very rare: the network simplex method works remarkably fast even on randomly generated problems.]

In practice, the network simplex method is a spectacular success. It takes just a few minutes to solve a typical problem with thousands of nodes and tens of thousands of arcs; even problems ten times as large are solved routinely. To attain this level of efficiency, the method must be properly implemented.

COMPUTER IMPLEMENTATIONS

Since the first outline of the network simplex method, we have revised our view of a typical iteration. Now we know that the numbers $y_1, y_2, \ldots, y_n$ can be easily updated by (19.12) or (19.13) rather than computed from (19.10) in each iteration. Consequently, we shall view each iteration as beginning with some T, $\mathbf{x}$, $\mathbf{y}$, and divided into two parts: (i) choose an entering arc; (ii) update T, $\mathbf{x}$, $\mathbf{y}$. The implementation of the first part involves questions that are radically different from the challenge presented in the second part.

To choose an entering arc, we have to have a pivoting rule. Simple pivoting rules, such as the first candidate, require a relatively short time *per iteration* whereas more complicated rules, such as the largest merit, tend to reduce the *number of iterations*. An efficient pivoting rule should attempt to combine both of these virtues, so as to minimize the total running time. For instance, we may occasionally scan the list of arcs until we find r candidates for the entering arc. In the subsequent s iterations, the entering arcs are chosen from this candidate list by the rule of the largest merit. After these s iterations, the scan begins again where it had previously left off. (Typical values of the two parameters might be $r = 40$ and $s = 20$.) Pivoting rules of such kind aim for a reasonable compromise between the largest merit, requiring a long search in each iteration, and the first candidate, disregarding the magnitudes of $y_j - y_i - c_{ij}$ altogether. This basic theme is open to endless variations. For example, we might form a relatively long candidate list by scanning 10% of all the arcs and putting aside every single arc ij such that $y_i + c_{ij} < y_j$. In each subsequent iteration, we consult only some 30 arcs on the list and choose the arc ij that maximizes $y_j - y_i - c_{ij}$ among these 30. As we move through the candidate list at the rate of 30 entries per iteration, we keep deleting those arcs ij that now show $y_i + c_{ij} \geq y_j$. Consequently, the candidate list shrinks until it eventually becomes empty. At that point, we scan the next 10% of all the arcs, and so on. A rigorous analysis of these stratagems seems impossible. The definitive choice of the actual procedure, and the setting of its various parameters, require a great deal of craftsmanship based on practical experience and/or experimental results. For more information, see J. M. Mulvey (1978) and G. H. Bradley et al. (1977).

The update of T, $\mathbf{x}$, and $\mathbf{y}$ begins by finding the path between the two endpoints of the entering arc. The time required to implement this operation depends heavily on the particular representation of T in the computer. For instance, if we had at our disposal only a list of the arcs of T, then finding the path might be a relatively cumbersome task. (To take a very small example, see how long it takes to find a path between S and T in the tree with arcs AL, BR, CM, CP, DG, EI, EQ, EU, FK, GT, GC, HI, HL, HS, IT, JK, KS, LN, OR, RT.) The path may be found faster if we choose a different representation of T. Consider an ordering $v_1, v_2, \ldots, v_n$ satisfying (19.3): for each node v_k other than the *root* v_1, there is precisely one arc with one endpoint equal to v_k and the other endpoint w equal to one of the nodes $v_1, v_2, \ldots, v_{k-1}$. We shall refer to w as the *predecessor* of v_k. When the tree is drawn as hanging down from the root, the predecessor of each node is its unique neighbor on the next higher level. For illustration, consider the tree in Figure 19.20. (In this context, the actual directions of the various arcs are irrelevant, and so we ignore them.) The predecessor $p(i)$ of each node i can be found in the following array.

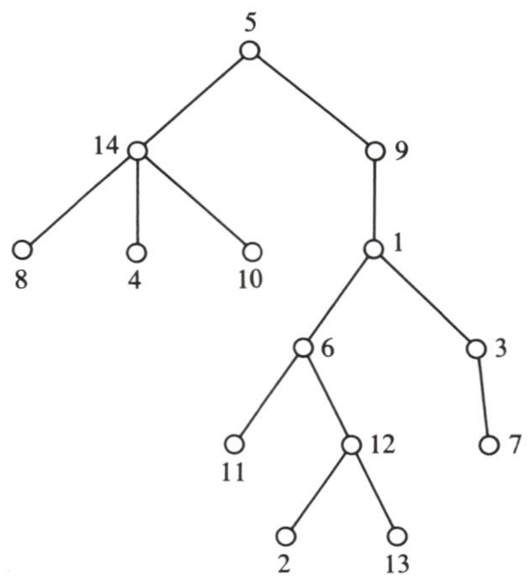

Figure 19.20

i	1	2	3	4	5	6	7	8	9	10	11	12	13	14
$p(i)$	9	12	1	14	—	1	3	14	5	14	6	6	12	5

The main virtue of this *predecessor array* is that it stores, in a compact and implicit fashion, all the paths between arbitrary nodes and the root. For example, to find the path from node 6 to root 5, we have to look up only $p(6) = 1$, $p(1) = 9$, $p(9) = 5$: the nodes of the desired path are 6, 1, 9, 5. From finding paths between prescribed nodes and the root, it is only a small step to finding paths between prescribed pairs of nodes. To find a path between u and v, we first find the path P from u to the root and the path Q from v to the root. Then, starting at the root, we move backward along P and Q at the same rate until we locate their join. The desired path from u to v consists of the portion of P between u and the join, and the portion of Q between the join and v.

Even though the path between the join and the root is totally useless, it gets traversed three times: first as a portion of P, then as a portion of Q, and finally in the backward scan. We can avoid looking at this path altogether as long as we maintain a certain auxiliary array. Each entry $d(i)$ in this array specifies the *depth* of a node i, defined as the number of nodes on the path between i and the root. The *depth array* for our example follows.

i	1	2	3	4	5	6	7	8	9	10	11	12	13	14
$d(i)$	3	6	4	3	1	4	5	3	2	3	5	5	6	2

The new array enables us to look for the join *while* we are constructing the two paths. Assuming $d(u) \geq d(v)$, we can describe the streamlined procedure as in Box 19.1.

Box 19.1 Finding the Path Between u and v

Step 0. [Initialize.] Set $i = 1, j = 1, u_1 = u, v_1 = v$.

Step 1. [The level of v reached?] If $d(u_i) = d(v)$ then go to step 3; otherwise proceed to step 2.

Step 2. [Climb up on the longer path.] Set $u_{i+1} = p(u_i)$; then replace i by $i + 1$ and return to step 1.

Step 3. [Join reached?] If $u_i = v_j$ then output the path $u_1, u_2, \ldots, u_i, v_{j-1}, \ldots, v_1$. Otherwise proceed to step 4.

Step 4. [Climb up on both paths.] Set $u_{i+1} = p(u_i)$ and $v_{j+1} = p(v_j)$. Then replace i by $i + 1$, replace j by $j + 1$ and return to step 3.

Incidentally, the leaving arc can be found *while* this procedure is being executed: we need only keep track of that reverse arc f that minimizes x_f among all the reverse arcs encountered so far. [Monitoring the values of x_e with $e \in T$ is easy as soon as $\mathbf{x}$ is represented by a certain array $x(\cdot)$ of length $n - 1$: for each node i other than the root, the value of $x(i)$ is either $x_{ip(i)}$ or $-x_{p(i)i}$, depending on the direction of the arc between i and $p(i)$. Cunningham's tie-breaking rule (19.16) can be incorporated in this scheme with no extra work.] As soon as the path between u and v and the leaving arc have been found, updating $\mathbf{x}$ presents no problem.

To update $\mathbf{y}$, we shall use formula (19.12) in case the root belongs to T_u and formula (19.13) in case the root belongs to T_v. It will be convenient to use S to denote that of the two trees T_u, T_v which does not contain the root. Now y_k changes only for $k \in S$: we have

$$\bar{y}_k = \begin{cases} y_k & (k \notin S) \\ y_k \pm \bar{c}_e & (k \in S) \end{cases} \tag{19.17}$$

with the $\pm$ sign interpreted as $+$ if e is directed away from the root in $T + e - f$, and as $-$ if e is directed toward the root.

A fast implementation of this formula requires a fast identification of S. The arrays $p(\cdot)$ and $d(\cdot)$ are not particularly well suited for this purpose; in order to identify S quickly, we shall make use of yet another array. When a rooted tree is drawn in the plane, imagine tracing a continuous line around its contours, beginning at the root and eventually returning there. As the line sweeps round the tree, it traverses each arc twice: first on its way down and later on its way in the opposite direction. This way of traversing the tree is known as *depth-first search.* In our example, depth-first search visits the nodes

5, 14, 8, 14, 4, 14, 10, 14, 5, 9, 1, 6, 11, 6, 12, 2, 12, 13, 12, 6, 1, 3, 7, 3, 1, 9, 5

in this order. Recording only the first visit to each node, we obtain an ordering of the nodes known as a *preorder.* In our example, the preorder is

5, 14, 8, 4, 10, 9, 1, 6, 11, 12, 2, 13, 3, 7.

A rooted tree may have many different preorders, depending on how it might be drawn in the plane. We choose a fixed preorder and, for each node i, record the successor $s(i)$ of i in this preorder. When i is the last node in the preorder, we let $s(i)$ be the root. The array $s(\cdot)$ is known as a *thread*; the following is the thread for our example.

i	1	2	3	4	5	6	7	8	9	10	11	12	13	14
$s(i)$	6	13	7	10	14	11	5	4	1	9	12	2	3	8

Now let f be the leaving arc in our tree T. As usual, $T - f$ splits into two disjoint trees. One of these trees contains the root; the other tree is easily seen to be the tree S featured in (19.17). Hence S can be identified easily by depth and thread: having traversed f downward, depth-first search proceeds to explore the entire S and, when finished, leaves by traversing f upward. The resulting implementation of (19.17) is described in Box 19.2.

BOX 19.2 Implementation of Formula (19.17)

Step 0. [Initialize.] Denote the two endpoints of f by f_1 and f_2 in such a way that $d(f_2) = d(f_1) + 1$. Set $k = f_2$ and proceed to step 1.

Step 1. [Update y_k.] Replace y_k by $y_k \pm \bar{c}_e$ and proceed to step 2.

Step 2. [Follow the thread through S.] Replace k by $s(k)$. If $d(k) > d(f_2)$ then return to step 1; otherwise stop.

The only remaining task is updating T, which means updating the three arrays $p(\cdot)$, $d(\cdot)$ and $s(\cdot)$. These arrays are merely tools for a quick update of $\mathbf{x}$ and $\mathbf{y}$; if they could not be updated fast in their own right, then the means would defeat the purpose. Fortunately, they can be updated fast. In fact, the update of $p(\cdot)$, $d(\cdot)$, and $s(\cdot)$ can be carried out concurrently with the update of $\mathbf{y}$.

Whereas the updated arrays $p(\cdot)$ and $d(\cdot)$ are uniquely determined by the new rooted tree $T + e - f$, the updated form of $s(\cdot)$ depends on the way we imagine $T + e - f$ drawn in the

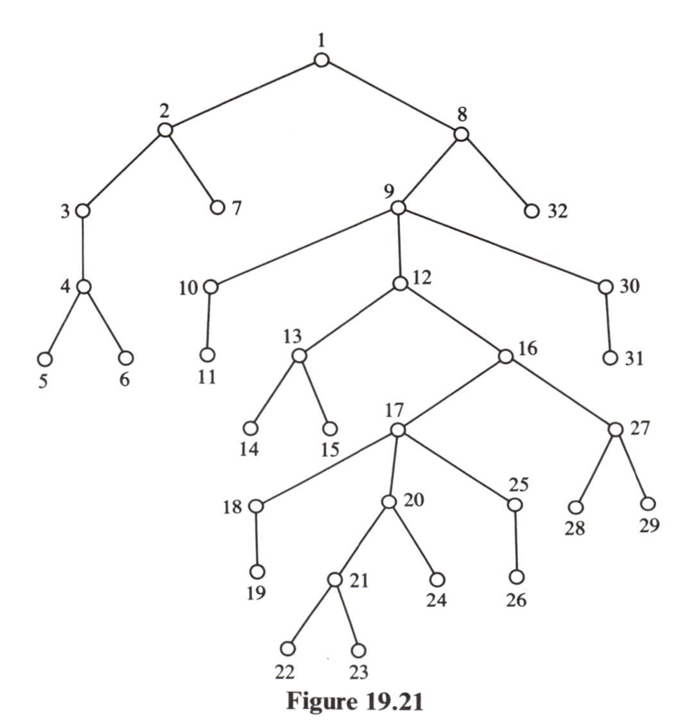

Figure 19.21

plane. In other words, $T + e - f$ may have many different preorders. We shall remove the ambiguity by choosing a preorder designed to minimize the work required in the update of $s(\cdot)$. A formal description of the new preorder is a little complicated; we shall illustrate it on the tree T shown in Figure 19.21. For clarity, the 32 nodes of T are labeled in such a way that $1, 2, \ldots, 32$ is a preorder. The entering arc e has endpoints 3 and 20; the leaving arc has endpoints 8 and 9.

The endpoints of the entering arc e will be labeled as e_1, e_2, and the endpoints of the leaving arc f will be labeled as f_1, f_2 in such a way that $e_1 \notin S$, $e_2 \in S$, $f_1 \notin S$, $f_2 \in S$. (The labels on f are easy to assign: as previously observed, $d(f_2) = d(f_1) + 1$. To assign the labels on e, observe that it is the path from e_2 to the join in T, rather than the path from e_1 to the join, which contains the leaving arc f.) The path between e_2 and f_2 in T is called the *pivot stem* or the *backpath*; we shall denote the sequence of its nodes by $v_1, v_2, \ldots, v_h$ in such a way that $v_1 = e_2$ and $v_h = f_2$. In our example, $v_1 = 20$, $v_2 = 17$, $v_3 = 16$, $v_4 = 12$, and $v_5 = 9$.

Now consider an arbitrary node $k \in S$ and find the smallest subscript t such that v_t appears on the path from k to the root: for this subscript t, we shall write $k \in S_t$. Thus S splits into pairwise disjoint subsets $S_1, S_2, \ldots, S_h$; we shall consider each S_t ordered by the preorder on T. In our example, we have

$$S_1 = \{20, 21, 22, 23, 24\}$$
$$S_2 = \{17, 18, 19, 25, 26\}$$
$$S_3 = \{16, 27, 28, 29\}$$
$$S_4 = \{12, 13, 14, 15\}$$
$$S_5 = \{9, 10, 11, 30, 31\}.$$

We shall denote by S^* the concatenation of the ordered sets $S_1, S_2, \ldots, S_h$. In our example, S^* is

$$20,\ 21,\ 22,\ 23,\ 24,\ 17,\ 18,\ 19,\ 25,\ 26,\ 16,\ 27,\ 28,\ 29,\ 12,\ 13,\ 14,\ 15,\ 9,\ 10,\ 11,\ 30,\ 31.$$

Finally, the preorder on $T + e - f$ is obtained from the preorder on T by first removing the block S and then inserting the block S^* directly after e_1. In our example, the corresponding drawing of $T + e - f$ is shown in Figure 19.22. This operation requires certain changes in the thread. When the block S is removed, the resulting gap has to be closed: if the first node f_2 in S was preceded by a node a, then the new successor of a will be the old successor z of the last node in S. Similarly, to insert S^* between e_1 and its old successor b, we make the first node e_2 of S^* the new successor of e_1, and we let b be the new successor of the last node in S^*. (In the singular case $e_1 = a$, we simply make e_2 the new successor of e_1, and we let z be the new successor of the last node in S^*.)

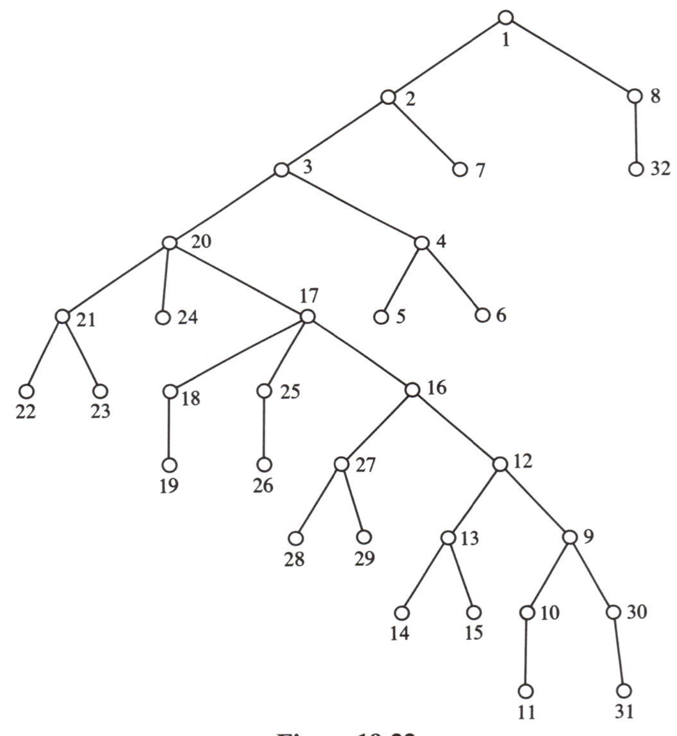

Figure 19.22

The transformation of S into S^* requires further changes in the thread. First, the new successor of the last node in each S_t with $t \leq h - 1$ will be the first node v_{t+1} of S_{t+1} (as the last node in S_h is the last node in S^*, its new successor will be b). Secondly, one additional node in each S_t with $t \geq 2$ may change its successor. In preparation for this change, note that each S_t with $t \geq 2$ splits into its *left part*, consisting of nodes appearing before v_{t-1} in the old preorder, and the remaining *right part*, consisting of nodes appearing after v_{t-1}. The left part always includes v_t, but the right part may be empty. (For instance, consider $t = 4$ in our example.) If the right part is nonempty, then the last node in the left part changes its successor from v_{t-1} to the first node in the right part.

Box 19.3 describes an implementation of these changes. The variable k scans through S_1, $S_2, \ldots, S_h$ in this order. While k is scanning through S_t, the variable i stands for v_t and, if $t \geq 2$, the variable j stands for v_{t-1}. The variable r is the first node in the right part of the set that, among all the sets $S_t, S_{t+1}, \ldots, S_h$ with nonempty right parts, has the smallest subscript; if there is no such set, then r is the old successor of the last node in S.

BOX 19.3 Updating the Thread

Step 0. [Initialize.] Set $a = f_1$ and, as long as $s(a) \neq f_2$, keep replacing a by $s(a)$. Set $b = s(e_1)$ and $i = e_2$.

Step 1. [Find the last node k in S_1 and initialize the value of r.] Set $k = i$ and, as long as $d(s(k)) > d(i)$, keep replacing k by $s(k)$. Then set $r = s(k)$.

Step 2. [If at the end of S^*, remove S and insert S^*.] If $i = f_2$ then set $s(a) = r$, $s(e_1) = e_2$, $s(k) = b$ in case $e_1 \neq a$, and $s(e_1) = e_2$, $s(k) = r$ in case $e_1 = a$, and stop.

Step 3 [Climb up the pivot stem and update $s(k)$.] Set $j = i$, replace i by $p(i)$, and then replace $s(k)$ by i.

Step 4. [Find the last node k in the left part of S_t.] Set $k = i$ and, as long as $s(k) \neq j$, keep replacing k by $s(k)$.

Step 5. [If the right part of S_t is nonempty then update $s(k)$, find the last node k in S_t, and update r.] If $d(r) > d(i)$ then replace $s(k)$ by r, keep replacing k by $s(k)$ as long as $d(s(k)) > d(i)$, and finally replace r by $s(k)$.

Step 6. [Iteration completed.] Return to step 2.

Now for the update of $p(\cdot)$ and $d(\cdot)$. Graphically, the transformation of T into $T + e - f$ may be seen as consisting of two steps: first, the link e between e_1 and e_2 is attached, and then the link f between f_1 and f_2 is cut. As a result, the pivot stem flips down: each of its nodes v_t, formerly hanging down from v_{t+1}, is now suspended from v_{t-1}. Nevertheless, each S_t continues to hang down from v_t in the same old way. This observation has two useful corollaries.

(i) The value of $p(k)$ changes only if k is on the pivot stem. The original values $p(v_1) = v_2$, $p(v_2) = v_3, \ldots, p(v_h) = f_1$ change into $p(v_1) = e_1$, $p(v_2) = v_1, \ldots, p(v_h) = v_{h-1}$.

(ii) For each $t = 1, 2, \ldots, h$ there is a constant c_t such that the new depth $d^*(k)$ of each $k \in S_t$ equals $d(k) + c_t$. Since $d^*(e_2) = d(e_1) + 1$, we have $c_1 = d(e_1) - d(e_2) + 1$. Since $d^*(v_t) = d^*(v_{t-1}) + 1$ and $d(v_{t-1}) = d(v_t) + 1$ whenever $t \geq 2$, we have $c_t = 2 + c_{t-1}$ whenever $t \geq 2$.

Now it becomes easy to incorporate the update of $p(\cdot)$ and $d(\cdot)$ into the procedure for updating $s(\cdot)$. The update of **y** may be incorporated into the same procedure rather than carried out separately.

For more information, including implementations based on data structures other than predecessor, depth and thread, the reader is referred to A. I. Ali et al. (1978); our exposition followed closely the lines of G. H. Bradley et al. (1977). □

PROBLEMS

△19.1 Solve the problems

$$\text{minimize} \quad \mathbf{cx} \qquad \text{subject to} \quad \mathbf{Ax} = \mathbf{b}, \mathbf{x} \geq \mathbf{0}$$

with

$$\mathbf{A} = \left[\begin{array}{cccccccccccc}
-1 & -1 & & 1 & & & & & & & & \\
1 & & -1 & & & & & 1 & & & & \\
 & & 1 & -1 & -1 & & & & & & & \\
 & 1 & & & & -1 & -1 & & & & & \\
 & & & & & 1 & & -1 & -1 & & & \\
 & & & & 1 & & & & & -1 & 1 & \\
 & & & & & & 1 & & & 1 & & -1 \\
 & & & & & & & & 1 & & -1 & 1
\end{array}\right]$$

and the following three choices of $\mathbf{b}$, $\mathbf{c}$:

(i) $\mathbf{b} = [-2, -1, 2, -6, 1, 0, \ 3, 3]^T, \mathbf{c} = [3, 1, \ 0, 0, 1, \ 2, 4, 2, 3, 1, 1, 0]$

(ii) $\mathbf{b} = [-2, -1, 2, -6, 1, 0, \ 3, 3]^T, \mathbf{c} = [2, 1, -3, 1, 1, -2, 4, 2, 3, 1, 2, 4]$

(iii) $\mathbf{b} = [-3, \ 1, 3, -1, 2, 0, -3, 1]^T, \mathbf{c} = [2, 1, -3, 1, 1, -2, 4, 2, 3, 1, 2, 4]$.

19.2 Decompose the problem

$$\text{minimize} \quad \mathbf{cx} \qquad \text{subject to} \quad \mathbf{Ax} = \mathbf{b}, \quad \mathbf{x} \geq \mathbf{0}$$

with

$$\mathbf{A} = \left[\begin{array}{cccccccccccc}
 & & & & & & 1 & 1 & -1 & & & \\
 & & -1 & -1 & & & & & & & 1 & \\
1 & & & 1 & & & & & & & & -1 \\
 & & & & 1 & -1 & & & & & -1 & \\
 & & & & -1 & & & -1 & & 1 & & \\
 & & & & & 1 & -1 & & & & & 1 \\
 & -1 & 1 & & & & & & & -1 & & \\
-1 & 1 & & & & & & & 1 & & &
\end{array}\right] \quad \text{and}$$

$$\mathbf{b} = [-1, 2, 3, -4, -5, 6, 7, -8]^T$$

into subproblems.

19.3 Update the arrays

i	1	2	3	4	5	6	7	8	9
p	9	5	—	1	9	5	3	1	3
d	3	4	1	4	3	4	2	4	2
s	4	6	7	8	2	3	9	5	1

representing a tree T into the arrays representing $T + 12 - 59$.

19.4 Prove that a network is not connected if and only if its nodes can be colored green and red in such a way that each color occurs at least once but the two endpoints of every arc have the same color.

19.5 Design an efficient algorithm that given any node v_1 in any network N with n nodes produces one of the following three outputs: (i) a labeling of the nodes as $v_1, v_2, \ldots, v_n$ such that (19.3) holds; (ii) a coloring of the nodes as in problem 19.4, showing that N is not connected; (iii) a cycle in N.

[*Hint*: Consider a sequence $v_1, v_2, \ldots, v_t$ which might conceivably be enlarged into output (i). Show that this sequence may be enlarged by v_{t+1} or else one of the outputs (ii), (iii) produced immediately.]

19.6 Prove: In an acyclic network, there is at most one path between any two nodes.

19.7 Prove: If $1 \leq k \leq n$ then every acyclic network with n nodes and $n - k$ arcs consists of k disjoint trees.

19.8 Prove that the network $T - e$ obtained by deleting an arc e from a tree T consists of two disjoint trees, each of which contains one endpoint of e.

19.9 Let $\mathbf{B}$ be a truncated incidence matrix of a network T with n nodes and $n - 1$ arcs. Prove that $\mathbf{B}$ is nonsingular if and only if T is a tree.

19.10 Construct an example of a transshipment problem with all arc costs positive and the following counterintuitive property: If the supply at a suitable source and the demand at a suitable sink are simultaneously lowered by one unit, then the optimal transshipment cost increases. (This can be done with as few as four nodes.)

19.11 Every transshipment problem

$$\text{minimize} \quad \mathbf{cx} \qquad \text{subject to} \quad \mathbf{Ax} = \mathbf{b}, \quad \mathbf{x} \geq \mathbf{0}$$

can be perturbed into

$$\text{minimize} \quad \mathbf{cx} \qquad \text{subject to} \quad \mathbf{Ax} = \tilde{\mathbf{b}}, \quad \mathbf{x} \geq \mathbf{0}$$

such that $\tilde{b}_w = b_w - (n - 1)\varepsilon$ for the root w and $\tilde{b}_i = b_i + \varepsilon$ for each remaining node i, with a very small positive ε (when $\mathbf{b}$ is integer valued, $\varepsilon = 1/n$ will do). Prove that each feasible tree solution in the perturbed problem is nondegenerate and that a tree is feasible in the perturbed problem if and only if it is strongly feasible in the original problem.

20

Applications of the Network Simplex Method

The scheduling problems that can be stated as transshipment problems, and solved by the network simplex method, extend beyond the obvious range involving shipments of goods. Furthermore, many combinatorial problems in pure mathematics can be seen as transshipment problems, and corresponding min-max theorems derived easily from an analysis of the network simplex method.

INEQUALITY CONSTRAINTS

The most straightforward applications of transshipment theory deal with actual physical shipments of some commodity. In this context, the total supply of the commodity often exceeds the total demand, and so the individual supplies at the various sources do not have to be fully exhausted. The resulting problem no longer fits the format described in Chapter 19, in which we assumed an *equality* constraint for each node. Nevertheless, the new variation can be incorporated into the old scheme.

For illustration, consider the network in Figure 20.1 with supplies of

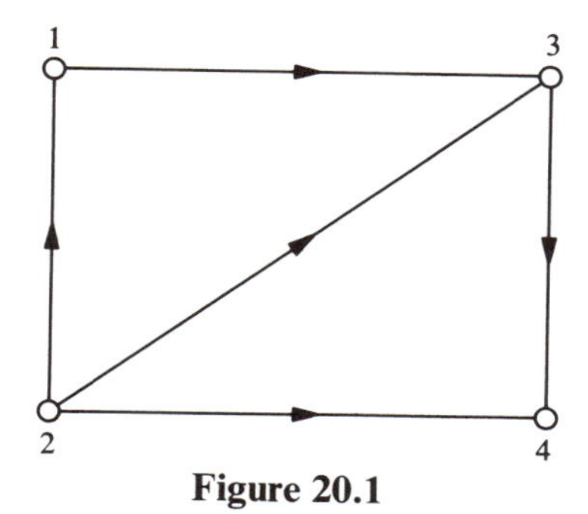

Figure 20.1

4 units at the source 1

9 units at the source 2

and demands of

7 units at the sink 3

2 units at the sink 4.

Let us assume that the demands at the sinks have to be satisfied exactly, but the supplies at the sources need not be exhausted fully. To accommodate this assumption, we introduce a new node with zero cost arcs leading into it from the sources as shown in Figure 20.2. This new node may be thought of as a dump into which the unused

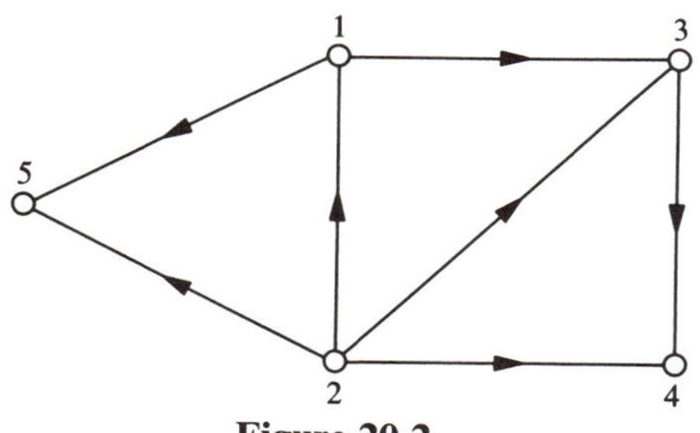

Figure 20.2

supplies may be diverted free of charge; the demand at the dump equals the excess supply, $(4 + 9) - (7 + 2) = 4$. Thus we have converted the original problem,

$$
\begin{array}{ll}
\text{minimize} & z = c_{13}x_{13} + c_{21}x_{21} + c_{23}x_{23} + c_{24}x_{24} + c_{34}x_{34} \\
\text{subject to} & -x_{13} + x_{21} \geq -4 \\
& -x_{21} - x_{23} - x_{24} \geq -9 \\
& x_{13} + x_{23} - x_{34} = 7 \\
& x_{24} + x_{34} = 2 \\
& x_{13}, x_{21}, x_{23}, x_{24}, x_{34} \geq 0
\end{array}
$$

into an equivalent transshipment problem,

$$\begin{array}{llcl}
\text{minimize} & z & & \\
\text{subject to} & -x_{13} + x_{21} \qquad\qquad\qquad\qquad - x_{15} & = & -4 \\
& - x_{21} - x_{23} - x_{24} \qquad\qquad\qquad - x_{25} & = & -9 \\
& x_{13} \qquad + x_{23} \qquad - x_{34} & = & 7 \\
& x_{24} + x_{34} & = & 2 \\
& x_{15} + x_{25} & = & 4 \\
& x_{13}, x_{21}, x_{23}, x_{24}, x_{34}, x_{15}, x_{25} & \geq & 0.
\end{array}$$

The new node 5 is called a *dummy node* and the new arcs 15, 25 are called *dummy arcs.*

In general, dummy arcs kd leading into a dummy node d convert inequality constraints

$$-\sum_j x_{kj} + \sum_i x_{ik} \geq b_k \tag{20.1}$$

into equations

$$-\sum_j x_{kj} + \sum_i x_{ik} - x_{kd} = b_k.$$

Similarly, dummy arcs dk leading out of a dummy node d convert inequality

$$-\sum_j x_{kj} + \sum_i x_{ik} \leq b_k \tag{20.2}$$

into equations

$$-\sum_j x_{kj} + \sum_i x_{ik} + x_{dk} = b_k.$$

Thus we may convert every problem with constraints of the types (20.1), (20.2) and

$$-\sum_j x_{kj} + \sum_i x_{ik} = b_k \tag{20.3}$$

into a proper transshipment problem, all of whose constraints have the form (20.3) and $x_{ij} \geq 0$. The only extra node required for this transformation is the dummy node d; the constraint corresponding to this node reads

$$-\sum_j x_{dj} + \sum_i x_{id} = -\sum_k b_k.$$

SCHEDULING PRODUCTION AND INVENTORY

There are other, less intuitive applications of transshipment theory in industry. Some of them are reminiscent of the inventory scheduling problems of Chapter 12. For example, consider a factory that produces goods whose predicted demand fluctuates over a period of time. There are three different ways of adjusting the volume of the output to meet the fluctuating demand:

(i) Change the level of the regular production.

(ii) Use overtime production to cover temporary shortages.

(iii) Store the present excess to cover future shortages.

We shall assume that there are no penalties on variations of the regular production level; however, this level is limited to at most r units per month. Similarly, the overtime production level is limited to at most s units per month. The overtime production cost, b dollars per unit, exceeds the regular production cost, a dollars per unit. There is no limit on the volume of the inventory, but there is a holding charge of c dollars per unit per month. For definiteness, we shall say that the predicted demand changes from one month to the next over a period of n months, and we shall denote the demand in the jth month by d_j.

The problem of meeting the fluctuating demand at the least possible cost can be formulated as a transshipment problem. The data corresponding to $n = 4$ are shown in Figure 20.3. For each of the n months, the network includes a regular production source with supply r, an overtime production source with supply s, and a sink with demand d_j. The sink can be supplied from the regular source at a dollars per unit, from the overtime source at b dollars per unit, or from the last month's inventory at c dollars per unit. (If the sum of the three exceeds the demand d_j, then the excess goes to the next sink in the form of inventory.) The unused production capacity at each source can be diverted to a dummy sink free of charge. [An essentially equivalent formulation has been pointed out by E. H. Bowman (1956).] Now the problem can be solved by the network simplex method. (Actually, even more straightforward solution procedures are available; we leave their design for problem 20.2.)

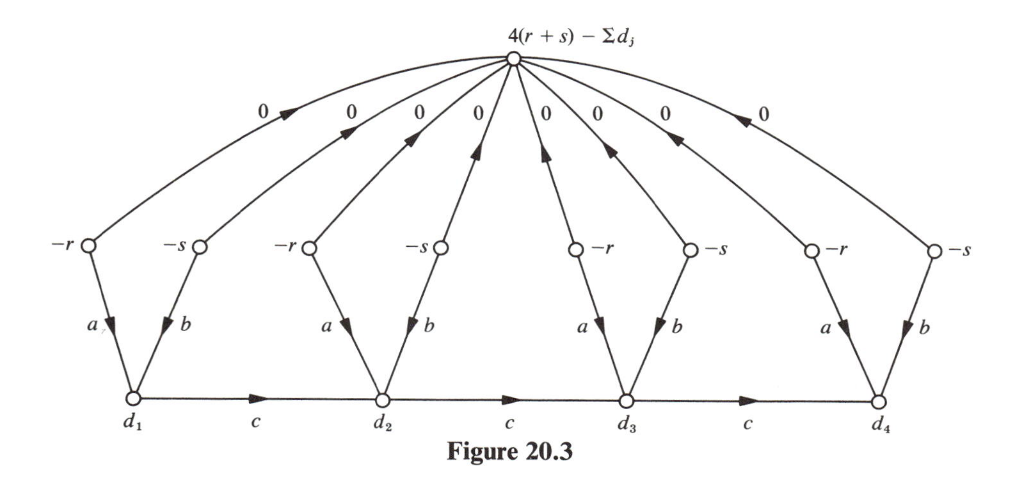

Figure 20.3

THE CATERER PROBLEM

A more intricate problem of a similar nature originated in scheduling aircraft engine overhauls. In a disguised description, it became known as the *caterer problem* [W. Jacobs (1954), J. W. Gaddum et al. (1954), W. Prager (1956)]. Imagine a caterer who has to provide fresh napkins over a period of n days; the number d_j of napkins required on the jth day is known in advance. To satisfy these requirements, the caterer can either buy new napkins (for a cents apiece) or have the used napkins laundered. The laundry provides a fast service (napkins returned q days later for b cents apiece) and a slow service (napkins returned p days later for c cents apiece). Naturally, $p > q$ and $a > b > c$. For illustration, we shall consider the case of $n = 10$,

$$\begin{aligned} &d_1 = 50, \quad d_2 = 60, \quad d_3 = 80, \quad d_4 = 70, \quad d_5 = 50, \\ &d_6 = 60, \quad d_7 = 90, \quad d_8 = 80, \quad d_9 = 50, \quad d_{10} = 100, \end{aligned}$$

and $p = 4$, $q = 2$, $a = 200$, $b = 75$, $c = 25$. One possible strategy is summarized in Table 20.1. The associated costs are $2 × 170 = $340 for purchase, $0.75 × 490 = $367.50 for fast laundry, and $0.25 × 30 = $7.50 for slow laundry; the total comes to $715.

Table 20.1 The Caterer Problem: A Feasible Solution

Day	Supply of fresh napkins		Dispatching used napkins		
	Laundered	Bought	Fast	Slow	Hold
1	—	50	50	—	—
2	—	60	60	—	—
3	50	30	50	30	—
4	60	10	60	—	10
5	50	—	60	—	—
6	60	—	60	—	—
7	90	—	50	—	40
8	60	20	100	—	20
9	50	—	—	—	70
10	100	—	—	—	170

The problem of finding the cheapest strategy can be formulated as a transshipment problem. The network and demands for our example are shown in Figure 20.4. Each of the left-hand side nodes may be thought of as the hamper at the end of the jth day, with the supply of used napkins equal to d_j. Each used napkin can be dispatched to

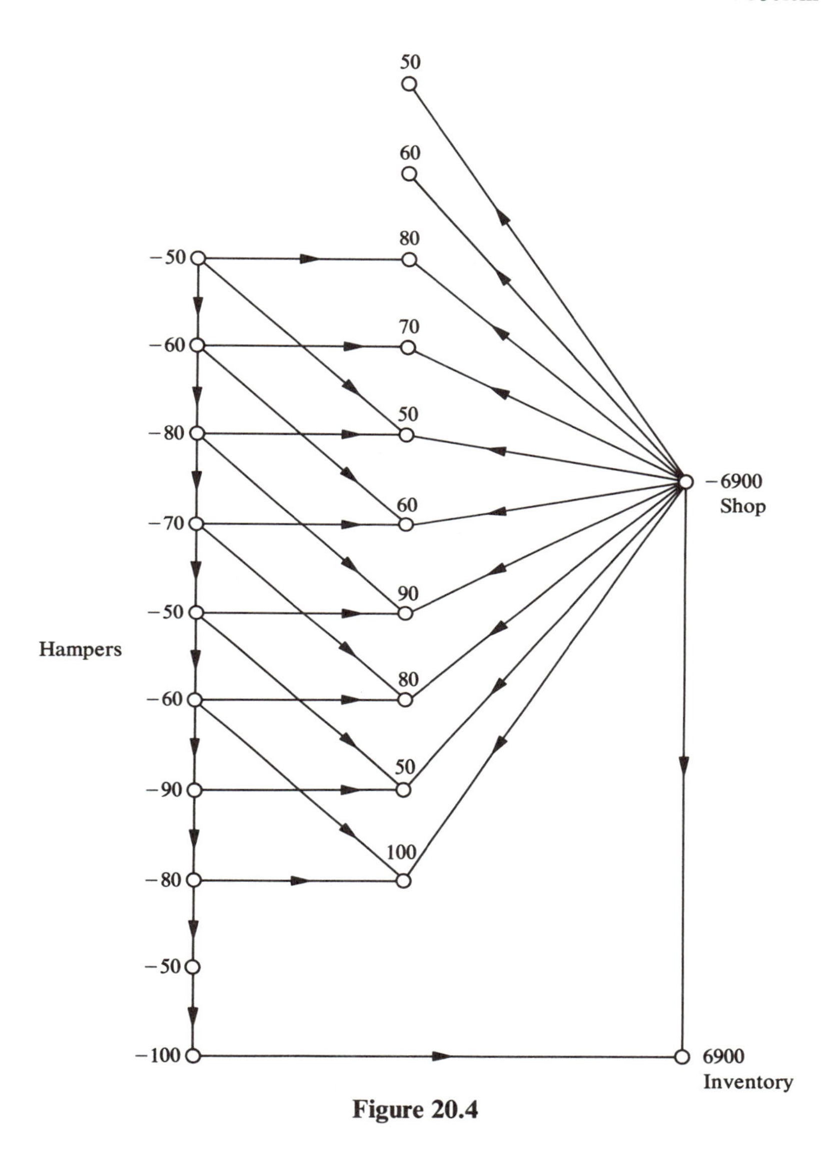

Figure 20.4

the fast laundry (horizontal arcs, cost b) or to the slow laundry (slanted arcs, cost c) or held till the next day (vertical arcs, zero cost). The required number d_j of napkins on day j can be obtained from the hamper of day $j - q$ by quick laundry, from the hamper of day $j - p$ by slow laundry, or from the shop (arcs of cost a). It is convenient to assume that the shop holds a sufficient supply of napkins: in an extreme case, all the napkins could be bought and none laundered. The used napkins from the last day's hamper are transferred to the final inventory at no extra charge, and so are the nap-

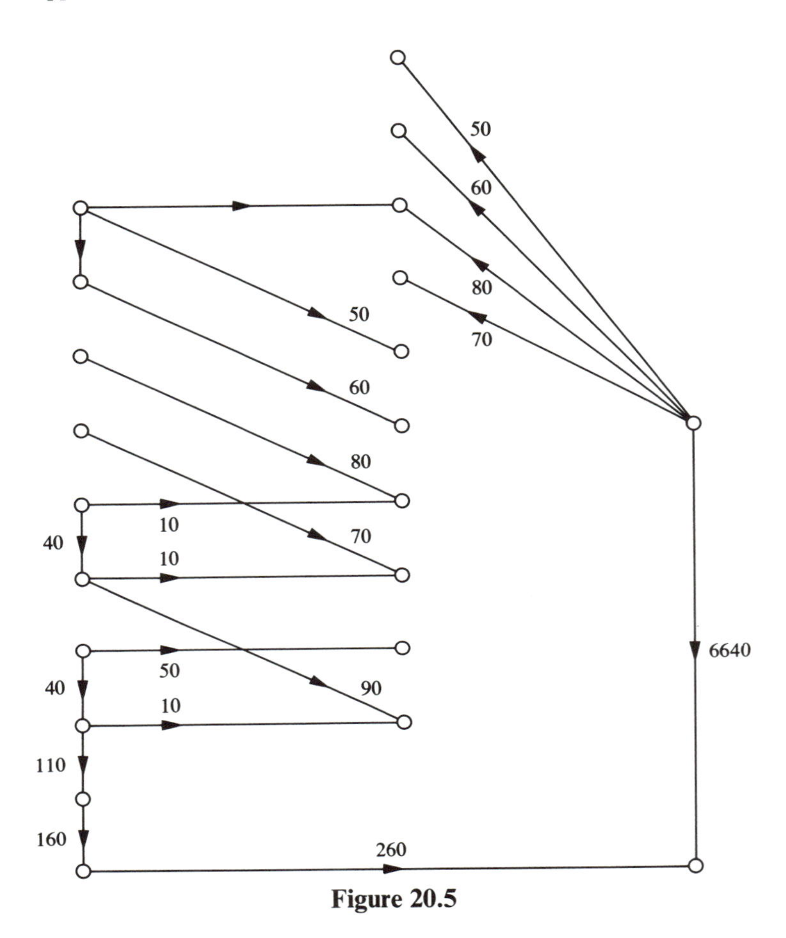

Figure 20.5

kins remaining at the store after the n-th day. The optimal solution of our illustrative case is shown in Figure 20.5, and the corresponding schedule is given in Table 20.2. The associated costs are $2 × 260 = $520 for purchase, $0.75 × 80 = 60 for fast laundry and $0.25 × 350 = $87.50 for slow laundry; the total comes to $667.50.

THE INTEGRALITY THEOREM

Our treatment of the caterer problem contains an apparent flaw: the transshipment formulation ignores the fact that the quantities of napkins bought, laundered, and stored are integers. An optimal solution of the transshipment problem would be of no use to the caterer if some of its variables had fractional values. Fortunately, variables never have fractional values in an optimal solution found by the network simplex method. More generally, we have the following theorem.

Table 20.2 The Caterer Problem: Optimal Solution

Day	Supply of fresh napkins		Dispatching used napkins		
	Laundered	Bought	Fast	Slow	Hold
1	—	50	—	50	—
2	—	60	—	60	—
3	—	80	—	80	—
4	—	70	—	70	—
5	50	—	10	—	40
6	60	—	10	90	—
7	90	—	50	—	40
8	80	—	10	—	110
9	50	—	—	—	160
10	100	—	—	—	260

THEOREM 20.1 (The Integrality Theorem). Consider a transshipment problem

$$\text{minimize} \quad \mathbf{cx} \qquad \text{subject to} \quad \mathbf{Ax} = \mathbf{b}, \quad \mathbf{x} \geq \mathbf{0} \tag{20.4}$$

such that all the components of the vector $\mathbf{b}$ are integers. If (20.4) has at least one feasible solution, then it has an integer-valued feasible solution. If (20.4) has an optimal solution, then it has an integer-valued optimal solution.

PROOF. The crucial point is that:

Every feasible tree solution of (20.4) is integer-valued. (20.5)

To justify (20.5), let $\mathbf{B}$ stand for the $(n-1) \times (n-1)$ truncated incidence matrix of that tree which defines the feasible tree solution. As observed in Chapter 19, the rows and columns of $\mathbf{B}$ may be permuted in such a way that the resulting matrix is upper triangular with each diagonal entry equal to ± 1. Hence each system $\mathbf{Bx}^* = \tilde{\mathbf{b}}$ may be solved by straightforward back substitution requiring no divisions whatsoever. In particular, if all the components of $\tilde{\mathbf{b}}$ are integers, then all the components of $\mathbf{x}^*$ are integers. Now (20.5) follows.

With (20.5) established, the theorem may be proved by induction on the number of nodes. If (20.4) has a feasible solution, then the network simplex method applied to the auxiliary problem finds either a decomposition of (20.4) into two smaller subproblems or a feasible tree solution. In the former case, the desired conclusions follow from the induction hypothesis applied to each of the subproblems. In the latter case, the feasible tree solution is integer-valued by virtue of (20.5) and, in case (20.4) has an optimal solution, the optimal tree solution found by the network simplex method is integer-valued by virtue of (20.5). ■

Alternatively, one might prove the Integrality Theorem by appealing to the mechanism of the network simplex method: if a feasible tree solution encountered in some iteration is integer-valued, then the next feasible tree solution is integer-valued again. Hence the network simplex method, once initialized by an integer-valued solution, encounters only integer-valued solutions.

Problems of the form

$$\text{minimize} \quad \mathbf{cx} \qquad \text{subject to} \quad \mathbf{Ax} = \mathbf{b}, \quad \mathbf{x} \geq \mathbf{0}, \quad \mathbf{x} \text{ integer-valued} \tag{20.6}$$

are known as *integer linear programming problems*. Such problems are notoriously hard to solve, both in theory and in practice. Yet if **A** happens to be the incidence matrix of a network, then (20.6) can be solved quite efficiently by the network simplex method. This fact is a direct consequence of the Integrality Theorem (or, rather, a direct consequence of the proof given here).

Furthermore, the Integrality Theorem has many applications in combinatorial mathematics. For instance, it readily implies a classical result [König (1916)] that may be stated frivolously as follows.

THEOREM 20.2 (König's Theorem). If, in a set of n girls and n boys, every girl knows exactly k boys and every boy knows exactly k girls, then n marriages can be arranged with everybody knowing her or his spouse.

PROOF. Consider a network with nodes $r_1, r_2, \ldots, r_n, s_1, s_2, \ldots, s_n$ in which $r_i s_j$ is an arc if and only if girl i and boy j know each other. Assign a unit supply to each girl-node r_i and a unit demand to each boy-node s_j. Regardless of the cost coefficients, the resulting transshipment problem has a feasible solution $\mathbf{x}$: we need only set $x_e = 1/k$ for every arc e. By the Integrality Theorem, the problem has an integer-valued solution $\mathbf{x}^*$. Clearly, we have $x_e^* = 0$ or $x_e^* = 1$ for every arc e. In fact, precisely one arc e with $x_e^* = 1$ leaves every girl-node and precisely one such arc enters every boy-node. These arcs point out the desired marriages. ■

DOUBLY STOCHASTIC MATRICES

The trick used in the proof of Theorem 20.2 applies in a more general setting of *doubly stochastic matrices.* An $n \times n$ matrix $\mathbf{X} = (x_{ij})$ is called doubly stochastic if

$$\begin{aligned} \sum_{i=1}^{n} x_{ij} &= 1 \qquad \text{for all} \quad j = 1, 2, \ldots, n \\ \sum_{j=1}^{n} x_{ij} &= 1 \qquad \text{for all} \quad i = 1, 2, \ldots, n \\ x_{ij} &\geq 0 \qquad \text{for all} \quad i \text{ and } j. \end{aligned} \tag{20.7}$$

(Thus, in terms of Chapter 15, a square matrix is doubly stochastic if and only if all of its rows and all of its columns are stochastic vectors.) For example,

$$\mathbf{X} = \begin{bmatrix} 0.86 & 0.07 & 0.07 & 0 & 0 \\ 0.07 & 0.93 & 0 & 0 & 0 \\ 0.07 & 0 & 0.07 & 0.86 & 0 \\ 0 & 0 & 0.56 & 0 & 0.44 \\ 0 & 0 & 0.30 & 0.14 & 0.56 \end{bmatrix}$$

is a doubly stochastic matrix. Doubly stochastic matrices with integer entries are particularly simple: exactly one entry in each row equals one, exactly one entry in each column equals one, and the remaining entries are zeros. These are precisely the *permutation matrices* defined in Chapter 6.

> ***THEOREM 20.3.*** For every $n \times n$-doubly stochastic matrix $\mathbf{X} = (x_{ij})$ there is an $n \times n$ permutation matrix $\mathbf{X}^* = (x_{ij}^*)$ such that $x_{ij}^* = 0$ whenever $x_{ij} = 0$.

(Note that Theorem 20.3 generalizes Theorem 20.2: we may set $x_{ij} = 1/k$ if girl i knows boy j, and $x_{ij} = 0$ otherwise. The permutation matrix $\mathbf{X}^*$ points out the desired marriages.)

PROOF. Consider a network with nodes $r_1, r_2, \ldots, r_n, s_1, s_2, \ldots, s_n$ in which $r_i s_j$ is an arc if and only if $x_{ij} > 0$. Assign a unit supply to each node r_i and a unit demand to each node s_j. Regardless of the cost coefficients, the resulting transshipment problem has a feasible solution: in fact, the entries x_{ij} of the matrix $\mathbf{X}$ constitute one. By the Integrality Theorem, the problem has an integer-valued solution $\mathbf{x}^*$. The numbers x_{ij}^* are the entries of the desired permutation matrix $\mathbf{X}^*$. ■

If $t_1, t_2, \ldots, t_k$ are nonnegative numbers such that $\sum t_i = 1$ then the matrix

$$\mathbf{M} = t_1\mathbf{M}_1 + t_2\mathbf{M}_2 + \cdots + t_k\mathbf{M}_k$$

is called a *convex combination* of the matrices $\mathbf{M}_1, \mathbf{M}_2, \ldots, \mathbf{M}_k$. It is a routine matter to verify that a convex combination of doubly stochastic matrices is a doubly stochastic matrix. In particular, *a convex combination of permutation matrices is a doubly stochastic matrix.* The converse of this claim, discovered independently by G. Birkhoff (1946) and J. von Neumann (1953), is commonly referred to as "the Birkhoff–von Neumann Theorem," even though it was published by D. König (1936) a decade earlier. [Other proofs were given by A. J. Hoffman and H. W. Wielandt (1953), and by J. M. Hammersley and J. G. Mauldon (1956).] For example, the doubly stochastic matrix **X** just shown may be expressed as

$$0.30\begin{bmatrix}1&0&0&0&0\\0&1&0&0&0\\0&0&0&1&0\\0&0&0&0&1\\0&0&1&0&0\end{bmatrix} + 0.56\begin{bmatrix}1&0&0&0&0\\0&1&0&0&0\\0&0&0&1&0\\0&0&1&0&0\\0&0&0&0&1\end{bmatrix}$$

$$+ 0.07\begin{bmatrix}0&1&0&0&0\\1&0&0&0&0\\0&0&1&0&0\\0&0&0&0&1\\0&0&0&1&0\end{bmatrix} + 0.07\begin{bmatrix}0&0&1&0&0\\0&1&0&0&0\\1&0&0&0&0\\0&0&0&0&1\\0&0&0&1&0\end{bmatrix}.$$

THEOREM 20.4 (The Birkhoff-von Neumann Theorem). Every doubly stochastic matrix is a convex combination of permutation matrices.

PROOF. The following algorithm expresses an arbitrary doubly stochastic matrix **M** as a convex combination of permutation matrices $\mathbf{M}_1, \mathbf{M}_2, \ldots, \mathbf{M}_k$.

Step 0. Set $t_1 = 1$, $\mathbf{X} = \mathbf{M}$, $k = 1$.

Step 1. Now we have permutation matrices $\mathbf{M}_1, \mathbf{M}_2, \ldots, \mathbf{M}_{k-1}$ a doubly stochastic matrix **X** and nonnegative numbers $t_1, t_2, \ldots, t_k$ such that

$$\mathbf{M} = t_1\mathbf{M}_1 + \cdots + t_{k-1}\mathbf{M}_{k-1} + t_k\mathbf{X}$$

and $t_1 + \cdots + t_k = 1$. If **X** is a permutation matrix, then set $\mathbf{M}_k = \mathbf{X}$ and stop. Otherwise use Theorem 20.3 to find a permutation matrix $\mathbf{X}^*$ such that $x_{ij}^* = 0$ whenever $x_{ij} = 0$.

Step 2. The n entries x_{ij} of $\mathbf{X}$ for which $x_{ij}^* = 1$ are all positive; denote the least of them by c. (Since $\mathbf{X}$ is not a permutation matrix, we have $c < 1$.) Set $t = t_k$, $t_k = ct$, $t_{k+1} = (1 - c)t$ and $\mathbf{M}_k = \mathbf{X}^*$. Then replace $\mathbf{X}$ by $(1/1 - c)(\mathbf{X} - c\mathbf{M}_k)$, replace k by $k + 1$ and return to Step 1.

With each iteration of this algorithm, the number of zero entries in $\mathbf{X}$ increases. Hence the algorithm must terminate. ∎

Readers acquainted with the problems at the end of Chapter 18 may appreciate the following effortless derivation of the Birkhoff–von Neumann Theorem. The n^2 entries of an $n \times n$ matrix may be thought of as the components of a vector of length n^2. Now every $n \times n$ matrix represents a point in the n^2-dimensional Euclidean space, and the system (20.7) defines a polyhedron in this space. The result of problem 18.5, combined with the easy observation that the polyhedron (20.7) is bounded, shows that every $n \times n$ doubly stochastic matrix represents a convex combination of vertices of (20.7). By the result of problem 18.4, each vertex of (20.7) is the unique optimal solution of some transshipment problem

$$\text{minimize} \quad \sum\sum c_{ij}x_{ij} \qquad \text{subject to (20.7).}$$

By the Integrality Theorem, this unique solution is integer-valued and therefore represented by a permutation matrix. (This argument has been pointed out, in conversation, by J. Edmonds.)

COVERS AND MATCHINGS IN BIPARTITE GRAPHS

Next, we shall describe an application of transshipment theory to graph theory. Loosely speaking, a *graph* is a network whose arcs have no directions (see Figure 20.6).

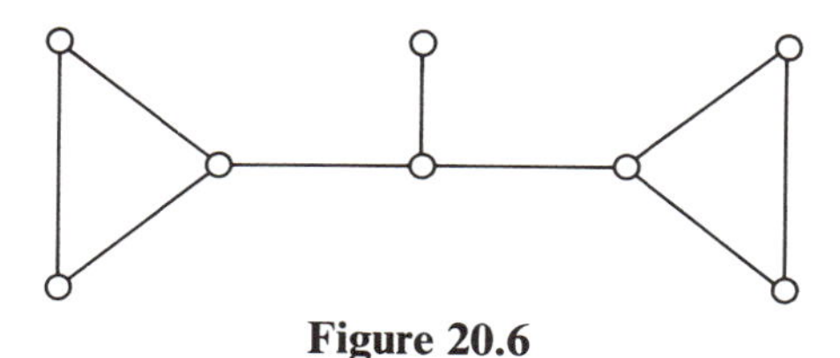

Figure 20.6

We shall investigate certain sets of nodes and certain sets of arcs in various graphs. A set of nodes is called a *cover* if it includes at least one endpoint of each arc. The dark nodes in Figure 20.7 constitute a cover. A set of arcs is called a *matching* if no two of its arcs share a common endpoint. The wiggly arcs in Figure 20.8 constitute

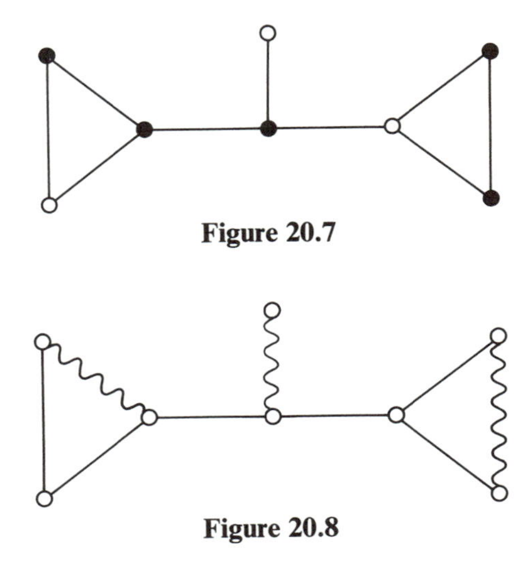

Figure 20.7

Figure 20.8

a matching. As usual, we shall denote the number of nodes in a cover C by $|C|$, and the number of arcs in a matching M by $|M|$. It is quite easy to see that

$$|C| \geq |M| \tag{20.8}$$

for every cover C and every matching M in the same graph: if M is a matching consisting of k arcs $v_1v_2, v_3v_4, \ldots, v_{2k-1}v_{2k}$, then the $2k$ nodes $v_1, v_2, \ldots, v_{2k}$ are all distinct. For each $i = 1, 2, \ldots, k$, a cover C must include at least one of the two nodes v_{2i-1}, v_{2i}. Hence C must include at least k nodes.

A graph is called *bipartite* if its nodes can be labeled "left" and "right" in such a way that each arc has one endpoint among the left nodes and the other endpoint among the right nodes. An example of a bipartite graph is shown in Figure 20.9.

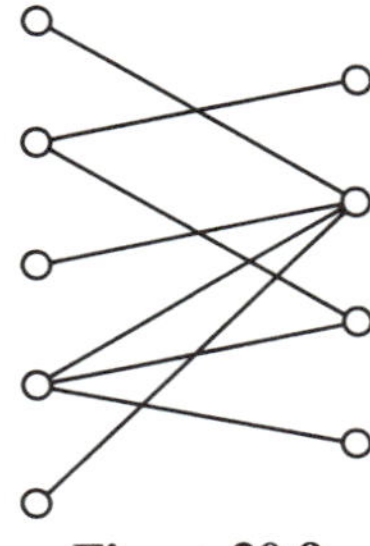

Figure 20.9

The problem of finding the largest matching in a bipartite graph can be turned into a transshipment problem. To begin, assign a unit supply to each left node, assign a unit demand to each right node, and direct each arc from its left endpoint to its right endpoint. Then add two extra nodes v and w with an arc wv between them, an arc iv for each left node i, and an arc wj for each right node j. The demand at v equals the number of left nodes and the supply at w equals the number of right nodes. Thus the bipartite graph in Figure 20.9 gives rise to the network and demands in Figure 20.10.

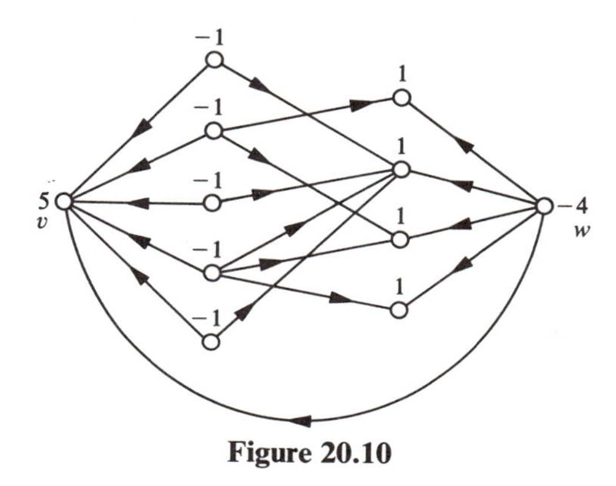

Figure 20.10

It is easy to see that every integer-valued feasible solution $\mathbf{x}$ in the network points out a matching in the original bipartite graph: each original arc e has either $x_e = 0$ or $x_e = 1$, and the arcs e with $x_e = 1$ form a matching. For instance, the feasible tree solution in our example, shown in Figure 20.11, points out the matching in Figure 20.12. In fact, the matching pointed out by $\mathbf{x}$ consists of precisely x_{wv} arcs:

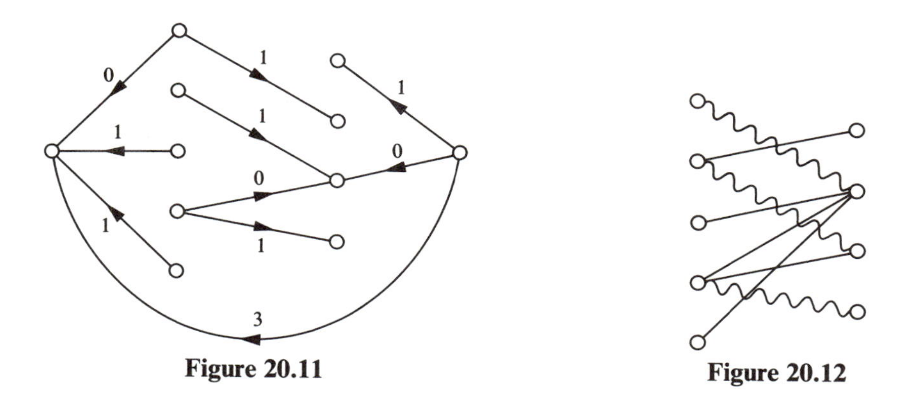

Figure 20.11

Figure 20.12

if the bipartite graph has r left nodes and s right nodes, then precisely $s - x_{wv}$ right nodes get supplied from w and the remaining x_{wv} right nodes must be supplied from the left nodes. Conversely, every matching consisting of k arcs points out an integer-valued feasible solution $\mathbf{x}$ with $x_{wv} = k$: in this solution, precisely $s - k$ right nodes are supplied from w and precisely $r - k$ left nodes supply v. Now denote the incidence matrix of the network by **A**, denote the demand vector by **b**, and define the cost vector **c** by $c_{ij} = 0$ everywhere except for $c_{wv} = -1$. The network simplex method applied to the transshipment problem

$$\text{minimize} \quad \mathbf{cx} \qquad \text{subject to} \quad \mathbf{Ax} = \mathbf{b}, \quad \mathbf{x} \geq \mathbf{0} \tag{20.9}$$

will find an integer-valued feasible solution **x** maximizing x_{wv}. Thus it will find a largest matching in the bipartite graph.

Next, we turn to the node numbers y_k associated with an optimal tree solution of (20.9). In our example, the feasible tree solution shown in Figure 20.11 is optimal; the corresponding numbers y_k are shown in Figure 20.13. In general, we may always set $y_v = 0$; then, as $c_{ij} = 0$ everywhere except for $c_{wv} = -1$, the value of each y_k will be zero or one. Now consider an arc ij coming from the original bipartite graph. We must have $y_i = 1$ or else $y_j = 0$: otherwise $y_i = c_{ij} = 0$ and $y_j = 1$, contradicting the optimality condition $y_i + c_{ij} \geq y_j$. To put it differently, all the left nodes i such that $y_i = 1$ along with all the right nodes j such that $y_j = 0$ form a cover in the bipartite graph. In our example, this cover consists of the three dark nodes in Figure 20.14.

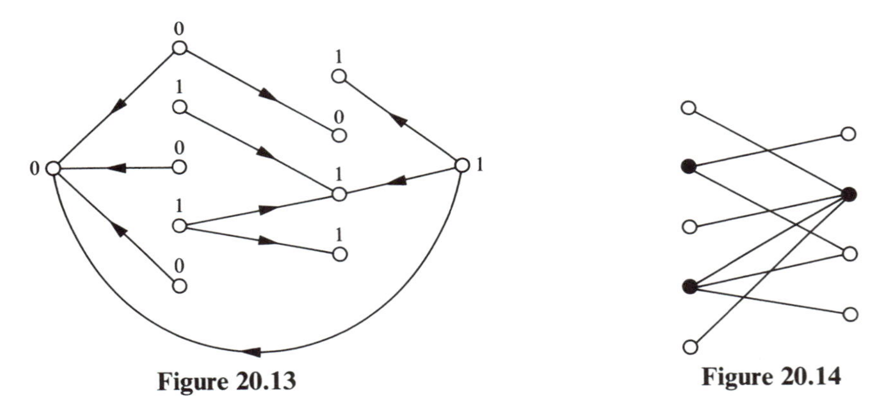

Figure 20.13 **Figure 20.14**

Now we are ready for a classical theorem discovered by D. König (1931) and J. Egerváry (1931).

> ***THEOREM 20.5 (The König–Egerváry Theorem).*** In every bipartite graph, the largest size of a matching equals the smallest size of a cover.

PROOF. We may assume that the bipartite graph has at least one arc (otherwise the theorem holds trivially). As we have just seen, the network simplex method applied to (20.9) finds a largest matching M^* and a certain cover C^* in the bipartite graph.

We propose to show that $|C^*| = |M^*|$. Since the inequality $|C^*| \geq |M^*|$ is a special case of (20.8), we need only establish the reverse inequality,

$$|M^*| \geq |C^*|. \tag{20.10}$$

For this purpose, denote the optimal tree by T, denote the optimal solution defined by T (and defining M^*) by $\mathbf{x}$, and denote the node numbers defined by T (and defining C^*) by y_k. (As before, we set $y_v = 0$; since $x_{wv} = |M^*| \geq 1$, we have $wv \in T$ and so $y_w = 1$.) Now consider a left node $i \in C^*$. As $iv \notin T$ (we have $y_i + c_{iv} = 1$ but $y_v = 0$), there must be a right node j with $x_{ij} = 1$. As $ij \in T$, we have $y_j = y_i = 1$ and so $j \notin C^*$. Similarly, consider a right node $j \in C^*$. As $wj \notin T$ (we have $y_w + c_{wj} = 1$, but $y_j = 0$), there must be a left node i with $x_{ij} = 1$. As $ij \in T$, we have $y_i = y_j = 0$ and so $i \notin C^*$. To summarize, every node in C^* is an endpoint of some arc in M^* whose other endpoint does not belong to C^*. Hence (20.10) follows.

So far we have established that the size of C^* equals the size of our largest matching M^*. The rest is easy: by (20.8) with M^* in place of M, every cover C satisfies $|C| \geq |M^*| = |C^*|$. Thus C^* is a smallest cover. ■

Consider an arbitrary bipartite graph with n nodes. Our proof of Theorem 20.5 shows that a smallest cover and a largest matching in this graph can be found simultaneously by applying the network simplex method to the transshipment problem (20.9). Now we are going to point out that the network simplex method does the job in at most $n(n + 1)/2$ iterations as long as it is initialized by a strongly feasible tree and as long as Cunningham's tie-breaking rule is used in the selection of leaving arcs. (An initial strongly feasible tree is readily available: it consists of all the arcs iv, all the arcs wj and the arc wv with the node w designated as the root. The only arc with $x_e = 0$ in the corresponding feasible tree solution $\mathbf{x}$ is $e = wv$.) It will suffice to show that:

(i) At most $n/2$ pivots are nondegenerate.

(ii) There are at most $n + 1$ degenerate pivots in a row.

The first claim is obvious: the value of x_{wv}, always between 0 and $n/2$, increases by one unit with each nondegenerate pivot. The second claim follows from a similar argument: the value of $\sum(y_k - y_w)$, always between $-(n + 1)$ and 0, decreases by at least one unit with each degenerate pivot. (For a justification, return to the proof of Theorem 19.1 and the subsequent discussion of Cunningham's rule in Chapter 19.)

Each individual iteration is bound to take a bit of time. Even though T, $\mathbf{x}$, $\mathbf{y}$ can be updated relatively quickly, finding an entering arc may require looking at all the out-of-tree arcs in an extreme case. Since the number of these arcs may be proportional to n^2, the time per iteration may be proportional to n^2. We conclude that the time required by the network simplex method to find a smallest cover and a largest matching in a bipartite graph with n nodes is at most proportional to n^4. From a theoretical point of view, this guarantee is quite satisfactory. Alternative algorithms for solving the same problem are guaranteed to terminate even earlier; we shall discuss them in Chapter 22.

In graphs that are not bipartite, the equality $\max|M| = \min|C|$ of the König–Egerváry Theorem may fail, even though (20.8) still implies $\max|M| \leq \min|C|$. In such graphs, the problem of finding the smallest cover may be difficult. It seems that there is no efficient algorithm for solving this problem; a fascinating theorem of S. A. Cook (1971) provides partial evidence supporting this belief. For more information, see M. R. Garey and D. S. Johnson (1979). On the other hand, the problem of finding the largest matching in a nonbipartite graph has been solved satisfactorily by an ingenious algorithm of J. Edmonds (1965). For more information, see E. L. Lawler (1976).

SYSTEMS OF DISTINCT REPRESENTATIVES

A sequence $x_1, x_2, \ldots, x_m$ is said to form a *system of distinct representatives* for a family of finite sets $S_1, S_2, \ldots, S_m$ if $x_i \in S_i$ for each i, and $x_i \neq x_j$ whenever $i \neq j$. For instance,

$$a, c, d, f, e \tag{20.11}$$

is a system of distinct representatives for the family

$$S_1 = \{a, b, c\}, \quad S_2 = \{c, d\}, \quad S_3 = \{c, d\}, \quad S_4 = \{c, e, f\}, \quad S_5 = \{e\}. \tag{20.12}$$

In this context, it is convenient to represent $S_1, S_2, \ldots, S_m$ by a bipartite graph: each of the left nodes represents one of the sets $S_1, S_2, \ldots, S_m$, each of the right nodes represents an element x of $S_1 \cup S_2 \cup \cdots \cup S_m$, and there is an arc between x and S_i if and only if $x \in S_i$. Thus (20.12) gives rise to the bipartite graph in Figure 20.15.

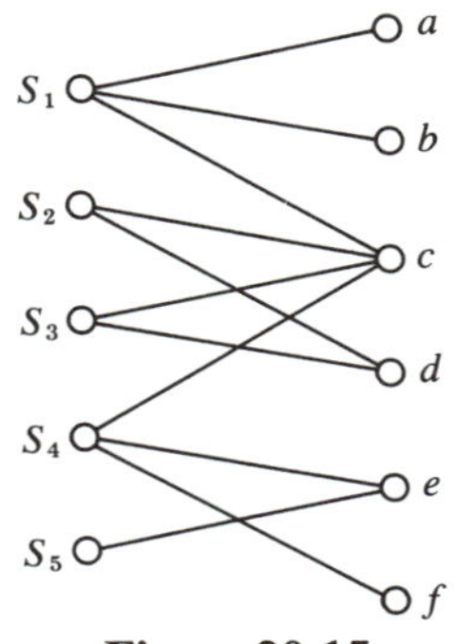

Figure 20.15

In this representation, systems of distinct representatives correspond to matchings of size m, and vice versa. For instance, the system of distinct representatives (20.11) corresponds to the matching in Figure 20.16. Therefore, our family has a system of

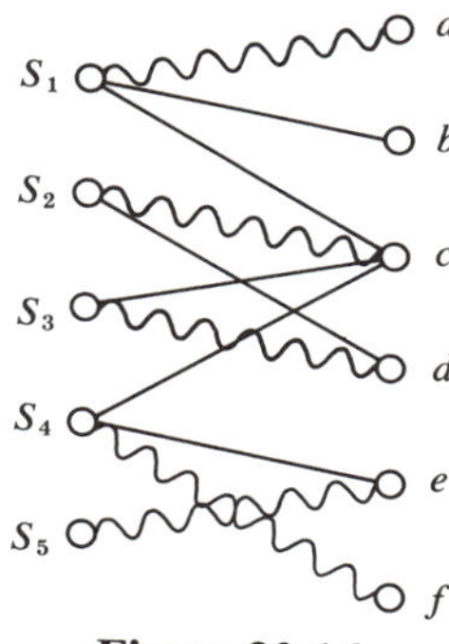

Figure 20.16

distinct representatives if and only if the corresponding bipartite graph has a matching of size m. That may or may not happen. In the bipartite graph representing

$$S_1 = \{a, b\}, \quad S_2 = \{b\}, \quad S_3 = \{b, c\}, \quad S_4 = \{c\}, \quad S_5 = \{c, d, e\}, \tag{20.13}$$

there is no matching of size five: the dark nodes in Figure 20.17 constitute a cover of size four. Hence there is no system of distinct representatives for the family (20.13). That can be concluded at once by looking at the three sets S_2, S_3, S_4: since their union $\{b, c\}$ has only two elements, it cannot accommodate three distinct representatives. More generally, we have a theorem of P. Hall (1935).

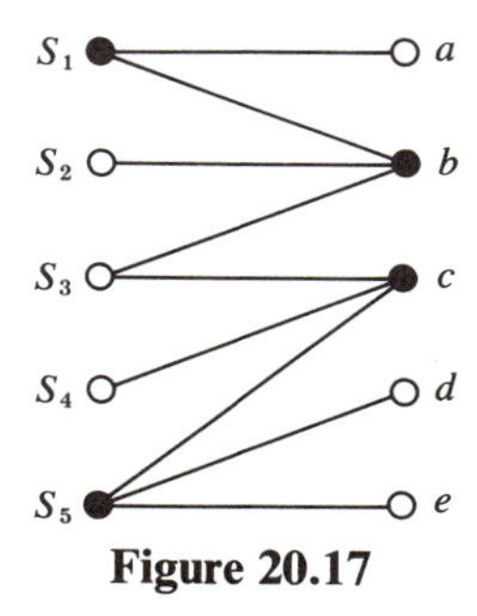

Figure 20.17

THEOREM 20.6 (Hall's Theorem). There is no system of distinct representatives for $S_1, S_2, \ldots, S_m$ if and only if there is a set $I \subseteq \{1, 2, \ldots, m\}$ such that

$$\left| \bigcup_{i \in I} S_i \right| < |I|. \tag{20.14}$$

PROOF. The "if" part is easy: if (20.14) holds then the union of all the sets S_i with $i \in I$ is not big enough to accomodate $|I|$ distinct representatives. To prove the "only if" part, consider a family $S_1, S_2, \ldots, S_m$ for which there is no system of distinct representatives. The corresponding bipartite graph has no matching of size m. Hence the largest matching M^* has size less than m and, by the König–Egerváry Theorem, there is a cover C^* such that $|C^*| = |M^*| < m$. Define a set $I \subseteq \{1, 2, \ldots, m\}$ by putting $i \in I$ if and only if the left node S_i does not belong to C^*. If $i \in I$, then each element x of S_i, viewed as a right node in our graph, must be in C^*: otherwise, C^* would not be a cover. We conclude that the union

$$\bigcup_{i \in I} S_i$$

is a subset of the right nodes in C^*. Since C^* has precisely $m - |I|$ left nodes, it has $|C^*| - (m - |I|) < |I|$ right nodes, and so (20.14) holds. ■

CHAINS AND ANTICHAINS IN PARTIALLY ORDERED SETS

In this section we shall investigate the case of a company that offers guided tours. Each of the tours has a fixed starting date and a fixed duration. If a tour x finishes sufficiently long before a tour y starts, then the guide in charge of x can be reassigned to y; whenever that is possible, we shall write $x \to y$. Let us note at once that:

$$\text{We never have} \quad x \to x. \tag{20.15}$$

Otherwise x would have to finish before it starts, a clear absurdity. Furthermore, note that

$$x \to y \quad \text{and} \quad y \to z \quad \text{together imply} \quad x \to z. \tag{20.16}$$

That is, if a guide can be reassigned from x to y and subsequently from y to z, then that guide can be reassigned from x to z directly. To take a miniature example, consider tours a, b, c, d, e, f, g, with all the possible reassignments listed below:

$$a \to c, \quad a \to d, \quad a \to f, \quad a \to g, \quad b \to c, \quad b \to g, \quad d \to g, \quad e \to f, \quad e \to g.$$

It does not take long to see that the seven trips can be taken care of by as few as three guides. We may assign

the first guide to a, d, g

the second guide to b, c (20.17)

the third guide to e, f.

In general, when the number of tours is large, a schedule minimizing the required number of guides may be far from obvious.

The problem of finding such a schedule may be turned into the problem of finding the largest matching in a certain bipartite graph. The transformation (in a slightly different form) is due to G. B. Dantzig and D. R. Fulkerson (1954). To construct the graph, we associate nodes x' and x'' with each trip x; each possible reassignment $x \to y$ gives rise to an arc with endpoints x' and y''. The graph resulting in our illustrative example is shown in Figure 20.18. The matching consisting of the arcs $a'd''$, $b'c''$, $d'g''$, and $e'f''$ in this graph suggests the reassignments $a \to d$, $b \to c$, $d \to g$ and $e \to f$, describing schedule (20.17). More generally, we are going to show that each matching M in the bipartite graph defines a schedule using only $n - |M|$ guides (with n standing for the total number of tours).

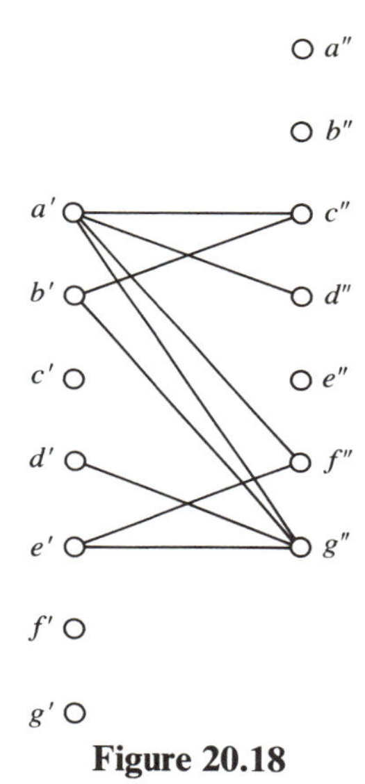

Figure 20.18

If x and y are tours with $x'y'' \in M$, then y will be called a *successor* of x and x will be called a *predecessor* of y. Note that each tour has at most one successor and at most one predecessor: otherwise, M would not be a matching. In particular, each tour t defines a *unique* sequence $t_0, t_1, \ldots$ such that $t_0 = t$ and each t_{i+1} is a successor of t_i. Similarly, t defines a *unique* sequence $\ldots, t_{-1}, t_0$ such that $t_0 = t$ and each t_{i-1} is a predecessor of t_i. Note that, in the concatenation

$$\ldots, t_{-1}, t_0, t_1, \ldots \tag{20.18}$$

of these two sequences, we have $t_i \to t_{i+1}$ for all i; thus, repeated applications of (20.16) yield $t_i \to t_j$ whenever $i < j$. Now it follows that the sequence (20.18) contains no repeated terms: $t_i = t_j$ with $i < j$ would contradict (20.15). In particular, as there are only a finite number of tours, the sequence (20.18) must be finite. To summarize, each tour z occurs as a term z_j in a unique sequence

$$z_1, z_2, \ldots, z_r \tag{20.19}$$

such that z_1 has no predecessor, z_r has no successor, and $z_i'z_{i+1}'' \in M$ for all $i = 1, 2, \ldots, r-1$. (Of course, the length r of this sequence may vary with the choice of z; if z has neither a successor nor a predecessor, then $r = 1$.) Furthermore, each arc $x'y'' \in M$ occurs as some $z_i'z_{i+1}''$ in one of the sequences (20.19). If the total number of these sequences is k and if the lengths of the sequences are $r_1, r_2, \ldots, r_k$, then clearly

$$\sum_{i=1}^{k} r_i = n \quad \text{and} \quad \sum_{i=1}^{k} (r_i - 1) = |M|.$$

These two equations imply $k = n - |M|$. This is the desired conclusion: as $z_1 \to z_2, z_2 \to z_3, \ldots, z_{r-1} \to z_r$, each sequence (20.19) can be taken care of by a single guide.

Conversely, it can be shown that each schedule using only k guides points out a matching with $n - k$ arcs in the bipartite graph; then it follows that the schedule defined by a largest matching M^* is optimal. Rather than going into the details, we shall establish the optimality of the schedule defined by M^* in a roundabout way. Our argument relies on the König–Egerváry theorem. Along with the matching M^*, the bipartite graph contains a cover C^* such that $|C^*| = |M^*|$. Define a set A^* of trips by setting $z \in A^*$ if, and only if, $z' \notin C^*$ and $z'' \notin C^*$. Note that no reassignment $x \to y$ with $x, y \in A^*$ is possible; otherwise, the bipartite graph would contain the arc $x'y''$ with $x' \notin C^*$, $y'' \notin C^*$, contradicting the fact that C^* is a cover. To put it differently, no guide can take care of two different trips in A^*. Hence each schedule must use at least $|A^*|$ different guides. But A^* has at least $n - |C^*|$ elements: each node in C^* makes only one trip ineligible for membership in A^*. To summarize, every schedule must use at least $|A^*| \geq n - |C^*| = n - |M^*|$ different guides. We conclude that the schedule defined by M^*, using only $n - |M^*|$ guides, is optimal.

Our findings can be reviewed in a more abstract context. A list of pairs $x \to y$ satisfying conditions (20.15), (20.16) is called a *partial order*; a set whose elements appear in a partial order is called a *partially ordered set*. A subset C of a partially ordered set is called a *chain* if

$$x \to y \quad \text{or} \quad y \to x \quad \text{whenever} \quad x, y \in C \quad \text{and} \quad x \neq y.$$

(To put it differently, C is a chain if and only if its elements can be enumerated as $z_1, z_2, \ldots, z_r$ in such a way that $z_1 \to z_2, z_2 \to z_3, \ldots, z_{r-1} \to z_r$.) A subset A of a partially ordered set is called an *antichain* if

$$x \to y \quad \text{for no} \quad x, y \in A.$$

The following theorem was discovered by R. P. Dilworth (1950).

***THEOREM 20.7* (*Dilworth's Theorem*).** In every partially ordered set P, the smallest number of chains whose union is P equals the largest size of an antichain.

PROOF [D. R. Fulkerson (1956)]. The partially ordered set P may be thought of as a set of tours with all possible reassignments. In this interpretation, chains are precisely those sets of tours that can be taken care of by a single guide. Hence the procedure described above produces the smallest number k of chains whose union equals P. Along with these chains $C_1, C_2, \ldots, C_k$, the procedure finds also an antichain A^* of size at least k. Since every antichain A shares at most one element with each of the chains $C_1, C_2, \ldots, C_k$, it must satisfy $|A| \leq k \leq |A^*|$. Now it follows that A^* is a largest antichain and (as A^* can be substituted for A) that its size is precisely k. ■

THE ASSIGNMENT PROBLEM

Now we shall turn to an application of the transshipment theory in which the Integrality Theorem again plays a crucial part. Consider a college with a small department offering five courses in the upcoming term. There are five teachers in the department; the performance of teacher i in charge of course j is roughly evaluated by a score s_{ij}, with a higher score indicating a better performance. (See Table 20.3.)

Table 20.3 Evaluation of Teachers' Performances

	Course 1	Course 2	Course 3	Course 4	Course 5
Teacher 1	7	5	4	4	5
Teacher 2	7	9	7	9	4
Teacher 3	4	6	5	8	5
Teacher 4	5	4	5	7	4
Teacher 5	4	5	5	8	9

The department chairman has to assign teachers to courses in some reasonable way. For instance, assigning

$$\begin{aligned}&\text{teacher 1 to course 1}\\&\text{teacher 2 to course 2}\\&\text{teacher 3 to course 3}\\&\text{teacher 4 to course 4}\\&\text{teacher 5 to course 5}\end{aligned}\tag{20.20}$$

is clearly better than assigning

$$\begin{aligned}&\text{teacher 1 to course 2}\\&\text{teacher 2 to course 1}\\&\text{teacher 3 to course 3}\\&\text{teacher 4 to course 4}\\&\text{teacher 5 to course 5.}\end{aligned}\tag{20.21}$$

[In passing from (20.21) to (20.20), the quality of teaching improves in course 2 and remains unaltered in the other courses.] However, comparing (20.20) with the assignment of

$$\begin{aligned}&\text{teacher 1 to course 1}\\&\text{teacher 2 to course 3}\\&\text{teacher 3 to course 2}\\&\text{teacher 4 to course 4}\\&\text{teacher 5 to course 5}\end{aligned}\tag{20.22}$$

is more difficult: in passing from (20.20) to (20.22), the quality of teaching improves in course 3 and deteriorates in course 2. The chairman decides to appraise each proposal on the basis of its *average* score. Under this criterion, assignment (20.20) with the average score of $\frac{1}{5}(7 + 9 + 5 + 7 + 9) = 7.4$ fares better than the assignment (20.22) with the average score of $\frac{1}{5}(7 + 6 + 7 + 7 + 9) = 7.2$.

Now the problem is clearly defined and we can state it in mathematical terms. Let us write $x_{ij} = 1$ if teacher i is assigned to course j, and $x_{ij} = 0$ otherwise. When there are n teachers and n courses, we have

$$\sum_{j=1}^{n} x_{ij} = 1 \quad \text{for all} \quad i = 1, 2, \ldots, n \tag{20.23}$$

(each teacher is in charge of precisely one course),

$$\sum_{i=1}^{n} x_{ij} = 1 \quad \text{for all} \quad j = 1, 2, \ldots, n \tag{20.24}$$

(each course is taught by precisely one teacher) and, of course,

$$x_{ij} = 0 \text{ or } 1 \quad \text{for all} \quad i \quad \text{and} \quad j. \tag{20.25}$$

Conversely, numbers x_{ij} satisfying (20.23), (20.24), (20.25) describe an assignment. The average score in this assignment equals

$$\frac{1}{n}\sum_{i=1}^{n}\sum_{j=1}^{n} s_{ij}x_{ij}.$$

Since maximizing the average score amounts to maximizing the total score, we are led to

$$\begin{aligned} \text{maximize} \quad & \sum_{i=1}^{n}\sum_{j=1}^{n} s_{ij}x_{ij} \\ \text{subject to} \quad & \sum_{j=1}^{n} x_{ij} = 1 \quad \text{for all} \quad i = 1, 2, \ldots, n \\ & \sum_{i=1}^{n} x_{ij} = 1 \quad \text{for all} \quad j = 1, 2, \ldots, n \\ & x_{ij} = 0 \text{ or } 1 \text{ for all } i \text{ and } j. \end{aligned} \tag{20.26}$$

This problem is called the *assignment problem.*

A related problem,

$$\begin{aligned} \text{minimize} \quad & \sum_{i=1}^{n}\sum_{j=1}^{n} (-s_{ij})x_{ij} \\ \text{subject to} \quad & \sum_{j=1}^{n} x_{ij} = 1 \qquad \text{for all} \quad i = 1, 2, \ldots, n \\ & \sum_{i=1}^{n} x_{ij} = 1 \qquad \text{for all} \quad j = 1, 2, \ldots, n \\ & x_{ij} \geq 0 \qquad \text{for all} \quad i \quad \text{and} \quad j \end{aligned} \tag{20.27}$$

can be solved by the network simplex method. The corresponding network has a source u_i with a unit supply for each teacher i, a sink v_j with a unit demand for each course j, and an arc from each source u_i to each sink v_j. The cost on each arc u_iv_j is $-s_{ij}$. Note that:

(i) Every feasible solution of (20.26) is a feasible solution of (20.27).

(ii) Every integer-valued feasible solution of (20.27) is a feasible solution of (20.26).

(iii) The network simplex method finds an integer-valued optimal solution of (20.27).

It follows that the optimal solution of (20.27) found by the network simplex method is, in fact, an optimal solution of (20.26). Thus the network simplex method, applied to the transshipment problem (20.27), solves the assignment problem (20.26).

Rather than maximizing the *average* score, the chairman might decide to maximize

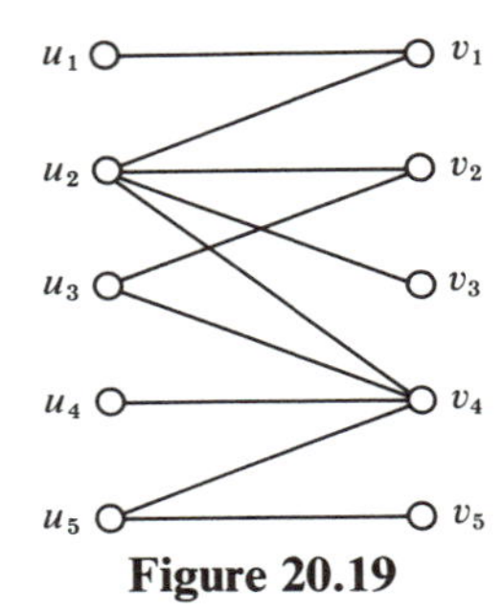

Figure 20.19

the *lowest* score: now the assignment (20.20) with the lowest score $s_{33} = 5$ fares worse than the assignment (20.22) with the lowest score $s_{32} = 6$. The new objective suits those situations where the overall performance of a team is measured by the worst performance of an individual member, just as the proverbial strength of a chain equals the strength of its weakest link. (For instance, if each of n workers performs one of n tasks on an assembly line, then the speed of the line equals the speed of the slowest worker.) The resulting *bottleneck assignment problem* can be solved by repeated applications of an algorithm for finding the largest matching in a bipartite graph. To begin, consider the problem of deciding whether there is an assignment in which every score exceeds a prescribed threshold t. This problem amounts to deciding whether a certain bipartite graph contains a matching of size n. This graph G_t has a node u_i for each teacher i, a node v_j for each course j, and an arc u_iv_j whenever $s_{ij} > t$. For instance, $t = 5$ in our example gives rise to the graph G_5 in Figure 20.19.

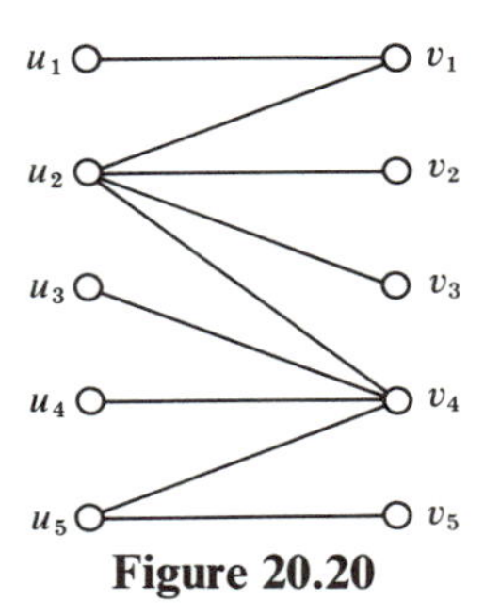

Figure 20.20

The matching u_1v_1, u_2v_3, u_3v_2, u_4v_4, u_5v_5 in this graph represents the assignment (20.22) with each individual score exceeding 5. Replacing $t = 5$ by $t = 6$, we obtain the graph G_6 with no matching of size five: the nodes u_1, u_2, u_5, v_4 form a cover of size four (see Figure 20.20). Hence (20.22) is an optimal solution of our bottleneck assignment problem. More generally, the following procedure suggests itself.

Step 0. Let M consist of the arcs $u_1v_1, u_2v_2, \ldots, u_nv_n$.

Step 1. Let t be the smallest of the n scores s_{ij} such that $u_iv_j \in M$.

Step 2. If G_t contains no matching M^* of size n, then stop: M represents an optimal solution. Otherwise, replace M by M^* and return to step 1.

This procedure can be speeded up by a more sophisticated search for the largest value of t such that G_t has a matching of size n. We shall not go into the details of such a search.

TRANSPORTATION PROBLEMS

The problem

$$\begin{aligned}
\text{minimize} \quad & \sum_{i=1}^{m} \sum_{j=1}^{n} c_{ij}x_{ij} \\
\text{subject to} \quad & \sum_{j=1}^{n} x_{ij} = r_i \qquad (i = 1, 2, \ldots, m) \\
& \sum_{i=1}^{m} x_{ij} = s_j \qquad (j = 1, 2, \ldots, n) \\
& x_{ij} \geq 0 \qquad \text{for all} \quad i \quad \text{and} \quad j
\end{aligned}$$

[studied by F. L. Hitchcock (1941)], is often referred to as the *Hitchcock transportation problem.* Problem (20.27) has this form with $m = n$, $c_{ij} = -s_{ij}$ for all i and j, $r_i = 1$ for all i, and $s_j = 1$ for all j. Clearly, the Hitchcock transportation problem is a transshipment problem. The corresponding network has sources $u_1, u_2, \ldots, u_m$, sinks $v_1, v_2, \ldots, v_n$ and an arc u_iv_j for each source u_i and sink v_j. The supply at u_i equals r_i, the demand at v_j equals s_j, and the cost on u_iv_j is c_{ij}.

More generally, a *transportation problem* is a transshipment problem with no intermediate nodes and with each arc leading from a source to a sink. For example, the transshipment problem shown in Figure 20.21 is actually a transportation problem (but not a Hitchcock transportation problem).

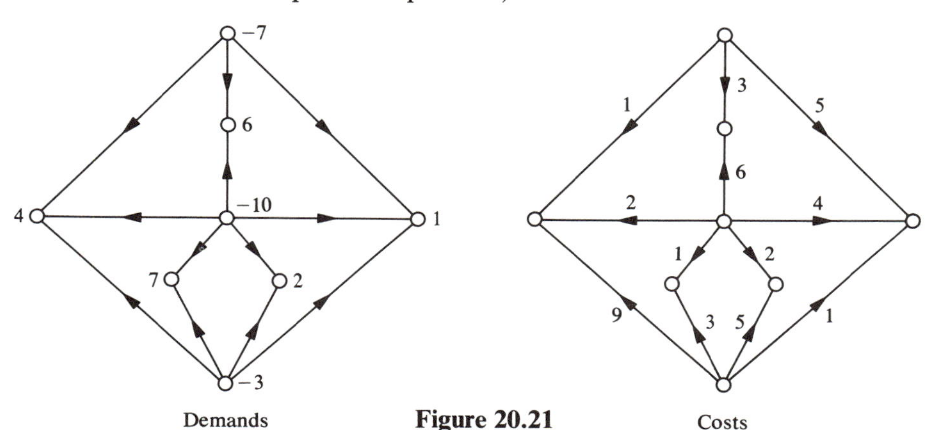

Figure 20.21

The special form of transportation problems does not lead to any essential simplifications of the network simplex method. However, it does make *manual solutions of small problems* more convenient, since the data and intermediate results can now be kept in a neat tabular form. We shall illustrate the point on the example shown in Figure 20.21. The data can be presented in the following array:

3	1	5	×	×	7
6	2	4	1	2	10
×	9	1	3	5	3
6	4	1	7	2	

Each row in this array represents a source and each column represents a sink. The cell in each row and column represents the arc between the corresponding source and sink; the cells that are crossed out indicate the absence of the corresponding arcs. The number beside each row is the supply at the corresponding source, the number below each column is the demand at the corresponding sink, and the number in each cell is the cost on the corresponding arc.

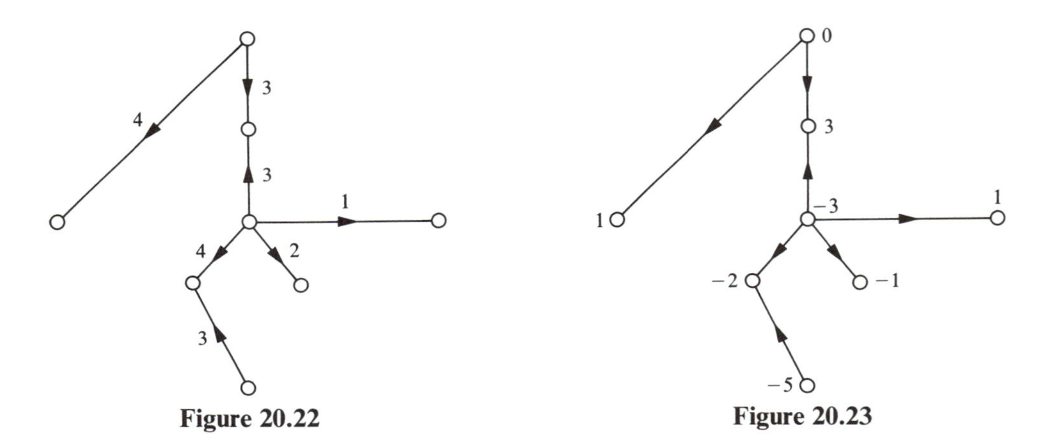

Figure 20.22 **Figure 20.23**

The intermediate results can be kept in a similar format. For instance, the feasible tree solution in Figure 20.22 and the corresponding node numbers in Figure 20.23 can be recorded as

	3	1	1	−2	−1
0	3	4		×	×
−3	3		1	4	2
−5	×			3	

The next set of intermediate results is

	3	1	-4	-2	-1
0	3	4		×	×
-3	3			5	2
-5	/		1	2	

and then

	3	1	-2	0	1
0	6	1		×	×
-1		3		5	2
-3	×		1	2	

which is the optimal solution. This format is particularly suitable for the Hitchcock transportation problems (20.27) arising from the assignment problem (20.26).

We shall use the tabular format in a discussion of a problem concerning recycled paper; the data come from C. R. Glassey and V. K. Gupta (1974). The principal sources of fiber for paper production are the virgin wood pulp and the secondary pulp recovered from waste paper. Paper and paperboard are broadly classified into 13 categories:

1. Newsprint.
2. Uncoated groundwood paper.
3. Coated paper.
4. Uncoated book paper.
5. Writing and related paper.
6. Bleached bristols.
7. Unbleached kraft packaging and industrial covering paper.
8. Other packaging and industrial covering paper.
9. Tissue paper.
10. Unbleached kraft linerboard and medium corrugated.
11. Bleached packaging.
12. Box board.
13. Building paper, board and construction paper, etc.

For each type j of new paper, the fixed level of production brings about a fixed demand for pulp, d_j thousands of tons (Table 20.4). The standards of quality may require that at least m_j thousands of tons come from virgin pulp and that the sources of secondary pulp be restricted to specified types of waste paper. The recovered waste paper is converted into secondary pulp by certain processes (de-inking, asphalt dispersion, bleaching) that cause a fiber loss. As a result, x tons of

Table 20.4 Recycled Paper Problem: Data

j	d_j	m_j	t_j	Allowed donors
1	3,475	0	0.85	1, 2
2	1,223	699	0.90	1, 2, 3, 4
3	2,260	1,077	0.85	2, 3, 4, 5, 6
4	2,700	1,285	0.85	2, 3, 4, 5, 6
5	2,950	1,965	0.90	5, 6
6	1,112	848	0.95	4, 5, 6
7	3,910	2,980	0.80	7, 8, 10
8	1,673	1,275	0.80	7, 8, 10
9	3,855	367	0.85	1, 2, 4
10	12,100	9,210	0.90	10
11	7,382	6,320	0.95	6
12	7,215	0	0.93	any
13	4,000	381	0.90	any

recovered waste paper contribute fewer than x tons of secondary pulp. The actual contribution $t_j x$ depends on the *recipient* (the new paper of type j that is being produced).

Finally, reliable predictions of the amount of waste paper actually recovered are difficult to obtain. Let us say that s_i thousands of tons of waste paper of type i will be recovered: thus s_i is a definite but as yet unknown number. For obvious reasons, we wish to minimize the actual consumption of virgin wood pulp. To achieve this objective, we shall judiciously dispatch the available waste paper to suitably chosen recipients. With the notation

v_j = amount of virgin wood pulp used to produce new paper of type j
w_{ij} = amount of waste paper of type i used to produce new paper of type j
$D(j)$ = set of allowable donors i in producing new paper of type j
$R(i)$ = set of allowable recipients j of waste paper of type i

our problem is to

$$\begin{array}{lll}\text{minimize} & \sum_{j=1}^{13} v_j & \\ \text{subject to} & t_j \sum_{i \in D(j)} w_{ij} + v_j = d_j & (j = 1, 2, \ldots, 13) \\ & \sum_{j \in R(i)} w_{ij} \leq s_i & (i = 1, 2, \ldots, 13) \\ & v_j \geq m_j & (j = 1, 2, \ldots, 13) \\ & w_{ij} \geq 0 & \text{for all } i \text{ and } j.\end{array}$$

Writing $x_{ij} = w_{ij}$, $c_{ij} = 0$ if $1 \leq i \leq 13$, and $x_{ij} = (v_j - m_j)/t_j$, $c_{ij} = t_j$ if $i = 14$, we convert this formulation into the problem

$$\text{minimize} \quad \sum_{i=1}^{14} \sum_{j=1}^{13} c_{ij} x_{ij}$$

$$\text{subject to} \quad \sum_{i=1}^{14} x_{ij} = (d_j - m_j)/t_j \qquad (j = 1, 2, \ldots, 13)$$

$$\sum_{j=1}^{13} x_{ij} \leq s_i \qquad (i = 1, 2, \ldots, 13)$$

$$x_{ij} \geq 0 \qquad \text{for all} \quad i \quad \text{and} \quad j.$$

(For convenience, we let the subscripts i and j run from 1 to 14 and 13, respectively, even though variables x_{ij} with $1 \leq i \leq 13$ and $i \notin D(j)$ are simply absent.) Note that a fourteenth inequality, $\sum x_{ij} \leq s_i$ with $i = 14$, may be added as long as the bound s_{14} is large enough to represent no real restriction. For instance, with r_j standing for $(d_j - m_j)/t_j$, every feasible solution satisfies

$$\sum_{j=1}^{13} \sum_{i=1}^{14} x_{ij} = \sum_{j=1}^{13} r_j$$

and so we may set $s_{14} = \sum r_j$. (Specifically, we have $r_1 = 4{,}088$, $r_2 = 582$, $r_3 = 1{,}392$, $r_4 = 1{,}665$, $r_5 = 1{,}094$, $r_6 = 278$, $r_7 = 1{,}162$, $r_8 = 498$, $r_9 = 4{,}104$, $r_{10} = 3{,}211$, $r_{11} = 1{,}118$, $r_{12} = 7{,}758$, $r_{13} = 4{,}021$, and so $s_{14} = 30{,}971$.) Finally, we let x_{ij} with $j = 14$ stand for the slack

$$s_i - \sum_{j=1}^{13} x_{ij}$$

and denote the sum of these slacks,

$$\sum_{i=1}^{14} \left(s_i - \sum_{j=1}^{13} x_{ij} \right) = \sum_{i=1}^{14} s_i - \sum_{i=1}^{14} \sum_{j=1}^{13} x_{ij} = \sum_{i=1}^{14} s_i - \sum_{j=1}^{13} r_j = \sum_{i=1}^{13} s_i$$

by r_{14}. Now our problem is converted into a proper transportation format,

$$\text{minimize} \quad \sum_{i=1}^{14} \sum_{j=1}^{13} c_{ij} x_{ij}$$

$$\text{subject to} \quad \sum_{i=1}^{14} x_{ij} = r_j \qquad (j = 1, 2, \ldots, 14)$$

$$\sum_{j=1}^{14} x_{ij} = s_i \qquad (i = 1, 2, \ldots, 14)$$

$$x_{ij} \geq 0 \qquad \text{for all} \quad i \quad \text{and} \quad j.$$

Its data in the tabular format (with zero costs indicated by blank cells) are shown in Table 20.5. Now certain simplifications become apparent. For instance, the third and fourth columns in the table look precisely the same, and so we can combine them into a single column whose demand is $1{,}392 + 1{,}665 = 3{,}057$. Similarly, the seventh and eighth column merge into a single column with demand $1{,}162 + 498 = 1{,}660$. (However, we cannot merge the twelfth and thirteenth columns, since the cost coefficients, 0.93 and 0.90, are not the same.) Analogous operations may be applied to the rows; Table 20.6 is the streamlined result.

As soon as the supplies s_i become known, this transportation problem can be easily solved.

Table 20.5 Recycled Paper Problem: Tabular Format

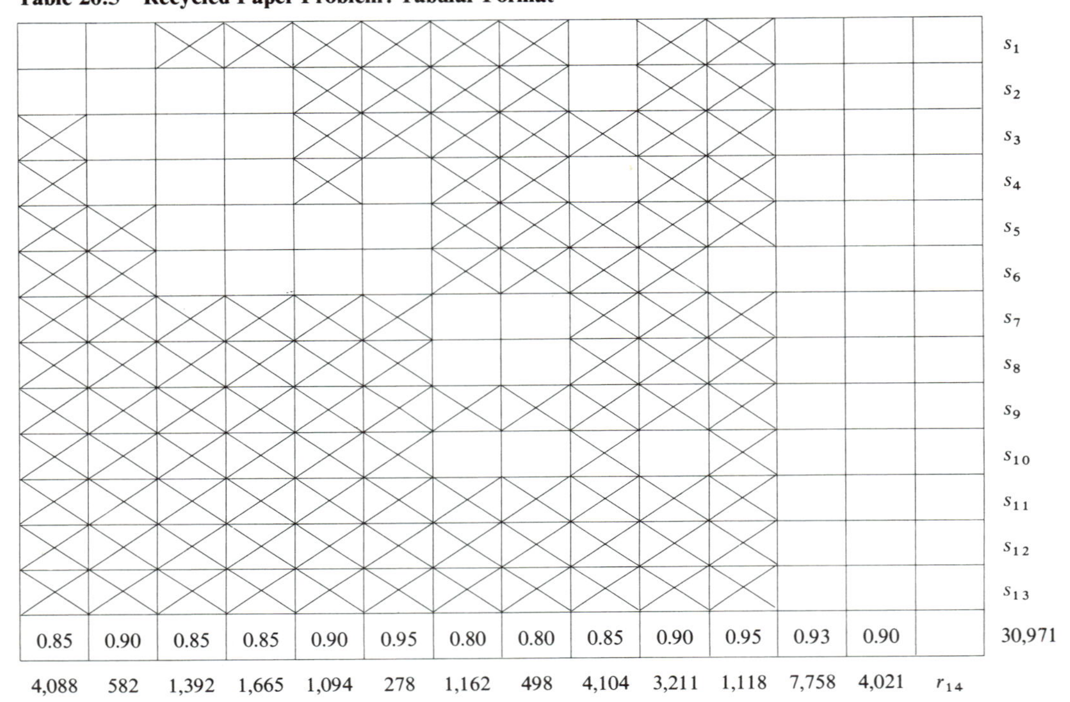

		X	X	X	X	X	X		X	X				s_1
				X	X	X	X		X	X				s_2
X				X	X	X	X	X	X	X				s_3
X				X		X	X		X	X				s_4
X	X					X	X	X	X	X				s_5
X	X					X	X	X	X					s_6
X	X	X	X	X	X			X	X	X				s_7
X	X	X	X	X	X			X	X	X				s_8
X	X	X	X	X	X	X	X	X	X	X				s_9
X	X	X	X	X	X			X		X				s_{10}
X	X	X	X	X	X	X	X	X	X	X				s_{11}
X	X	X	X	X	X	X	X	X	X	X				s_{12}
X	X	X	X	X	X	X	X	X	X	X				s_{13}
0.85	0.90	0.85	0.85	0.90	0.95	0.80	0.80	0.85	0.90	0.95	0.93	0.90		30,971
4,088	582	1,392	1,665	1,094	278	1,162	498	4,104	3,211	1,118	7,758	4,021	r_{14}	

Table 20.6 Recycled Paper Problem: Streamlined Tabular Format

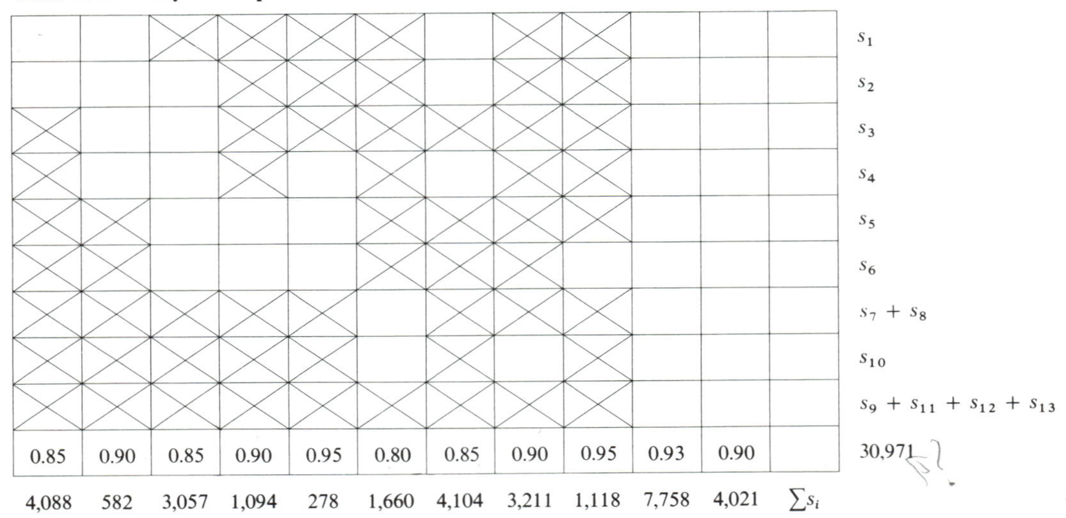

		X	X	X	X		X	X				s_1
			X	X	X		X	X				s_2
X			X	X	X	X	X	X				s_3
X			X		X	X	X	X				s_4
X	X				X	X	X	X				s_5
X	X				X	X	X					s_6
X	X	X	X	X		X	X	X				$s_7 + s_8$
X	X	X	X	X		X		X				s_{10}
X	X	X	X	X	X	X	X	X				$s_9 + s_{11} + s_{12} + s_{13}$
0.85	0.90	0.85	0.90	0.95	0.80	0.85	0.90	0.95	0.93	0.90		30,971
4,088	582	3,057	1,094	278	1,660	4,104	3,211	1,118	7,758	4,021	$\sum s_i$	

PROBLEMS

△**20.1** Minimize $2x_1 + 3x_2 + 4x_3 + x_4 + 3x_5 + x_6 + 3x_7 + 2x_8 + x_9$

subject to

$$\begin{aligned} -x_1 + x_6 - x_7 &\leq -1 \\ x_1 - x_2 - x_8 &\geq 0 \\ x_2 - x_3 - x_9 &\leq -2 \\ x_3 - x_4 + x_7 &= -4 \\ x_4 - x_5 + x_8 &\geq 4 \\ x_5 - x_6 + x_9 &\geq 4 \\ x_1, x_2, \ldots, x_9 &\geq 0. \end{aligned}$$

20.2 Develop a simple algorithm for scheduling production and inventory to meet a fluctuating demand. (*Hint*: Assume that an optimal solution for the first $n - 1$ months has been found.)

20.3 Solve the caterer problem with $n = 10$,

$$d_1 = 3, \quad d_2 = 1, \quad d_3 = 4, \quad d_4 = 1, \quad d_5 = 5, \quad d_6 = 9, \quad d_7 = 2, \quad d_8 = 6,$$
$$d_9 = 8, \quad d_{10} = 1$$

and

$$p = 3, \quad q = 1, \quad a = 3, \quad b = 2, \quad c = 1.$$

20.4 Illustrate the Birkhoff–von Neumann Theorem on the following doubly stochastic matrix:

$$\begin{bmatrix} 0.3 & 0.3 & 0 & 0.3 & 0.1 \\ 0.1 & 0.5 & 0.2 & 0.1 & 0.1 \\ 0.2 & 0 & 0.3 & 0.5 & 0 \\ 0 & 0.2 & 0.5 & 0 & 0.3 \\ 0.4 & 0 & 0 & 0.1 & 0.5 \end{bmatrix}.$$

20.5 Illustrate the König–Egerváry theorem on the bipartite graph with left nodes a, b, c, d, e, right nodes f, g, h, i, j, and arcs $ag, ai, bf, bg, bh, bj, cg, df, dh, di, dj, eg, ei$.

20.6 Illustrate Hall's theorem on the following family of sets:

$$\begin{aligned} S_1 &= \{1, 3, 5, 9\} \\ S_2 &= \{1, 3, 5, 8\} \\ S_3 &= \{3, 5, 6, 8\} \\ S_4 &= \{2, 4, 7\} \\ S_5 &= \{6, 8\} \\ S_6 &= \{2, 4, 7, 9\} \\ S_7 &= \{3, 5, 8\} \\ S_8 &= \{5, 6, 8\} \\ S_9 &= \{1, 3, 5\}. \end{aligned}$$

20.7 Illustrate Dilworth's theorem on the partial order $a \to d, a \to e, a \to g, a \to i, b \to i, c \to e, c \to f, c \to g, c \to h, d \to i, e \to g, f \to h$.

20.8 Solve the assignment and the bottleneck assignment problems with

$$(s_{ij}) = \begin{bmatrix} 6 & 6 & 7 & 8 & 6 & 8 \\ 6 & 5 & 3 & 6 & 4 & 5 \\ 9 & 8 & 6 & 9 & 7 & 9 \\ 4 & 5 & 5 & 6 & 3 & 4 \\ 5 & 3 & 2 & 6 & 4 & 3 \\ 9 & 8 & 8 & 9 & 6 & 8 \end{bmatrix}.$$

20.9 Solve the paper recycling problem with $s_1 = 4{,}160$, $s_2 = 620$, $s_3 = 840$, $s_4 = 680$, $s_5 = 980$, $s_6 = 500$, $s_7 = 1{,}440$, $s_8 = 620$, $s_9 = 160$, $s_{10} = 4{,}380$, $s_{11} = 3{,}280$, $s_{12} = 3{,}080$, $s_{13} = 720$.

20.10 Turn problem 1.6 into a transportation problem and solve it by the network simplex method.

20.11 Construct an unbounded transportation problem or show that there is none.

20.12 Eastern Provincial Airways flights served by Boeing 737 are listed in Table 20.7.

Table 20.7 Flight Timetable

Number	Departure	Arrival	Frequency
101	St. John's 6:55	Montreal 11:15	Daily
102	Montreal 16:10	St. John's 23:10	Daily
	St. John's 23:35	Gander 0:05	MWF Sat
103	Gander 8:15	St. John's 8:45	Th Su
	St. John's 9:25	Wabush 12:20	Th F Su
104	Wabush 13:35	St. John's 17:55	Th F Su
107	Charlottetown 6:45	Montreal 7:55	Daily
108	Montreal 18:55	Halifax 21:50	Daily
109	St. John's 18:40	Halifax 21:10	Except W
	St. John's 19:45	Halifax 22:15	W
110	Halifax 9:40	St. John's 12:20	Except Su
	Halifax 9:00	St. John's 12:20	Su
111	St. John's 13:00	Montreal 17:10	Daily
112	Montreal 9:00	St. John's 16:10	Daily
115	Gander 8:15	Wabush 12:30	Tu
116	Wabush 13:15	St. John's 17:55	Tu Sat
117	Chatham 15:05	Montreal 15:25	Except Sat
118	Montreal 11:55	Chatham 14:45	Except Sat
119	St. John's 7:35	Wabush 11:15	M
120	Wabush 15:50	Gander 21:05	M
123	Gander 7:40	Wabush 12:30	Sat
124	Gander 6:00	St. John's 6:30	Daily
125	St. John's 7:20	Wabush 10:00	W
126	Wabush 14:35	St. John's 19:00	W
127	St. John's 17:10	Halifax 18:15	Daily
134	Halifax 19:15	Gander 22:00	Daily
151	St. John's 15:10	Toronto 18:05	Daily
152	Toronto 9:15	St. John's 14:25	Daily
153	Halifax 7:05	Toronto 8:30	Daily
154	Toronto 19:15	Charlottetown 23:10	Daily

Find the smallest number of planes capable of implementing this timetable (and justify your answer). Can you state and prove a relevant minimax theorem?

21

Upper-Bounded Transshipment Problems

The network simplex method can be easily modified to handle problems

$$\text{minimize} \quad \mathbf{cx} \qquad \text{subject to} \quad \mathbf{Ax} = \mathbf{b}, \quad \mathbf{l} \le \mathbf{x} \le \mathbf{u}$$

such that $\mathbf{A}$ is the incidence matrix of a network, each component l_{ij} of the vector $\mathbf{l}$ is either a number or the symbol $-\infty$ (meaning that no lower bound is imposed on x_{ij}) and each component u_{ij} of the vector $\mathbf{u}$ is either a number or the symbol $+\infty$ (meaning that no upper bound is imposed on x_{ij}). These modifications are simply a particular instance of the upper-bounding technique described in Chapter 8; we shall develop them from scratch so as to make our presentation of the network simplex method self-contained. To avoid fussy formalism, we shall restrict ourselves to so-called *upper-bounded transshipment problems* (or *minimum cost network flow problems*),

$$\text{minimize} \quad \mathbf{cx} \qquad \text{subject to} \quad \mathbf{Ax} = \mathbf{b}, \quad \mathbf{0} \le \mathbf{x} \le \mathbf{u}.$$

The generalization to arbitrary lower bounds is straightforward.

AN EXAMPLE

For illustration, let us consider the data in Figure 21.1. The key to understanding the modifications is the realization that the new constraints $x_{ij} \leq u_{ij}$ should be treated analogously to the constraints $x_{ij} \geq 0$. For instance, a feasible tree solution in a transshipment problem had the property that $x_{ij} = 0$ whenever $ij \notin T$. In the context of upper-bounded transshipment problems, we shall require that

$$x_{ij} = 0 \quad \text{or} \quad x_{ij} = u_{ij} \qquad \text{whenever} \quad ij \notin T.$$

The arcs ij with $x_{ij} = u_{ij}$ are sometimes called *saturated.* An example of a feasible tree solution is shown in Figure 21.2: the in-tree arcs are solid, whereas the saturated out-of-tree arcs are dashed. We shall initialize the network simplex method by this feasible tree solution. The node numbers y_i are defined by the equations

$$y_i + c_{ij} = y_j \qquad \text{for all} \quad ij \in T$$

just as they were in Chapter 19 (see Figure 21.3). In transshipment problems, each entering arc $ij \notin T$ has the property that

$$y_i + c_{ij} < y_j \quad \text{and} \quad x_{ij} = 0. \tag{21.1}$$

There is no harm in choosing such an arc in the present context either (although we shall see later in this section that certain arcs $ij \notin T$ with $x_{ij} = u_{ij}$ are eligible as well). For instance, arc 13 will do. As we used to do, we now set $x_{13} = t$ and adjust the values of x_{ij} for $ij \in T$ (see Figure 21.4). The largest value of t for which feasibility is maintained is $t = 8$. The resulting feasible tree solution is shown in Figure 21.5 along with the corresponding node numbers.

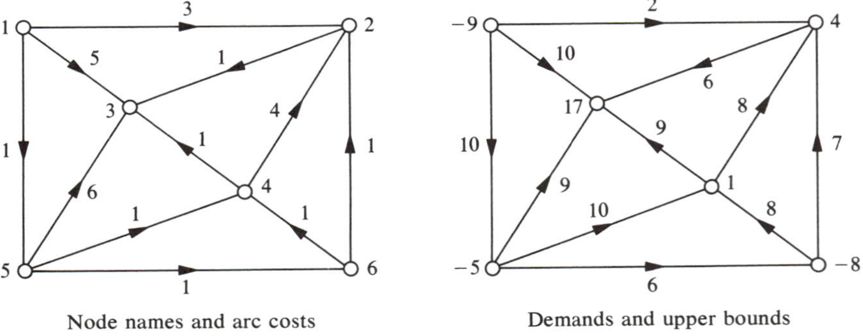

Figure 21.1

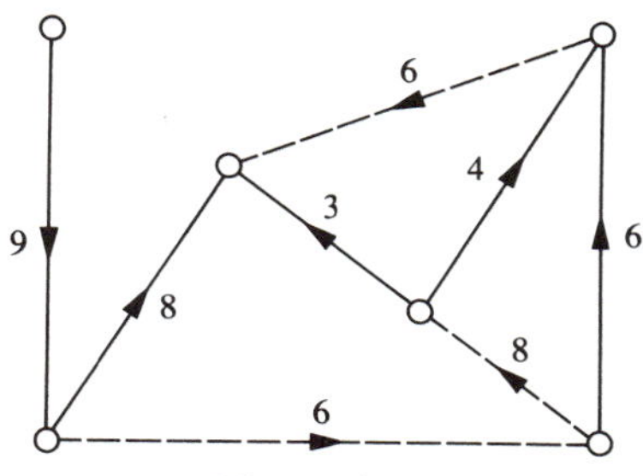

Figure 21.2

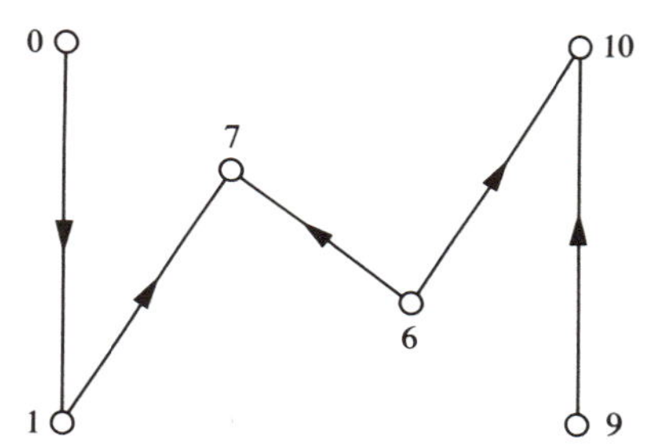

Figure 21.3

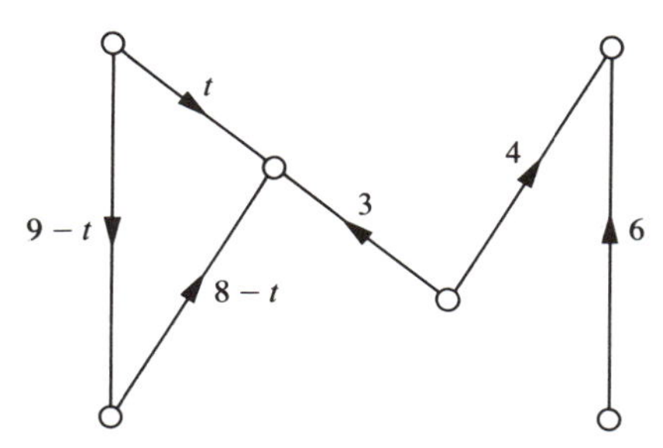

Figure 21.4

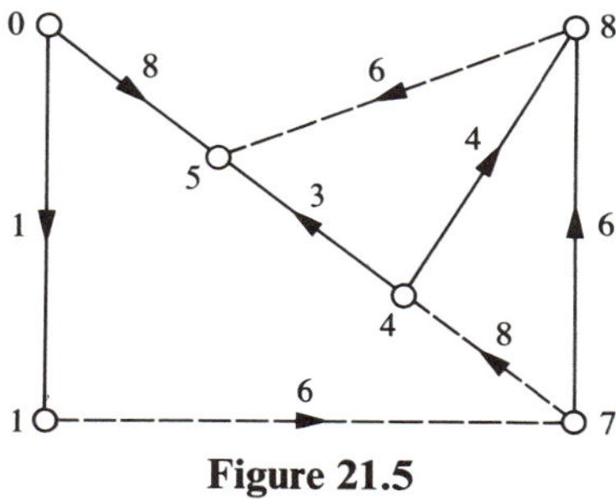

Figure 21.5

Now we choose 54 for the entering arc (note that this arc satisfies (21.1) in place of ij). Setting $x_{54} = t$ and adjusting the values of x_{ij} for $ij \in T$, we obtain Figure 21.6.

Now the constraint $x_{43} \leq u_{43} = 9$ dictates $t \leq 6$. Setting $t = 6$, we obtain the feasible tree solution and new node numbers shown in Figure 21.7. Note that arc 4 3 leaves the tree and becomes saturated.

Now the only arc satisfying (21.1) in place of ij is arc 12. We choose this arc for our next entering arc. With $x_{12} = t$ we are led to consider Figure 21.8. The constraint $x_{12} \leq u_{12} = 2$ forces $t \leq 2$; in fact, the solution corresponding to $t = 2$ remains feasible. Thus the entering arc becomes saturated before any arc is forced to leave the tree. Consequently, the tree does not change even though the solution does. The new solution is shown in Figure 21.9; since the tree does not change, the node numbers do not change.

At this moment, no arc $ij \notin T$ satisfies (21.1) and yet our solution is *not* optimal. Consider, for instance, arc 23. Under the current prices y_i, shipping through this arc is unprofitable ($y_2 + c_{23} > y_3$) and yet we are doing it; in fact, we are using this arc to its full capacity. It seems sensible to *decrease* the value of x_{23}, and that is precisely what we are going to do. Setting $x_{23} = u_{23} - t$, and adjusting the values of x_{ij} with $ij \in T$ accordingly, we obtain Figure 21.10. Since the constraint $x_{42} \geq 0$

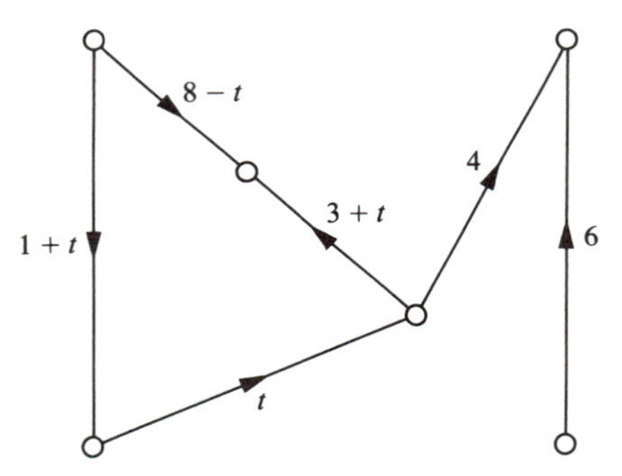

Figure 21.6

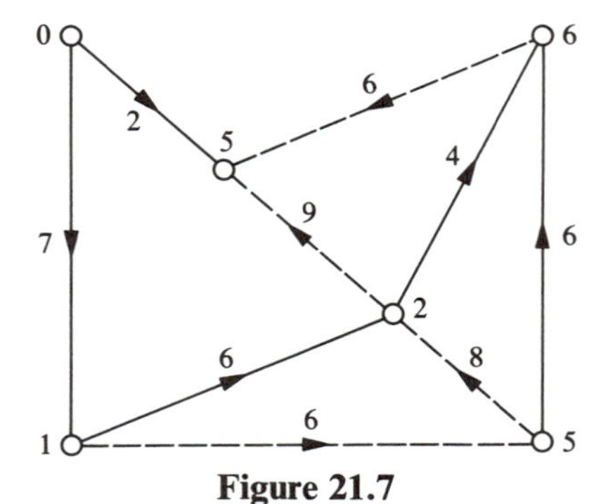

Figure 21.7

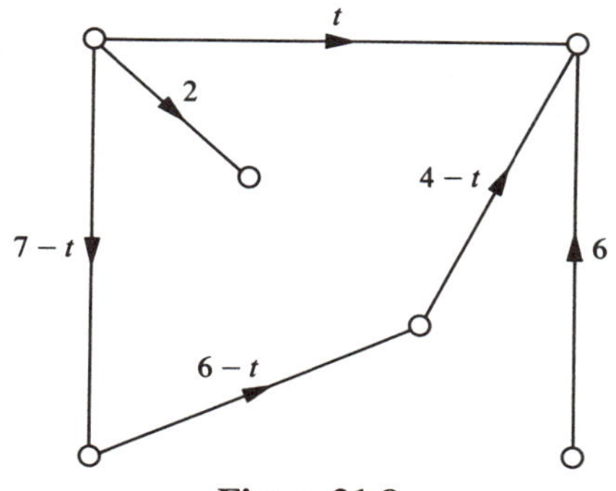

Figure 21.8

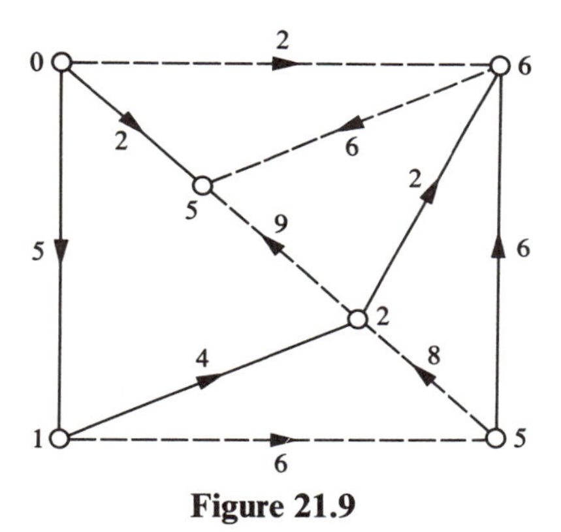

Figure 21.9

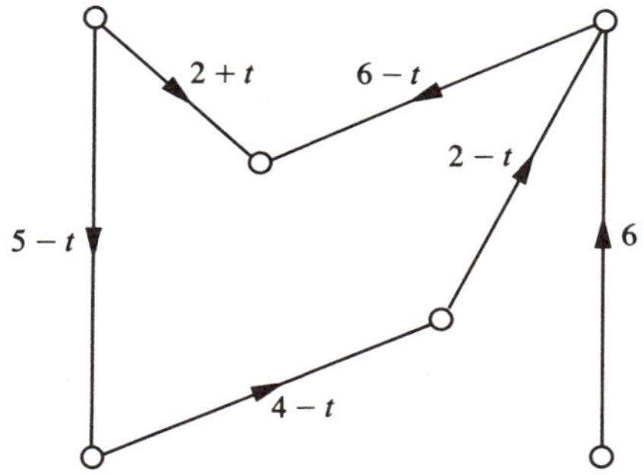

Figure 21.10

forces $t \leq 2$, and since $t = 2$ does yield a feasible solution, we enter the next iteration with the feasible tree solution and node numbers shown in Figure 21.11.

More generally, any arc $ij \notin T$ such that

$$y_i + c_{ij} > y_j \quad \text{and} \quad x_{ij} = u_{ij} \tag{21.2}$$

can be chosen for an entering arc: we shall reduce the total cost by decreasing the value of x_{ij} to $u_{ij} - t$ and adjusting the values of x_{ij} with $ij \in T$ accordingly. At present, the arc 64 satisfies (21.2) in place of ij; choosing this arc for an entering arc we obtain Figure 21.12. As t increases, arc 62 becomes saturated and leaves the tree. The resulting feasible tree solution, along with the corresponding node numbers, is shown in Figure 21.13.

Now arc 56 satisfies (21.2) in place of ij. Choosing this arc for the entering arc we are led to Figure 21.14. As t increases, the entering arc 56 obtains $x_{56} = 0$ before any arc is forced to leave the tree. Hence the tree does not change and the node numbers do not change either. The next feasible tree solution and its node numbers are shown in Figure 21.15.

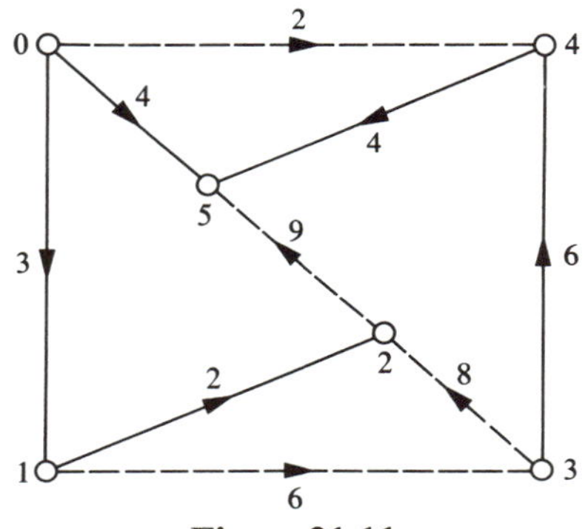

Figure 21.11

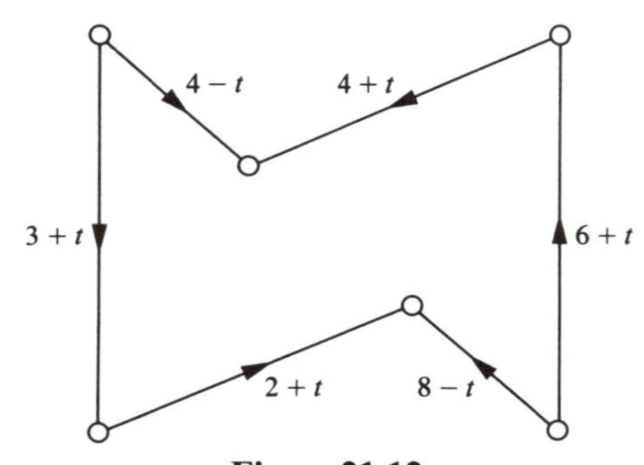

Figure 21.12

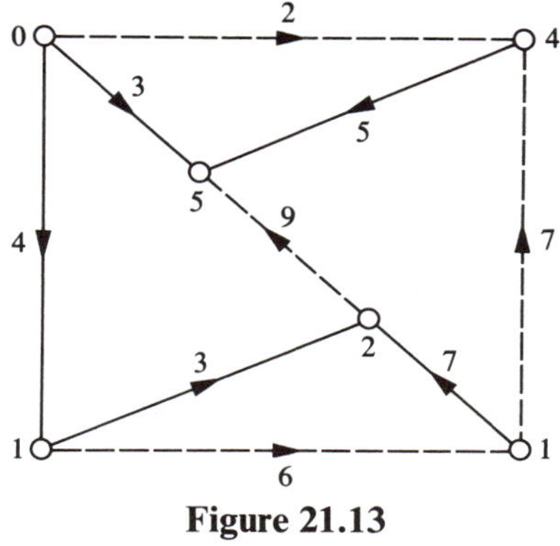

Figure 21.13

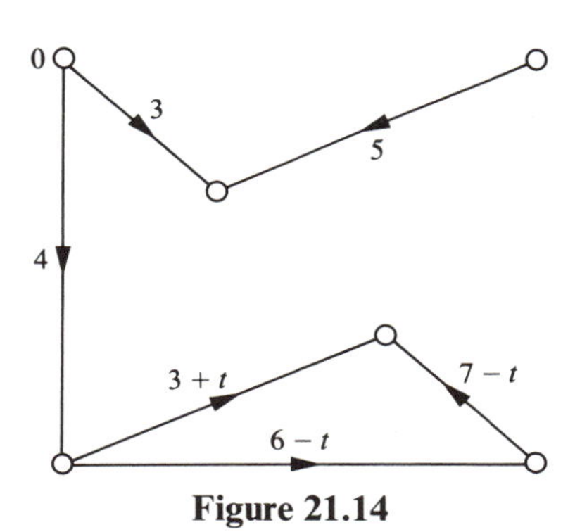

Figure 21.14

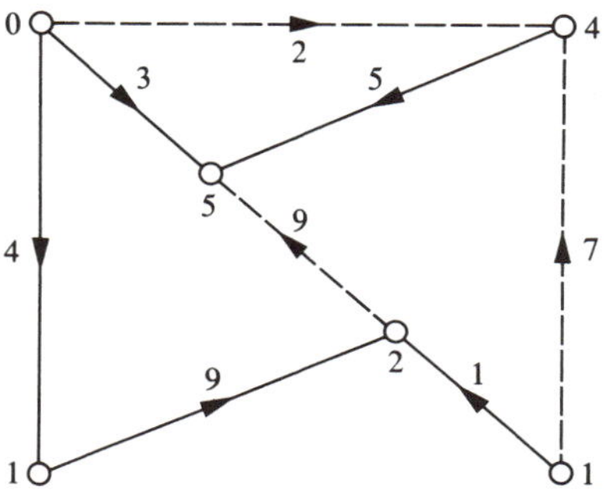

Figure 21.15

Now we have

$$x_{ij} = u_{ij} \qquad \text{whenever} \quad y_i + c_{ij} < y_j \tag{21.3}$$

and

$$x_{ij} = 0 \qquad \text{whenever} \quad y_i + c_{ij} > y_j. \tag{21.4}$$

Thus no arc ij satisfies (21.1) or (21.2). As we are about to prove, this circumstance indicates that the current solution is optimal.

ANALYSIS

Let us show at once that a feasible solution $\mathbf{x}$ of an upper-bounded transshipment problem

$$\text{minimize} \quad \mathbf{cx} \qquad \text{subject to} \quad \mathbf{Ax} = \mathbf{b}, \quad \mathbf{0} \le \mathbf{x} \le \mathbf{u} \tag{21.5}$$

is optimal whenever there are numbers $y_1, y_2, \ldots, y_n$ satisfying (21.3) and (21.4). The proof of this claim is simple: by virtue of (21.3) and (21.4), any feasible solution $\tilde{\mathbf{x}}$ of (21.5) satisfies

$$(c_{ij} + y_i - y_j)\tilde{x}_{ij} \ge (c_{ij} + y_i - y_j)x_{ij}$$

for all arcs ij. If $\bar{\mathbf{c}}$ denotes the vector with components $\bar{c}_{ij} = c_{ij} + y_i - y_j$ then $\mathbf{c} = \bar{\mathbf{c}} + \mathbf{yA}$ and so

$$\mathbf{c}\tilde{\mathbf{x}} = \bar{\mathbf{c}}\tilde{\mathbf{x}} + \mathbf{yA}\tilde{\mathbf{x}} = \bar{\mathbf{c}}\tilde{\mathbf{x}} + \mathbf{yb} \ge \bar{\mathbf{c}}\mathbf{x} + \mathbf{yb} = \bar{\mathbf{c}}\mathbf{x} + \mathbf{yAx} = \mathbf{cx}$$

which is the desired conclusion.

Along similar lines, we can explain the choice of the entering arc e as one satisfying $\bar{c}_e < 0$, $x_e = 0$ or $\bar{c}_e > 0$, $x_e = u_e$. In the corresponding pivot, the current feasible solution $\mathbf{x}$ is replaced by a feasible solution $\tilde{\mathbf{x}}$ such that

$$\tilde{x}_{ij} = x_{ij} \qquad \text{for all arcs } ij \notin T \text{ except the entering arc } e.$$

Since

$$\bar{c}_{ij} = 0 \qquad \text{whenever } ij \in T$$

we have

$$\bar{\mathbf{c}}\tilde{\mathbf{x}} = \bar{\mathbf{c}}\mathbf{x} + \bar{c}_e(\tilde{x}_e - x_e).$$

Now $\mathbf{c} = \bar{\mathbf{c}} + \mathbf{yA}$, $\mathbf{Ax} = \mathbf{A}\tilde{\mathbf{x}} = \mathbf{b}$ imply

$$\mathbf{c}\tilde{\mathbf{x}} = \mathbf{cx} + \bar{c}_e(\tilde{x}_e - x_e).$$

Since we set, for some nonnegative t,

$$\tilde{x}_e = x_e + t \qquad \text{in case} \quad \bar{c}_e < 0, \quad x_e = 0$$

and

$$\tilde{x}_e = x_e - t \qquad \text{in case} \quad \bar{c}_e > 0, \quad x_e = u_e$$

it follows that

$$\bar{c}_e(\tilde{x}_e - x_e) = -|\bar{c}_e| \cdot t.$$

In particular, if t is positive, then $\mathbf{c}\tilde{\mathbf{x}} < \mathbf{cx}$, and so the pivot improves the value of the objective function. Such pivots are called *nondegenerate*; pivots with $t = 0$ are called *degenerate*.

To obtain the new solution $\tilde{\mathbf{x}}$, we locate the unique cycle C in $T + e$ and then adjust x_{ij} on all the arcs $ij \in C$. Each of these arcs is either *forward* (directed in the same sense as e) or *reverse* (directed in the opposite sense). In case $\bar{c}_e < 0$, we set

$$\tilde{x}_{ij} = \begin{cases} x_{ij} + t & \text{for all forward arcs } ij \\ x_{ij} - t & \text{for all reverse arcs } ij. \end{cases}$$

In case $\bar{c}_e > 0$, we set

$$\tilde{x}_{ij} = \begin{cases} x_{ij} - t & \text{for all forward arcs } ij \\ x_{ij} + t & \text{for all reverse arcs } ij. \end{cases}$$

In either case, the adjustments cancel each other out at every node of C, and so $\mathbf{A}\tilde{\mathbf{x}} = \mathbf{b}$. As we have just observed, the value of $\mathbf{c}\tilde{\mathbf{x}}$ decreases when t increases, and so we are led to choosing t as large as possible. The constraints that prevent us from increasing t beyond every bound arise from the need to maintain feasibility. More precisely, the requirement $\mathbf{0} \leq \tilde{\mathbf{x}} \leq \mathbf{u}$ amounts to

$$\left.\begin{array}{ll} x_{ij} + t \leq u_{ij} & \text{for all forward arcs } ij \\ x_{ij} - t \geq 0 & \text{for all reverse arcs } ij \end{array}\right\} \text{in case } \bar{c}_e < 0$$

and

$$\left.\begin{array}{ll} x_{ij} - t \geq 0 & \text{for all forward arcs } ij \\ x_{ij} + t \leq u_{ij} & \text{for all reverse arcs } ij \end{array}\right\} \text{in case } \bar{c}_e > 0.$$

It may happen that these formal constraints represent no real restrictions on the value of t. Clearly, this is the case if and only if

$$\tilde{x}_{ij} = x_{ij} + t \quad \text{and} \quad u_{ij} = \infty \quad \text{for all arcs } ij \in C. \tag{21.6}$$

Let us examine this situation more closely. To begin, (21.6) implies $\tilde{x}_e = x_e + t$ for the entering arc e and, since e is a forward arc, we have $\bar{c}_e < 0$. Now it follows that every arc $ij \in C$ must be forward. The change of $\mathbf{x}$ into $\tilde{\mathbf{x}}$ amounts to sending t extra units around the cycle C; since $\mathbf{c}\tilde{\mathbf{x}} < \mathbf{cx}$, sending a unit around C must cost a negative amount. In other words, the sum of the cost coefficients c_{ij} over all the arcs $ij \in C$ must be negative. To summarize, the cycle C has the following three properties:

(i) All the arcs $ij \in C$ are directed in the same sense.

(ii) There is no finite upper bound u_{ij} on any x_{ij} such that $ij \in C$.

(iii) $\sum_{ij \in C} c_{ij} < 0.$

We shall refer to such cycles as *negative cycles*. The presence of a negative cycle indicates that the problem is unbounded: by sending larger and larger amounts around this cycle, we can make the objective function negative and arbitrarily large in magnitude.

If C is not a negative cycle, then there is at least one upper bound on the value of t, and some arc $f \in C$ provides the most stringent bound. We set t at the corresponding level ($u_f - x_f$ if f is forward, x_f if f is reverse) and define $\tilde{\mathbf{x}}$ as above. If $f \neq e$, then the tree that goes with $\tilde{\mathbf{x}}$ is $T + e - f$. It may happen, as it did in our example, that $e = f$. In that case, $\mathbf{x}$ changes into $\tilde{\mathbf{x}}$, but the tree T and the node numbers $\mathbf{y}$ remain intact. To summarize, each iteration may be of one of the following six types:

(i) $\bar{c}_e < 0, \quad x_e = 0 \quad$ and $\quad e \neq f, \quad \tilde{x}_f = 0.$

(ii) $\bar{c}_e < 0, \quad x_e = 0 \quad$ and $\quad e \neq f, \quad \tilde{x}_f = u_f.$

(iii) $\bar{c}_e < 0, \quad x_e = 0 \quad$ and $\quad e = f, \quad \tilde{x}_e = u_e.$

(iv) $\bar{c}_e > 0, \quad x_e = u_e \quad$ and $\quad e \neq f, \quad \tilde{x}_f = 0.$

(v) $\bar{c}_e > 0, \quad x_e = u_e \quad$ and $\quad e \neq f, \quad \tilde{x}_f = u_f.$

(vi) $\bar{c}_e > 0, \quad x_e = u_e \quad$ and $\quad e = f, \quad \tilde{x}_e = 0.$

Each of these six types has been illustrated on one of the six iterations in our example.

It may happen that some arc $f \in C$ limits the increase of t to zero. In that case, we have

$$f \in T \quad \text{and either} \quad x_f = 0 \quad \text{or} \quad x_f = u_f.$$

Feasible tree solutions $\mathbf{x}$ with this property are called *degenerate*. The corresponding pivot, replacing T by $T + e - f$ but leaving $\mathbf{x}$ unchanged, is also called *degenerate*.

In the presence of degeneracy, the network simplex method may cycle: we have seen an example of this phenomenon even in the case without upper bounds ($u_{ij} = +\infty$ for every arc ij). Nevertheless, Cunningham's cycling-prevention rule, discussed in Chapter 19, extends to the upper-bounded case. We used to call a feasible tree solution *strongly feasible* whenever *every arc* $ij \in T$ *with* $x_{ij} = 0$ *was directed away from the root*. Now we shall also require that *every arc* $ij \in T$ *with* $x_{ij} = u_{ij}$ *be directed towards the root*. Again, there is an easy procedure that either replaces a feasible tree solution by a strongly feasible one or else decomposes the problem into smaller subproblems. In Chapter 19, the decomposition was induced by a set S of nodes such that:

(i) There was no arc ij with $i \notin S, j \in S$.

(ii) We had $\sum_{k \in S} b_k = 0$.

Intuitively, S was thought of as an autonomous region whose home supply matched the home demand, and which possessed no import channels. In the upper-bounded case, these two conditions get replaced by

$$\sum_{k \in S} b_k = \sum_{\substack{i \notin S \\ j \in S}} u_{ij}.$$

Now S may be thought of as an autonomous region whose total net demand matches the upper bound on the volume of import. Hence in every feasible solution, the region S will import all it can and export nothing:

$$x_{ij} = u_{ij} \quad \text{whenever} \quad i \notin S, \quad j \in S, \qquad \text{and } x_{ij} = 0 \quad \text{whenever} \quad i \in S, \quad j \notin S.$$

It is an easy exercise to couch this argument in algebraic terms.

Having initialized the algorithm by a strongly feasible solution, we break ties in each choice of a leaving arc by the following rule:

> Choose the first candidate encountered when C is traversed, beginning at the join, in the direction of the entering arc e in case $\overline{c}_e < 0$, or in the opposite direction in case $\overline{c}_e > 0$.

Again, it can be proved that this procedure transforms each strongly feasible solution into another strongly feasible solution. Furthermore, it can be proved that in each degenerate pivot of this kind, the quantity $\sum(y_k - y_w)$, for a fixed w and with the summation running through all k, decreases. Hence no tree appears in two different iterations. Now cycling is purged, and so the algorithm terminates.

The only remaining problem is finding a feasible tree solution to begin with. As we did in Chapter 19, we shall get around this difficulty by solving a related *auxiliary problem*. To obtain the auxiliary problem, we first designate some node w as a root; then we add an artificial arc iw for each node i such that $b_i < 0$ and an artificial arc wj for each node j such that $b_j \geq 0$. The demands at the nodes and the upper bounds on the original (nonartificial) arcs remain unchanged, but we replace the cost coefficient c_{ij} on each original arc ij by $p_{ij} = 0$. The artificial arcs ij receive $p_{ij} = 1$ and $u_{ij} = +\infty$. In the resulting auxiliary problem, a feasible tree solution is readily apparent: the tree consists of the $m - 1$ artificial arcs. Hence we can use the network simplex method to find a feasible tree solution $\mathbf{x}^*$ minimizing the new objective function $\mathbf{px}$. If $x_{ij}^* = 0$ for every artificial arc ij, then the remaining components of $\mathbf{x}^*$ describe a feasible solution of the original problem. (In order to initialize the network simplex method on the original problem, we also need a tree T that goes with $\mathbf{x}^*$. If the optimal tree T^* in the auxiliary problem includes no artificial arcs, then we can set $T = T^*$. If T^* does include artificial arcs ij, but we still have $x_{ij}^* = 0$ for all such arcs, then an easy procedure decomposes the original problem into independent subproblems and finds a feasible tree solution in each of them. The details follow the lines of Chapter 19 and will not be repeated here.) On the other hand, if $x_{ij}^* > 0$ for some artificial arc ij then no feasible solution $\mathbf{x}$ of the original problem can exist: such an $\mathbf{x}$, extended by $x_{ij} = 0$ for every artificial arc ij, would yield $\mathbf{px} = 0 < \mathbf{px}^*$, contradicting optimality of $\mathbf{x}^*$. A further analysis of this case leads to the following result.

***THEOREM 21.1* [*D. Gale* (1957)].** An upper-bounded transshipment problem has no feasible solution if and only if there is a set S of nodes such that

$$\sum_{k \in S} b_k > \sum_{\substack{i \notin S \\ j \in S}} u_{ij}. \tag{21.7}$$

PROOF. The "if" part is easy: the left-hand side of (21.7) represents the total net demand of the region S, whereas the right-hand side represents an upper bound on the total volume of import into S. To prove the more difficult "only if" part, consider an optimal solution $\mathbf{x}^*$ of our auxiliary problem. Since the original problem has no feasible solution, there is an artificial arc uv such that $x_{uv}^* > 0$. The corresponding node numbers $y_1, y_2, \ldots, y_m$ must satisfy

$$y_v = y_u + p_{uv} = y_u + 1.$$

Let S consist of all the nodes k such that $y_k \geq y_v$. We propose to show that

$$\sum_{\substack{i \notin S \\ j \in S}} x_{ij}^* \geq x_{uv}^* + \sum_{\substack{i \notin S \\ j \in S}} u_{ij} \tag{21.8}$$

with the right-hand side summation restricted to original arcs ij, and that

$$\sum_{\substack{i \in S \\ j \notin S}} x_{ij}^* = 0. \tag{21.9}$$

Once these two facts are established, the desired inequality (21.7) will follow: it is easy to see intuitively, and to justify rigorously, that

$$\sum_{k \in S} b_k = \sum_{\substack{i \notin S \\ j \in S}} x_{ij}^* - \sum_{\substack{i \in S \\ j \notin S}} x_{ij}^*$$

for an arbitrary set S. [See also equation (19.8).]

To prove (21.8), we need only show that $x_{ij}^* = u_{ij}$ for each original arc ij with $i \notin S$, $j \in S$. But this fact follows directly from the optimality conditions since $y_i + p_{ij} =$

$y_i < y_j$. To prove (21.9), we need only show that $x^*_{ij} = 0$ for each arc ij, original or artificial, such that $i \in S, j \notin S$. Again, this fact follows directly from the optimality conditions since $y_i + p_{ij} \geq y_i > y_j$. ■

In closing, let us note that Theorem 20.1 generalizes easily to the context of upper bounded transshipment problems.

THEOREM 21.2 (The Integrality Theorem). Consider an upper-bounded transshipment problem

$$\text{minimize} \quad \mathbf{cx} \qquad \text{subject to} \quad \mathbf{Ax} = \mathbf{b}, \quad \mathbf{0} \leq \mathbf{x} \leq \mathbf{u} \tag{21.10}$$

such that all the components of the vector **b** and all the finite components of the vector **u** are integers. If (21.10) has at least one feasible solution, then it has an integer-valued feasible solution. If (21.10) has an optimal solution, then it has an integer-valued optimal solution. ■

PROBLEMS

△ **21.1** Solve the problems

$$\text{minimize} \quad \mathbf{cx} \qquad \text{subject to} \quad \mathbf{Ax} = \mathbf{b}, \quad \mathbf{0} \leq \mathbf{x} \leq \mathbf{u}$$

with

$$\mathbf{A} = \begin{bmatrix} -1 & -1 & & & & & 1 & & \\ 1 & & -1 & & & & & 1 & \\ & 1 & 1 & -1 & & & & & \\ & & & 1 & -1 & -1 & & & \\ & & & & 1 & & -1 & & 1 \\ & & & & & 1 & & -1 & -1 \end{bmatrix}, \quad \mathbf{b} = \begin{bmatrix} 0 \\ 1 \\ 2 \\ -7 \\ 0 \\ 4 \end{bmatrix}$$

and the following choices of **c**, **u**:

(i) $\mathbf{c} = [2, 4, 0, 2, 1, 3, 0, 1, 3]$, $\mathbf{u} = [2, 5, +\infty, +\infty, 5, +\infty, 5, +\infty, 3]^T$.

(ii) $\mathbf{c} = [4, 1, 3, -5, -1, 1, 4, 0, 5]$, $\mathbf{u} = [2, 5, +\infty, +\infty, 5, +\infty, 5, +\infty, 3]^T$.

(iii) $\mathbf{c} = [4, 1, 3, -5, -1, 1, 4, 0, 5]$, $\mathbf{u} = [3, +\infty, +\infty, 5, +\infty, 4, 1, 2, +\infty]^T$.

21.2 Prove in detail that Cunningham's rule prevents cycling when extended to the context of upper-bounded transshipment problems along the lines described in the text.

21.3 Decompose the problem

$$\text{minimize} \quad \mathbf{cx} \qquad \text{subject to} \quad \mathbf{Ax} = \mathbf{b}, \quad \mathbf{0} \leq \mathbf{x} \leq \mathbf{u}$$

with

$$\mathbf{A} = \begin{bmatrix} 1 & & & & 1 & & 1 & & \\ & -1 & -1 & & & & & & -1 \\ -1 & 1 & & -1 & & & & & \\ & & & 1 & -1 & -1 & & & \\ & & 1 & & & & -1 & -1 & \\ & & & & & 1 & & 1 & 1 \end{bmatrix}, \quad \mathbf{b} = \begin{bmatrix} 8 \\ -9 \\ -3 \\ -2 \\ 4 \\ 2 \end{bmatrix}$$

and

$$\mathbf{u} = [+\infty, 1, +\infty, 6, 7, +\infty, 2, 5, 3]^T$$

into two subproblems.

21.4 Generalize Theorem 21.1 to the context of problems

$$\text{minimize} \quad \mathbf{cx} \qquad \text{subject to} \quad \mathbf{Ax} = \mathbf{b}, \quad \mathbf{l} \leq \mathbf{x} \leq \mathbf{u}$$

such that $\mathbf{A}$ is the incidence matrix of a network.

21.5 [D. Gale (1957), H. J. Ryser (1957).] Find necessary and sufficient conditions, in terms of positive integers $r_1, r_2, \ldots, r_m$ and $s_1, s_2, \ldots, s_n$, for the nonexistence of an $m \times n$ zero–one matrix $A = (a_{ij})$ such that

$$\sum_{j=1}^{n} a_{ij} = r_i \qquad \text{for all} \quad i = 1, 2, \ldots, m$$

and

$$\sum_{i=1}^{m} a_{ij} = s_j \qquad \text{for all} \quad j = 1, 2, \ldots, n.$$

21.6 [H. G. Landau (1953).] A vector $s = [s_1, s_2, \ldots, s_n]$ is called a *score vector* if each s_i specifies the number of wins scored by the ith player in a tournament where every two participants played against each other precisely once, with draws excluded by the rules of the game. Find necessary and sufficient conditions under which s is not a score vector.

21.7 [G. B. Dantzig (1953), pp. 380–382.] Given an upper bounded transshipment problem, add two new nodes r, s with $b_s = u_{ij}$, $b_r = -u_{ij}$ for each arc ij with a finite u_{ij}. Then replace ij by three arcs is, rs, rj with $u_{is} = u_{rs} = u_{rj} = +\infty$ and $c_{is} = c_{ij}$, $c_{rs} = c_{rj} = 0$. How is the resulting problem related to the original one?

22

Maximum Flows Through Networks

The subject of this chapter is a restricted class of upper-bounded transshipment problems, known as the *maximum-flow problems*. Even though these problems are quite special, they still retain a wide range of applications: we shall see that (i) the task of finding a feasible solution in an upper-bounded transshipment problem, (ii) the task of finding a largest matching in a bipartite graph, and (iii) an important strip mining problem, known as the *open-pit problem*, can all be turned into maximum flow problems. What makes maximum-flow problems appealing is the fact that they can be solved very fast by specialized algorithms. We shall describe first the classical *augmenting path method* developed by L. R. Ford, Jr. and D. R. Fulkerson (1957) and then its more recent efficient variations.

THE TRAVELERS' EXAMPLE

Consider a hypothetical group of people that must get from San Francisco to New York City on a short notice. Even though all direct flights have been booked up, it is still possible for a traveler starting from San Francisco to reach New York City on the same day by changing planes at other airports. The seat availability

Table 22.1 The Travelers' Example: Seat Availability

From	To	Number of seats
San Francisco	Denver	5
San Francisco	Houston	6
Denver	Atlanta	4
Denver	Chicago	2
Houston	Atlanta	5
Atlanta	New York	7
Chicago	New York	4

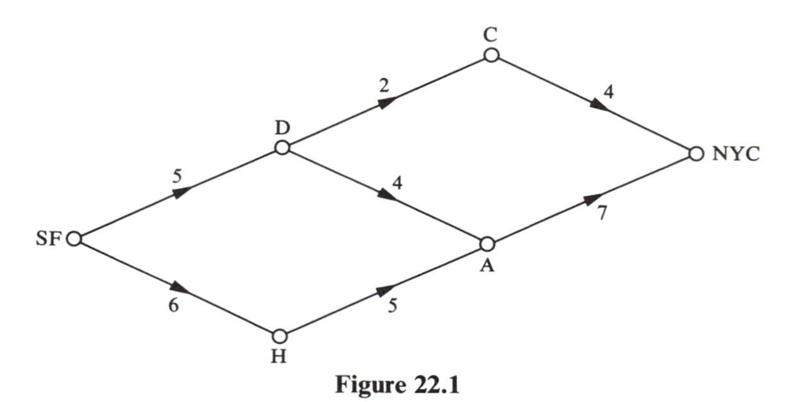

Figure 22.1

on potentially useful flights is shown in Table 22.1 and in Figure 22.1. We assume that there is sufficient time between the arrival of each flight at an airport and the departure of another flight from the same airport.

Thus, for instance, four people may travel via the route SF–Denver–Atlanta–NYC. Once these seats are committed, an extra passenger may take the northern route (SF–Denver–Chicago–NYC) and an additional three passengers may take the southern route (SF–Houston–Atlanta–NYC). The resulting schedule is represented in Figure 22.2.

Now we have eight people on their way; can we do better than that? As the reader may have noticed, our problem may be presented in an upper-bounded transshipment form. The trick is to enlarge our network by adding a *return arc* that enables every passenger to get from New York City back to San Francisco. The people using this arc will be precisely those passengers that arrived in New York City; since we wish to maximize their number, we assign the cost coefficient $c_{ij} = -1$ to the return arc ij and set $c_{ij} = 0$ for all the remaining arcs ij. (As usual, we are going to *minimize* $\mathbf{cx}$.) In this closed system, people neither emerge out of the blue nor disappear;

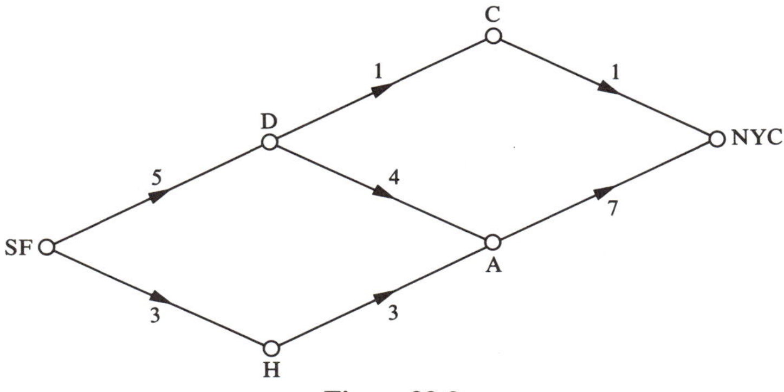

Figure 22.2

every passenger eventually returns home. Hence the demand b_i at each node i equals zero. The upper bounds, naturally, are as given in Figure 22.1, with $u_{ij} = \infty$ on the return arc ij. The resulting upper-bounded transshipment problem may be solved routinely by the network simplex method. The point of this chapter is that highly special problems of this kind may be solved even faster by certain specialized algorithms.

WHAT IS A MAXIMUM-FLOW PROBLEM?

In the present chapter, a *network* will mean an ordinary network with n nodes, one of which is designated as a *source* s and another as a *sink* t. (In our example, the source is San Francisco and the sink is New York City.) The remaining $n - 2$ nodes will be called *intermediate*. For simplicity of exposition, we shall assume that *no arc enters the source and no arc leaves the sink*. Each arc ij will carry a positive upper bound u_{ij}, possibly $u_{ij} = +\infty$. In a contrast with the general upper-bounded transshipment problem, there will be no demands b_i associated with nodes i and no cost coefficients c_{ij} associated with arcs ij. By a *flow* $\mathbf{x}$ through our network, we shall mean an assignment of numbers x_{ij} to the various arcs ij such that

$$\sum_i x_{ij} = \sum_k x_{jk} \tag{22.1}$$

for all intermediate nodes j. In our example, this *conservation law* reflects the fact the passengers arriving at an airport j are precisely the passengers leaving that airport. In the sum of the $n - 2$ equations (22.1), the terms corresponding to arcs between intermediate nodes cancel each other out, and we obtain

$$\sum_j x_{sj} = \sum_j x_{jt}. \tag{22.2}$$

In our example, this identity states that the number of passengers leaving San Francisco equals the number of passengers arriving in New York City. The common value of the two sides in (22.2) will be referred to as the *volume* of the flow. Finally, a flow $\mathbf{x}$ will be called *feasible* if

$$0 \leq x_{ij} \leq u_{ij}$$

for each arc ij. A feasible flow of the largest possible volume is referred to as a *maximum flow* through the network; the *maximum-flow problem* is the problem of finding a maximum flow through a prescribed network.

THE MAX-FLOW MIN-CUT THEOREM

By a *cut* in a network, we shall mean any set C of nodes that includes the source s but does not include the sink t. If C is a cut and if $\mathbf{x}$ is a flow of volume v, then the sum of the identity $v = \sum x_{sk}$ and all the equations (22.1) with $j \neq s, j \in C$ reads

$$v + \sum_{\substack{i \in C \\ j \in C}} x_{ij} + \sum_{\substack{i \notin C \\ j \in C}} x_{ij} = \sum_{\substack{j \in C \\ k \in C}} x_{jk} + \sum_{\substack{j \in C \\ k \notin C}} x_{jk}.$$

After simplifications, we obtain

$$v = \sum_{\substack{j \in C \\ k \notin C}} x_{jk} - \sum_{\substack{i \notin C \\ j \in C}} x_{ij}. \tag{22.3}$$

In intuitive terms, this identity says that the volume of $\mathbf{x}$ equals the net export out of the region C. In particular, the identities $v = \sum x_{sk}$ and $v = \sum x_{jt}$ arise from (22.3) by letting C consist of the single node s and of all nodes except t, respectively.

The *capacity* of a cut C is defined to be the quantity

$$\sum_{\substack{j \in C \\ k \notin C}} u_{jk}.$$

(If at least one of the summands u_{jk} is $+\infty$, then the capacity of C is said to be infinite.) Identity (22.3) implies at once that, for every feasible flow $\mathbf{x}$ and for every cut C, the volume of $\mathbf{x}$ is at most the capacity of C. The fundamental theorem of this chapter asserts that the gap between the two can always be closed by a suitable choice of $\mathbf{x}$ and C.

THEOREM 22.1 (The Max-Flow Min-Cut Theorem). Every maximum-flow problem has precisely one of the following two properties:

(i) There are feasible flows of arbitrarily large voiumes and every cut has an infinite capacity.

(ii) There is a maximum flow and its volume equals the minimum capacity of a cut.

PROOF. As previously illustrated, every maximum-flow problem can be turned into an upper-bounded transshipment problem: we need only add the return arc ts with $u_{ts} = +\infty$ and $c_{ts} = -1$, to set $c_{ij} = 0$ for all the original arcs ij, and to set $b_k = 0$ for all the nodes k. Every upper-bounded transshipment problem has precisely one of the following three properties: (i) it is unbounded, (ii) it has an optimal solution, (iii) it has no feasible solution. Our particular problem cannot have the property (iii), since a feasible solution can be obtained by setting $x_{ij} = 0$ for all arcs ij. Hence it is unbounded or it has an optimal solution. In the first case, there are feasible flows of arbitrarily large volumes; furthermore, since the capacity of each cut is bounded from below by the volume of every flow, this capacity must be infinite. In the second case, the network simplex method finds an optimal solution $\mathbf{x}$ along with node numbers y_k such that

$$x_{ij} = 0 \qquad \text{whenever} \quad y_i + c_{ij} > y_j \tag{22.4}$$

and

$$x_{ij} = u_{ij} \qquad \text{whenever} \quad y_i + c_{ij} < y_j. \tag{22.5}$$

In particular, since $u_{ts} = +\infty$ and $c_{ts} = -1$, we must have $y_t - 1 \geq y_s$. Hence the set C of all nodes k such that $y_k \leq y_s$ is a cut. Since $c_{ij} = 0$ for all original arcs ij, fact (22.5) implies $x_{ij} = u_{ij}$ for each original arc ij with $i \in C$, $j \notin C$ and fact (22.4) implies $x_{ij} = 0$ for each original arc ij with $i \notin C$, $j \in C$. Now (22.3) implies that the volume of the flow defined by $\mathbf{x}$ equals the capacity of C. To complete the proof, we need only observe that $\mathbf{x}$ defines a maximum flow (the volume of any feasible flow is at most the capacity of C) and that C is a cut of the minimum capacity (the capacity of any cut is at least the volume of the flow defined by $\mathbf{x}$). ■

The Max-Flow Min-Cut Theorem was obtained independently by P. Elias, A. Feinstein and C. E. Shannon (1956), by L. R. Ford, Jr. and D. R. Fulkerson (1956) and, in the special case where each u_{ij} is a positive integer, by A. Kotzig (1956). Theorem 22.2 is a companion theorem, often useful in applications.

THEOREM 22.2 (The Integral Flow Theorem). If each finite u_{ij} is a positive integer and if there is a maximum flow then there is an integer-valued maximum flow. ■

This fact follows easily from the reduction of a maximum-flow problem into an upper-bounded transshipment problem combined with Theorem 21.2. The details are left for an exercise (problem 22.3). Alternative proofs of Theorems 22.1 and 22.2 follow from an analysis of the specialized algorithms presented later in this chapter.

APPLICATIONS

The problem of finding a feasible solution in an upper bounded transshipment problem can be turned into a maximum-flow problem: we need only create a fictitious source capable of supplying up to $-b_j$ units to each real source j and a fictitious sink capable of absorbing up to b_i units from each real sink i. More precisely, we begin with the original network of the transshipment problem, discard the cost coefficients c_{ij}, and retain the upper bounds u_{ij}. Then we introduce a source s and a sink t, join s to each node j having $b_j < 0$ by an arc sj having $u_{sj} = -b_j$ and join each node i having $b_i > 0$ to t by an arc it having $u_{it} = b_i$. Since $\sum b_k = 0$, we have $\sum u_{sj} = \sum u_{it}$. Since the common value w of the two sides in this equation is the capacity of the cut consisting of the single node s (as well as the capacity of the cut consisting of all nodes but t), no feasible flow can have a volume greater than w. In fact, the following three properties of a feasible flow $\mathbf{x}$ are easily seen to be equivalent:

(i) The volume of $\mathbf{x}$ is w.

(ii) $x_{sj} = u_{sj}$ for all arcs sj.

(iii) $x_{it} = u_{it}$ for all arcs it.

Now it follows that every feasible flow of volume w represents a feasible solution in the upper bounded transshipment problem (we can ignore all the arcs sj and it) and, conversely, every feasible solution in the upper-bounded transshipment problem represents a feasible flow of volume w (we can write $x_{sj} = u_{sj}$ for all arcs sj, and $x_{it} = u_{it}$ for all arcs it). In short, the upper-bounded transshipment problem has a feasible solution if and only if the maximum flow has volume w. (This reduction, along with Theorem 22.1, yields an alternative proof of Theorem 21.1; we leave the details for problem 22.4.)

Along similar lines, the problem of finding a largest matching in a bipartite graph may be turned into a problem of finding a maximum integer-valued flow. We direct every arc of the bipartite graph away from its left endpoint towards its right endpoint, and assign it an infinite upper bound. Then we add a source s with an arc si for every left node i and a sink t with an arc jt for every right node j. The new arcs have upper bounds $u_{si} = u_{jt} = 1$. It is not difficult to see that every *integer-valued* feasible flow $\mathbf{x}$ of volume k points out a matching M of size k ($ij \in M$ if and only if $x_{ij} = 1$) and, conversely, that every matching M of size k points out a feasible flow $\mathbf{x}$ of volume k ($x_{si} = x_{ij} = x_{jt} = 1$ whenever $ij \in M$). In particular, a maximum integer-valued flow $\mathbf{x}$ points out a largest matching in the graph. (This reduction, along with Theorem 22.1, yields an alternative proof of Theorem 20.5; we leave the details for problem 22.5.)

A rather unexpected application of the maximum flow theory involves a problem of some practical significance in strip mining: given a reliable survey of underground ore deposits, design the open pit that will be created. The walls of the pit must slope gently, for otherwise they might cave in. Consequently, before the deposit can be tapped, much of the earth lying above it must be removed. It may even be advisable to ignore part of the deposit altogether when the profit obtained from the ore does not make up for the attendant cost of earth removal. Thus, finding the most profitable contour of the pit may be a nontrivial task. Customarily, the volume under consideration is partitioned into relatively small blocks with sides measuring about 30 feet. (The length and width of each block, usually equal to each other, increase with the distance between successive bore-holes in the geological survey; the height increases with the angle at which the walls are permitted to slope.) A typical number of these blocks is several thousand or more. With each block i, we associate the net profit w_i resulting when i is included in the pit: to obtain w_i, we subtract the cost of excavating i (*not* the accumulated cost of excavating i along with the blocks above it) from the profit (if any) brought in by the ore found in i. Thus the total net profit of creating a pit P equals $\sum_{i \in P} w_i$. Of course, not every set P of blocks constitutes a feasible pit: whenever we decide to excavate a particular block i, we are forced to excavate a number of other blocks, usually forming a vertical cone above i. For each of these blocks j, we shall write $i \to j$. (It is not difficult to see that the list of all the pairs $i \to j$ forms a partial order, as defined in Chapter 20. Nevertheless, this fact is irrelevant to our discussion.) Now a set P of blocks constitutes a feasible pit if and only if

$$j \in P \quad \text{whenever} \quad i \in P \quad \text{and} \quad i \to j.$$

In the abstract version of the open pit problem, the data consist of a network and a set of numbers w_i, one for each node i. A set C of nodes is called *closed* if $j \in C$ for every arc ij such that $i \in C$. The objective is to find a closed set C maximizing $\sum_{i \in C} w_i$. J.-C. Picard (1976), generalizing an earlier work of M. L. Balinski (1970) and J. Rhys (1970), exhibited an ingenious way of converting this problem into maximum-flow (or rather, minimum-cut) form. For each node i of the network, write $i \in A$ if $w_i \geq 0$ and $i \in B$ if $w_i < 0$. Extend the network by adding a source s and a sink t, with arcs si for each $i \in A$ and it for each $i \in B$. The upper bounds on the new arcs are defined by $u_{si} = w_i$ for each $i \in A$ and $u_{it} = -w_i$ for each $i \in B$; the upper bounds on the original arcs are infinite. It is not difficult to see that a set C of nodes is closed if and only if $C \cup \{s\}$ has a finite capacity. Furthermore, if the capacity of $C \cup \{s\}$ is finite then it equals

$$\sum_{i \in A - C} u_{si} + \sum_{i \in B \cap C} u_{it} = \sum_{i \in A - C} w_i + \sum_{i \in B \cap C} (-w_i) = \sum_{i \in A} w_i - \sum_{i \in C} w_i.$$

Since $\sum_{i \in A} w_i$ is a constant, minimizing the capacity of $C \cup \{s\}$ amounts to maximizing $\sum_{i \in C} w_i$. We conclude that a minimum cut in our network points out an optimal closed set C.

THE AUGMENTING PATH METHOD

A classical algorithm for finding a maximum flow through a network was developed by L. R. Ford, Jr. and D. R. Fulkerson (1957) from earlier contributions of H. W. Kuhn (1955) and J. Egerváry (1931). We shall refer to it as the *augmenting path method.* Rudiments of this algorithm appear in our treatment of the travelers' example. There, we began with four passengers going via Denver and Atlanta, then we added an extra passenger taking the northern route, and finally an additional three passengers taking the southern route. In effect, we have moved through the sequence of flows (with a steadily increasing volume) shown in Figure 22.3.

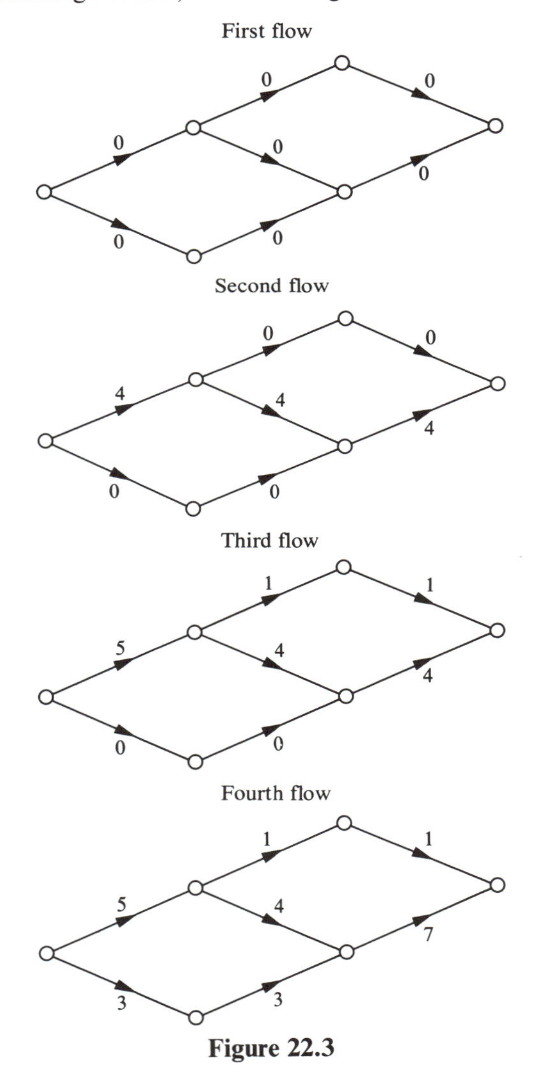

Figure 22.3

More generally, each iteration of the augmenting path method begins with some feasible flow; if we find a path along which an extra passenger (or passengers) starting at the source may reach the sink then we adjust the flow accordingly and proceed to the next iteration with a feasible flow of a larger volume. This happened in each of the first three iterations (Figure 22.3a–c). However, this crude trick fails after the third iteration: now the flight from San Francisco to Denver is full, and so the extra passenger would have to proceed via Houston to Atlanta, only to get stuck there. Fortunately, a more sophisticated technique is available. Imagine that we do send the extra passenger to Atlanta, thereby creating a passenger surplus at that airport (eight passengers getting in and only seven coming out). The surplus may be transferred to Denver by cancelling one of the seats on the Denver–Atlanta flight. But a surplus in Denver can be easily taken care of by routing the extra passenger via Chicago to New York City. Altogether, the proposed modification may be represented as in Figure 22.4. The resulting feasible flow is shown in Figure 22.5.

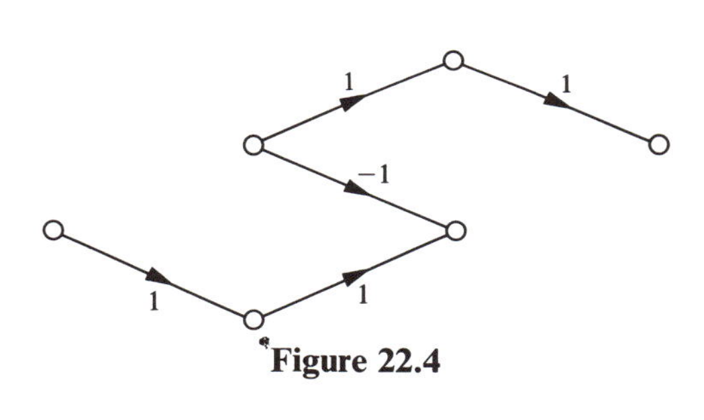

Figure 22.4

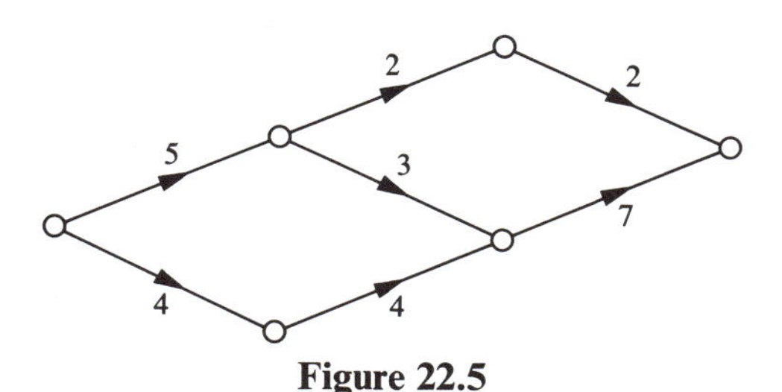

Figure 22.5

Now we arrive at the fundamental notion of an *augmenting path*. Consider a feasible flow **x** through a network. A path with nodes $v_0, v_1, \ldots, v_r$ will be called **x**-*alterable* if $x_{ij} < u_{ij}$ for each of its forward arcs ij (an arc ij is forward if $i = v_k$ and $j = v_{k+1}$ for some k) and if $x_{ij} > 0$ for each of its reverse arcs ij (an arc ij is reverse if $i = v_{k+1}$ and $j = v_k$ for some k). An **x**-alterable path from the source v_0 to the sink v_r will be called **x**-*augmenting*. For instance, if **x** is the fourth flow in our illustrative example then the path show in Figure 22.4, San Francisco–Houston–Atlanta–Denver–Chicago–New York is **x**-augmenting. Each **x**-augmenting path points out a simple way of augmenting the volume of **x** by uniformly increasing **x** on the forward arcs and by decreasing **x** on the reverse arcs. More precisely, let us set

$$\bar{x}_{ij} = \begin{cases} x_{ij} + d & \text{if } ij \text{ is a forward arc} \\ x_{ij} - d & \text{if } ij \text{ is a reverse arc} \\ x_{ij} & \text{if } ij \text{ is not on the path.} \end{cases} \tag{22.6}$$

It is not difficult to see that $\bar{\mathbf{x}}$ is a flow (the conservation law remains preserved since the new contributions cancel each other out at every intermediate node) and that its volume exceeds the volume of **x** by d units. Furthermore, $\bar{\mathbf{x}}$ is a *feasible* flow as long as $d \le u_{ij} - x_{ij}$ for each forward arc ij and $d \le x_{ij}$ for each reverse arc ij. Each of these constraints imposes a *positive* upper bound on d; we may simply let d be the smallest of these upper bounds. (A somewhat unusual situation occurs if all the arcs ij on the augmenting path are forward and if all of them have $u_{ij} = \infty$. In that case, the problem is unbounded: by choosing arbitrarily large values of d, we may obtain feasible flows $\bar{\mathbf{x}}$ of arbitrarily large volumes.)

Each iteration of the augmenting path mathod is initialized by some feasible flow **x** and amounts to an efficient search for an **x**-augmenting path. If such a path is found, then the volume of **x** is augmented as in (22.6) and the next iteration entered with $\bar{\mathbf{x}}$ in place of **x**. On the other hand, the search is organized in such a way that a failure to find an **x**-augmenting path reveals a cut C such that

$$x_{jk} = u_{jk} \qquad \text{for each arc } jk \quad \text{with} \quad j \in C, \quad k \notin C \tag{22.7}$$

and

$$x_{ij} = 0 \qquad \text{for each arc } ij \quad \text{with} \quad i \notin C, \quad j \in C. \tag{22.8}$$

In this case, the method may terminate at once, for **x** is a maximum flow and C is a minimum cut. Indeed, a substitution from (22.7) and (22.8) into (22.3) shows that the volume of **x** equals the capacity of C.

The search consists of successive enlargements of a set C of nodes j such that the existence of an **x**-alterable path from the source to j has been established; it starts with the source s being the only node in C and it may stop as soon as the sink t finds its way into C. New members j of C are recruited by old members i in two different ways:

(i) If there is an arc ij with $i \in C, j \notin C$ and $x_{ij} < u_{ij}$, then we may set $j \in C$.

(ii) If there is an arc ji with $j \notin C$, $i \in C$ and $x_{ji} > 0$, then we may set $j \in C$.

In either case, setting $j \in C$ is justified: the **x**-alterable path $v_0, v_1, \ldots, v_{r-1}$ with $v_0 = s$ and $v_{r-1} = i$, whose existence is guaranteed as $i \in C$, extends into the **x**-alterable path $v_0, v_1, \ldots, v_{r-1}, v_r$ with $v_r = j$. (It is convenient to set $p(j) = ij$ in (i) and $p(j) = ji$ in (ii): the resulting array $p(\cdot)$ makes it easy to reconstruct the **x**-alterable path v_0,

$v_1, \ldots, v_r$ in the reverse order, from v_r to v_0.) In particular, the search may stop as soon as the sink finds its way into C, for now an **x**-augmenting path may be reconstructed. On the other hand, the search may also stop when neither (i) nor (ii) yield any new members of C and yet the sink remains outside C. But then C is a cut satisfying (22.7) and (22.8); therefore, **x** is a maximum flow and C is a minimum cut in this case.

Now our discussion of an individual iteration is completed; it only remains to be shown that the method can always be initialized and that it always terminates. Initialization is easy: we can simply enter the first iteration with $x_{ij} = 0$ for all the arcs ij. On the other hand, it is easy to show that the method always terminates under certain assumptions that are only mildly restrictive. More precisely, suppose that

$$\text{each finite } u_{ij} \text{ is a positive integer} \tag{22.9}$$

and that

$$\text{at least one cut has a finite capacity.} \tag{22.10}$$

If the augmenting path method is initialized by an integer-valued flow, then it will produce an integer-valued flow in every iteration: assumption (22.9) guarantees that the quantity d featured in (22.6) will always be a positive integer. In particular, the volume of the flow will increase by at least one unit with each new iteration. Hence the volume of the flow initializing the kth iteration will be at least $k - 1$. On the other hand, the volume of every feasible flow is at most the finite capacity M of the cut featured in the assumption (22.10). We conclude that the kth iteration can take place only if $k - 1 \leq M$. To put it differently, the method will terminate after no more than $M + 1$ iterations.

Assumptions (22.9) and (22.10) can be easily satisfied in nearly all practical applications (in particular, upper bounds u_{ij} recorded in floating-point may be converted into integers by a change of scale). However, if either of them is violated, then the method may fail to terminate. For example, consider the data shown in Figure 22.6.

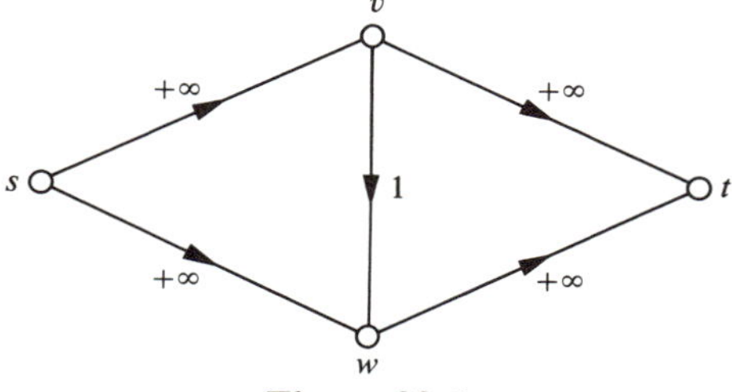

Figure 22.6

If the flow, initially $x_{ij} = 0$ for all the arcs ij, is augmented along the path $svwt$ in each odd iteration and along the path $swvt$ in each even iteration, then the method never terminates. Similarly, the augmenting path method may fail to terminate if all the upper bounds u_{ij} are finite but some of them are irrational. (The relevant example, constructed by L. R. Ford, Jr. and D. R. Fulkerson (1962), is more complicated; we shall present its simplified version in problem 22.2.) Nevertheless, these examples present no cause for alarm: we are about to describe a natural implementation of the augmenting path method, whose fast termination is guaranteed even if the upper bounds u_{ij} are infinite and/or irrational.

Implementations

Our description of the augmenting path method does not specify a way of searching for the arcs ij such that $i \in C$, $j \notin C$, $x_{ij} < u_{ij}$, and the arcs ji such that $j \notin C$, $i \in C$, $x_{ji} > 0$. Ford and Fulkerson did specify a way of doing that. In their terminology, nodes in C are called *labeled* and nodes outside C are called *unlabeled*; the labeled nodes are divided further into *scanned* and *unscanned*. Initially, the source s is labeled but unscanned and all the remaining nodes are unlabeled. Scanning a labeled node i means *examining all the arcs* ij *and, whenever such an arc satisfies* $x_{ij} < u_{ij}$, $j \notin C$, *setting* $j \in C$, $p(j) = ij$ and *examining all the arcs* ji *and, whenever such an arc satisfies* $x_{ji} > 0, j \notin C$, *setting* $j \in C$, $p(j) = ji$. The search may be described as in Box 22.1.

BOX 22.1 Search for an Augmenting Path

Step 0. Mark s as labeled unscanned; mark the remaining nodes as unlabeled.

Step 1. If all the labeled nodes are scanned then stop [the set C of labeled nodes satisfies (22.7) and (22.8)]; otherwise, choose a labeled unscanned node i.

Step 2 Scan i. If t has become labeled then stop (an $\mathbf{x}$-augmenting path has been found); otherwise return to step 1.

The amount of time required for scanning a node depends heavily on the representation of the network. No generality is lost in assuming that the network, whose nodes are named $1, 2, \ldots, n$ and whose arcs are named $1, 2, \ldots, m$, is presented by two arrays $T(\cdot)$ and $H(\cdot)$ of length m such that $T(e)$ is the tail and $H(e)$ is the head of an arc e. For instance, our travelers' example with

1 = San Francisco
2 = Denver
3 = Houston
4 = Chicago
5 = Atlanta
6 = New York City

may be presented by the following arrays:

e	1	2	3	4	5	6	7
$T(e)$	1	1	2	2	3	4	5
$H(e)$	2	3	4	5	5	6	6

If the network is represented in this way, then scanning a node may involve a certain amount of wasted work: every single one of the m arcs e has to be examined, only to be discarded at once if neither $T(e)$ nor $H(e)$ is the node being scanned. To avoid this waste, we need only weave threads through the list of arcs by certain arrays $FT(\cdot)$, $FH(\cdot)$ of length n and $NT(\cdot)$, $NH(\cdot)$ of length m. Each $FT(i)$ is either the smallest e such that $T(e) = i$, or zero if there is no such e. Each $NT(e)$ is either the smallest e' such that $e' > e$ and $T(e') = T(e)$, or zero if there is no such e'. The values of $FH(\cdot)$ and $NH(\cdot)$ are defined similarly, with $H(\cdot)$ in place of $T(\cdot)$. In our travelers' example, these four arrays are as follows:

i	1	2	3	4	5	6
$FT(i)$	1	3	5	6	7	0
$FH(i)$	0	1	2	3	4	6

e	1	2	3	4	5	6	7
$NT(e)$	2	0	4	0	0	0	0
$NH(e)$	0	0	0	5	0	7	0

It is an easy exercise (problem 22.7) to design an algorithm that, given the arrays $T(\cdot)$ and $H(\cdot)$, produces the arrays $FT(\cdot)$, $FH(\cdot)$, $NT(\cdot)$, $NH(\cdot)$ in a number of steps proportional to m. As soon as these arrays are available, scanning i takes time proportional only to the number of arcs e whose head or tail is i; these are the only arcs examined while i is being scanned. (To list all the arcs e with $T(e) = i$, it suffices to begin with $e = FT(i)$ and then to keep replacing e by $NT(e)$, until $e = 0$ indicates that the end of the list has been reached. All the arcs e with $H(e) = i$ may be listed in an analogous way.) Thus, during the entire search for an $\mathbf{x}$-augmenting path, each of the m arcs ij is examined at most twice (while i is being scanned and while j is being scanned). It follows that the total amount of time spent on all the executions of step 2 is at most proportional to m. On the other hand, each execution of step 1 takes only a constant amount of time: we need only maintain a pool of labeled unscanned nodes to which a node is added as soon as it gets labeled and from which a node is deleted as soon as it gets scanned. We conclude that each iteration of the augmenting path method can be carried out in a number of steps that is proportional only to m. (It will be convenient to adopt the so-called "big oh" notation, commonly used in the analysis of algorithms: a quantity x is denoted by $O(y)$ if it is at most proportional to y. In particular, each iteration of the augmenting path method takes only $O(m)$ steps.)

The pool of labeled unscanned nodes can be managed in many different ways. A particularly interesting case arises when the nodes are removed from the pool in the same order in which they have arrived. This version of the augmenting path method is known under the name *first-labeled, first-scanned.* In a pioneering paper, J. Edmonds and R. M. Karp (1972) proved that the first-labeled, first-scanned version of the augmenting path method finds a maximum flow after no more than $mn/2$ iterations regardless of the upper bounds u_{ij} (which may be infinite or irrational). Since each iteration takes only $O(m)$ steps, it follows that this method finds a maximum flow in only $O(m^2n)$ steps. We are not going to prove the Edmonds–Karp theorem, for there are even better things to come: improvements leading up to an algorithm that finds a maximum flow in only $O(n^3)$ steps. (On networks that are sparse, in the sense that m is much smaller than n^2, this bound can be improved even further: algorithms designed by Z. Galil (1978) and D. D. K. Sleator (1980) find a maximum flow in only $O(n^{5/3}m^{2/3})$ and $O(nm \log n)$ steps, respectively. However, these algorithms are too complicated to be discussed here.)

The Core and Blocking Flows

Edmonds and Karp first announced their results at a conference in 1969; independently of them, E. A. Dinic (1970) designed an even faster algorithm for finding a maximum flow. (Actually, these two developments are related: each iteration of Dinic's algorithm may be seen as several iterations of the first-labeled, first-scanned method lumped together in a way that avoids unnecessary duplication of work.) Even though Dinic's algorithm, too, has been surpassed by faster methods, its overall scheme is still the basis for all these subsequent improvements. In order to outline this scheme, we shall need notions that may seem artificial until the moment Theorem 22.3 provides the hindsight to demonstrate their worth.

When an iteration of the first-labeled, first-scanned method has terminated, the array $p(\cdot)$ points out an **x**-alterable path from the source s to each labeled node j. The number of arcs in this path will be denoted by $d(j)$. To find the values of $d(j)$, it suffices to set $d(s) = 0$ in step 0 of the first-labeled, first-scanned method and to set $d(j) = d(i) + 1$ whenever j gets labeled in the process of scanning i. Now each iteration may be seen as divided into stages: in the qth stage, all the nodes i with $d(i) = q - 1$ are scanned and all the nodes j with $d(j) = q$ are labeled. Of course, the very last stage may stop short of completion as soon as the sink t gets labeled; nevertheless, each node i with $d(i) \leq d(t) - 2$ gets completely scanned. In particular,

$$\text{if} \quad x_{ij} < u_{ij} \quad \text{and} \quad d(i) \leq d(t) - 2, \quad \text{then} \quad d(j) \leq d(i) + 1 \tag{22.11}$$

and

$$\text{if} \quad x_{ji} > 0 \quad \text{and} \quad d(i) \le d(t) - 2, \quad \text{then} \quad d(j) \le d(i) + 1. \tag{22.12}$$

This observation will be required later on.

By the *core defined by* **x**, we mean the network consisting of all the labeled nodes and all the arcs ij such that $d(j) = d(i) + 1$ and either (i) $x_{ij} < u_{ij}$, or (ii) $x_{ji} > 0$. (Note that the arcs ij of the second kind do not necessarily appear in the original network.) In the core, each arc ij carries an upper bound u^*_{ij} defined as $u_{ij} - x_{ij}$ in the first case (i) and as x_{ji} in the second case (ii). For illustration, consider Figure 22.7 with each arc ij labeled by u_{ij} outside and x_{ij} inside the parentheses. The core resulting in this case is shown in Figure 22.8, with each node j labeled by $d(j)$ and each arc ij labeled by u^*_{ij}. The number $d(t)$ will be referred to as the *length* of the core.

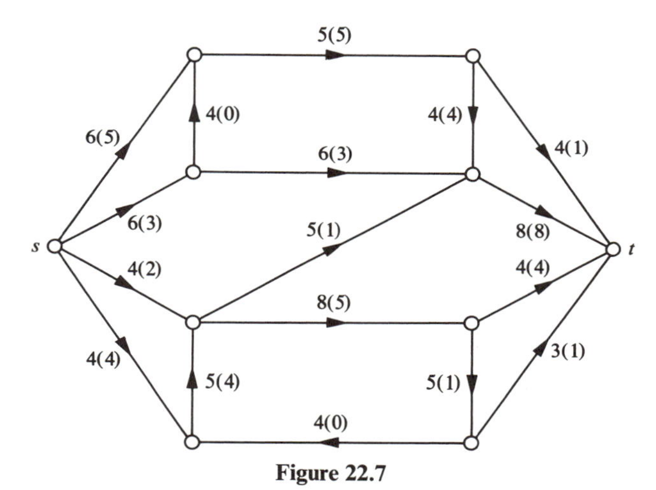

Figure 22.7

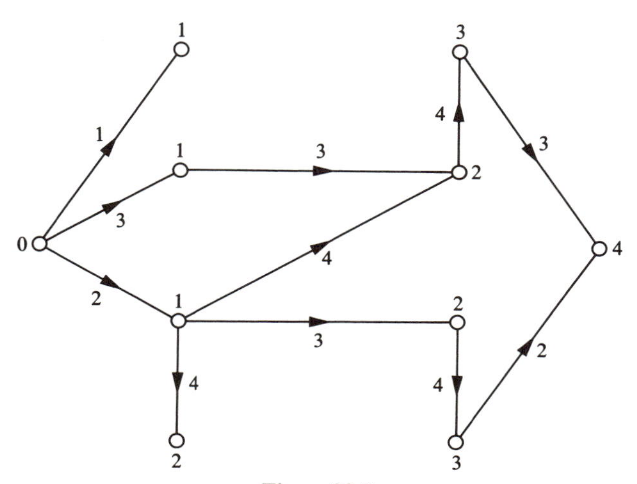

Figure 22.8

Each feasible flow $\mathbf{x}^*$ in the core points out a way of replacing the original flow $\mathbf{x}$ by a new flow $\bar{\mathbf{x}}$: it suffices to set $\bar{x}_{ij} = x_{ij} + x^*_{ij}$ for each arc ij of type (i) and $\bar{x}_{ji} = x_{ji} - x^*_{ij}$ for each arc ij of type (ii). For illustration, a feasible flow $\mathbf{x}^*$ in the core and the resulting flow $\bar{\mathbf{x}}$ in the original network are shown in Figure 22.9.

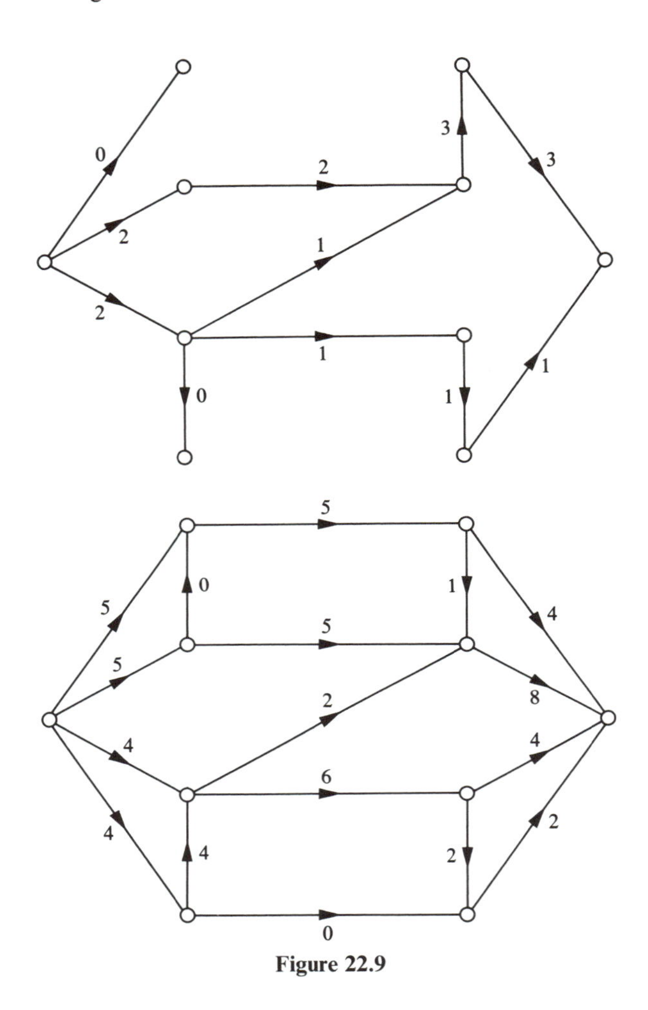

Figure 22.9

It is not difficult to verify that, in general, $\bar{\mathbf{x}}$ is indeed a feasible flow in the original network and that its volume equals the volume of $\mathbf{x}$ plus the volume of $\mathbf{x}^*$. This new flow $\bar{\mathbf{x}}$ will be referred to as $\mathbf{x}$ *augmented by* $\mathbf{x}^*$. Note that $\bar{x}_{ij} > x_{ij}$ only if $d(j) = d(i) + 1$ and that $\bar{x}_{ji} < x_{ji}$ only if $d(j) = d(i) + 1$. Again, this observation will be required in the future; it will be convenient to record it with i and j interchanged as

$$\bar{x}_{ji} > x_{ji} \qquad \text{only if} \quad d(i) = d(j) + 1 \tag{22.13}$$

and

$$\bar{x}_{ij} < x_{ij} \quad \text{only if} \quad d(i) = d(j) + 1. \tag{22.14}$$

Finally, a feasible flow $\mathbf{x}^*$ (in the core) will be called *blocking* if there is no $\mathbf{x}^*$-augmenting path whose arcs are all forward. (In particular, every maximum flow is blocking. The converse does not always hold: for instance, the flow $\mathbf{x}^*$ shown in Figure 22.9 is blocking but not maximum.) Now we are ready for a crucial fact.

THEOREM 22.3. If $\mathbf{x}$ is a feasible flow, if $\mathbf{x}^*$ is a blocking flow in the core defined by $\mathbf{x}$, and if $\bar{\mathbf{x}}$ is $\mathbf{x}$ augmented by $\mathbf{x}^*$, then the core defined by $\bar{\mathbf{x}}$ is longer than the core defined by $\mathbf{x}$.

PROOF. Let numbers $d(j)$ be defined by the old flow $\mathbf{x}$ and let r be the length of the new core defined by $\bar{\mathbf{x}}$. By definition, there is an $\bar{\mathbf{x}}$-augmenting path $w_0, w_1, \ldots, w_r$. Let us prove at once that

$$d(w_k) \le 1 + d(w_{k-1}) \quad \text{whenever} \quad d(w_{k-1}) \le d(t) - 2. \tag{22.15}$$

Writing $i = w_{k-1}$ and $j = w_k$, we observe that either $\bar{x}_{ij} < u_{ij}$ or $\bar{x}_{ji} > 0$. In the first case, $d(j) \le 1 + d(i)$ follows from (22.11) if $x_{ij} < u_{ij}$ and from (22.14) if $x_{ij} = u_{ij}$. In the second case, $d(j) \le 1 + d(i)$ follows from (22.12) if $x_{ji} > 0$ and from (22.13) if $x_{ji} = 0$. Thus (22.15) is proved.

To prove the theorem, we have to show that $r > d(t)$. This task will be accomplished indirectly, by deriving a contradiction from the assumption that

$$r \le d(t). \tag{22.16}$$

Assuming (22.16), we shall first prove that

$$d(w_k) = k \quad \text{for all} \quad k = 0, 1, \ldots, r. \tag{22.17}$$

For this purpose, note that (22.15) and (22.16) imply

$$d(w_k) \le 1 + d(w_{k-1}) \quad \text{whenever} \quad d(w_{k-1}) \le r - 2. \tag{22.18}$$

Repeated applications of (22.18) show that $d(w_k) \le k$ for all $k = 0, 1, \ldots, r - 1$; on the other hand, if we had $d(w_i) < i$ for some i such that $i < r$, then repeated applications of (22.18) with $k = i + 1, i + 2, \ldots, r$ would yield $d(w_r) < r$, contradicting (22.16). Hence $d(w_k) = k$ for all $k = 0, 1, \ldots, r - 1$. Finally, the inequality $d(w_r) \ge r$ is nothing but (22.16), whereas the reversed inequality $d(w_r) \le 1 + d(w_{r-1})$ follows from (22.15) with $k = r$. Thus (22.17) is proved.

Now we are ready for the final contradiction. Consider an arbitrary subscript k such that $1 \le k \le r$, again write $i = w_{k-1}$, $j = w_k$, and observe again that either $\bar{x}_{ij} < u_{ij}$ or $\bar{x}_{ji} > 0$. By (22.17), we have $d(j) = d(i) + 1$; hence (22.14) shows that $x_{ij} \le \bar{x}_{ij}$ in the first case and (22.13) shows that $x_{ij} \ge \bar{x}_{ji}$ in the second case. Thus each forward arc ij on the path $w_0, w_1, \ldots, w_r$ appears in the old core defined by $\mathbf{x}$, with

$$u^*_{ij} = u_{ij} - x_{ij} > \bar{x}_{ij} - x_{ij} = x^*_{ij}$$

and each reverse arc ji on this path yields an arc ij in the old core with

$$u^*_{ij} = x_{ji} > x_{ji} - \bar{x}_{ji} = x^*_{ij}.$$

But this is the desired contradiction: since all the arcs on the path $w_0, w_1, \ldots, w_r$ in the old core are forward, and since each of these arcs ij has $x^*_{ij} < u^*_{ij}$, the flow $\mathbf{x}^*$ is not blocking. ■

The scheme of Dinic's algorithm evolves from this theorem. Each iteration, entered with some feasible flow **x**, amounts to finding a blocking flow **x*** in the core defined by **x**; the next iteration is entered with **x** augmented by **x***. Unless **x** is a maximum flow, the length of the core defined by **x** is a positive integer smaller than n; by Theorem 22.3, this length increases with each iteration. Hence a maximum flow is found after fewer than n iterations.

Evidently, the efficiency of this scheme depends on the efficiency of the procedure used to find the blocking flow in each iteration. (Constructing the core is a relatively easy task: it takes only $O(m)$ steps. This fact is implicit in our definition of the core; the details are left for problem 22.8.) The procedure originally suggested by Dinic (see problem 22.14) finds the blocking flow in $O(mn)$ steps; we are going to describe a procedure, developed more recently by V. M. Malhotra, M. P. Kumar, and S. N. Maheshwari (1978), that finds the blocking flow in only $O(n^2)$ steps.

A Fast Way to Find a Blocking Flow

The core has a special structure: its nodes can be arranged in a sequence such that all the arcs point from left to right. To see this, recall that the first-labeled, first-scanned method labels the nodes in a sequence $v_1, v_2, \ldots, v_k$ such that $d(v_1) \leq d(v_2) \leq \cdots \leq d(v_k)$; these labeled nodes are precisely the nodes of the core and each arc ij of the core has $d(j) = d(i) + 1$. This special structure is crucial in the procedure we are about to describe. The sequence $v_1, v_2, \ldots, v_k$ can be represented by a pair of arrays $A(\cdot)$, $B(\cdot)$ such that $A(v_i)$ is the node v_{i+1} coming just after v_i, and $B(v_i)$ is the node v_{i-1} coming just before v_i. (We set $A(v_k) = 0$ for the sink v_k and $B(v_1) = 0$ for the source v_1.) In addition, we shall make use of arrays $T^*(\cdot)$, $H^*(\cdot)$, $FT^*(\cdot)$, $FH^*(\cdot)$, $NT^*(\cdot)$, $NH^*(\cdot)$ representing the core in the same way as the arrays $T(\cdot)$, $H(\cdot)$, $FT(\cdot)$, $FH(\cdot)$, $NT(\cdot)$, $NH(\cdot)$ represent the original network. All of these arrays can be constructed in only $O(m)$ steps (see problem 22.8).

The blocking flow **x*** is found by beginning with $x_{ij}^* = 0$ for all arcs ij and then successively increasing the various values of x_{ij}^*. It will be convenient to write $x_e^* = x_{ij}^*$ and $u_e^* = u_{ij}^*$ with $i = T^*(e), j = H^*(e)$. We shall store and update a certain array $E(\cdot)$ such that each $E(e)$ represents the "extra amount of the commodity that can be sent through e." In particular, an arc e with $E(e) = 0$ cannot appear on any **x***-augmenting path whose arcs are all forward. Initially, we shall have $x_e^* = 0$ and $E(e) = u_e^*$ for all arcs e. At all times, the sum of all the numbers $E(e)$ with $H^*(e) = i$ will be referred to as the *left capacity* $LC(i)$ of a node i, and the sum of all the numbers $E(e)$ with $T^*(e) = i$ will be referred to as the *right capacity* $RC(i)$ of a node i. The *capacity* $C(i)$ of the node i will be defined by

$$C(i) = \begin{cases} \min(LC(i), RC(i)) & \text{if } i \text{ is an intermediate node} \\ RC(i) & \text{if } i \text{ is the source} \\ LC(i) & \text{if } i \text{ is the sink.} \end{cases}$$

(Whenever $LC(i)$ or $RC(i)$ changes, we shall tacitly assume that $C(i)$ is updated at once.)

The capacity of a node represents the extra amount of the commodity that can be sent through this node. In particular, nodes of capacity zero are useless (they cannot appear on any **x***-augmenting path whose arcs are all forward) and might just as well be removed. The removal of an intermediate node w can be implemented by the following subroutine, referred to as REMOVE (w):

1. Set $e = FT^*(w)$ and, as long as $e \neq 0$, keep repeating the following operations: reduce $LC(H^*(e))$ by $E(e)$, replace $E(e)$ by zero, replace e by $NT^*(e)$.
2. Set $e = FH^*(e)$ and, as long as $e \neq 0$, keep repeating the following operations: reduce $RC(T^*(e))$ by $E(e)$, replace $E(e)$ by zero, replace e by $NH^*(e)$.
3. Replace $A(B(w))$ by $A(w)$ and replace $B(A(w))$ by $B(w)$.

The removal of a node w may result in reducing the capacities of other nodes. In particular, it may make the capacities of certain nodes drop to zero. In turn, these nodes may now be removed, possibly causing the removal of yet other nodes, and so on. When this chain reaction stops, all the nodes have positive capacities.

Now the idea is to send some extra units all the way from the source to the sink, making the capacity of some node w drop to zero, and subsequently removing w. The natural candidate for w is the node of the smallest capacity; let us write $d = C(w)$. To ensure that the d extra units do indeed pass through w, we push them first from w to the sink and then from the source to w. (The second stage is carried out in reverse, beginning at w and working back towards the source.) Informally, we imagine that the d extra units have suddenly materialized at w. These units are immediately dispatched to the right via the various arcs wj, thus creating a temporary surplus,

$$S(j) = \sum_i x_{ij}^* - \sum_i x_{ji}^*$$

at their heads j. Then we sweep from w to the sink and, at each node j that we encounter, restore the flow balance by dispatching the extra units further to the right. [This can be done since $S(j) \leq d = C(w) \leq RC(j)$.] Eventually, we arrive at the sink, which has by now absorbed the d extra units. The second stage is a mirror image of the first. We begin by supplying the d extra units to w via the various arcs iw, thus creating a temporary shortage, $-S(i)$, at their tails i. Then we sweep from w toward the source and, at each node i that we encounter, restore the flow balance by supplying the extra units from nodes further to the left. [This can be done since $-S(i) \leq d = C(w) \leq LC(i)$.] Eventually, we arrive at the source, which is by now supplying the d extra units.

A formal presentation begins with a detailed account of what is involved in increasing the value of x_e^* by c units. The following sequence of instructions will be referred to as INCREASE (e, c):

1. Increase x_e^* by c units.
2. Reduce $E(e)$ by c units.
3. Reduce $\mathrm{RC}(T^*(e))$ by c units.
4. Reduce $\mathrm{LC}(H^*(e))$ by c units.
5. Increase $S(H^*(e))$ by c units.
6. Reduce $S(T^*(e))$ by c units.

The intuitive notion of "dispatching c units from j to the right" may be implemented by the following subroutine, referred to as PUSHRIGHT (j, c):

1. If $c = 0$ then stop; otherwise, set $e = FT^*(j)$.
2. As long as $c \geq E(e)$, keep repeating the following operations: reduce c by $E(e)$, call INCREASE $(e, E(e))$, replace $FT^*(j)$ by $NT^*(e)$.
3. Call INCREASE (e, c) and stop.

Replacing j, FT^* and NT^* in PUSHRIGHT (j, c) by i, FH^*, and NH^*, respectively, we obtain a subroutine referred to as PUSHLEFT $(i. c)$. Now the entire procedure for finding a blocking flow in the core may be described as in Box 22.2.

Proving that the flow $\mathbf{x}^*$ delivered by this procedure is indeed blocking is an easy exercise, left for problem 22.12. To prove that the running time of the procedure is $O(n^2)$, observe that the loop formed by steps 1–4 is executed fewer than n times: each of its executions ends by a removal of a node. Since the time required to execute step 1 once is $O(n)$, and since the total time spent on executions of step 4 is $O(m)$, we need only prove that the total time spent on executions of step 2 is $O(n^2)$. In fact, since steps 2.1 and 2.2 are mirror images of each other, we need only prove that

Box 22.2 Finding a Blocking Flow in the Core

Step 1. [Locate a bottleneck.] Set $k = w =$ source and, as long as $A(k) \neq 0$, keep repeating the following operations: replace k by $A(k)$ and, in case $C(k) < C(w)$, replace w by k.

Step 2. [Send an extra d units from the source to the sink through the bottleneck.] If $C(w) > 0$ then set $d = C(w)$ and execute the following steps:

2.1. [From the bottleneck to the sink.] Set $j = w$, $c = d$ and, as long as $A(j) \neq 0$, keep repeating the following operations: call PUSHRIGHT (j, c), replace j by $A(j)$, replace c by $S(j)$.

2.2. [From the source to the bottleneck.] Set $i = w$, $c = d$ and, as long as $B(i) \neq 0$, keep repeating the following operations: call PUSHLEFT (i, c), replace i by $B(i)$, replace c by $-S(i)$.

Step 3. [Blocking flow obtained?] If w is the source or the sink then stop.

Step 4. [Remove the bottleneck.] Call REMOVE (w) and return to Step 1.

the total time spent on executions of step 2.1 is $O(n^2)$. For this purpose, note that the loop of step 2.1 is executed fewer than n^2 times (at most once for each combination of w and j) and that, in each of these executions, only the loop (line 2) of PUSHRIGHT (j, c) may require more than a constant amount of time. But, *during the entire run of the procedure*, line 2 of PUSHRIGHT is executed at most once for each arc e: as soon as $FT^*(j)$ is replaced by $NT^*(e)$, the arc e is guaranteed to never come up again in an execution of PUSHRIGHT. □

PROBLEMS

22.1 Find a maximum flow and a minimum cut in Figure 22.10 by (i) the network simplex method, (ii) the first-labeled, first-scanned method, (iii) Dinic's algorithm with the Malhotra–Kumar–Maheshwari procedure for finding a blocking flow in the core.

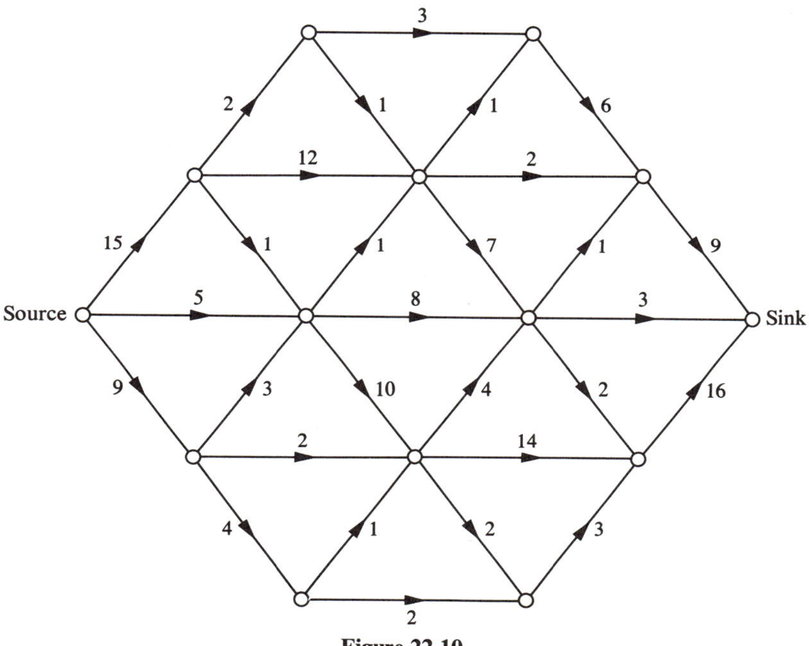

Figure 22.10

22.2 Write $c = (-1 + \sqrt{5})/2$ and consider the network in Figure 22.11 with upper bounds of

c on v_1v_2, $\quad$ 1 on v_3v_6, $\quad$ c on v_4v_6

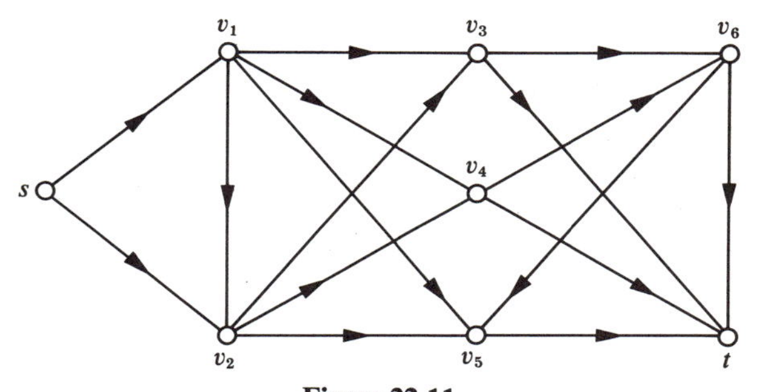

Figure 22.11

Table 22.2 An Endless Sequence of Iterations

		Resulting flow through			
Iteration	Augmenting path	v_1v_2	v_3v_6	v_4v_6	v_6v_5
$6k+1$	$sv_1v_2v_3v_6t$	c	$1-c^{3k+2}$	$c-c^{3k+1}$	0
$6k+2$	$sv_2v_1v_3v_6v_5t$	$c-c^{3k+2}$	1	$c-c^{3k+1}$	c^{3k+2}
$6k+3$	$sv_1v_2v_4v_6t$	c	1	$c-c^{3k+3}$	c^{3k+2}
$6k+4$	$sv_2v_1v_4v_6v_3t$	$c-c^{3k+3}$	$1-c^{3k+3}$	c	c^{3k+2}
$6k+5$	$sv_1v_2v_5v_6t$	c	$1-c^{3k+3}$	c	c^{3k+4}
$6k+6$	$sv_2v_1v_5v_6v_4t$	$c-c^{3k+4}$	$1-c^{3k+3}$	$c-c^{3k+4}$	0

and large upper bounds on the remaining arcs. Verify that the augmenting path method may go through an endless sequence of iterations characterized in Table 22.2.

22.3 Derive Theorem 22.2 from Theorem 21.2.

22.4 Derive Theorem 21.1 from Theorem 22.1.

22.5 Derive Theorem 20.5 from Theorem 22.1.

22.6 Suppose our definition of a network is relaxed so as to allow arcs to enter the source and/or leave the sink; now the volume of a flow $\mathbf{x}$ is properly defined as

$$\sum_j x_{sj} - \sum_i x_{is}.$$

In this generalized context, prove that for every feasible flow of volume v there is a feasible flow $\mathbf{x}$ of volume at least v such that $x_{is} = 0$ for all arcs is entering the source and $x_{tj} = 0$ for all arcs tj leaving the sink. What is the significance of this observation?

22.7 Design an algorithm that, given a network represented by the arrays $T(\cdot)$ and $H(\cdot)$, constructs the arrays $FT(\cdot)$, $FH(\cdot)$, $NT(\cdot)$, and $NH(\cdot)$ in only $O(m)$ steps.

22.8 Design an algorithm that, given a feasible flow in a network, constructs the corresponding core in only $O(m)$ steps. The output must include the arrays $T^*(\cdot)$, $H^*(\cdot)$, $FT^*(\cdot)$, $FH^*(\cdot)$, $NT^*(\cdot)$, $NH^*(\cdot)$, and $A(\cdot)$, $B(\cdot)$ as well as the upper bounds u_{ij}^*.

22.9 A free-lance photographer can take on a variety of assignments during the next month, the kth assignment yielding a profit of p_k dollars and requiring a set S_k of pieces of equipment. Since she owns no equipment, she must rent at prescribed monthly rates whatever she needs. The problem of maximizing her net profit is complicated by the fact that distinct sets S_i and S_j may overlap. Turn this problem into the open pit problem.

22.10 If each vertex of a graph G carries a positive weight, then the weight of a set S of vertices is defined as the sum of weights of all the vertices in S and the *optimum cover problem* requires finding a cover of the smallest weight. Show that this problem can be turned into the minimum cut problem whenever G is bipartite.

22.11 Prove that a flow $\mathbf{x}^*$ in a network with upper bounds u_{ij}^* is blocking if and only if there is a cut C such that $x_{ij}^* = u_{ij}^*$ for all arcs ij such that $i \in C$ and $j \notin C$.

22.12 Modify the Malhotra–Kumar–Maheshwari procedure so as to make it deliver a cut C with the property specified in problem 22.11.

22.13 If some upper bounds u^*_{ij} in the core are infinite, then some or all of the capacities $C(w)$ defined in the Malhotra–Kumar–Maheshwari procedure may be infinite. What modifications are required in this case?

22.14 In the notation used toward the end of this chapter, Dinic's procedure for finding a blocking flow in the core may be described as follows.

Step 0. [Initialize.] Set $x^*_e = 0$ for each arc e. Set $i = s$, $k = 0$, and $e = FT^*(s)$.

Step 1. [Extend the path as far as you can.] As long as $e \neq 0$, keep repeating the following operations: replace k by $k + 1$, set $e_k = e$ and $d_k = u^*_k - x^*_k$, replace i by $H^*(e)$, replace e by $FT^*(i)$.

Step 2. [Still at the source?] If $i = s$ then stop: the current $\mathbf{x}^*$ is a blocking flow.

Step 3. [A blind alley found?] If $i \neq t$ then replace i by $T^*(e_k)$, replace e by $NT^*(e_k)$, replace $FT^*(i)$ by e, replace k by $k - 1$, and return to step 1.

Step 4. [An augmenting path found.] Find the minimum d of $d_1, d_2, \ldots, d_k$. If $d = +\infty$ then stop: there are flows of arbitrarily large volumes. Otherwise do the following for $j = 1, 2, \ldots, k$: set $e = e_j$, increase x^*_e by d and, in case the new value of x^*_e equals u^*_e, replace $FT^*(T^*(e))$ by $NT^*(e)$. Then return to Step 1.

Prove that the flow $\mathbf{x}^*$ delivered by this procedure is indeed blocking and that the procedure terminates in $O(mn)$ steps.

23

The Primal-Dual Method

In this chapter, we shall present an alternative algorithm for solving the upper-bounded transshipment problems

$$\text{minimize} \quad \mathbf{cx} \qquad \text{subject to} \quad \mathbf{Ax} = \mathbf{b}, \quad \mathbf{0} \le \mathbf{x} \le \mathbf{u}. \tag{23.1}$$

This algorithm, developed by L. R. Ford, Jr. and D. R. Fulkerson (1958a) from earlier contributions of J. Egerváry (1931) and H. W. Kuhn (1955), is known as the *primal-dual method.* The network simplex method and the primal-dual method are similar in the sense that both update a pair of vectors $\mathbf{x}$, $\mathbf{y}$ in each iteration, until a pair satisfying

$$\mathbf{Ax} = \mathbf{b} \tag{23.2}$$

$$\mathbf{0} \le \mathbf{x} \le \mathbf{u} \tag{23.3}$$

and

$$\begin{aligned} x_{ij} &= 0 \qquad \text{whenever} \quad y_i + c_{ij} > y_j \\ x_{ij} &= u_{ij} \qquad \text{whenever} \quad y_i + c_{ij} < y_j \end{aligned} \tag{23.4}$$

is found. The two methods are different in the sense that the network simplex method maintains (23.2) and (23.3), terminating when (23.4) becomes satisfied, whereas the primal-dual method maintains (23.3) and (23.4), terminating when (23.2) becomes satisfied.

In an informal experiment conducted by Dantzig, Ford, and Fulkerson in the late 1950s, the primal-dual method was found to be about twice as fast as the network simplex method. However, this experiment involved only very small examples solved by hand; in experiments conducted by F. Glover et al. (1974), involving computer solutions of large problems, the primal-dual method turned out to be slower than the network simplex method. It is conceivable that the two methods are on a par and that the balance can be swung in favor of one or the other by ingenuity in computer implementations.

From a theoretical point of view, the primal-dual method is more important than the network simplex method. J. Edmonds and R. M. Karp (1972) have shown that a theoretically satisfactory algorithm (in the sense of Chapter 4) for solving (23.1) consists of successive applications of the primal-dual method to problems

$$\text{minimize} \quad \mathbf{cx} \quad \text{subject to} \quad \mathbf{Ax} = \tilde{\mathbf{b}}, \quad \mathbf{0} \leq \mathbf{x} \leq \tilde{\mathbf{u}}$$

approximating (23.1) with greater and greater precision. No comparable scheme using the network simplex method is known.

SHORTEST PATHS

As we shall see in the next section, each iteration of the primal-dual method requires a solution of a problem that generalizes the commonplace problem of finding the shortest route between two cities on a road map. The generalization may be described in terms of a network whose arcs ij carry nonnegative *lengths* l_{ij}. A *directed path* from a node u to a node v is a sequence of nodes $v_0, v_1, \ldots, v_k$ such that $v_0 = u$, $v_k = v$ and such that each $v_{i-1}v_i$ $(i = 1, 2, \ldots, k)$ is an arc (that is, a directed path is simply a path whose arcs are all forward); the *length* of this path is simply the sum of the lengths of its arcs. It will be convenient to admit $k = 0$ in this definition, so that every single node constitutes a directed path (from itself to itself) of length zero. E. Dijkstra (1959) proposed an efficient algorithm for finding shortest directed paths from a fixed node to all the remaining nodes in the network. A slightly modified version of this algorithm, presented in Box 23.1, finds shortest directed paths from a fixed set W of nodes to all the remaining nodes (a directed path from W being a directed path from any u such that $u \in W$).

The variables in this algorithm are labels d_v assigned to nodes v, a set R of nodes and an array $p(\cdot)$. Each label d_v with $v \in R$ is permanent; it specifies the shortest length of a directed path $v_0, v_1, \ldots, v_k$ with $v_0 \in W$ and $v_k = v$. (Such a path exists whenever $v \in R$.) Each label d_v with $v \notin R$ is temporary; it specifies the shortest length of a directed path $v_0, v_1, \ldots, v_k$ with $v_0 \in W$, $v_k = v$, and $v_j \in R$ whenever $j < k$. (If there is no such path, then $d_v = +\infty$.) Whenever d_v is finite, the array $p(\cdot)$ represents the relevant path $v_0, v_1, \ldots, v_k$ by $p(v_j) = v_{j-1}v_j$ for all $j = 1, 2, \ldots, k$. It is an easy exercise (left for problem 23.2) to show that these properties of the variables remain preserved throughout the execution of the algorithm.

BOX 23.1 The Shortest Paths Algorithm

Step 0. [Initialize.] Set $R = \emptyset$, $d_v = 0$ for all $v \in W$, and $d_v = +\infty$ for all $v \notin W$.

Step 1. [Make a temporary label permanent.] If $d_v = +\infty$ whenever $v \notin R$, then stop; otherwise, find a node i such that $i \notin R$ and $d_i \leq d_v$ for all $v \notin R$. Add i to R.

Step 2. [Update temporary labels.] For every arc ij such that $j \notin R$ and $d_i + l_{ij} < d_j$, replace d_j by $d_i + l_{ij}$ and set $p(j) = ij$. Then return to step 1.

Another easy exercise (left for problem 23.3) is verifying that, upon termination of the algorithm, the output has the following properties:

$$W \subseteq R \tag{23.5}$$

$$\text{there is no arc } ij \text{ such that} \quad i \in R \quad \text{and} \quad j \notin R \tag{23.6}$$

$$d_i + l_{ij} \geq d_j \qquad \text{for all arcs} \quad ij \tag{23.7}$$

$$d_i + l_{ij} = d_j \qquad \text{whenever} \quad ij = p(j). \tag{23.8}$$

To estimate the running time of the algorithm, we shall use the "big oh" notation introduced in Chapter 22 and assume that the network is represented by the arrays $T(\cdot)$, $H(\cdot)$, $FT(\cdot)$, and $NT(\cdot)$ of Chapter 22 (the arrays $FH(\cdot)$ and $NH(\cdot)$ are useless in the present context). Now the total amount of time spent on executions of step 2 is only $O(m)$; on the other hand, step 1 is executed at most n times and each of its executions takes only $O(n)$ steps. Thus the entire algorithm terminates in only $O(n^2)$ steps. (As shown by D. B. Johnson (1977), this bound can be improved to $O(m \log_k n)$ with $k = \max(2, m/n)$. The details are suggested in problem 23.17.)

THE PRIMAL-DUAL METHOD

Each iteration of the primal-dual method begins with a pair of vectors $\mathbf{x}$ and $\mathbf{y}$ such that

$$0 \leq x_{ij} \leq u_{ij} \qquad \text{for all arcs} \quad ij \tag{23.9}$$

and such that

$$\begin{aligned} x_{ij} &= 0 \qquad \text{whenever} \quad y_i + c_{ij} > y_j \\ x_{ij} &= u_{ij} \qquad \text{whenever} \quad y_i + c_{ij} < y_j. \end{aligned} \tag{23.10}$$

(Finding an initial pair **x**, **y** to begin with is easy as long as all the arc costs c_{ij} are nonnegative, for then (23.9) and (23.10) can be satisfied by setting $x_{ij} = 0$ for all arcs ij and $y_k = 0$ for all nodes k. The case of arbitrary arc costs is more intricate and will be discussed later.) If $\mathbf{Ax} = \mathbf{b}$, then the method terminates at once (as proved in Chapter 21, **x** is an optimal solution); otherwise the pair **x**, **y** is replaced by a pair $\bar{\mathbf{x}}$, $\bar{\mathbf{y}}$ that (i) satisfies (23.9) and (23.10) in place of **x**, **y**, (ii) makes $\mathbf{A}\bar{\mathbf{x}}$ "closer to **b**" than **Ax**. To elaborate on the second point, let us call a node j *wet*, *balanced*, or *dry* if the net flow

$$\sum_i x_{ij} - \sum_k x_{jk}$$

of the commodity into j is greater than, equal to, or less than the demand b_j, respectively. The change of **x** into $\bar{\mathbf{x}}$, preceded by the change of **y** into $\bar{\mathbf{y}}$, is carried out in such a way that the net flow into some wet node decreases (without this node becoming dry), the net flow into some dry node increases (without this node becoming wet), and the net flows into all other nodes remain unchanged. Our presentation of the details begins with a brief sketch followed by an illustrative example; an analysis will be provided afterwards.

To change **y** into $\bar{\mathbf{y}}$, we apply the shortest path algorithm to an *auxiliary network*, having

(i) an arc $e(+) = ij$ of length $y_i - y_j + c_{ij}$ for each original arc $e = ij$ with $x_e < u_e$

(ii) an arc $e(-) = ji$ of length $y_j - y_i - c_{ij}$ for each original arc $e = ij$ with $x_e > 0$

and to the set W consisting of all the wet nodes. If all the labels d_j returned by the shortest paths algorithm are finite, we set

$$\bar{y}_j = y_j + d_j \qquad \text{for every node } j.$$

To change **x** into $\bar{\mathbf{x}}$, we choose a dry node v and reconstruct the path $v_0, v_1, \ldots, v_k$ with $v_0 \in W$, $v_k = v$ and $p(v_j) = v_{j-1}v_j$ for all $j = 1, 2, \ldots, k$; then we set

$$\bar{x}_e = x_e + t \qquad \text{when some } p(v_j) \text{ is } e(+)$$

$$\bar{x}_e = x_e - t \qquad \text{when some } p(v_j) \text{ is } e(-)$$

with t as large as possible subject to the constraints that (i) the wet node v_0 does not become dry, (ii) the dry node v_k does not become wet, (iii) $0 \leq \bar{x}_{ij} \leq u_{ij}$ for each arc ij on the path.

For illustration, we return to the example from Chapter 21, whose data (originally presented in Figure 21.1) are reproduced in Figure 23.1a. The primal-dual method, initialized by $\mathbf{x} = \mathbf{0}$ and $\mathbf{y} = \mathbf{0}$, takes eight iterations to find the optimal solution. Each of these iterations is recorded in figure 23.1b–i by (i) the auxiliary network with arcs ij labeled by l_{ij} (occasionally, there is a pair of arcs ij, ji with $l_{ij} = l_{ji} = 0$) and nodes v labeled d_v, and (ii) the updated pair **x**, **y**.

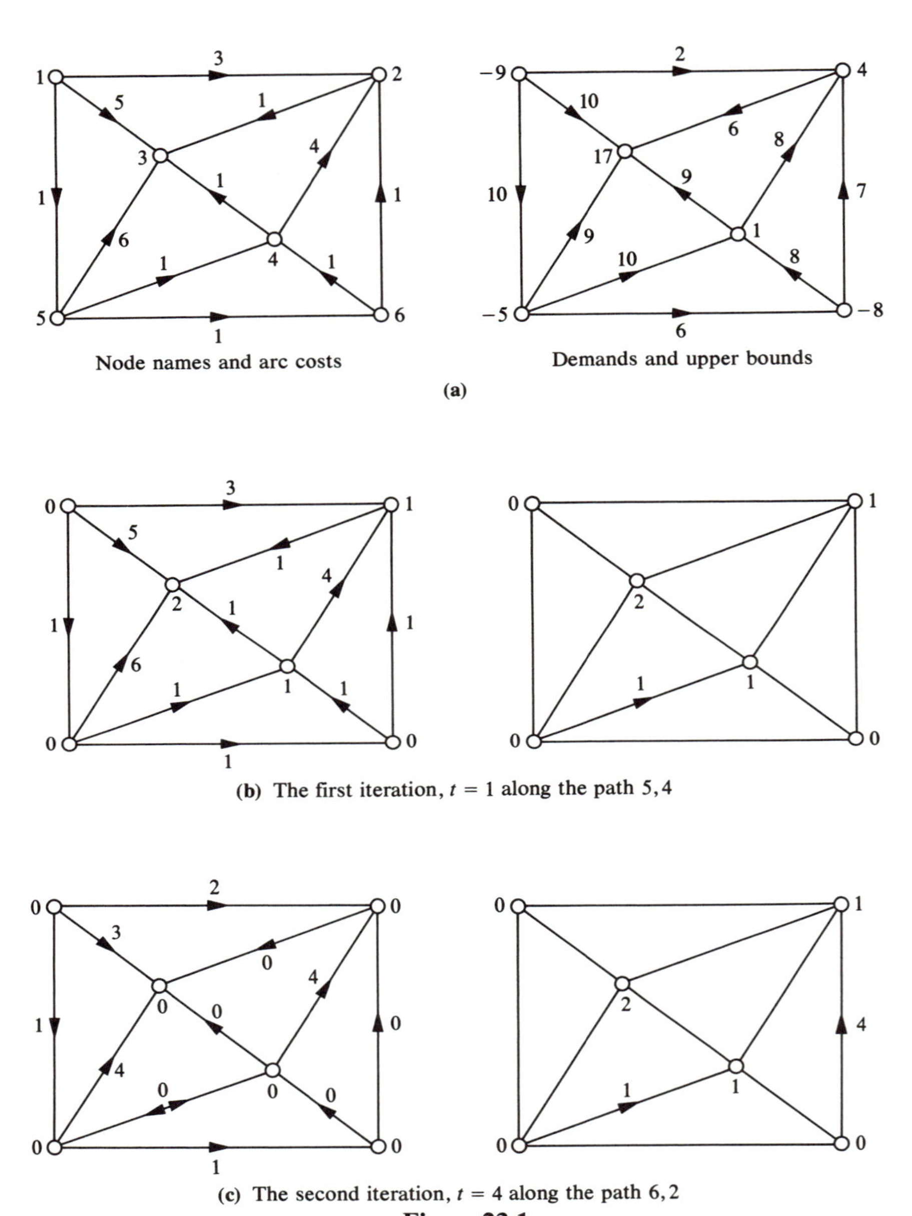
Node names and arc costs
Demands and upper bounds
(a)
(b) The first iteration, $t = 1$ along the path 5, 4
(c) The second iteration, $t = 4$ along the path 6, 2

Figure 23.1

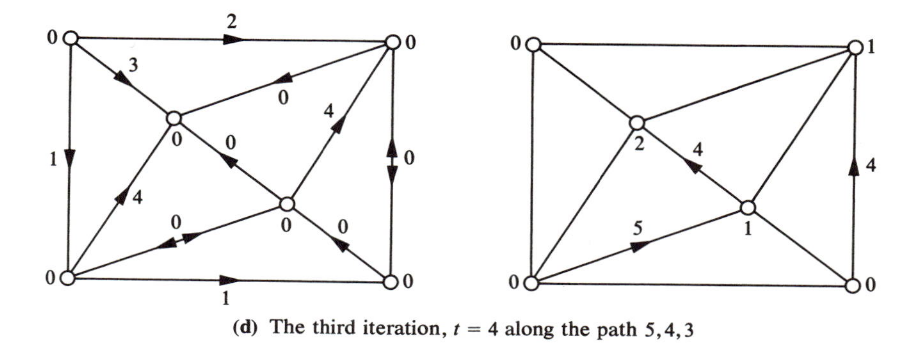

(d) The third iteration, $t = 4$ along the path 5, 4, 3

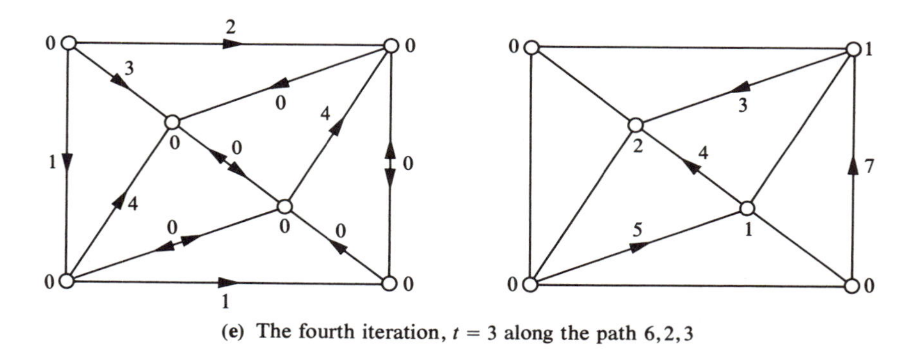

(e) The fourth iteration, $t = 3$ along the path 6, 2, 3

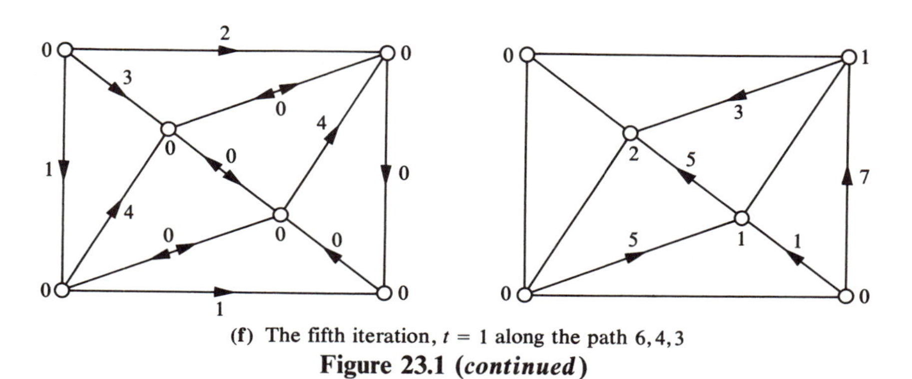

(f) The fifth iteration, $t = 1$ along the path 6, 4, 3

Figure 23.1 *(continued)*

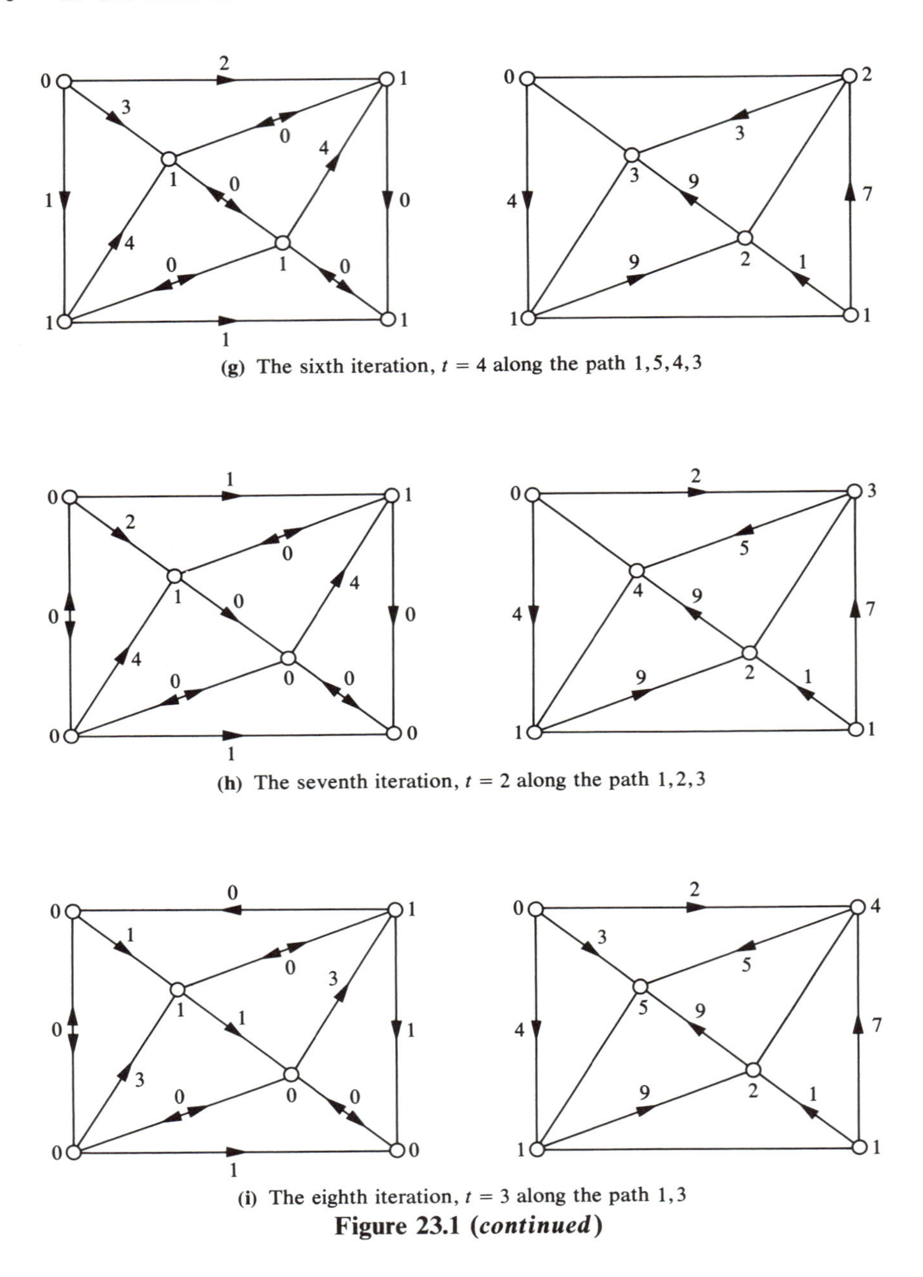

(g) The sixth iteration, $t = 4$ along the path 1, 5, 4, 3

(h) The seventh iteration, $t = 2$ along the path 1, 2, 3

(i) The eighth iteration, $t = 3$ along the path 1, 3

Figure 23.1 ***(continued)***

To prove that the method is sound, we have to clarify a number of points.

(i) Is there always a wet node to start off the shortest-paths algorithm and a dry node to start off the reconstruction of the path $v_0, v_1, \ldots, v_k$?

(ii) The shortest paths algorithm works under the assumption that the arc lengths are nonnegative; is this assumption satisfied in the auxiliary network?

(iii) What is to be done if $d_j = +\infty$ for one or more nodes j?

(iv) Does the updated pair $\bar{\mathbf{x}}$, $\bar{\mathbf{y}}$ satisfy (23.9) and (23.10) in place of $\mathbf{x}$, $\mathbf{y}$?

(v) Does the method always terminate?

(vi) How can we initialize if the arc costs are arbitrary?

The first two questions are easily answered in the affirmative; we leave the details for Problems 23.4 and 23.5, respectively.

To answer question (iii), we write $j \in S$ if $d_j = +\infty$; the remaining nodes form the set R returned by the shortest-paths algorithm. By (23.5), every node in S is dry or balanced; that is,

$$\sum_i x_{ij} - \sum_k x_{jk} \leq b_j$$

whenever $j \in S$. Summing up all these inequalities, one for each $j \in S$, we find that

$$\sum_{\substack{i \in R \\ j \in S}} x_{ij} - \sum_{\substack{j \in S \\ k \in R}} x_{jk} \leq \sum_{j \in S} b_j.$$

By (23.6), the auxiliary network has no arcs ij with $i \in R$ and $j \in S$. In terms of the original network, this means $x_{ij} = u_{ij}$ whenever $i \in R$, $j \in S$; and $x_{jk} = 0$ whenever $j \in S$, $k \in R$. Substituting into the above inequality, we conclude that

$$\sum_{j \in S} b_j \geq \sum_{\substack{i \in R \\ j \in S}} u_{ij}.$$

If this inequality holds as strict then, by the easy "if" part of Theorem 21.1, the problem has no feasible solution (the net demand within the region S exceeds the upper bound on the import to S). If the inequality holds as an equation, then, as in Chapter 21, the problem decomposes into two subproblems, which can be solved separately. [In fact, the subproblem whose node set is S is already solved (all the nodes in S are balanced), and so only the subproblem whose node set is R remains to be solved.]

To answer question (iv), we distinguish between the arcs on the path $v_0, v_1, \ldots, v_k$ and the arcs off this path. For each arc ij off the path, we have $\bar{x}_{ij} = x_{ij}$, so that (23.9) is preserved trivially and the preservation of (23.10) is guaranteed by (23.7). For each arc ij on the path, (23.9) is preserved by our choice of t and the preservation of (23.10) is guaranteed by (23.8). The tedious details are left for problems 23.6 and 23.7.

To answer question (v), we note that $x_{ij} < u_{ij}$ for each forward arc ij on the path $v_0, v_1, \ldots, v_k$ and that $x_{ij} > 0$ for each reverse arc on this path. Hence each of the upper bounds imposed on t is strictly positive, and so $t > 0$. The progress towards termination can be marked by a quantity called *displacement* of $\mathbf{x}$ and defined as the sum of the n absolute values $|e_j|$, each e_j being the difference between the net flow into j and the demand b_j. (In the notation of Chapter 14, the displacement of $\mathbf{x}$ is simply $\|\mathbf{Ax} - \mathbf{b}\|_1$.) Since the net flows into the nodes $v_1, v_2, \ldots, v_{k-1}$ do not change when $\mathbf{x}$ is replaced by $\bar{\mathbf{x}}$ (the contributions involving t cancel each other out), the displacement decreases by $2t$ in each iteration (the net flow into the wet node v_0 drops by t units, and the net flow into the dry node v_k rises by t units).

THEOREM 23.1. If all the demands b_j, all the finite upper bounds u_{ij}, and all the components x_{ij} of the initial $\mathbf{x}$ are integers, then the primal-dual method terminates within $D/2$ iterations with D standing for the displacement of $\mathbf{x}$.

PROOF. In view of the remarks made above, it suffices to show that $t \geq 1$ in each iteration. But this fact follows at once from observing that all the components $\bar{x}_{ij}$ of each updated $\bar{\mathbf{x}}$ remain integer, and so t is a positive integer. ■

The assumption of Theorem 23.1 is hardly restrictive: when the numbers b_j and u_{ij} are presented as finite decimal expansions, they can be converted into integers by a simple change of scale. Furthermore, if the primal-dual method can be initialized at all, then it can be initialized by an $\mathbf{x}$ such that each x_{ij} is either zero or u_{ij}.

As for question (vi), it suffices to find numbers y_v such that

$$\text{there is no arc } ij \text{ with} \quad u_{ij} = +\infty \quad \text{and} \quad y_i + c_{ij} < y_j: \tag{23.11}$$

then (23.9) and (23.10) can be satisfied by setting

$$x_{ij} = \begin{cases} 0 & \text{if} \quad y_i + c_{ij} \geq y_j \\ u_{ij} & \text{if} \quad y_i + c_{ij} < y_j. \end{cases}$$

We are going to describe a procedure that, given an upper-bounded transshipment problem

$$\text{minimize} \quad \mathbf{cx} \qquad \text{subject to} \quad \mathbf{Ax} = \mathbf{b}, \quad \mathbf{0} \leq \mathbf{x} \leq \mathbf{u}, \tag{23.12}$$

attempts to find numbers y_v satisfying (23.11); in case of failure, this procedure returns a proof that (23.12) has no optimal solution. The procedure amounts to an application of the primal-dual method to the problem

$$\text{minimize} \quad \mathbf{cx} \qquad \text{subject to} \quad \mathbf{Ax} = \mathbf{0}, \quad \mathbf{0} \le \mathbf{x} \le \tilde{\mathbf{u}} \tag{23.13}$$

obtained from (23.12) by replacing each demand b_j by zero and setting

$$\tilde{u}_{ij} = \begin{cases} 0 & \text{if} \quad u_{ij} \text{ is finite} \\ 1 & \text{if} \quad u_{ij} \text{ is infinite.} \end{cases}$$

Since (23.13) has at least one feasible solution (namely, $\mathbf{x} = \mathbf{0}$), the primal-dual method (initialized by $\mathbf{y} = \mathbf{0}$, $x_{ij} = 0$ whenever $c_{ij} \ge 0$, and $x_{ij} = \tilde{u}_{ij}$ whenever $c_{ij} < 0$) will find an optimal solution $\mathbf{x}^*$ of (23.13) along with numbers y_v such that

$$\begin{aligned} x_{ij}^* &= 0 \qquad \text{whenever} \quad y_i + c_{ij} > y_j \\ x_{ij}^* &= \tilde{u}_{ij} \qquad \text{whenever} \quad y_i + c_{ij} < y_j. \end{aligned} \tag{23.14}$$

If these numbers y_v satisfy (23.11) then we are done; otherwise (23.14) implies $(\mathbf{c} - \mathbf{yA})\mathbf{x}^* < 0$ [at least one of the numbers $(c_{ij} + y_i - y_j)x_{ij}^*$ is negative and none of them is positive], and so $\mathbf{cx}^* < 0$. But then (23.12) cannot have an optimal solution: if $\mathbf{x}$ is a feasible solution of (23.12) then $\mathbf{x} + \mathbf{x}^*$ is another and $\mathbf{c}(\mathbf{x} + \mathbf{x}^*) < \mathbf{cx}^*$.

Practical Improvements

The vector $\mathbf{y}$ may remain unchanged in some iterations of the primal-dual method, such as the second, third, fourth, and fifth iterations in the previous example. In such iterations, much of the time spent on computing the numbers d_v is wasted: when $\mathbf{y}$ does not change, the only progress toward termination is made by the change of $\mathbf{x}$. This waste can be avoided by lumping consecutive iterations (such as the first five iterations in the previous example) into a single one; in the resulting streamlined version of the primal-dual method, both $\mathbf{y}$ and $\mathbf{x}$ change in every iteration. To change $\mathbf{y}$, one has to solve a shortest path problem; to change $\mathbf{x}$, one has to solve a certain maximum flow problem.

More precisely, when $\mathbf{y}$ has been changed into $\bar{\mathbf{y}}$ by

$$\bar{y}_v = y_v + d_v \qquad \text{for all nodes } v, \tag{23.15}$$

we still have

$$\begin{aligned} x_{ij} &= 0 \qquad \text{whenever} \quad \bar{y}_i + c_{ij} > \bar{y}_j \\ x_{ij} &= u_{ij} \qquad \text{whenever} \quad \bar{y}_i + c_{ij} < \bar{y}_j \end{aligned} \tag{23.16}$$

(see problem 23.6). Since the updated vector $\bar{\mathbf{x}}$ must satisfy (23.16) in place of $\mathbf{x}$, changes of x_{ij} are allowed only if $\bar{y}_i + c_{ij} = \bar{y}_j$. The novelty of the streamlined version comes from no longer restricting these changes to a single path between a wet node and a dry node, but allowing more extensive changes as well. Such changes may be described in terms of a *working network* whose arcs ij carry upper bounds v_{ij}. In addition to all the original nodes, the working network has an extra source s, an extra sink t, and:

(i) An arc $e(+) = ij$ with $v_{e(+)} = u_e - x_e$ for each original arc $e = ij$ with $\bar{y}_i + c_{ij} = \bar{y}_j$.

(ii) An arc $e(-) = ji$ with $v_{e(-)} = x_e$ for each original arc $e = ij$ with $\bar{y}_i + c_{ij} = \bar{y}_j$.

(iii) An arc si, with v_{si} equal to the difference between the net flow into i and the demand b_i, for each wet node i.

(iv) An arc jt, with v_{jt} equal to the difference between the demand b_j and the net flow into j, for each dry node j.

Every flow $\mathbf{x}^*$ in the working network defines a vector $\bar{\mathbf{x}}$ by

$$\bar{x}_e = x_e + x^*_{e(+)} - x^*_{e(-)} \tag{23.17}$$

If $\mathbf{x}^*$ is a feasible flow in the working network, then clearly $0 \leq \bar{x}_{ij} \leq u_{ij}$ for all arcs ij; furthermore, it is an easy exercise (problem 23.8) to show that the displacement of $\bar{\mathbf{x}}$ equals the displacement of $\mathbf{x}$ minus twice the volume of $\mathbf{x}^*$. Hence the natural course of action is to find a maximum flow $\mathbf{x}^*$ in the working network by any of the efficient algorithms of Chapter 22 and then to enter the next iteration with $\bar{\mathbf{x}}$ defined by (23.17) and $\bar{\mathbf{y}}$ defined by (23.15). (Another easy exercise, left for problem 23.9, shows that a change of $\bar{\mathbf{y}}$ will occur in this next iteration.)

Theoretical Improvements

If the primal-dual method can be initialized at all then it can be initialized by $\mathbf{y}$ and $\mathbf{x}$ such that each x_{ij} is either zero or u_{ij}. It is an easy exercise (left for problem 23.10) to show that the displacement of $\mathbf{x}$ is at most $\sum|b_k| + 2\sum u_{ij}$, with the first sum running through all the nodes k and the second sum running through all the arcs ij such that u_{ij} is finite; hence Theorem 23.1 guarantees that an optimal solution will be found in at most

$$\frac{1}{2}\sum|b_k| + \sum u_{ij}$$

iterations. [In particular, the primal-dual method will solve the assignment problem (20.26) in no more than n iterations. Since the time required to execute each iteration is $O(n^2)$, the total running time comes to $O(n^3)$. It was in this context that the first version of the primal-dual method was developed by H. W. Kuhn (1955); the upper bound $O(n^3)$ on its running time was established by J. Munkres (1957). This may be the first instance of a combinatorial algorithm proved theoretically satisfactory in the sense of Chapter 4.] When the integers $|b_k|$ and u_{ij} are very large, this bound is not very satisfactory. However, a device proposed by J. Edmonds and R. M. Karp (1972) and known as *scaling*, limits the number of iterations to at most

$$d(5n + 10m)$$

with d standing for the number of digits in the largest of the integers $|b_k|$ and u_{ij}. From the theoretical point of view, the latter bound is quite satisfactory.

The idea is to apply the primal-dual method to a sequence of problems

$$\text{minimize} \quad \mathbf{cx} \qquad \text{subject to} \quad \mathbf{Ax} = \tilde{\mathbf{b}}, \quad \mathbf{0} \leq \mathbf{x} \leq \tilde{\mathbf{u}} \tag{23.18}$$

approximating the original problem

$$\text{minimize} \quad \mathbf{cx} \qquad \text{subject to} \quad \mathbf{Ax} = \mathbf{b}, \quad \mathbf{0} \leq \mathbf{x} \leq \mathbf{u}$$

with greater and greater precision: the overall number of iterations is kept low because the optimal solution of each approximation yields a good initial $\mathbf{x}$ in the next approximation. There are d approximations altogether; in the ith of them, each finite upper bound $\tilde{u}_{ij}$ is $u_{ij}/10^{d-i}$ rounded up to the nearest integer, and each demand $\tilde{b}_k$ is $b_k/10^{d-i}$ rounded up or down to the nearest integer in such a way that $\sum \tilde{b}_k = 0$. In particular, each $|\tilde{b}_k|$ and each $\tilde{u}_{ij}$ in the very first approximation is an integer between zero and ten; hence this iteration can be solved in at most $5n + 10m$ iterations of the primal-dual method. Having found an optimal pair $\mathbf{x}^*, \mathbf{y}^*$ in the ith approximation $(i = 1, 2, \ldots, d - 1)$, we may initialize the $(i + 1)$th approximation by $\mathbf{x}$ and $\mathbf{y}^*$ such that each x_{ij} is the minimum of $10x^*_{ij}$ and the new upper bound $\tilde{u}_{ij}$. It is an easy exercise (left for problem 23.11) to show that the displacement of $\mathbf{x}$ in the $(i + 1)$th approximation is at most $10n + 18m$; hence the $(i + 1)$th approximation can be solved in at most $5n + 9m$ iterations of the primal-dual method. Thus the overall number of iterations is limited by $d(5n + 10m)$, as claimed above.

Each reduction of the displacement by t units in the ith approximation amounts to a reduction by $10^{d-i}t$ units in the original problem. From this point of view, scaling may be seen as a way of forcing large improvements early on. In practice, such improvements tend to occur whether forced or not; for this reason, the practical importance of scaling is negligible. □

PROBLEMS

23.1 Solve problems 19.1 and 21.1 by the primal-dual method.

23.2 Prove that, throughout the execution of the shortest-paths algorithm, the variables d_v, R, and $p(\cdot)$ maintain the properties specified in the text.

23.3 Prove that the output of the shortest-paths algorithm satisfies (23.5), (23.6), (23.7), and (23.8).

23.4 Prove: If there is at least one wet node, then there is at least one dry node and vice versa.

23.5 Prove that all the arcs in the auxiliary network have nonnegative lengths.

23.6 Prove that $\mathbf{x}$ and $\bar{\mathbf{y}}$ satisfy (23.10) in place of $\mathbf{x}$ and $\mathbf{y}$.

23.7 Prove that $\bar{\mathbf{x}}$ and $\bar{\mathbf{y}}$ satisfy (23.10) in place of $\mathbf{x}$ and $\mathbf{y}$.

23.8 Prove: if $\mathbf{x}^*$ is a feasible flow in the working network and if $\bar{\mathbf{x}}$ is defined by (23.17) then the displacement of $\bar{\mathbf{x}}$ equals the displacement of $\mathbf{x}$ minus twice the volume of $\mathbf{x}^*$.

23.9 Prove: If $\mathbf{x}^*$ is a maximum flow in the working network and if $\bar{\mathbf{x}}$ is defined by (23.17) then, in the next iteration of the primal-dual method, there will be no dry node v with $d_v = 0$.

23.10 Prove: If each x_{ij} is zero or u_{ij}, then the displacement of **x** is at most $\sum |b_k| + 2\sum u_{ij}$, with the first sum running through all the nodes k and the second sum running through all the arcs ij with finite u_{ij}.

23.11 Let $\mathbf{x}^*$ be a feasible solution of the ith approximation (23.18) and let each x_{ij} be the minimum of $10x_{ij}^*$ and the upper bound $\tilde{u}_{ij}$ featured in the $(i + 1)$th approximation. Prove that the displacement of **x** in the $(i + 1)$th approximation is at most $10n + 18m$.

23.12 How much time is required to initialize the primal-dual method?

23.13 How can you find numbers y_v satisfying (23.11) if (23.13) decomposes into subproblems?

△ **23.14** Our exposition of scaling disregards the possibility of one or more approximations being infeasible, even though the original problem is feasible. Find a way of getting around this obstacle.

△ **23.15** In Chapter 21, a negative cycle was defined as a cycle such that all of its arcs are forward, all of them have $u_{ij} = +\infty$ and the sum of their costs c_{ij} is negative. Design an efficient algorithm which, given a feasible solution $\mathbf{x}^*$ of (23.13) such that $\mathbf{cx}^* < 0$, finds a negative cycle in (23.12).

23.16 Show that the problem of finding shortest paths from a fixed node to all the remaining nodes can be turned into a transshipment problem.

23.17 Let k be an integer greater than one. A sequence of numbers $x_0, x_1, \ldots, x_N$ is called a *k-heap* if

$$x_i \le x_j \qquad \text{whenever} \quad ki < j \le k(i+1).$$

Assuming that $x_0, x_1, \ldots, x_N$ is a k-heap, prove that

(i) If the value of some x_j drops then the resulting sequence can be rearranged into a k-heap in only $O(\log_k N)$ steps.

(ii) If the value of some x_i rises then the resulting sequence can be rearranged into a k-heap in only $O(k \log_k N)$ steps.

and use (ii) to show that

(iii) The sequence $x_N, x_2, \ldots, x_{N-1}$ can be rearranged into a k-heap in only $O(k \log_k N)$ steps.

Then use (iii) and (i) to implement the shortest paths algorithm in such a way that the total time spent on the execution of steps 1 and 2 is only $O(nk \log_k n)$ and $O(m \log_k n)$, respectively.

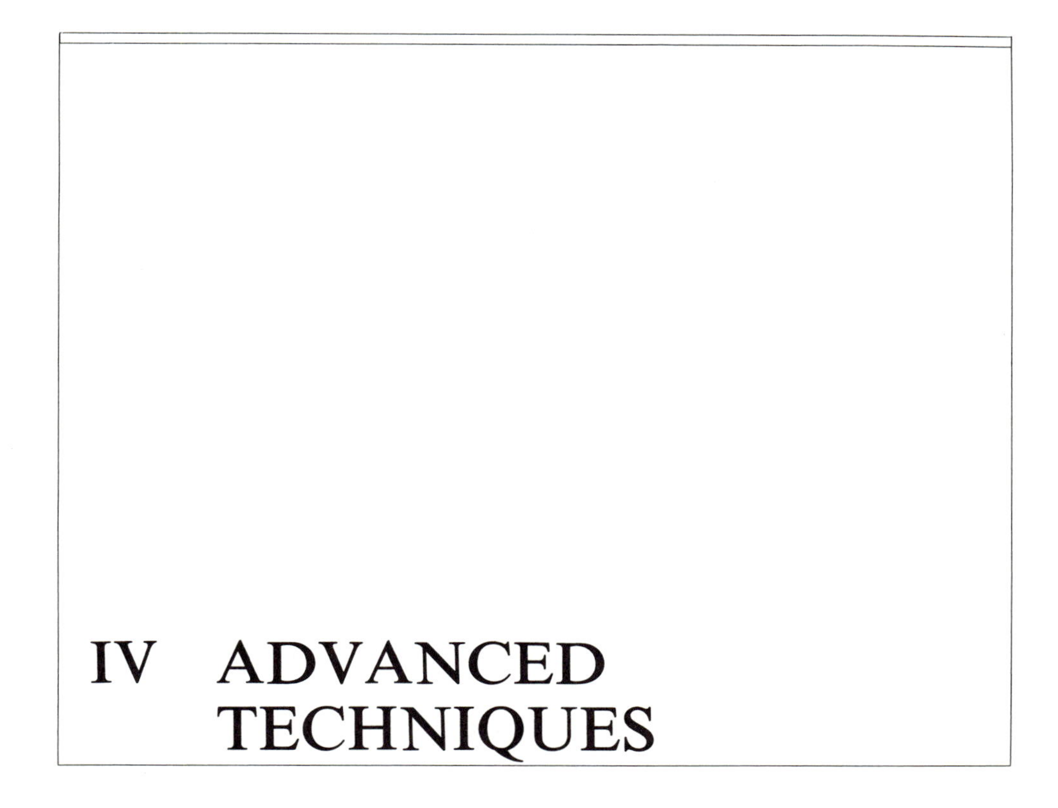

IV ADVANCED TECHNIQUES

24

Updating a Triangular Factorization of the Basis

A device facilitating solutions of the two systems $\mathbf{yB} = \mathbf{c}_B$ and $\mathbf{Bd} = \mathbf{a}$ in each iteration of the revised simplex method was suggested in Chapter 7. It consists of a triangular factorization

$$\mathbf{L}_m\mathbf{P}_m \cdots \mathbf{L}_1\mathbf{P}_1\mathbf{B} = \mathbf{U}_m\mathbf{U}_{m-1} \cdots \mathbf{U}_1\mathbf{E}_1\mathbf{E}_2 \cdots \mathbf{E}_k \tag{24.1}$$

such that each $\mathbf{L}_j$ is a lower triangular eta matrix whose eta column is in the jth position, each $\mathbf{P}_j$ is a permutation matrix, each $\mathbf{U}_j$ is an upper triangular eta matrix whose eta column is in the jth position, and each $\mathbf{E}_i$ is an eta matrix. (All the diagonal entries in each $\mathbf{U}_j$ are ones.) In this chapter, we shall consider devices consisting of factorizations

$$\mathbf{L}_s\mathbf{P}_s \cdots \mathbf{L}_1\mathbf{P}_1\mathbf{B} = \mathbf{Q}\mathbf{U}_m\mathbf{U}_{m-1} \cdots \mathbf{U}_1\mathbf{R} \tag{24.2}$$

such that the matrices $\mathbf{L}_j$, $\mathbf{P}_j$, $\mathbf{U}_j$ with $1 \le j \le m$ are as in (24.1), each $\mathbf{L}_j$ with $m < j \le s$ is an eta-matrix with at most one non zero off-diagonal entry, each $\mathbf{P}_j$ with $m \le j \le s$ is a permutation matrix, and $\mathbf{Q}$ and $\mathbf{R}$ are permutation matrices. The use of (24.2) is analogous to the use of (24.1). More precisely, the system $\mathbf{yB} = \mathbf{c}_B$ may be solved by first solving the system

$$(((\mathbf{v}\mathbf{U}_m)\mathbf{U}_{m-1}) \cdots)\mathbf{U}_1 = \mathbf{c}_B\mathbf{R}^{-1}$$

and then computing

$$\mathbf{y} = ((((\mathbf{v}\mathbf{Q}^{-1})\mathbf{L}_s)\mathbf{P}_s) \cdots \mathbf{L}_1)\mathbf{P}_1.$$

Similarly, the system $\mathbf{Bd} = \mathbf{a}$ may be solved by first computing

$$\mathbf{w} = \mathbf{Q}^{-1}(\mathbf{L}_s(\mathbf{P}_s \cdots (\mathbf{L}_1(\mathbf{P}_1\mathbf{a})))) \tag{24.3}$$

and then solving the system

$$\mathbf{U}_m(\mathbf{U}_{m-1} \cdots (\mathbf{U}_1(\mathbf{Rd}))) = \mathbf{w}.$$

Since the files of nonzero entries in (24.2) tend to be shorter than the files of nonzero entries in (24.1), fewer arithmetical operations are required when (24.2) is used to solve the two systems. Even though the update of (24.2) requires a little extra work, the use of (24.2) leads to faster implementations of the revised simplex method.

R. H. Bartels and G. H. Golub (1969) were the first to propose the use of the triangular factorizations (24.2) and a way of updating them in each iteration of the simplex method. Their aim was to improve numerical accuracy in solving the two systems of equations; on this count, the Bartels–Golub method is remarkably successful. However, its straightforward implementations become very slow as soon as the nonzero entries of $\mathbf{U}_m, \mathbf{U}_{m-1}, \ldots, \mathbf{U}_1$ overflow the core memory of a computer and have to be stored on peripheral devices. In a later proposal by J. J. H. Forrest and J. A. Tomlin (1972), this obstacle is circumnavigated at the cost of abandoning the measures to improve numerical accuracy. Subsequent implementations of the Bartels–Golub method, designed by J. K. Reid (1982) and M. A. Saunders (1976b), combine the virtues of accuracy and speed. In Reid's implementation, judicious choices of $\mathbf{Q}$ and $\mathbf{R}$ curb fill-in in the matrices $\mathbf{U}_j$ to such an extent that the file representing these matrices can be packed in core. In Saunders's implementation, a large part of this file is handled as in the Forrest–Tomlin method, and the remaining part is small enough to fit in core.

GENERAL SCHEME

When the basis matrix $\mathbf{B}$ is replaced by a new $\bar{\mathbf{B}}$, the factorization (24.2) is updated into

$$\bar{\mathbf{L}}_m\bar{\mathbf{P}}_m \cdots \bar{\mathbf{L}}_1\bar{\mathbf{P}}_1\mathbf{L}_s\mathbf{P}_s \cdots \mathbf{L}_1\mathbf{P}_1\bar{\mathbf{B}} = \bar{\mathbf{Q}}\bar{\mathbf{U}}_m\bar{\mathbf{U}}_{m-1} \cdots \bar{\mathbf{U}}_1\bar{\mathbf{R}}. \tag{24.4}$$

Here, $\mathbf{L}_j$ and $\mathbf{P}_j$ are the matrices featured in (24.2), each $\bar{\mathbf{L}}_j$ differs from the identity matrix in at most one (off-diagonal) entry, each $\bar{\mathbf{P}}_j$ is a permutation matrix, each $\bar{\mathbf{U}}_j$ is an upper triangular eta matrix whose eta column is in the jth position, and $\bar{\mathbf{Q}}$ and $\bar{\mathbf{R}}$ are permutation matrices.

For an easy understanding of the technical details, it is crucial to see through the relationship between a square matrix $\mathbf{A} = (a_{ij})$ and the matrix $\mathbf{QAR}$ such that $\mathbf{Q}$ and $\mathbf{R}$ are permutation matrices. The matrix $\mathbf{QAR}$ arises from $\mathbf{A}$ by permuting rows and columns: *the rows of* **A** (*with their entries permuted*) *appear in* **QAR** *in the same order in which the rows of* **I** *appear in* **Q**, and *the columns of* **A** (*with their entries permuted*) *appear in* **QAR** *in the same order in which the columns of* **I** *appear in* **R**. A special case arises when $\mathbf{R} = \mathbf{Q}^{-1}$: since the rows of **I** appear in **Q** in the same order in which the columns of **I** appear in $\mathbf{Q}^{-1}$, *the diagonal entries of* **A** *remain on the diagonal of* $\mathbf{QAQ}^{-1}$ *but appear in the same order in which the rows of* **I** *appear in* **Q**. We shall sometimes describe the transformation of **A** into **QAR** informally by saying that, for instance, "the entry a_{ij} is moved to the bottom right corner of **A**." This description means moving the ith row of **A** into the last position, keeping the relative order of the remaining rows undisturbed, moving the jth column of **A** into the last position, and keeping the relative order of the remaining columns undisturbed.

The update of (24.2) into (24.4) relies on the fact that $\bar{\mathbf{B}} = \mathbf{BE}$ with an eta matrix **E**, whose eta column is the vector **d** computed in step 3 of the simplex iteration. Let us say that the entering column **a** replaces the leaving column of **B** in the pth position, so that **d** is the pth column of **E**. Furthermore, let us say that the pth row of **I** is the qth row of the permutation matrix **R** featured in (24.2). Finally, let us write $\mathbf{U} = \mathbf{U}_m\mathbf{U}_{m-1} \cdots \mathbf{U}_1$. As noted in Chapter 7, the jth column of **U** is the eta column of $\mathbf{U}_j$.

Observe that $\mathbf{RER}^{-1}$ is an eta matrix whose eta column **Rd** appears in the qth position. Hence the matrix

$$\mathbf{V} = \mathbf{U}(\mathbf{RER}^{-1})$$

differs from **U** only in its qth column. This column of **V** is **URd**, which is nothing but the vector **w** featured in (24.3). Since this vector can be saved in step 3 of the simplex iteration, all the entries of **V** are readily available. As we shall see in the next four sections, it is easy to find permutation matrices $\tilde{\mathbf{Q}}$ and $\tilde{\mathbf{R}}$ such that $\tilde{\mathbf{Q}}\mathbf{V}\tilde{\mathbf{R}}$ differs from an upper triangular matrix in only one row. In turn, it is easy to find a triangular factorization

$$\tilde{\mathbf{L}}_m\tilde{\mathbf{P}}_m \cdots \tilde{\mathbf{L}}_1\tilde{\mathbf{P}}_1(\tilde{\mathbf{Q}}\mathbf{V}\tilde{\mathbf{R}}) = \bar{\mathbf{U}}$$

such that each $\tilde{\mathbf{L}}_j$ differs from the identity matrix in at most one (off-diagonal) entry, each $\tilde{\mathbf{P}}_j$ is a permutation matrix, and $\bar{\mathbf{U}}$ is an upper triangular matrix all of whose diagonal entries are one. If $\bar{\mathbf{U}}_j$ stands for the eta matrix obtained when the jth column of **I** is replaced by the jth column of **U**, then again $\bar{\mathbf{U}} = \bar{\mathbf{U}}_m\bar{\mathbf{U}}_{m-1} \cdots \bar{\mathbf{U}}_1$.

Now (24.4) can be obtained by setting

$$\bar{\mathbf{Q}} = \mathbf{Q}\tilde{\mathbf{Q}}^{-1}, \qquad \bar{\mathbf{R}} = \tilde{\mathbf{R}}^{-1}\mathbf{R}$$

and

$$\bar{\mathbf{L}}_j = \bar{\mathbf{Q}}\tilde{\mathbf{L}}_j\bar{\mathbf{Q}}^{-1}, \qquad \bar{\mathbf{P}}_j = \bar{\mathbf{Q}}\tilde{\mathbf{P}}_j\bar{\mathbf{Q}}^{-1} \qquad \text{for all} \quad j = 1, 2, \ldots, m$$

since

$$\bar{\mathbf{L}}_m\bar{\mathbf{P}}_m \cdots \bar{\mathbf{L}}_1\bar{\mathbf{P}}_1 = \bar{\mathbf{Q}}\tilde{\mathbf{L}}_m\tilde{\mathbf{P}}_m \cdots \tilde{\mathbf{L}}_1\tilde{\mathbf{P}}_1\bar{\mathbf{Q}}^{-1}$$

and so

$$\begin{aligned}
\bar{\mathbf{L}}_m\bar{\mathbf{P}}_m \cdots \bar{\mathbf{L}}_1\bar{\mathbf{P}}_1\mathbf{L}_s\mathbf{P}_s \cdots \mathbf{L}_1\mathbf{P}_1\bar{\mathbf{B}} &= (\bar{\mathbf{L}}_m\bar{\mathbf{P}}_m \cdots \bar{\mathbf{L}}_1\bar{\mathbf{P}}_1)(\mathbf{L}_s\mathbf{P}_s \cdots \mathbf{L}_1\mathbf{P}_1\mathbf{BE}) \\
&= (\bar{\mathbf{Q}}\tilde{\mathbf{L}}_m\tilde{\mathbf{P}}_m \cdots \tilde{\mathbf{L}}_1\tilde{\mathbf{P}}_1\bar{\mathbf{Q}}^{-1})(\mathbf{QURE}) \\
&= \bar{\mathbf{Q}}\tilde{\mathbf{L}}_m\tilde{\mathbf{P}}_m \cdots \tilde{\mathbf{L}}_1\tilde{\mathbf{P}}_1\tilde{\mathbf{Q}}\mathbf{VR} \\
&= \bar{\mathbf{Q}}(\tilde{\mathbf{L}}_m\tilde{\mathbf{P}}_m \cdots \tilde{\mathbf{L}}_1\tilde{\mathbf{P}}_1\tilde{\mathbf{Q}}\mathbf{V}\tilde{\mathbf{R}})\bar{\mathbf{R}} \\
&= \bar{\mathbf{Q}}\bar{\mathbf{U}}\bar{\mathbf{R}}.
\end{aligned}$$

The four methods discussed next follow this general pattern (even though this is not quite the way the Bartels–Golub and Reid methods were originally presented). They differ from each other only in their choices of the permutation matrices $\tilde{\mathbf{Q}}$, $\tilde{\mathbf{R}}$ and $\tilde{\mathbf{P}}_1, \tilde{\mathbf{P}}_2, \ldots, \tilde{\mathbf{P}}_m$.

In order to assess properly the efficiency of these methods, it is crucial to consider not only the mere numbers of arithmetical operations per simplex iteration but also the ways of handling the data on a computer. As noted in Chapter 7, data can be stored either in the core memory, which is fast to access but limited in capacity, or on peripheral devices such as disks, drums, and tapes, which are virtually unlimited in capacity but slow to access. Any location in the core memory can be accessed instantaneously, but jumps between two locations in peripheral memory are instantaneous only if these locations are adjacent. This difference is important when it comes to handling sequential files that have to be scanned over and over again. In core, the individual items $x_1, x_2, \ldots, x_n$ may be scattered through arbitrary locations and the scans facilitated by "pointers" specifying, for each item x_i on the file, the locations of the preceding item x_{i-1} and the next term x_{i+1}. Consequently, deletions and insertions of items are easy to implement. A deletion amounts to changing a couple of pointers; an insertion of a new item into a prescribed position in the file involves placing this item in any available location, properly setting its pointers, and adjusting a couple of old pointers. The situation changes drastically when the file is stored in peripheral memory: to facilitate frequent scans, consecutive items have to be stored in adjacent locations. Even though deletions can still be implemented with a relative ease (simply by marking the appropriate items as deleted), inserting a new item into the middle of a file becomes very awkward.

These general remarks pertain to handling the file of nonzero entries of $\mathbf{U}_m$, $\mathbf{U}_{m-1}, \ldots, \mathbf{U}_1$. Following Forrest and Tomlin (1972), we shall refer to this file as the *backward file* and call the file representing $\mathbf{P}_1, \mathbf{L}_1, \ldots, \mathbf{P}_s, \mathbf{L}_s$ the *forward file.* The update of the forward file presents no difficulty as neither deletions nor insertions of items into the middle of this file are ever required: items $\mathbf{P}_{s+1} = \bar{\mathbf{P}}_1, \mathbf{L}_{s+1} = \bar{\mathbf{L}}_1, \mathbf{P}_{s+2} = \bar{\mathbf{P}}_2, \ldots$ added to the file are simply tacked on to its end. However, insertions into the middle of a backward file may be required if $\mathbf{U}$ acquires new nonzero entries in arbitrary locations. Such insertions are extremely time-consuming unless the backward file is kept in core.

THE BARTELS–GOLUB METHOD

In this method, (i) the permutation matrices $\tilde{\mathbf{Q}}$ and $\tilde{\mathbf{R}} = \tilde{\mathbf{Q}}^{-1}$ are chosen so as to move the qth diagonal element w_q of $\mathbf{V}$ to the bottom right corner and (ii) the permutation matrices $\tilde{\mathbf{P}}_1, \tilde{\mathbf{P}}_2, \ldots, \tilde{\mathbf{P}}_m$ are determined by the rules of partial pivoting.

To illustrate (i), consider the case $m = 5$ and $q = 2$:

$$\mathbf{V} = \begin{bmatrix} 1 & w_1 & u_{13} & u_{14} & u_{15} \\ & w_2 & u_{23} & u_{24} & u_{25} \\ & w_3 & 1 & u_{34} & u_{35} \\ & w_4 & & 1 & u_{45} \\ & w_5 & & & 1 \end{bmatrix}.$$

In this case, setting

$$\tilde{\mathbf{Q}} = \begin{bmatrix} 1 & & & & \\ & & 1 & & \\ & & & 1 & \\ & & & & 1 \\ & 1 & & & \end{bmatrix} \qquad \text{and} \qquad \tilde{\mathbf{R}} = \tilde{\mathbf{Q}}^{-1}$$

$$\text{we obtain} \quad \tilde{\mathbf{Q}}\mathbf{V}\tilde{\mathbf{R}} = \begin{bmatrix} 1 & u_{13} & u_{14} & u_{15} & w_1 \\ & 1 & u_{34} & u_{35} & w_3 \\ & & 1 & u_{45} & w_4 \\ & & & 1 & w_5 \\ & u_{23} & u_{24} & u_{25} & w_2 \end{bmatrix}.$$

As for (ii), partial pivoting yields remarkably accurate results: even artificially concocted nasty examples can be solved with only negligible rounding errors, and backward analysis of this updating procedure provides a guarantee of its numerical accuracy. For more information, see R. H. Bartels (1971).

It had been thought that the Bartels–Golub update would suffer from considerable fill-in throughout $\mathbf{U}$. The experimental results of J. A. Tomlin (1972) provided the first evidence to the contrary: on two problems originating from the oil industry (with the matrix $\mathbf{A}$ having size $822 \times 1{,}571$ in one, $2{,}978 \times 6{,}333$ in the other, and fewer than ten nonzeros per column on the average in both), the number of nonzeros in the triangular factorization (24.2) of $\mathbf{B}$ after 30 iterations was less than 50% of the number of nonzeros in (24.1). Furthermore, fill-in in $\mathbf{U}$ can be reduced by mild violations of the rigid rules of partial pivoting. Recall that each $\tilde{\mathbf{P}}_j$ is either the identity matrix $\mathbf{I}$ or the permutation matrix $\mathbf{P}$ arising from $\mathbf{I}$ by interchanging the jth row with the last row. The actual choice of $\tilde{\mathbf{P}}_j$ depends on two entries in the jth column of the matrix

$$\tilde{\mathbf{L}}_{j-1}\tilde{\mathbf{P}}_{j-1} \cdots \tilde{\mathbf{L}}_1\tilde{\mathbf{P}}_1 (\tilde{\mathbf{Q}}\mathbf{V}\tilde{\mathbf{R}}), \tag{24.5}$$

namely, the entry x in the jth row and the entry y in the last row. Partial pivoting requires $\tilde{\mathbf{P}}_j = \mathbf{I}$ whenever $|x| > |y|$ and $\tilde{\mathbf{P}}_j = \mathbf{P}$ whenever $|y| > |x|$. If these requirements are relaxed by

insisting on $\tilde{\mathbf{P}}_j = \mathbf{I}$ only if $|x| > c|y|$

and

insisting on $\tilde{\mathbf{P}}_j = \mathbf{P}$ only if $|y| > c|x|$

for some constant c, then one is free to choose between $\tilde{\mathbf{P}}_j = \mathbf{I}$ and $\tilde{\mathbf{P}}_j = \mathbf{P}$ whenever $|x| \le c|y|$ and $|y| \le c|x|$. This freedom may be used to produce a sparser $\mathbf{U}_{k+1}$ by setting $\tilde{\mathbf{P}}_j = \mathbf{I}$ *if the jth row of* (24.5) *is sparser than its last row* and $\tilde{\mathbf{P}}_j = \mathbf{P}$ *if the last row of* (24.5) *is sparser than its jth row*. Even though these choices of $\tilde{\mathbf{P}}_j$ may violate the strict rules of partial pivoting, numerical accuracy of the results remains satisfactory as long as c is reasonably small (say, 10 or 100).

Nevertheless, even very slow fill-in may create serious difficulties if the backward file is stored in peripheral memory: as noted in the preceding section, inserting a new item into the middle of such a file may be a very awkward task.

THE FORREST–TOMLIN METHOD

In this method, (i) the permutation matrices $\tilde{\mathbf{Q}}$ and $\tilde{\mathbf{R}}$ are chosen as in the Bartels–Golub update but (ii) each of the permutation matrices $\tilde{\mathbf{P}}_j$ is the identity matrix. Thus, partial pivoting, the original motivation behind the Bartels–Golub proposal, is abandoned altogether; in return, fill-in is restricted to the last column of $\mathbf{U}$ and the backward file can be updated easily even if it resides in peripheral memory. The resulting implementation of the simplex method is very fast. The results of experiments by Forrest and Tomlin (1972) show that a simplex iteration with their update of (24.2) takes, on the average, about 70% of the time required when the product form (24.1) of the basis is used.

For illustration of the details, we return to the case $m = 5$ and $q = 2$; now

$$\mathbf{V} = \begin{bmatrix} 1 & w_1 & u_{13} & u_{14} & u_{15} \\ & w_2 & u_{23} & u_{24} & u_{25} \\ & w_3 & 1 & u_{34} & u_{35} \\ & w_4 & & 1 & u_{45} \\ & w_5 & & & 1 \end{bmatrix} \quad \text{and} \quad \tilde{\mathbf{Q}}\mathbf{V}\tilde{\mathbf{R}} = \begin{bmatrix} 1 & u_{13} & u_{14} & u_{15} & w_1 \\ & 1 & u_{34} & u_{35} & w_3 \\ & & 1 & u_{45} & w_4 \\ & & & 1 & w_5 \\ & u_{23} & u_{24} & u_{25} & w_2 \end{bmatrix}$$

as in the Bartels–Golub method, but

$$\bar{\mathbf{U}} = \tilde{\mathbf{L}}_5\tilde{\mathbf{L}}_4\tilde{\mathbf{L}}_3\tilde{\mathbf{L}}_2(\tilde{\mathbf{Q}}\mathbf{V}\tilde{\mathbf{R}}) = \begin{bmatrix} 1 & u_{13} & u_{14} & u_{15} & w_1 \\ & 1 & u_{34} & u_{35} & w_3 \\ & & 1 & u_{45} & w_4 \\ & & & 1 & w_5 \\ & & & & 1 \end{bmatrix}.$$

The transformation of $\mathbf{U}$ into $\bar{\mathbf{U}}$ may be seen as consisting of three phases:

$$\mathbf{U} \to \begin{bmatrix} 1 & 0 & u_{13} & u_{14} & u_{15} \\ & 0 & 0 & 0 & 0 \\ & & 1 & u_{34} & u_{35} \\ & & & 1 & u_{45} \\ & & & & 1 \end{bmatrix} \to \begin{bmatrix} 1 & u_{13} & u_{14} & u_{15} & 0 \\ & 1 & u_{34} & u_{35} & \\ & & 1 & u_{45} & \\ & & & 1 & \\ & 0 & 0 & 0 & 0 \end{bmatrix} \to \bar{\mathbf{U}}.$$

In general, each of these three phases is easy to implement even if the backward file is stored in peripheral memory. In the first phase, only deletions of items are required. In the second phase, the row and column permutations are recorded without disturbing the file. (A column permutation is a change in the basis heading, and a row permutation is a similar change in a "row heading.") In the third phase, the new column of $\bar{\mathbf{U}}$ is simply tacked on to the end of the file.

As for the update of the forward file, this requires computing the matrices $\tilde{\mathbf{L}}_q$, $\tilde{\mathbf{L}}_{q+1}, \ldots, \tilde{\mathbf{L}}_m$ featured in the triangular factorization

$$\tilde{\mathbf{L}}_m \cdots \tilde{\mathbf{L}}_{q+1}\tilde{\mathbf{L}}_q(\tilde{\mathbf{Q}}\mathbf{V}\tilde{\mathbf{R}}) = \bar{\mathbf{U}}.$$

To compute (the off-diagonal entry in) each $\tilde{\mathbf{L}}_j$, we need only know the jth column of $\tilde{\mathbf{L}}_{j-1} \cdots \tilde{\mathbf{L}}_q(\tilde{\mathbf{Q}}\mathbf{V}\tilde{\mathbf{R}})$. This column is $\tilde{\mathbf{L}}_{j-1} \cdots \tilde{\mathbf{L}}_q\tilde{\mathbf{u}}$ with $\tilde{\mathbf{u}}$ standing for the jth column of $\tilde{\mathbf{Q}}\mathbf{V}\tilde{\mathbf{R}}$; if $q \le j < m$ then $\tilde{\mathbf{u}}$ is $\tilde{\mathbf{Q}}\mathbf{u}$ with $\mathbf{u}$ standing for the $(j+1)$th column of $\mathbf{U}$. Thus the update of the two files may be implemented as follows.

1. Mark all the nonzero entries in the qth column of $\mathbf{U}$ as deleted.

2. Repeat the following operations for $j = q, q+1, \ldots, m$: Read the $(j+1)$th column $\mathbf{u}$ of $\mathbf{U}$, compute $\tilde{\mathbf{L}}_j$ (which requires computing $\tilde{\mathbf{L}}_{j-1} \cdots \tilde{\mathbf{L}}_q\tilde{\mathbf{Q}}\mathbf{u}$) and add it to the end of the forward file. If u_q is a nonzero then mark it as deleted.

3. Compute $\tilde{\mathbf{L}}_m$ (which requires computing $\tilde{\mathbf{L}}_{m-1} \cdots \tilde{\mathbf{L}}_q\tilde{\mathbf{Q}}\mathbf{w}$) and add it to the end of the forward file. Add $\mathbf{w}$ (with w_q replaced by 1) to the end of the backward file.

4. Adjust the row and column headings.

(The items in the backward file that are marked as deleted form only a small fraction of the total number of nonzeros in the two files. In experiments carried out by Forrest and Tomlin, this fraction was less than 4% after 70 iterations.) Solving the system $\mathbf{v}\bar{\mathbf{U}} = \mathbf{c}_B\bar{\mathbf{R}}^{-1}$ featured in Step 1 of the upcoming simplex iteration may be incorporated into this procedure. [To obtain $\mathbf{v}$, we set $\mathbf{v} = \mathbf{c}_B\bar{\mathbf{R}}^{-1}$ initially, and then, for $j = 1, 2, \ldots, m$, keep replacing $\mathbf{v}$ by the solution $\mathbf{y}$ of $\mathbf{y}\bar{\mathbf{U}}_j = \mathbf{v}$ with $\bar{\mathbf{U}}_j$ standing for the eta matrix whose eta column (in the jth position) is the jth column of $\bar{\mathbf{U}}$. The first $q - 1$ iterations may be carried out before the execution of line 1, each of the subsequent $m - q$ iterations after each execution of line 2, and the last iteration after the execution of line 3.]

Note that the procedure for updating the two files and solving the system $\mathbf{v}\bar{\mathbf{U}} = \mathbf{c}_B\bar{\mathbf{R}}^{-1}$ requires scanning the backward file from $\mathbf{U}_1$ to $\mathbf{U}_m$. When the procedure has terminated, the heads scanning the backward and forward files are positioned at $\mathbf{U}_m$ and $\mathbf{L}_s$, respectively. In Step 1 of the next simplex iteration, the forward file is scanned from $\mathbf{L}_s$ to $\mathbf{P}_1$; in Step 3, the forward file is scanned from $\mathbf{P}_1$ to $\mathbf{L}_s$ and the backward file from $\mathbf{U}_m$ to $\mathbf{U}_1$. Therefore, neither file is ever scanned without being read.

REID'S METHOD

In this method, (i) the permutation matrices $\tilde{\mathbf{Q}}$ and $\tilde{\mathbf{R}}$ are chosen by a simple procedure designed to minimize fill-in in $\mathbf{U}$ and (ii) the permutation matrices $\tilde{\mathbf{P}}_1, \tilde{\mathbf{P}}_2, \ldots, \tilde{\mathbf{P}}_m$ are as in the Bartels–Golub method. Fill-in in $\mathbf{U}$ is curbed to such an extent that the entire backward file can be kept in core, where insertions of new items, even into the middle of the file, presents no difficulty whatsoever. The efficiency of the resulting method is comparable to that of the Forrest–Tomlin method.

A rudiment of the idea of Reid's method appears in an earlier proposal by M. A. Saunders (1976a): if nonzero components of the vector $\mathbf{w}$ extend only to the rth row (rather than all the way to the bottom), then the number of subdiagonal elements in $\tilde{\mathbf{Q}}\mathbf{V}\tilde{\mathbf{R}}$ may be reduced by moving the qth diagonal entry w_q of $\mathbf{V}$ only to the rth diagonal position (rather than all the way to the bottom diagonal position). For instance, if $m = 5$, $q = 2$, and $r = 4$, then we obtain

$$\tilde{\mathbf{Q}}\mathbf{V}\tilde{\mathbf{R}} = \begin{bmatrix} 1 & u_{13} & u_{14} & w_1 & u_{15} \\ & 1 & u_{34} & w_3 & u_{35} \\ & & 1 & w_4 & u_{45} \\ & u_{23} & u_{24} & w_2 & u_{25} \\ & & & & 1 \end{bmatrix} \quad \text{rather than} \quad \begin{bmatrix} 1 & u_{13} & u_{14} & u_{15} & w_1 \\ & 1 & u_{34} & u_{35} & w_3 \\ & & 1 & u_{45} & w_4 \\ & & & 1 & \\ & u_{23} & u_{24} & u_{25} & w_2 \end{bmatrix}.$$

More generally, the matrix $\tilde{\mathbf{Q}}\mathbf{V}\tilde{\mathbf{R}}$ now assumes the "block-triangular" form

$$\begin{bmatrix} \mathbf{V}_{11} & \mathbf{V}_{12} & \mathbf{V}_{13} \\ & \mathbf{V}_{22} & \mathbf{V}_{23} \\ & & \mathbf{V}_{33} \end{bmatrix}$$

such that the submatrices $\mathbf{V}_{11}$, $\mathbf{V}_{22}$, $\mathbf{V}_{23}$ are square and, in fact, $\mathbf{V}_{11}$ and $\mathbf{V}_{33}$ are upper triangular. The submatrix $\mathbf{V}_{22}$ is called the *bump*.

Reid's procedure attempts to reduce the size of the bump further by additional row and column permutations. These permutations are conveniently described in terms of *row singletons* and *column singletons*, a row singleton being any nonzero entry that is the only nonzero entry in a row of the bump and a column singleton defined analogously. The procedure consists of three phases.

Phase I. Whenever a column singleton is found in any column but the last, it is moved to the top left corner of the bump (and so the bump shrinks).

Phase II. Whenever a row singleton is found in any row but the last, it is moved to the bottom right corner of the bump (and so the bump shrinks).

Phase III. Whenever a column singleton is found in the last column, it is moved to the top left corner of the bump (and so the bump shrinks).

An example can be found in problem 24.1. When the procedure terminates, the bump in the resulting $\tilde{\mathbf{Q}}\mathbf{V}\tilde{\mathbf{R}}$ still differs from an upper triangular matrix only in its last row (the proof is left for an exercise; see problem 24.2). If the bump extends from row and column i to row and column j, then the triangular factorization of $\tilde{\mathbf{Q}}\mathbf{V}\tilde{\mathbf{R}}$ becomes

$$\tilde{\mathbf{L}}_j\tilde{\mathbf{P}}_j \cdots \tilde{\mathbf{L}}_i\tilde{\mathbf{P}}_i(\tilde{\mathbf{Q}}\mathbf{V}\tilde{\mathbf{R}}) = \bar{\mathbf{U}}.$$

Thus a smaller bump means a slower growth of the forward and backward files. Despite its simplicity, the procedure is remarkably successful in reducing the sizes of bumps in sparse matrices $\mathbf{V}$. In particular, it will permute $\mathbf{V}$ into an upper triangular $\tilde{\mathbf{Q}}\mathbf{V}\tilde{\mathbf{R}}$ whenever possible (see problem 24.4).

SAUNDERS'S METHOD

In this method, (i) the permutation matrices $\tilde{\mathbf{Q}}$, $\tilde{\mathbf{R}}$, as well as (ii) the permutation matrices $\tilde{\mathbf{P}}_1, \tilde{\mathbf{P}}_2, \ldots, \tilde{\mathbf{P}}_m$ are just as in the Bartels–Golub method. The only novelty is the initial ordering of the columns of $\mathbf{B}$ just after refactorization: the basis heading is permuted in such a way that the spikes come last. This apparently insignificant detail has far-reaching consequences; in short, it confines fill-in in $\mathbf{U}$ to a small area in the bottom right corner.

This conclusion is easy to justify. To begin, if the spikes in $\mathbf{U}$ are restricted to the last t columns, then the off-diagonal entries in $\tilde{\mathbf{Q}}\mathbf{V}\tilde{\mathbf{R}}$ are restricted to the last $t + 1$ columns. Consequently, in computing the triangular factorization

$$\tilde{\mathbf{L}}_m\tilde{\mathbf{P}}_m \cdots \tilde{\mathbf{L}}_{m-t}\tilde{\mathbf{P}}_{m-t}(\tilde{\mathbf{Q}}\mathbf{V}\tilde{\mathbf{R}}) = \bar{\mathbf{U}}$$

fill-in is confined to the last $t + 1$ rows of $\mathbf{U}$ and spikes in $\bar{\mathbf{U}}$ are restricted to the last $t + 1$ columns. Now an easy induction on k shows that, with s standing for the number of spikes in $\mathbf{U}$ just after refactorization, fill-in is confined to the last $s + k$ rows in the first k iterations.

Saunders exploits this observation by keeping the bottom $s + k$ rows of $\mathbf{U}$ in core and storing the rest of the backward file in peripheral memory. Now both parts can be updated easily even though partial pivoting is used. The success of this proposal is due to the empirical fact that the initial s is always small, and so the lower part of $\mathbf{U}$ does indeed fit in core. A typical value of s is about 100 or less even for very large problems; if the basis is refactorized periodically after every r iterations, then the lower part of $\mathbf{U}$ has no more than $(s + r)(s + r + 1)/2$ entries.

Since Saunders's method combines the arithmetic of Bartels–Golub with the logistic of Forrest–Tomlin, it has the accuracy of the former and the efficiency of the latter. [Incidentally, note that the bump in $\tilde{\mathbf{Q}}\mathbf{V}\tilde{\mathbf{R}}$ always comes from the last $s + k$ rows residing in core; thus, Reid's bump-reducing procedure can be incorporated in Saunders's method. This has been actually done by D. M. Gay (1978).]

PROBLEMS

24.1 Illustrate Reid's bump-reducing procedure on

$$\mathbf{V} = \begin{bmatrix} u_{11} & w_1 & & u_{14} & & & u_{17} & & u_{19} \\ & & u_{23} & & & u_{26} & & & \\ & & u_{33} & u_{34} & & u_{36} & & & \\ & & & u_{44} & & & u_{47} & & u_{49} \\ & & & & u_{55} & & u_{57} & u_{58} & \\ & & & & & u_{66} & & u_{68} & \\ & & & & & & u_{77} & & u_{79} \\ & w_8 & & & & & & u_{88} & \\ & & & & & & & & u_{99} \end{bmatrix}.$$

24.2 Prove that (a) throughout the application of Reid's bump-reducing procedure, the bump differs from an upper triangular matrix only in its last row and that (b) when the procedure terminates, there are no column singletons in the bump.

24.3 Consider a block-diagonal matrix

$$\mathbf{V} = \begin{bmatrix} \mathbf{V}_{11} & \mathbf{V}_{12} & \mathbf{V}_{13} \\ & \mathbf{V}_{22} & \mathbf{V}_{23} \\ & & \mathbf{V}_{33} \end{bmatrix}$$

such that $\mathbf{V}_{11}$ and $\mathbf{V}_{33}$ are nonsingular upper triangular. Prove that $\mathbf{V}$ is a permuted triangular matrix if and only if $\mathbf{V}_{22}$ is a permuted triangular matrix.

24.4 Combine the results of problem 24.2b and problem 24.3 to prove that Reid's bump-reducing procedure permutes $\mathbf{V}$ into an upper triangular matrix $\tilde{\mathbf{Q}}\mathbf{V}\tilde{\mathbf{R}}$ whenever such a permutation is possible.

24.5 Use (a) the Bartels–Golub method, (b) the Forrest–Tomlin method, and (c) Reid's method to solve problem (4.3). In each iteration, let the variable with the largest coefficient in $\mathbf{c}_N - \mathbf{y}\mathbf{A}_N$ enter the basis.

25

Generalized Upper Bounding

The simplex method can often be streamlined or modified in order to work faster on linear programming problems

$$\text{maximize} \quad \mathbf{cx} \qquad \text{subject to} \quad \mathbf{Ax} = \mathbf{b}, \quad \mathbf{l} \leq \mathbf{x} \leq \mathbf{u} \tag{25.1}$$

whose matrices **A** exhibit some kind of a special structure. Most notably, (25.1) can be solved quickly by the network simplex method if **A** is the incidence matrix of a network (see Chapter 19). Another kind of special structure occurring frequently in applications is illustrated by the following matrix:

$$\mathbf{A} = \begin{bmatrix} 3 & 2 & 4 & 3 & 5 & 9 & 6 & 7 & 4 & 8 & 2 & 5 & 4 \\ 2 & 5 & 6 & 8 & 6 & 5 & 7 & 9 & 2 & 3 & 4 & 6 & 7 \\ 1 & 1 & 1 & & & & & & & & & & \\ & & & 1 & 1 & & & & & & & & \\ & & & & & 1 & 1 & 1 & 1 & & & & \\ & & & & & & & & & 1 & 1 & & \end{bmatrix}.$$

A matrix is said to exhibit a *generalized upper bounding*, or GUB, structure if a relatively large number m'' of its m rows constitute a matrix with at most one nonzero entry in each column, the nonzero entry being equal to one. In this chapter, we shall present an implementation of the revised simplex method tailored for problems (25.1) whose matrices $\mathbf{A}$ exhibit a GUB structure. In this implementation, developed by G. B. Dantzig and R. M. Van Slyke (1967), and known as *generalized upper bounding*, each of the two systems $\mathbf{yB} = \mathbf{c}_B$ and $\mathbf{Bd} = \mathbf{a}$ solved in each iteration is reduced to a system of m' equations in m' variables such that $m' = m - m''$. It has been said that generalized upper bounding becomes faster than straightforward implementations of the revised simplex method when $m' \doteq 0.7m$ and nearly ten times as fast when $m' \doteq 0.2m$. Thus even very large problems (25.1) may be solved in practice as soon as they have a GUB structure with a reasonably small m'. For instance, in 1972, D. M. Hirshfeld reported having solved a problem with $m = 50{,}215$, $m' = 627$, and 282,468 variables in only $2\frac{1}{2}$ hours of total elapsed time on an IBM 370/165.

HOW TO SOLVE THE TWO SYSTEMS

We may assume that the m rows of $\mathbf{A}$ are ordered as just illustrated, with the last m'' components of each column forming either a column of the $m'' \times m''$ identity matrix or the zero column of length m''. Now consider the basis matrix $\mathbf{B}_k$ obtained after k iterations of the revised simplex method. For each $j = 1, 2, \ldots, m''$, this matrix must contain a column whose last m'' components form the jth column of the $m'' \times m''$ identity matrix; otherwise, all the entries in the $(m' + j)$th row of $\mathbf{B}_k$ would be zeros, contradicting the fact that $\mathbf{B}_k$ is nonsingular. Thus, the m columns of $\mathbf{B}_k$ may be permuted so that the resulting matrix contains the $m'' \times m''$ identity matrix in its lower right corner; as observed in Chapter 7, permutations of columns of $\mathbf{B}_k$ are implemented easily by permutations of the basis heading. For instance,

$$\mathbf{B}_k = \begin{bmatrix} 7 & 5 & 3 & 2 & 3 & 5 \\ 9 & 6 & 8 & 4 & 2 & 6 \\ & & & & 1 & \\ & & 1 & & & 1 \\ 1 & & & & & \\ & & & 1 & & \end{bmatrix} \quad \text{with} \quad \mathbf{x}_B = \begin{bmatrix} x_8 \\ x_{12} \\ x_4 \\ x_{11} \\ x_1 \\ x_5 \end{bmatrix}$$

may be presented as

$$\mathbf{B}_k = \begin{bmatrix} 3 & 5 & 3 & 5 & 7 & 2 \\ 8 & 6 & 2 & 6 & 9 & 4 \\ & & 1 & & & \\ 1 & & & 1 & & \\ & & & & 1 & \\ & & & & & 1 \end{bmatrix} \quad \text{with} \quad \mathbf{x}_B = \begin{bmatrix} x_4 \\ x_{12} \\ x_1 \\ x_5 \\ x_8 \\ x_{11} \end{bmatrix}.$$

Now we may write

$$\mathbf{B}_k = \left[\begin{array}{c|c} \mathbf{R}_k & \mathbf{S}_k \\ \hline \mathbf{T}_k & \mathbf{I} \end{array}\right] \begin{array}{l} m' \text{ rows} \\ \\ m'' \text{ rows} \end{array}$$

m' columns, m'' columns

When $\mathbf{z}$ is a vector of length m, let $\mathbf{z}'$ stand for the vector consisting of the first m' components of $\mathbf{z}$, and let $\mathbf{z}''$ stand for the vector consisting of the last m'' components of $\mathbf{z}$. In this notation, $\mathbf{y}\mathbf{B}_k = \mathbf{c}_B$ may be written as

$$[\mathbf{y}', \mathbf{y}''] \cdot \begin{bmatrix} \mathbf{R}_k & \mathbf{S}_k \\ \mathbf{T}_k & \mathbf{I} \end{bmatrix} = [\mathbf{c}'_B, \mathbf{c}''_B]$$

and then broken down into

$$\mathbf{y}'\mathbf{R}_k + \mathbf{y}''\mathbf{T}_k = \mathbf{c}'_B, \qquad \mathbf{y}'\mathbf{S}_k + \mathbf{y}'' = \mathbf{c}''_B.$$

Substituting for $\mathbf{y}''$ from the second equation into the first, we obtain

$$\mathbf{y}'(\mathbf{R}_k - \mathbf{S}_k\mathbf{T}_k) = \mathbf{c}'_B - \mathbf{c}''_B\mathbf{T}_k. \tag{25.2}$$

Thus $\mathbf{y}$ may be obtained by first solving system (25.2) and then computing

$$\mathbf{y}'' = \mathbf{c}''_B - \mathbf{y}'\mathbf{S}_k.$$

Similarly, $\mathbf{B}_k\mathbf{d} = \mathbf{a}$ may be written as

$$\begin{bmatrix} \mathbf{R}_k & \mathbf{S}_k \\ \mathbf{T}_k & \mathbf{I} \end{bmatrix} \cdot \begin{bmatrix} \mathbf{d}' \\ \mathbf{d}'' \end{bmatrix} = \begin{bmatrix} \mathbf{a}' \\ \mathbf{a}'' \end{bmatrix}$$

and then broken down into

$$\mathbf{R}_k\mathbf{d}' + \mathbf{S}_k\mathbf{d}'' = \mathbf{a}', \qquad \mathbf{T}_k\mathbf{d}' + \mathbf{d}'' = \mathbf{a}''.$$

Substituting for $\mathbf{d}''$ from the second equation into the first, we obtain

$$(\mathbf{R}_k - \mathbf{S}_k\mathbf{T}_k)\mathbf{d}' = \mathbf{a}' - \mathbf{S}_k\mathbf{a}''. \tag{25.3}$$

Thus $\mathbf{d}$ may be obtained by first solving system (25.3) and then computing

$$\mathbf{d}'' = \mathbf{a}'' - \mathbf{T}_k\mathbf{d}'.$$

As usual, systems (25.2) and (25.3) are easier to solve if a suitable factorization of the "working basis" $\mathbf{R}_k - \mathbf{S}_k\mathbf{T}_k$ is available. The key point of generalized upper bounding is a clever way of updating

$$\mathbf{B}_k = \begin{bmatrix} \mathbf{R}_k & \mathbf{S}_k \\ \mathbf{T}_k & \mathbf{I} \end{bmatrix} \quad \text{into} \quad \mathbf{B}_{k+1} = \begin{bmatrix} \mathbf{R}_{k+1} & \mathbf{S}_{k+1} \\ \mathbf{T}_{k+1} & \mathbf{I} \end{bmatrix}$$

which permits an easy factorization of $\mathbf{R}_{k+1} - \mathbf{S}_{k+1}\mathbf{T}_{k+1}$ in terms of $\mathbf{R}_k - \mathbf{S}_k\mathbf{T}_k$.

HOW TO UPDATE A FACTORIZATION OF THE WORKING BASIS

There are three different cases to consider. We shall illustrate them on the problem

$$\text{maximize} \quad \mathbf{cx} \qquad \text{subject to} \quad \mathbf{Ax} = \mathbf{b}, \quad \mathbf{x} \geq \mathbf{0}$$

with

$$\mathbf{A} = \begin{bmatrix} 8 & 3 & 6 & 1 & 1 & 2 & 3 & 5 & 2 & 3 \\ 5 & 3 & 4 & 4 & 4 & 3 & 4 & 4 & 3 & 4 \\ 9 & 2 & 4 & 2 & 3 & 3 & 4 & 3 & 1 & 4 \\ 1 & 1 & 1 & & & & & & & \\ & & & 1 & 1 & 1 & & & & \\ & & & & & & 1 & 1 & & \\ & & & & & & & & 1 & 1 \end{bmatrix}, \quad \mathbf{b} = \begin{bmatrix} 14.49 \\ 15.47 \\ 12.50 \\ 1 \\ 1 \\ 1 \\ 1 \end{bmatrix}$$

and

$$\mathbf{c} = [5 \quad 3 \quad 4 \quad 44 \quad 43 \quad 3 \quad 16 \quad 4 \quad 3 \quad 14].$$

Let us initialize by

$$\mathbf{B}_0 = \begin{bmatrix} 6 & 2 & 3 & 3 & 1 & 5 & 2 \\ 4 & 3 & 4 & 3 & 4 & 4 & 3 \\ 4 & 3 & 4 & 2 & 2 & 3 & 1 \\ 1 & & & 1 & & & \\ & 1 & & & 1 & & \\ & & & & & 1 & \\ & & 1 & & & & 1 \end{bmatrix} \quad \text{with} \quad \mathbf{x}_B = \begin{bmatrix} x_3 \\ x_6 \\ x_{10} \\ x_2 \\ x_4 \\ x_8 \\ x_9 \end{bmatrix} = \begin{bmatrix} 0.79 \\ 0.22 \\ 0.90 \\ 0.21 \\ 0.78 \\ 1 \\ 0.10 \end{bmatrix}$$

and note that

$$\mathbf{R}_0 - \mathbf{S}_0\mathbf{T}_0 = \begin{bmatrix} 6 & 2 & 3 \\ 4 & 3 & 4 \\ 4 & 3 & 4 \end{bmatrix} - \begin{bmatrix} 3 & 1 & 5 & 2 \\ 3 & 4 & 4 & 3 \\ 2 & 2 & 3 & 1 \end{bmatrix} \cdot \begin{bmatrix} 1 & & \\ & 1 & \\ & & \\ & & 1 \end{bmatrix} = \begin{bmatrix} 3 & 1 & 1 \\ 1 & -1 & 1 \\ 2 & 1 & 3 \end{bmatrix}.$$

The first iteration is as follows.

Step 1. Solving the system

$$\mathbf{y}'(\mathbf{R}_0 - \mathbf{S}_0\mathbf{T}_0) = [c_3 - c_2, c_6 - c_4, c_{10} - c_9] = [1, -41, 11]$$

we obtain $\mathbf{y}' = [-7, 30, -4]$. Then we compute

$$\mathbf{y}'' = [3, 44, 4, 3] - [-7, 30, -4] \cdot \begin{bmatrix} 3 & 1 & 5 & 2 \\ 3 & 4 & 4 & 3 \\ 2 & 2 & 3 & 1 \end{bmatrix} = [-58, -61, -69, -69].$$

Step 2. Since $c_1 - \mathbf{y}[8, 5, 9, 1, 0, 0, 0]^T = 5$, we may let x_1 enter the basis.

Step 3. Solving the system

$$(\mathbf{R}_0 - \mathbf{S}_0\mathbf{T}_0) \cdot \mathbf{d}' = \begin{bmatrix} a_{11} - a_{12} \\ a_{21} - a_{22} \\ a_{31} - a_{32} \end{bmatrix} = \begin{bmatrix} 5 \\ 2 \\ 7 \end{bmatrix} \quad \text{we get } \mathbf{d}' = \begin{bmatrix} 1 \\ 0.5 \\ 1.5 \end{bmatrix}.$$

Then we compute

$$\mathbf{d}'' = \begin{bmatrix} a_{41} - 1 \\ a_{51} - 0.5 \\ a_{61} \\ a_{71} - 1.5 \end{bmatrix} = \begin{bmatrix} 0 \\ -0.5 \\ 0 \\ -1.5 \end{bmatrix}.$$

Step 4. The variable x_6 has to leave the basis and the entering variable x_1 will assume the value of $t = 0.44$.

Step 5. Now we shall illustrate the general case where the leaving pth column of $\mathbf{B}_k$ happens to be one of its first m' columns. To obtain $\mathbf{B}_{k+1}$ in this case, we simply replace the leaving column by the entering column $\mathbf{a}$. In our example,

$$\mathbf{B}_1 = \begin{bmatrix} 6 & 8 & 3 & 3 & 1 & 5 & 2 \\ 4 & 5 & 4 & 3 & 4 & 4 & 3 \\ 4 & 9 & 4 & 2 & 2 & 3 & 1 \\ 1 & 1 & & 1 & & & \\ & & & & 1 & & \\ & & & & & 1 & \\ & & 1 & & & & 1 \end{bmatrix} \quad \text{with} \quad \mathbf{x}_B = \begin{bmatrix} x_3 \\ x_1 \\ x_{10} \\ x_2 \\ x_4 \\ x_8 \\ x_9 \end{bmatrix} = \begin{bmatrix} 0.79 - t \\ t \\ 0.90 - 1.5t \\ 0.21 \\ 0.78 + 0.5t \\ 1 \\ 0.10 + 1.5t \end{bmatrix} = \begin{bmatrix} 0.35 \\ 0.44 \\ 0.24 \\ 0.21 \\ 1 \\ 1 \\ 0.76 \end{bmatrix}.$$

Note that $\mathbf{S}_{k+1} = \mathbf{S}_k$, and so

$$\mathbf{B}_k = \begin{bmatrix} \mathbf{R}_k & \mathbf{S}_k \\ \mathbf{T}_k & \mathbf{I} \end{bmatrix} \quad \text{is updated into} \quad \mathbf{B}_{k+1} = \begin{bmatrix} \mathbf{R}_{k+1} & \mathbf{S}_k \\ \mathbf{T}_{k+1} & \mathbf{I} \end{bmatrix}.$$

The matrices $\mathbf{R}_k$ and $\mathbf{R}_{k+1}$ differ only in their pth columns, as do $\mathbf{T}_k$ and $\mathbf{T}_{k+1}$. Hence, $\mathbf{R}_k - \mathbf{S}_k\mathbf{T}_k$ and $\mathbf{R}_{k+1} - \mathbf{S}_{k+1}\mathbf{T}_{k+1} = \mathbf{R}_{k+1} - \mathbf{S}_k\mathbf{T}_{k+1}$ differ only in their pth columns; the pth column of $\mathbf{R}_{k+1} - \mathbf{S}_k\mathbf{T}_{k+1}$ is $\mathbf{a}' - \mathbf{S}_k\mathbf{a}'' = (\mathbf{R}_k - \mathbf{S}_k\mathbf{T}_k)\mathbf{d}'$. Thus

$$(\mathbf{R}_{k+1} - \mathbf{S}_{k+1}\mathbf{T}_{k+1}) = (\mathbf{R}_k - \mathbf{S}_k\mathbf{T}_k)\mathbf{F}_{k+1}$$

with $\mathbf{F}_{k+1}$ standing for the $m' \times m'$ eta matrix whose eta column, in the pth position, is $\mathbf{d}'$. In our example,

$$\mathbf{R}_1 - \mathbf{S}_1\mathbf{T}_1 = (\mathbf{R}_0 - \mathbf{S}_0\mathbf{T}_0) \cdot \begin{bmatrix} 1 & 1 & \\ & 0.5 & \\ & 1.5 & 1 \end{bmatrix}.$$

The second iteration goes as follows.

Step 1. Solving the system

$$\mathbf{y}'(\mathbf{R}_1 - \mathbf{S}_1\mathbf{T}_1) = [c_3 - c_2, c_1 - c_2, c_{10} - c_9] = [1, 2, 11]$$

we obtain $\mathbf{y}' = [-6, 23, -2]$. Then we compute

$$\mathbf{y}'' = [3, 44, 4, 3] - [-6, 23, -2] \cdot \begin{bmatrix} 3 & 1 & 5 & 2 \\ 3 & 4 & 4 & 3 \\ 2 & 2 & 3 & 1 \end{bmatrix} = [-44, -38, -52, -52].$$

Step 2. Since $c_7 - \mathbf{y}[3, 4, 4, 0, 0, 1, 0]^T = 2$, we may let x_7 enter the basis.

Step 3. Solving the system

$$(\mathbf{R}_1 - \mathbf{S}_1\mathbf{T}_1) \cdot \mathbf{d}' = \begin{bmatrix} a_{17} - a_{18} \\ a_{27} - a_{28} \\ a_{37} - a_{38} \end{bmatrix} = \begin{bmatrix} -2 \\ 0 \\ 1 \end{bmatrix} \quad \text{we get } \mathbf{d}' = \begin{bmatrix} -1 \\ 0 \\ 1 \end{bmatrix}.$$

Then we compute

$$\mathbf{d}'' = \begin{bmatrix} a_{47} + 1 \\ a_{57} \\ a_{67} \\ a_{77} - 1 \end{bmatrix} = \begin{bmatrix} 1 \\ 0 \\ 1 \\ -1 \end{bmatrix}.$$

Step 4. The leaving variable is x_2 and the entering variable x_7 will assume the value of $t = 0.21$.

Step 5. Now we shall illustrate the general case where the leaving column $\mathbf{f}$ of $\mathbf{B}_k$ happens to be one of its last m'' columns and $\mathbf{f}'' = \mathbf{g}''$ for some other column $\mathbf{g}$ of $\mathbf{B}_k$. Note that $\mathbf{g}$ is necessarily one of the first m' columns of $\mathbf{B}_k$. To obtain $\mathbf{B}_{k+1}$ in this case, we first replace $\mathbf{f}$ by $\mathbf{g}$ (to preserve the $m'' \times m''$ identity matrix in the lower right corner) and then insert the entering column $\mathbf{a}$ into the position formerly occupied by $\mathbf{g}$. In our example, letting the first column of $\mathbf{B}_1$ play the role of $\mathbf{g}$, we obtain

$$\mathbf{B}_2 = \begin{bmatrix} 3 & 8 & 3 & 6 & 1 & 5 & 2 \\ 4 & 5 & 4 & 4 & 4 & 4 & 3 \\ 4 & 9 & 4 & 4 & 2 & 3 & 1 \\ & 1 & & 1 & & & \\ & & & & 1 & & \\ 1 & & & & & 1 & \\ & & 1 & & & & 1 \end{bmatrix} \quad \text{with} \quad \mathbf{x}_B = \begin{bmatrix} x_7 \\ x_1 \\ x_{10} \\ x_3 \\ x_4 \\ x_8 \\ x_9 \end{bmatrix} = \begin{bmatrix} t \\ 0.44 \\ 0.24 - t \\ 0.35 + t \\ 1 \\ 1 - t \\ 0.76 + t \end{bmatrix} = \begin{bmatrix} 0.21 \\ 0.44 \\ 0.03 \\ 0.56 \\ 1 \\ 0.79 \\ 0.97 \end{bmatrix}.$$

An analysis of the relationship between $\mathbf{R}_k - \mathbf{S}_k\mathbf{T}_k$ and $\mathbf{R}_{k+1} - \mathbf{S}_{k+1}\mathbf{T}_{k+1}$ gets a little complicated in this case. It will be convenient to view the update of $\mathbf{B}_k$ into $\mathbf{B}_{k+1}$ as consisting of two stages. In the first stage, a matrix $\tilde{\mathbf{B}}_k$ is created by interchanging the two columns **f** and **g** in $\mathbf{B}_k$; in the second stage, the matrix $\mathbf{B}_{k+1}$ is created by replacing **f** by **a** in $\tilde{\mathbf{B}}_k$. Thus we shall consider three matrices,

$$\mathbf{B}_k = \begin{bmatrix} \mathbf{R}_k & \mathbf{S}_k \\ \mathbf{T}_k & \mathbf{I} \end{bmatrix}, \quad \tilde{\mathbf{B}}_k = \begin{bmatrix} \tilde{\mathbf{R}}_k & \mathbf{S}_{k+1} \\ \mathbf{T}_k & \mathbf{I} \end{bmatrix}, \quad \text{and} \quad \mathbf{B}_{k+1} = \begin{bmatrix} \mathbf{R}_{k+1} & \mathbf{S}_{k+1} \\ \mathbf{T}_{k+1} & \mathbf{I} \end{bmatrix}.$$

For definiteness, let us say that **f** is the $(m' + i)$th column of $\mathbf{B}_k$, and let **r** denote the ith row of $\mathbf{T}_k$. Furthermore, let us say that **g** is the qth column of $\mathbf{B}_k$ and let $\mathbf{J}_{k+1}$ denote the $m' \times m'$ identity matrix whose qth row has been replaced by $-\mathbf{r}$. In our example,

$$\mathbf{J}_2 = \begin{bmatrix} -1 & -1 & 0 \\ & 1 & \\ & & 1 \end{bmatrix}.$$

We claim that

$$(\tilde{\mathbf{R}}_k - \mathbf{S}_{k+1}\mathbf{T}_k) = (\mathbf{R}_k - \mathbf{S}_k\mathbf{T}_k)\mathbf{J}_{k+1}. \tag{25.4}$$

This claim can be verified by comparing the two sides in (25.4) column by column; we leave the tedious details for problem 25.2. Next, denote the qth column of the $m' \times m'$ identity matrix by $\mathbf{e}_q$; write

$$\mathbf{z} = \begin{cases} \mathbf{J}_{k+1}\mathbf{d}' & \text{if } \mathbf{a}'' \neq \mathbf{f}'' \\ \mathbf{J}_{k+1}\mathbf{d}' + \mathbf{e}_q & \text{if } \mathbf{a}'' = \mathbf{f}'' \end{cases}$$

and let $\mathbf{F}_{k+1}$ stand for the $m' \times m'$ identity matrix whose qth column has been replaced by **z**. In our example,

$$\mathbf{z} = \begin{bmatrix} -1 & -1 & 0 \\ & 1 & \\ & & 1 \end{bmatrix} \cdot \begin{bmatrix} -1 \\ 0 \\ 1 \end{bmatrix} = \begin{bmatrix} 1 \\ 0 \\ 1 \end{bmatrix} \quad \text{and so} \quad \mathbf{F}_2 = \begin{bmatrix} 1 & & \\ 0 & 1 & \\ 1 & & 1 \end{bmatrix}.$$

We claim that

$$(\mathbf{R}_{k+1} - \mathbf{S}_{k+1}\mathbf{T}_{k+1}) = (\tilde{\mathbf{R}}_k - \mathbf{S}_{k+1}\mathbf{T}_k)\mathbf{F}_{k+1}. \tag{25.5}$$

Again, this claim can be verified by comparing the two sides in (25.5) column by column. Since $\mathbf{R}_{k+1} - \mathbf{S}_{k+1}\mathbf{T}_{k+1}$ differs from $\tilde{\mathbf{R}}_k - \mathbf{S}_{k+1}\mathbf{T}_k$ only in its qth column, which is $\mathbf{a}' - \mathbf{S}_{k+1}\mathbf{a}''$, verifying (25.5) amounts to verifying

$$(\tilde{\mathbf{R}}_k - \mathbf{S}_{k+1}\mathbf{T}_k)\mathbf{z} = \mathbf{a}' - \mathbf{S}_{k+1}\mathbf{a}''.$$

Since $\mathbf{J}_{k+1}$ is its own inverse, we have

$$(\tilde{\mathbf{R}}_k - \mathbf{S}_{k+1}\mathbf{T}_k)\mathbf{J}_{k+1}\mathbf{d}' = (\mathbf{R}_k - \mathbf{S}_k\mathbf{T}_k)\mathbf{d}' = \mathbf{a}' - \mathbf{S}_k\mathbf{a}''.$$

In addition,

$$\mathbf{a}' - \mathbf{S}_{k+1}\mathbf{a}'' = (\mathbf{a}' - \mathbf{S}_k\mathbf{a}'') + (\mathbf{S}_k - \mathbf{S}_{k+1})\mathbf{a}''.$$

Now it only remains to observe that $(\mathbf{S}_k - \mathbf{S}_{k+1})\mathbf{a}''$ is the zero vector if $\mathbf{a}'' \neq \mathbf{f}''$ and that

$$(\mathbf{S}_k - \mathbf{S}_{k+1})\mathbf{a}'' = (\tilde{\mathbf{R}}_k - \mathbf{S}_{k+1}\mathbf{T}_k)\mathbf{e}_q \quad \text{if} \quad \mathbf{a}'' = \mathbf{f}''.$$

(Both sides of the last equation equal $\mathbf{f}' - \mathbf{g}'$.) Combining (25.4) and (25.5) we obtain

$$(\mathbf{R}_{k+1} - \mathbf{S}_{k+1}\mathbf{T}_{k+1}) = (\mathbf{R}_k - \mathbf{S}_k\mathbf{T}_k)\mathbf{J}_{k+1}\mathbf{F}_{k+1}.$$

In our example,

$$\mathbf{R}_2 - \mathbf{S}_2\mathbf{T}_2 = (\mathbf{R}_1 - \mathbf{S}_1\mathbf{T}_1) \cdot \begin{bmatrix} -1 & -1 & 0 \\ & 1 & \\ & & 1 \end{bmatrix} \cdot \begin{bmatrix} 1 & & \\ 0 & 1 & \\ 1 & & 1 \end{bmatrix}.$$

The third iteration is as follows.

Step 1. Solving the system

$$\mathbf{y}'(\mathbf{R}_2 - \mathbf{S}_2\mathbf{T}_2) = [c_7 - c_8, c_1 - c_3, c_{10} - c_9] = [12, 1, 11]$$

we obtain $\mathbf{y}' = [-6.8, 22.6, -1.6]$. Then we compute

$$\begin{aligned} \mathbf{y}'' &= [4, 44, 4, 3] - [-6.8, 22.6, -1.6] \begin{bmatrix} 6 & 1 & 5 & 2 \\ 4 & 4 & 4 & 3 \\ 4 & 2 & 3 & 1 \end{bmatrix} \\ &= [-39.2, -36.4, -47.6, -49.6]. \end{aligned}$$

Step 2. Since $c_5 - \mathbf{y}[1, 4, 3, 0, 1, 0, 0]^T = 0.6$, we may let x_5 enter the basis.

Step 3. Solving the system

$$(\mathbf{R}_2 - \mathbf{S}_2\mathbf{T}_2) \cdot \mathbf{d}' = \begin{bmatrix} a_{15} - a_{14} \\ a_{25} - a_{24} \\ a_{35} - a_{34} \end{bmatrix} = \begin{bmatrix} 0 \\ 0 \\ 1 \end{bmatrix} \quad \text{we get } \mathbf{d}' = \begin{bmatrix} 0.2 \\ 0.4 \\ -0.4 \end{bmatrix}.$$

Then we compute

$$\mathbf{d}'' = \begin{bmatrix} a_{45} - 0.4 \\ a_{55} \\ a_{65} - 0.2 \\ a_{75} + 0.4 \end{bmatrix} = \begin{bmatrix} -0.4 \\ 1 \\ -0.2 \\ 0.4 \end{bmatrix}.$$

Step 4. The variable x_4 has to leave the basis and the entering variable x_5 will assume the value of $t = 1$.

Step 5. Now we shall illustrate the general case where the leaving column $\mathbf{f}$ of $\mathbf{B}_k$ happens to be one of its last m'' columns but $\mathbf{B}_k$ has no other column $\mathbf{g}$ with $\mathbf{g}'' = \mathbf{f}''$. Since $\mathbf{B}_{k+1}$ must contain a column $\mathbf{h}$ with $\mathbf{h}'' = \mathbf{f}''$, we necessarily have

$\mathbf{h} = \mathbf{a}$ and so $\mathbf{a}'' = \mathbf{f}''$. To obtain $\mathbf{B}_{k+1}$, we simply replace the leaving column $\mathbf{f}$ by the entering column $\mathbf{a}$. In our example,

$$\mathbf{B}_3 = \begin{bmatrix} 3 & 8 & 3 & 6 & 1 & 5 & 2 \\ 4 & 5 & 4 & 4 & 4 & 4 & 3 \\ 4 & 9 & 4 & 4 & 3 & 3 & 1 \\ & 1 & & 1 & & & \\ & & & & 1 & & \\ 1 & & & & & 1 & \\ & & 1 & & & & 1 \end{bmatrix} \quad \text{with} \quad \mathbf{x}_B = \begin{bmatrix} x_7 \\ x_1 \\ x_{10} \\ x_3 \\ x_5 \\ x_8 \\ x_9 \end{bmatrix} = \begin{bmatrix} 0.21 - 0.2t \\ 0.44 - 0.4t \\ 0.03 + 0.4t \\ 0.56 + 0.4t \\ t \\ 0.79 + 0.2t \\ 0.97 - 0.4t \end{bmatrix} = \begin{bmatrix} 0.01 \\ 0.04 \\ 0.43 \\ 0.96 \\ 1 \\ 0.99 \\ 0.57 \end{bmatrix}.$$

Note that $\mathbf{R}_{k+1} = \mathbf{R}_k$ and $\mathbf{T}_{k+1} = \mathbf{T}_k$; thus,

$$\mathbf{B}_k = \begin{bmatrix} \mathbf{R}_k & \mathbf{S}_k \\ \mathbf{T}_k & \mathbf{I} \end{bmatrix} \quad \text{is updated into} \quad \mathbf{B}_{k+1} = \begin{bmatrix} \mathbf{R}_k & \mathbf{S}_{k+1} \\ \mathbf{T}_k & \mathbf{I} \end{bmatrix}.$$

Furthermore, if the leaving column $\mathbf{f}$ appeared in the $(m' + i)$th position of $\mathbf{B}_k$, then $\mathbf{S}_{k+1}$ differs from $\mathbf{S}_k$ only in its ith column, and all the entries in the ith row of $\mathbf{T}_k$ are zeros (no column $\mathbf{g}$ in the first m' positions of $\mathbf{B}_k$ satisfied $\mathbf{g}'' = \mathbf{f}''$). We conclude that $\mathbf{S}_{k+1}\mathbf{T}_k = \mathbf{S}_k\mathbf{T}_k$, and so

$$\mathbf{R}_{k+1} - \mathbf{S}_{k+1}\mathbf{T}_{k+1} = \mathbf{R}_k - \mathbf{S}_k\mathbf{T}_k.$$

We leave the example here and review our general findings. The update of each $\mathbf{B}_k$ falls into one of three cases.

Case 1. The leaving column is one of the first m' columns of $\mathbf{B}_k$. In this case, we simply replace the leaving column by the entering column. Consequently,

$$\mathbf{R}_{k+1} - \mathbf{S}_{k+1}\mathbf{T}_{k+1} = (\mathbf{R}_k - \mathbf{S}_k\mathbf{T}_k)\mathbf{F}_{k+1}$$

with a readily available eta matrix $\mathbf{F}_{k+1}$.

Case 2. The leaving column $\mathbf{f}$ is one of the last m'' columns of $\mathbf{B}_k$, and some other column $\mathbf{g}$ of $\mathbf{B}_k$ satisfies $\mathbf{g}'' = \mathbf{f}''$. In this case, we replace $\mathbf{f}$ by $\mathbf{g}$ and then insert the entering column into the position formerly occupied by $\mathbf{g}$. Consequently,

$$\mathbf{R}_{k+1} - \mathbf{S}_{k+1}\mathbf{T}_{k+1} = (\mathbf{R}_k - \mathbf{S}_k\mathbf{T}_k)\mathbf{J}_{k+1}\mathbf{F}_{k+1}$$

with a readily available matrix $\mathbf{J}_{k+1}$, differing from $\mathbf{I}$ in only one row, and an easily obtainable eta matrix $\mathbf{F}_{k+1}$.

Case 3. The leaving column $\mathbf{f}$ is one of the last m'' columns of $\mathbf{B}_k$ and no other column $\mathbf{g}$ of $\mathbf{B}_k$ satisfies $\mathbf{g}'' = \mathbf{f}''$. In this case, we simply replace the leaving column by the entering column. Consequently,

$$\mathbf{R}_{k+1} - \mathbf{S}_{k+1}\mathbf{T}_{k+1} = \mathbf{R}_k - \mathbf{S}_k\mathbf{T}_k.$$

After k iterations, $\mathbf{R}_k - \mathbf{S}_k\mathbf{T}_k$ may be represented as

$$\mathbf{R}_k - \mathbf{S}_k\mathbf{T}_k = (\mathbf{R}_0 - \mathbf{S}_0\mathbf{T}_0)\mathbf{J}_1\mathbf{F}_1\mathbf{J}_2\mathbf{F}_2 \cdots \mathbf{J}_k\mathbf{F}_k$$

possibly with some of the matrices $\mathbf{J}_i$ and $\mathbf{F}_i$ missing. This factorization facilitates the solutions of $\mathbf{y}'(\mathbf{R}_k - \mathbf{S}_k\mathbf{T}_k) = \mathbf{c}'_B - \mathbf{c}''_B\mathbf{T}_k$ and $(\mathbf{R}_k - \mathbf{S}_k\mathbf{T}_k)\mathbf{d}' = \mathbf{a}' - \mathbf{S}_k\mathbf{a}''$ in the usual way (not illustrated here).

PROBLEMS

25.1 Maximize $2x_1 + x_2 + 4x_3 + 4x_4 + 5x_5 + 3x_6$

subject to

$$\begin{aligned} 2x_1 - x_2 + x_3 + 2x_4 - x_5 + x_6 &\leq 15 \\ -x_1 + x_2 + 2x_3 + x_4 + x_5 + x_6 &\leq 21 \\ x_1 + x_2 &\leq 4 \\ x_3 + x_4 &\leq 5 \\ x_1, x_2, x_3, x_4 \geq 0, \quad 0 \leq x_5 \leq 5, \quad 0 \leq x_6 &\leq 9. \end{aligned}$$

25.2 Verify (25.4).

25.3 How many multiplications are required to compute $\mathbf{c}''_B\mathbf{T}_k$, $\mathbf{S}_k\mathbf{a}''$, $\mathbf{T}_k\mathbf{d}'$, and $\mathbf{J}_{k+1}\mathbf{d}'$?

26

The Dantzig–Wolfe Decomposition Principle

The subject of this chapter consists of two parts. First, every linear programming problem

$$\text{maximize} \quad \mathbf{cx} \qquad \text{subject to} \quad \mathbf{Ax} = \mathbf{b}, \quad \mathbf{l} \le \mathbf{x} \le \mathbf{u} \tag{26.1}$$

is equivalent to a so-called *master problem*,

$$\text{maximize} \quad \tilde{\mathbf{c}}\tilde{\mathbf{x}} \qquad \text{subject to} \quad \tilde{\mathbf{A}}\tilde{\mathbf{x}} = \tilde{\mathbf{b}}, \quad \tilde{\mathbf{x}} \ge \mathbf{0} \tag{26.2}$$

with $\tilde{\mathbf{A}}$ having fewer rows but typically many more columns than $\mathbf{A}$. Secondly, the master problem may be solved even when $\tilde{\mathbf{A}}$ and $\tilde{\mathbf{c}}$ are not explicitly available: a technique of *delayed column generation*, similar to that of Chapter 13, may be used to find the entering column of $\tilde{\mathbf{A}}$ and the corresponding component of $\tilde{\mathbf{c}}$ in each iteration of the revised simplex method. The resulting *decomposition algorithm* may be applied advantageously to a variety of LP problems exhibiting a special structure; in addition, it has an interesting economic interpretation.

THE MASTER PROBLEM

Let us partition the $m \times n$ matrix $\mathbf{A}$ arbitrarily into an $m' \times n$ matrix $\mathbf{A}'$ and an $m'' \times n$ matrix $\mathbf{A}''$. Now (26.1) may be written as

$$\text{maximize} \quad \mathbf{cx} \qquad \text{subject to} \quad \mathbf{A}'\mathbf{x} = \mathbf{b}', \quad \mathbf{A}''\mathbf{x} = \mathbf{b}'', \quad \mathbf{l} \le \mathbf{x} \le \mathbf{u}. \tag{26.3}$$

Then we enumerate the normal basic feasible solutions of

$$\mathbf{A}''\mathbf{x} = \mathbf{b}'', \quad \mathbf{l} \le \mathbf{x} \le \mathbf{u} \tag{26.4}$$

as $\mathbf{v}^1, \mathbf{v}^2, \ldots, \mathbf{v}^M$ and the basic feasible directions as $\mathbf{w}^1, \mathbf{w}^2, \ldots, \mathbf{w}^N$. By Theorem 18.4, a vector $\mathbf{x}$ satisfies (26.4) if and only if

$$\mathbf{x} = \sum_{k=1}^{M} r_k \mathbf{v}^k + \sum_{k=1}^{N} s_k \mathbf{w}^k \tag{26.5}$$

for some choice of nonnegative numbers $r_1, r_2, \ldots, r_M, s_1, s_2, \ldots, s_N$ such that $\sum r_k = 1$. Substituting for $\mathbf{x}$ from (26.5), we may write (26.3) as

$$\begin{aligned}
\text{maximize} \quad & \mathbf{c}\left(\sum_{k=1}^{M} r_k \mathbf{v}^k + \sum_{k=1}^{N} s_k \mathbf{w}^k\right) \\
\text{subject to} \quad & \mathbf{A}'\left(\sum_{k=1}^{M} r_k \mathbf{v}^k + \sum_{k=1}^{N} s_k \mathbf{w}^k\right) = \mathbf{b}' \\
& \sum_{k=1}^{M} r_k = 1 \\
& r_k \ge 0 \quad (1 \le k \le M), \quad s_k \ge 0 \quad (1 \le k \le N)
\end{aligned}$$

or, after simplification, as

$$\begin{aligned}
\text{maximize} \quad & \sum_{k=1}^{N} (\mathbf{c}\mathbf{v}^k) r_k + \sum_{k=1}^{N} (\mathbf{c}\mathbf{w}^k) s_k \\
\text{subject to} \quad & \sum_{k=1}^{M} r_k (\mathbf{A}'\mathbf{v}^k) + \sum_{k=1}^{N} s_k (\mathbf{A}'\mathbf{w}^k) = \mathbf{b}' \\
& \sum_{k=1}^{M} r_k = 1 \\
& r_k \ge 0 \quad (1 \le k \le M), \quad s_k \ge 0 \quad (1 \le k \le N).
\end{aligned}$$

This is the *master problem* associated with (26.3). For illustration, we shall consider

$$\mathbf{A} = \begin{bmatrix} 2 & 1 & -2 & -1 & 2 & -1 & -2 & -3 \\ 1 & -3 & 2 & 3 & -1 & 2 & 1 & 1 \\ -1 & 0 & 1 & 0 & 1 & 0 & 0 & 0 \\ 1 & -1 & 0 & 1 & 0 & 0 & 0 & 0 \\ 0 & 1 & -1 & 0 & 0 & 1 & -1 & 0 \\ 0 & 0 & 0 & -1 & 0 & -1 & 0 & 1 \\ 0 & 0 & 0 & 0 & -1 & 0 & 1 & -1 \end{bmatrix}, \quad \mathbf{b} = \begin{bmatrix} 4 \\ -2 \\ -3 \\ 1 \\ 4 \\ 3 \\ -5 \end{bmatrix},$$

$$\mathbf{c} = [\;9 \quad -1 \quad -4 \quad -2 \quad 8 \quad -2 \quad -8 \quad -12]$$

and $l_j = 0$, $u_j = +\infty$ for all j. Let us partition $\mathbf{A}$ and $\mathbf{b}$ into

$$\mathbf{A}' = \begin{bmatrix} 2 & 1 & -2 & -1 & 2 & -1 & -2 & -3 \\ 1 & -3 & 2 & 3 & -1 & 2 & 1 & 1 \end{bmatrix}, \quad \mathbf{b}' = \begin{bmatrix} 4 \\ -2 \end{bmatrix}$$

and

$$\mathbf{A}'' = \begin{bmatrix} -1 & 0 & 1 & 0 & 1 & 0 & 0 & 0 \\ 1 & -1 & 0 & 1 & 0 & 0 & 0 & 0 \\ 0 & 1 & -1 & 0 & 0 & 1 & -1 & 0 \\ 0 & 0 & 0 & -1 & 0 & -1 & 0 & 1 \\ 0 & 0 & 0 & 0 & -1 & 0 & 1 & -1 \end{bmatrix}, \quad \mathbf{b}'' = \begin{bmatrix} -3 \\ 1 \\ 4 \\ 3 \\ -5 \end{bmatrix}.$$

Now a vector $\mathbf{x}$ satisfies $\mathbf{A}''\mathbf{x} = \mathbf{b}''$, $\mathbf{x} \geq \mathbf{0}$ if and only if

$$\mathbf{x} = r_1 \begin{bmatrix} 3 \\ 4 \\ 0 \\ 2 \\ 0 \\ 0 \\ 0 \\ 5 \end{bmatrix} + r_2 \begin{bmatrix} 3 \\ 2 \\ 0 \\ 0 \\ 0 \\ 2 \\ 0 \\ 5 \end{bmatrix} + r_3 \begin{bmatrix} 5 \\ 4 \\ 0 \\ 0 \\ 2 \\ 0 \\ 0 \\ 3 \end{bmatrix} + s_1 \begin{bmatrix} 1 \\ 1 \\ 1 \\ 0 \\ 0 \\ 0 \\ 0 \\ 0 \end{bmatrix} + s_2 \begin{bmatrix} 0 \\ 0 \\ 0 \\ 0 \\ 0 \\ 1 \\ 1 \\ 1 \end{bmatrix} + s_3 \begin{bmatrix} 1 \\ 1 \\ 0 \\ 0 \\ 1 \\ 0 \\ 1 \\ 0 \end{bmatrix} + s_4 \begin{bmatrix} 0 \\ 1 \\ 0 \\ 1 \\ 0 \\ 0 \\ 1 \\ 1 \end{bmatrix}$$

for some choice of nonnegative numbers $r_1, r_2, r_3, s_1, s_2, s_3, s_4$ such that $r_1 + r_2 + r_3 = 1$. Hence the master problem assumes the form

$$\begin{aligned} &\text{maximize} \quad -41r_1 - 39r_2 + 21r_3 + 4s_1 - 22s_2 + 8s_3 - 23s_4 \\ &\text{subject to} \quad r_1 \begin{bmatrix} -7 \\ 2 \end{bmatrix} + r_2 \begin{bmatrix} -9 \\ 6 \end{bmatrix} + r_3 \begin{bmatrix} 9 \\ -6 \end{bmatrix} \\ &\qquad + s_1 \begin{bmatrix} 1 \\ 0 \end{bmatrix} + s_2 \begin{bmatrix} -6 \\ 4 \end{bmatrix} + s_3 \begin{bmatrix} 3 \\ -2 \end{bmatrix} + s_4 \begin{bmatrix} -5 \\ 2 \end{bmatrix} = \begin{bmatrix} 4 \\ -2 \end{bmatrix} \\ &\qquad r_1 + r_2 + r_3 = 1 \\ &\qquad r_1, r_2, r_3, s_1, s_2, s_3, s_4 \geq 0 \end{aligned}$$

which may be written as (26.2) with

$$\tilde{\mathbf{A}} = \begin{bmatrix} -7 & -9 & 9 & 1 & -6 & 3 & -5 \\ 2 & 6 & -6 & 0 & 4 & -2 & 2 \\ 1 & 1 & 1 & 0 & 0 & 0 & 0 \end{bmatrix}, \quad \tilde{\mathbf{b}} = \begin{bmatrix} 4 \\ -2 \\ 1 \end{bmatrix},$$

$$\tilde{\mathbf{c}} = [-41 \quad -39 \quad 21 \quad 4 \quad -22 \quad 8 \quad -23]$$

and

$$\tilde{\mathbf{x}} = [r_1, \quad r_2, \quad r_3, \quad s_1, \quad s_2, \quad s_3, \quad s_4]^T.$$

In general, constructing the master problem could be a formidable task: even when $\mathbf{A}$ is reasonably small, the number $M + N$ of the columns of $\tilde{\mathbf{A}}$ may be enormous. Nevertheless, one can solve the master problem by the revised simplex method without constructing $\tilde{\mathbf{A}}$ and $\tilde{\mathbf{c}}$ in advance. (The right-hand side column vector $\tilde{\mathbf{b}}$ is readily available; it consists of $\mathbf{b}'$ augmented by an extra component equal to one.) We are about to explain the details. (A similar trick was used in Chapter 13: the cutting-stock problem can be solved by the revised simplex method without tabulating all the data in advance.)

DELAYED COLUMN GENERATION

Each iteration of the revised simplex method applied to the master problem begins with a matrix $\mathbf{B}$, a row vector $\tilde{\mathbf{c}}_B$, and a column vector $\tilde{\mathbf{x}}_B$. The matrix $\mathbf{B}$ is nonsingular and consists of $m' + 1$ columns of $\tilde{\mathbf{A}}$, the vector $\tilde{\mathbf{c}}_B$ constitutes the corresponding portion of $\tilde{\mathbf{c}}$, and the vector $\tilde{\mathbf{x}}_B$ satisfies $\mathbf{B}\tilde{\mathbf{x}}_B = \tilde{\mathbf{b}}$, $\tilde{\mathbf{x}}_B \geq 0$. Now step 1, solving the system $\mathbf{yB} = \tilde{\mathbf{c}}_B$, can be carried out quite routinely. In step 2, we have to find the entering column. This can be any column $\tilde{\mathbf{a}}$ of $\tilde{\mathbf{A}}$ such that $\mathbf{y}\tilde{\mathbf{a}}$ is strictly less than the corresponding component of $\tilde{\mathbf{c}}$. We claim that this step can be implemented by solving the *subproblem*

$$\text{maximize} \quad (\mathbf{c} - \mathbf{y}'\mathbf{A}')\mathbf{x} \qquad \text{subject to} \quad \mathbf{A}''\mathbf{x} = \mathbf{b}'', \mathbf{l} \leq \mathbf{x} \leq \mathbf{u}$$

with $\mathbf{y}'$ standing for the row vector $[y_1, y_2, \ldots, y_{m'}]$, which is obtained from $\mathbf{y}$ by deleting the last component $y_{m'+1}$.

To justify this claim, recall first that every feasible solution of the master problem yields a feasible solution of the original problem and, therefore, a feasible solution of the subproblem. In particular, the basic feasible solution $\tilde{\mathbf{x}}_B$ of the master problem guarantees that the subproblem has a feasible solution. Now we may restrict our considerations to three outcomes:

(i) The subproblem has an optimal solution $\mathbf{x}^*$ such that $(\mathbf{c} - \mathbf{y}'\mathbf{A}')\mathbf{x}^* > y_{m'+1}$.

(ii) The subproblem is unbounded.

(iii) The subproblem has an optimal solution $\mathbf{x}^*$ such that $(\mathbf{c} - \mathbf{y}'\mathbf{A}')\mathbf{x}^* \leq y_{m'+1}$.

In the first case, the simplex method applied to the subproblem finds a normal basic feasible solution $\mathbf{v}$ such that $(\mathbf{c} - \mathbf{y}'\mathbf{A}')\mathbf{v} > y_{m'+1}$. This vector $\mathbf{v}$, being one of the vectors $\mathbf{v}^k$ in (26.5), gives rise to a column

$$\tilde{\mathbf{a}} = \begin{bmatrix} \mathbf{A}'\mathbf{v} \\ 1 \end{bmatrix} \tag{26.6}$$

of $\tilde{\mathbf{A}}$, with the corresponding component of $\tilde{\mathbf{c}}$ equal to $\mathbf{cv}$. Since $\mathbf{y}\tilde{\mathbf{a}} = \mathbf{y}'\mathbf{A}'\mathbf{v} + y_{m'+1} < \mathbf{cv}$, we may see use $\tilde{\mathbf{a}}$ as the entering column. In the second case, the simplex method finds a basic feasible direction $\mathbf{w}$ such that $(\mathbf{c} - \mathbf{y}'\mathbf{A}')\mathbf{w} > 0$. This vector $\mathbf{w}$, being one of the vectors $\mathbf{w}^k$ in (26.5), gives rise to a column

$$\tilde{\mathbf{a}} = \begin{bmatrix} \mathbf{A}'\mathbf{w} \\ 0 \end{bmatrix} \tag{26.7}$$

of $\tilde{\mathbf{A}}$, with the corresponding component of $\tilde{\mathbf{c}}$ equal to $\mathbf{cw}$. Since $\mathbf{y}\tilde{\mathbf{a}} = \mathbf{y}'\mathbf{A}'\mathbf{w} < \mathbf{cw}$, we may use $\tilde{\mathbf{a}}$ as the entering column. In the last case, every basic feasible solution $\mathbf{v}$ must satisfy $(\mathbf{c} - \mathbf{y}'\mathbf{A}')\mathbf{v} \leq y_{m'+1}$ and every basic feasible direction $\mathbf{w}$ must satisfy $(\mathbf{c} - \mathbf{y}'\mathbf{A}')\mathbf{w} \leq 0$. Since every column $\tilde{\mathbf{a}}$ of $\tilde{\mathbf{A}}$ has either the form (26.6) for some basic feasible solution $\mathbf{v}$ or the form (26.7) for some basic feasible direction $\mathbf{w}$, we conclude that $\mathbf{y}\tilde{\mathbf{A}} \geq \tilde{\mathbf{c}}$. Hence, the current solution of the master problem is optimal.

Once the entering column has been found, the remaining steps of the iteration can be carried out routinely without any reference to $\tilde{\mathbf{A}}$ and $\tilde{\mathbf{c}}$.

THE DECOMPOSITION ALGORITHM

For illustration, let us return to our example and pretend that the master problem has not been constructed. We shall consider the problem of finding the initial $\mathbf{B}$ and $\tilde{\mathbf{c}}_B$ later; for the moment, let us assume that

$$\mathbf{B} = \begin{bmatrix} 1 & 3 & -7 \\ 0 & -2 & 2 \\ 0 & 0 & 1 \end{bmatrix}, \quad \tilde{\mathbf{c}}_B = [4, 8, -41], \quad \text{and} \quad \tilde{\mathbf{x}}_B = \begin{bmatrix} 5 \\ 2 \\ 1 \end{bmatrix}$$

have been obtained somehow, with the three columns of $\mathbf{B}$ arising from

$$\begin{bmatrix} 1 \\ 1 \\ 1 \\ 0 \\ 0 \\ 0 \\ 0 \\ 0 \end{bmatrix}, \quad \begin{bmatrix} 1 \\ 1 \\ 0 \\ 0 \\ 1 \\ 0 \\ 1 \\ 0 \end{bmatrix}, \quad \text{and} \quad \begin{bmatrix} 3 \\ 4 \\ 0 \\ 2 \\ 0 \\ 0 \\ 0 \\ 5 \end{bmatrix},$$

respectively. Now the first iteration begins.

Step 1. Solving the system $\mathbf{yB} = \tilde{\mathbf{c}}_B$ we obtain $\mathbf{y} = [4, 2, -17]$, so that $\mathbf{y}' = [4, 2]$.

Step 2. We first calculate

$$\mathbf{c} - \mathbf{y}'\mathbf{A}' = [-1, 1, 0, -4, 2, -2, -2, -2]$$

and then we solve the subproblem. Since the basic feasible solution

$$\mathbf{v} = [5, 4, 0, 0, 2, 0, 0, 3]^T$$

satisfies $(\mathbf{c} - \mathbf{y}'\mathbf{A}')\mathbf{v} = -3 > -17 = y_3$, it gives rise to the entering column $\tilde{\mathbf{a}} = [9, -6, 1]^T$ with the corresponding component of $\tilde{\mathbf{c}}$ equal to $\mathbf{cv} = 21$.

Step 3. Solving the system $\mathbf{Bd} = \tilde{\mathbf{a}}$ we obtain $\mathbf{d} = [4, 4, 1]^T$.

Step 4. Comparing the ratios 5/4, 2/4, and 1/1 we obtain $t = 0.5$ and find that the second column of **B** has to leave.

Step 5. Now we have

$$\mathbf{B} = \begin{bmatrix} 1 & 9 & -7 \\ 0 & -6 & 2 \\ 0 & 1 & 1 \end{bmatrix}, \quad \tilde{\mathbf{c}}_B = [4, 21, -41], \quad \text{and} \quad \tilde{\mathbf{x}}_B = \begin{bmatrix} 5 - 4t \\ t \\ 1 - t \end{bmatrix} = \begin{bmatrix} 3 \\ 0.5 \\ 0.5 \end{bmatrix}.$$

The second iteration begins.

Step 1. Solving the system $\mathbf{yB} = \tilde{\mathbf{c}}_B$ we obtain $\mathbf{y} = [4, 0.25, -13.5]$, so that $\mathbf{y}' = [4, 0.25]$.

Step 2. We first calculate

$$\mathbf{c} - \mathbf{y}'\mathbf{A}' = [0.75, -4.25, 3.5, 1.25, 0.25, 1.5, -0.25, -0.25]$$

and then we solve the subproblem. Since the basic feasible direction

$$\mathbf{w} = [0, 0, 0, 0, 0, 1, 1, 1]^T$$

satisfies $(\mathbf{c} - \mathbf{y}'\mathbf{A}')\mathbf{w} = 1 > 0$, it gives rise to the entering column $\tilde{\mathbf{a}} = [-6, 4, 0]^T$ with the corresponding component of $\tilde{\mathbf{c}}$ equal to $\mathbf{cw} = -22$.

Step 3. Solving the system $\mathbf{Bd} = \tilde{\mathbf{a}}$ we obtain $\mathbf{d} = [2, -0.5, 0.5]^T$.

Step 4. Comparing the ratios 3/2 and 0.5/0.5 we obtain $t = 1$ and find that the third column of **B** has to leave.

Step 5. Now we have

$$\mathbf{B} = \begin{bmatrix} 1 & 9 & -6 \\ 0 & -6 & 4 \\ 0 & 1 & 0 \end{bmatrix}, \quad \tilde{\mathbf{c}}_B = [4, 21, -22], \quad \text{and} \quad \tilde{\mathbf{x}}_B = \begin{bmatrix} 3 - 2t \\ 0.5 + 0.5t \\ t \end{bmatrix} = \begin{bmatrix} 1 \\ 1 \\ 1 \end{bmatrix}.$$

The third iteration begins.

Step 1. Solving the system $\mathbf{yB} = \tilde{\mathbf{c}}_B$ we obtain $\mathbf{y} = [4, 0.5, -12]$, so that $\mathbf{y}' = [4, 0.5]$.

Step 2. We first calculate

$$\mathbf{c} - \mathbf{y}'\mathbf{A}' = [0.5, -3.5, 3, 0.5, 0.5, 1, -0.5, -0.5]$$

and then we solve the subproblem. Since the basic feasible solution

$$\mathbf{v} = [3, 2, 0, 0, 0, 2, 0, 5]^T$$

satisfies $(\mathbf{c} - \mathbf{y}'\mathbf{A}')\mathbf{v} = -6 > -12 = y_3$, it gives rise to the entering column $\tilde{\mathbf{a}} = [-9, 6, 1]^T$ with the corresponding component of $\tilde{\mathbf{c}}$ equal to $\mathbf{cv} = -39$.

Step 3. Solving the system $\mathbf{Bd} = \tilde{\mathbf{a}}$ we obtain $\mathbf{d} = [0, 1, 3]^T$.

Step 4. Comparing the ratios 1/1 and 1/3 we obtain $t = \frac{1}{3}$ and find that the third column of $\mathbf{B}$ has to leave.

Step 5. Now we have

$$\mathbf{B} = \begin{bmatrix} 1 & 9 & -9 \\ 0 & -6 & 6 \\ 0 & 1 & 1 \end{bmatrix}, \quad \tilde{\mathbf{c}}_B = [4, 21, -39], \quad \text{and} \quad \tilde{\mathbf{x}}_B = \begin{bmatrix} 1 \\ 1-t \\ t \end{bmatrix} = \begin{bmatrix} 1 \\ \frac{2}{3} \\ \frac{1}{3} \end{bmatrix}.$$

The fourth iteration begins.

Step 1. Solving the system $\mathbf{yB} = \tilde{\mathbf{c}}_B$ we obtain $\mathbf{y} = [4, 1, -9]$, so that $\mathbf{y}' = [4, 1]$.

Step 2. We first calculate

$$\mathbf{c} - \mathbf{y}'\mathbf{A}' = [0, -2, 2, -1, 1, 0, -1, -1]$$

and then we solve the subproblem. Since its optimum value equals $-9 = y_3$, we conclude that the current solution of the master problem is optimal. Recalling that the three columns of $\mathbf{B}$ arise from

$$\begin{bmatrix} 1 \\ 1 \\ 1 \\ 0 \\ 0 \\ 0 \\ 0 \\ 0 \end{bmatrix}, \quad \begin{bmatrix} 5 \\ 4 \\ 0 \\ 0 \\ 2 \\ 0 \\ 0 \\ 3 \end{bmatrix}, \quad \text{and} \quad \begin{bmatrix} 3 \\ 2 \\ 0 \\ 0 \\ 0 \\ 2 \\ 0 \\ 5 \end{bmatrix},$$

respectively, we transform our optimal solution $\tilde{\mathbf{x}}_B = [1, \frac{2}{3}, \frac{1}{3}]^T$ of the master problem into an optimal solution

$$\mathbf{x} = \begin{bmatrix} 1 \\ 1 \\ 1 \\ 0 \\ 0 \\ 0 \\ 0 \\ 0 \end{bmatrix} + \frac{2}{3}\begin{bmatrix} 5 \\ 4 \\ 0 \\ 0 \\ 2 \\ 0 \\ 0 \\ 3 \end{bmatrix} + \frac{1}{3}\begin{bmatrix} 3 \\ 2 \\ 0 \\ 0 \\ 0 \\ 2 \\ 0 \\ 5 \end{bmatrix} = \begin{bmatrix} \frac{16}{3} \\ \frac{13}{3} \\ 1 \\ 0 \\ \frac{4}{3} \\ \frac{2}{3} \\ 0 \\ \frac{11}{3} \end{bmatrix}$$

of the original problem.

Initialization

The initial $\mathbf{B}$ and $\tilde{\mathbf{c}}_B$ can be found by applying the two-phase simplex method directly to the master problem. During the first phase, each iteration begins with a matrix $\mathbf{B}$ and a vector $\tilde{\mathbf{x}}_B$ such that $\mathbf{B}\tilde{\mathbf{x}}_B = \tilde{\mathbf{b}}$. Some of the components of $\tilde{\mathbf{x}}_B$ are the values of artificial variables; the remaining components must be nonnegative. Initially $\mathbf{B} = \mathbf{I}$ and all the basic variables are artificial; when all the artificial variables have been driven out of the basis (or simply to zero), the second phase may begin. In each iteration, we aim to *decrease* the value of an ad hoc objective function

$$\sum_{j \in Q} \tilde{x}_j - \sum_{j \in P} \tilde{x}_j$$

with $j \in Q$ if and only if $\tilde{x}_j$ is an artificial variable whose value is positive and $j \in P$ if and only if $\tilde{x}_j$ is an artificial variable whose value is negative. Accordingly, $\tilde{\mathbf{c}}_B$ is defined by

$$\tilde{c}_j = \begin{cases} 1 & \text{if } j \in P \\ -1 & \text{if } j \in Q \\ 0 & \text{otherwise} \end{cases}$$

in each iteration, and $\mathbf{c}$ is replaced by the zero vector in each subproblem. For illustration, we return once again to our example. The first phase is initialized by

$$\mathbf{B} = \begin{bmatrix} 1 & & \\ & 1 & \\ & & 1 \end{bmatrix} \quad \text{and} \quad \tilde{\mathbf{x}}_B = \begin{bmatrix} 4 \\ -2 \\ 1 \end{bmatrix}.$$

The first iteration begins.

Step 1. Solving the system $\mathbf{yB} = [-1, 1, -1]$ we obtain $\mathbf{y} = [-1, 1, -1]$.

Step 2. We first calculate

$$-\mathbf{y}'\mathbf{A}' = [1, 4, -4, -4, 3, -3, -3, -4]$$

and then we solve the subproblem. Since the basic feasible direction

$$\mathbf{w} = [1, 1, 0, 0, 1, 0, 1, 0]^T$$

satisfies $(-\mathbf{y}'\mathbf{A}')\mathbf{w} > 0$, it gives rise to the entering column $\tilde{\mathbf{a}} = [3, -2, 0]^T$.

Step 3. Solving the system $\mathbf{Bd} = \tilde{\mathbf{a}}$ we obtain $\mathbf{d} = [3, -2, 0]^T$.

Step 4. The basic variables are about to assume the values $4 - 3t$, $-2 + 2t$, and 1, respectively. Hence, $t = 1$ and the second column of $\mathbf{B}$ leaves.

Step 5. Now we have

$$\mathbf{B} = \begin{bmatrix} 1 & 3 & \\ & -2 & \\ & 0 & 1 \end{bmatrix} \quad \text{and} \quad \tilde{\mathbf{x}}_B = \begin{bmatrix} 1 \\ 1 \\ 1 \end{bmatrix}.$$

Only the first and third of the three basic variables are artificial.

The second iteration begins.

Step 1. Solving the system $\mathbf{yB} = [-1, 0, -1]$ we obtain $\mathbf{y} = [-1, -1.5, -1]$.

Step 2. We first calculate

$$-\mathbf{y}'\mathbf{A}' = [3.5, -3.5, 1, 3.5, 0.5, 2, -0.5, -1.5]$$

and then we solve the subproblem. Since the basic feasible direction

$$\mathbf{w} = [1, 1, 1, 0, 0, 0, 0, 0]^T$$

satisfies $(-\mathbf{y}'\mathbf{A}')\mathbf{w} > 0$, it gives rise to the entering column $\tilde{\mathbf{a}} = [1, 0, 0]^T$. Since the entering column is identical with the first column of $\mathbf{B}$, the remaining steps of this iteration may be skipped; we shall maintain

$$\mathbf{B} = \begin{bmatrix} 1 & 3 & \\ 0 & -2 & \\ 0 & 0 & 1 \end{bmatrix} \quad \text{and} \quad \tilde{\mathbf{x}}_B = \begin{bmatrix} 1 \\ 1 \\ 1 \end{bmatrix}.$$

However, the first basic variable is no longer artificial.

The third iteration begins.

Step 1. Solving the system $\mathbf{yB} = [0, 0, -1]$ we obtain $\mathbf{y} = [0, 0, -1]$.

Step 2. We first calculate

$$-\mathbf{y}'\mathbf{A}' = [0, 0, 0, 0, 0, 0, 0, 0]$$

and then we solve the subproblem. Since the basic feasible solution

$$\mathbf{v} = [3, 4, 0, 2, 0, 0, 0, 5]^T$$

satisfies $(-\mathbf{y}'\mathbf{A}')\mathbf{v} > y_3$, it gives rise to the entering column $\tilde{\mathbf{a}} = [-7, 2, 1]^T$.

Step 3. Solving the system $\mathbf{Bd} = \tilde{\mathbf{a}}$ we obtain $\mathbf{d} = [-4, -1, 1]^T$.

Step 4. The basic variables are about to assume the values $1 + 4t$, $1 + t$, and $1 - t$, respectively. Hence $t = 1$ and the third column of $\mathbf{B}$ leaves.

Step 5. Now we have

$$\mathbf{B} = \begin{bmatrix} 1 & 3 & -7 \\ 0 & -2 & 2 \\ 0 & 0 & 1 \end{bmatrix} \quad \text{and} \quad \tilde{\mathbf{x}}_B = \begin{bmatrix} 5 \\ 2 \\ 1 \end{bmatrix}.$$

Since none of the three basic variables is artificial, the second phase may begin. To obtain $\tilde{\mathbf{c}}_B$, we recall that the three columns of $\mathbf{B}$ arise from

$$\begin{bmatrix}1\\1\\1\\0\\0\\0\\0\\0\end{bmatrix} \quad \begin{bmatrix}1\\1\\0\\0\\1\\0\\1\\0\end{bmatrix} \quad \text{and} \quad \begin{bmatrix}3\\4\\0\\2\\0\\0\\0\\5\end{bmatrix},$$

respectively. Hence

$$\tilde{\mathbf{c}}_B = \mathbf{c} \cdot \begin{bmatrix}1 & 1 & 3\\1 & 1 & 4\\1 & 0 & 0\\0 & 0 & 2\\0 & 1 & 0\\0 & 0 & 0\\0 & 1 & 0\\0 & 0 & 5\end{bmatrix} = [4, 8, -41].$$

Block-Angular Problems

The decomposition algorithm has obvious advantages and disadvantages when compared with the revised simplex method applied directly to the original problem. Each of the two systems $\mathbf{yB} = \tilde{\mathbf{c}}_B$ and $\mathbf{Bd} = \tilde{\mathbf{a}}$ solved in each iteration of the decomposition algorithm consists of only $m' + 1$ equations in $m' + 1$ variables rather than m equations in m variables. However, the execution of step 2 in each iteration of the decomposition algorithm may take up considerable time since it requires solving a subproblem with m'' equations and n variables. Clearly, the key to the success of the decomposition algorithm lies in a judicious partition of $\mathbf{A}$ into $\mathbf{A}'$ and $\mathbf{A}''$. Ideally, $\mathbf{A}'$ should have relatively few rows and yet each subproblem should be relatively easy to solve. These two requirements may be satisfied at the same time only if $\mathbf{A}$ has a rather special structure. The example just solved exhibits one kind of such a structure; by removing only a few rows of $\mathbf{A}$ we obtain the incidence matrix $\mathbf{A}''$ of a network. Problems with this structure are amenable to the decomposition algorithm: each subproblem can be solved easily by the network simplex method.

Another class of problems amenable to the decomposition algorithm is the class of *block-angular* problems. In these problems, each subproblem breaks down into independent parts that can be solved separately from each other. For instance, the problem

$$\text{maximize} \quad \mathbf{cx} \quad \text{subject to} \quad \mathbf{Ax} = \mathbf{b}, \quad \mathbf{x} \geq \mathbf{0} \tag{26.8}$$

with

$$\mathbf{A} = \begin{bmatrix} 1 & 2 & 0 & 0 & 2 & 3 & 0 & 0 & 1 & 0 \\ 1 & 1 & 0 & 0 & 1 & 1 & 0 & 0 & 0 & 1 \\ 2 & 3 & 1 & 0 & & & & & & \\ 1 & 1 & 0 & 1 & & & & & & \\ & & & & 1 & 3 & 1 & 0 & & \\ & & & & 2 & 1 & 0 & 1 & & \end{bmatrix}, \quad \mathbf{b} = \begin{bmatrix} 225 \\ 114 \\ 120 \\ 50 \\ 150 \\ 100 \end{bmatrix}$$

and

$$\mathbf{c} = [3 \quad 4 \quad 0 \quad 0 \quad 4 \quad 7 \quad 0 \quad 0 \quad 0 \quad 0]$$

is block-angular: the matrix $\mathbf{A}'$ consists of the first two rows of $\mathbf{A}$ and the matrix $\mathbf{A}''$ consists of the remaining four rows. When (26.8) is solved by the decomposition algorithm, each subproblem

$$\text{maximize} \quad \mathbf{c}'\mathbf{x} \quad \text{subject to} \quad \mathbf{A}''\mathbf{x} = \mathbf{b}'', \quad \mathbf{x} \geq \mathbf{0}$$

breaks down into independent parts,

$$\begin{aligned} \text{maximize} \quad & c'_1x_1 + c'_2x_2 + c'_3x_3 + c'_4x_4 \\ \text{subject to} \quad & 2x_1 + 3x_2 + x_3 = 120 \\ & x_1 + x_2 + x_4 = 50 \\ & x_1, x_2, x_3, x_4 \geq 0 \end{aligned}$$

and

$$\begin{aligned} \text{maximize} \quad & c'_5x_5 + c'_6x_6 + c'_7x_7 + c'_8x_8 \\ \text{subject to} \quad & x_5 + 3x_6 + x_7 = 150 \\ & 2x_5 + x_6 + x_8 = 100 \\ & x_5, x_6, x_7, x_8 \geq 0 \end{aligned}$$

which can be solved separately from each other. The constraints $\mathbf{A}'\mathbf{x} = \mathbf{b}'$ in block-angular problems are referred to as *linking contraints*, and the matrix $\mathbf{A}''$ is said to have a *block-diagonal structure*.

It may be interesting to note that an important class of problems combines the two kinds of special structure pointed out in our discussion. In so-called *multicommodity flow problems*, the matrix $\mathbf{A}''$ is *both* the incidence matrix of a network and block-diagonal. It was the work of L. R. Ford, Jr. and D. R. Fulkerson (1958b) and W. S. Jewell (1958) on these problems that inspired the general development of the decomposition principle by G. B. Dantzig and P. Wolfe (1960).

An Economic Interpretation of the Decomposition Algorithm

To take a fictitious example, consider a company manufacturing furniture from wood and metal. The company has two branches, one specializing in chairs and desks, and the other specializing in bedframes and bookcases. In the first branch:

- Each chair requires 1 unit of wood, 1 unit of metal, 2 hours in workshop A, 1 hour in workshop B, and brings in a net profit of $3.
- Each desk requires 2 units of wood, 1 unit of metal, 3 hours in workshop A, 1 hour in workshop B, and brings in a net profit of $4.
- 120 hours in workshop A and 50 hours in workshop B are available per day.

In the second branch:

- Each bedframe requires 2 units of wood, 1 unit of metal, 1 hour in workshop C, 2 hours in workshop D, and brings in a net profit of $4.
- Each bookcase requires 3 units of wood, 1 unit of metal, 3 hours in workshop C, 1 hour in workshop D, and brings in a net profit of $7.
- 150 hours in workshop C and 100 hours in workshop D are available per day.

The operations of the two branches are nearly but not quite independent; only 225 units of wood and 114 units of metal per day are available to the entire company. The problem of maximizing the total profit of the company (under the assumption that all the furniture can be sold) is nothing but the block-angular problem (26.8). More generally, block-angular problems often arise from scheduling groups of activities that are nearly but not quite independent. (Another example is provided by problem 1.7: mixing Avgas A and mixing Avgas B are two nearly independent groups of activities linked together only by their demand for the same four types of raw gasoline.)

The decomposition algorithm applied to such problems can be given an interesting economic interpretation, apparently first pointed out by C. Almon [see Dantzig (1963), Section 23.3]. In this interpretation, each iteration of the decomposition algorithm involves an exchange of memoranda between the manager in charge of the resources and managers in charge of production at their respective branches. The resource manager requires no information about the kind of goods that are produced and the constraints limiting their production; all he has to know is how many units of each resource are available to the company. In this sense, the decomposition algorithm amounts to planning *without complete information at the center*. The actual decisions as to the quantities produced are left to the branch managers (who have no idea of the amounts of the resources available to the entire company); in this sense, the planning is *decentralized*.

More precisely, each iteration begins with a set of proposals P_j submitted earlier by the branch managers. Each of these proposals P_j, *obtained by combining the memoranda received from the various branches*, is described only by the amount $\tilde{a}_{ij}$

of each resource i that it requires and by the net profit $\tilde{c}_j$ it brings in. The present tentative program P formulated by the resource manager is a mixture of these proposals, with each P_j used at only $x_j \cdot 100\%$ of its full intensity; in symbols, $P = \sum x_j P_j$. Of course, the numbers x_j have to be chosen in such a way that the resulting total demand for the resources can be satisfied. In particular, the plant itself has to be regarded as the $(m' + 1)$th resource; since each P_j uses up the full production capacity, the resource manager has to maintain $\sum x_j \leq 1$. (In other words, if the $(m' + 1)$th resource is measured in plants, then $\tilde{a}_{ij} = 1$ for $i = m' + 1$ and all j.)

The resource manager's aim is to design a more lucrative program by introducing a new constituent P_r into the current mixture P. This change has to be done in such a way that the resulting demand for the $m' + 1$ resources remains satisfied. This requirement presents an imminent constraint only for the resources that are fully used up by P; we shall refer to them as *bottleneck resources.* To be able to introduce P_r into the mixture P at some positive level $t \cdot 100\%$, the manager will have to release $t\tilde{a}_{ir}$ units of each bottleneck resource i by appropriate adjustments of the current concentrations x_j. In other words, P_r will be *substituted* for some suitable mixture $\sum d_j P_j$. The advantage or disadvantage of this substitution can be measured by *shadow prices* y_i assigned to bottleneck resources i in such a way that each constituent P_j of P pays for the resources it consumes; in symbols,

$$\sum_i y_i \tilde{a}_{ij} = \tilde{c}_j.$$

The loss of profit suffered when the concentrations x_j are adjusted equals the price of the bottleneck resources released by this change. Hence, the substitution will be profitable if and only if the net profit $t\tilde{c}_r$ brought in by tP_r exceeds the price $\sum_i y_i(t\tilde{a}_{ir})$ of the resources consumed by tP_r. To put it differently, P_r is worth introducing at some positive level if and only if it overpays for the resources it consumes. To obtain such a proposal P_r, the resource manager appeals to the production managers, asking them to describe their most profitable strategies under the assumption that they get charged $\$y_i$ for each unit of each bottleneck resource i with $1 \leq i \leq m'$. (To comply with this request, each production manager has to solve a linear programming problem.) Having received all the replies, the resource manager combines them into a single proposal P_r, described by the amounts $\tilde{a}_{ir}$ of each resource i consumed and by the net profit $\tilde{c}_r$. (While computing $\tilde{c}_r$, the resource manager has to keep in mind that each production manager has subtracted the phony charges for bottleneck resources from his or her own profit figure.)

If P_r does overpay for the resources it consumes, then the rest is straightforward. First, the resource manager determines the numbers d_j such that P_r can be substituted for the mixture $\sum d_j P_j$. Since the mixture must use up the same amounts of bottleneck resources as P_r, the numbers d_j can be found by solving the system of equations

$$\sum_j d_j \tilde{a}_{ij} = \tilde{a}_{ir}$$

(one equation for each bottleneck resource i). Then the resource manager finds the largest value of t such that tP_r can be substituted for $t\sum d_j P_j$. The gradual increase of t from zero to larger and larger values has to stop when the concentration of some P_j drops to zero or some resource previously available in excess gets fully used up. Finally, the resource manager updates the current program P by actually performing the substitution. Note that the production managers find only P_r; the remaining computations are done by the resource manager alone. This process is repeated until the most recent P_r does not overpay for the resources it consumes. At that moment, P is optimal: the net profit of any program is bounded from above by the total price of the bottleneck resources (otherwise, the production managers would have come up with something better than P_r) and this upper bound is attained by the net profit of P.

In our example, the execution of the decomposition algorithm can be interpreted as follows.

Iteration 1. Since the resource manager begins with no proposals whatsoever, his initial program P is to leave the entire plant idle. Consequently, there are no bottleneck resources and the memo sent to the branch managers asks simply for their most profitable strategies. Having come up with 30 chairs and 20 desks, the first manager replies:

70 units of wood and 50 units of metal used, profit of \$170 realized.

Having come up with 30 bedframes and 40 bookcases, the second manager replies:

180 units of wood and 70 units of metal used, profit of \$400 realized.

The resource manager combines these replies into the proposal

P_1: 250 units of wood and 120 units of metal used, profit of \$570 realized.

The largest value of t for which tP_1 remains feasible is 0.9 (at this point, wood gets fully used up). The updated program is $0.9P_1$.

Iteration 2. Now wood is the only bottleneck resource and its shadow price, \$2.28 per unit, is announced to the branch managers. Having come up with 50 chairs, the first manager replies:

50 units of wood and 50 units of metal used, net profit of \$36 realized.

Having come up with 50 bookcases, the second manager replies:

150 units of wood and 50 units of metal used, net profit of \$8 realized.

Discounting the phony charges of \$456 for 200 units of wood, the resource manager combines these replies into the proposal

P_2: 200 units of wood and 100 units of metal used, profit of \$500 realized.

This proposal can be substituted for $0.8P_1$. The largest value of t for which (0.9 −

$0.8t)P_1 + tP_2$ remains feasible equals 0.5 (at this point, the plant capacity gets fully used up). The updated program is $0.5P_1 + 0.5P_2$.

Iteration 3. Now the bottleneck resources are wood and the plant itself, with shadow prices \$1.40 and \$220.00 per unit, respectively. The shadow price of wood is announced to the branch managers. Having come up with 50 chairs, the first manager replies:

50 units of wood and 50 units of metal used, net profit of \$80 realized.

Having come up with 30 bedframes and 40 bookcases, the second manager replies:

180 units of wood and 70 units of metal used, net profit of \$148 realized.

Discounting the phony charges of \$322 for 230 units of wood, the resource manager combines these replies into the proposal

P_3: 230 units of wood and 120 units of metal used, profit of \$550 realized.

Note that this proposal does overpay for the resources it consumes (\$322 worth of wood plus \$220 worth of plant capacity); hence, the computations go on. The new proposal P_3 can be substituted for $0.6P_1 + 0.4P_2$. The largest value of t for which $(0.5 - 0.6t)P_1 + (0.5 - 0.4t)P_2 + tP_3$ remains feasible equals 0.5 (at this point, metal gets fully used up). The updated proposal is $0.2P_1 + 0.3P_2 + 0.5P_3$.

Iteration 4. Now the bottleneck resources are all three—wood, metal, and the plant itself—with shadow prices \$1, \$1, and \$200 per unit, respectively. The shadow prices of wood and metal are announced to the branch managers. Having come up with 50 chairs, the first manager replies:

50 units of wood and 50 units of metal used, net profit of \$50 realized.

Having come up with 30 bedframes and 40 bookcases, the second manager replies:

180 units of wood and 70 units of metal used, net profit of \$150 realized.

Discounting the phony charges of \$230 for 230 units of wood and \$120 for 120 units of metal, the resource manager combines these replies into a proposal that turns out to be identical to P_3. Since P_3 is being used in the current mixture, it does not overpay for the resources it consumes. Hence, the process stops here; the resource manager instructs the production managers to implement 20% of the first proposal plus 30% of the second proposal plus 50% of the third proposal. (An alternative way of telling the production managers what to do is suggested in problem 26.5.)

A Revelation

Our version of the decomposition algorithm differs from that which is commonly found in the literature: it is customary (i) to view the decomposition algorithm as a specialized tool for solving block-angular problems and (ii) to develop it from a master problem different from ours.

As for (i), the decomposition principle is quite independent of the concept of block-angular problems. Furthermore, the scope of the decomposition algorithm extends beyond this class of problems: all that is required is a partition of the set of constraints into a small set of (possibly "hard") constraints $\mathbf{A}'\mathbf{x} = \mathbf{b}'$ and a (possibly large) set of "easy" constraints $\mathbf{A}''\mathbf{x} = \mathbf{b}''$, $\mathbf{l} \leq \mathbf{x} \leq \mathbf{u}$. This broader point of view, related to many "relaxation procedures" of nonlinear and integer programming, has been suggested by A. F. Perold.

To explain (ii), we record a block-angular problem as

$$\begin{array}{lll} \text{maximize} & \displaystyle\sum_{j=1}^{k} \mathbf{c}^j\mathbf{x}^j & \\ \text{subject to} & \displaystyle\sum_{j=1}^{k} \mathbf{L}^j\mathbf{x}^j = \mathbf{b}^0 & \\ & \mathbf{S}^j\mathbf{x}^j = \mathbf{b}^j & (j = 1, 2, \ldots, k) \\ & \mathbf{l}^j \leq \mathbf{x}^j \leq \mathbf{u}^j & (j = 1, 2, \ldots, k) \end{array} \tag{26.9}$$

with matrices $\mathbf{L}^j$, $\mathbf{S}^j$, row vectors $\mathbf{c}^j$, and column vectors $\mathbf{b}^j$, $\mathbf{x}^j$, $\mathbf{l}^j$, $\mathbf{u}^j$. In particular,

$$\sum_{j=1}^{k} \mathbf{L}^j\mathbf{x}^j = \mathbf{b}^0$$

are the m' linking constraints (denoted earlier by $\mathbf{A}'\mathbf{x} = \mathbf{b}'$) and the k matrices $\mathbf{S}^j$ are blocks in the block-diagonal matrix (denoted earlier by $\mathbf{A}''$). In the master problem

$$\text{maximize} \quad \tilde{\mathbf{c}}\tilde{\mathbf{x}} \qquad \text{subject to} \quad \tilde{\mathbf{A}}\tilde{\mathbf{x}} = \tilde{\mathbf{b}}, \quad \tilde{\mathbf{x}} \geq \mathbf{0}$$

commonly associated with (26.9), the matrix $\tilde{\mathbf{A}}$ has $m' + k$ rather than $m' + 1$ rows. Each basic feasible solution $\mathbf{v}$ of $\mathbf{S}^j\mathbf{x}^j = \mathbf{b}^j$, $\mathbf{l}^j \leq \mathbf{x}^j \leq \mathbf{u}^j$ gives rise to a column

$$\begin{bmatrix} \mathbf{L}^j\mathbf{v} \\ \mathbf{e}^j \end{bmatrix}$$

of $\tilde{\mathbf{A}}$, with $\mathbf{e}^j$ standing for the jth column of the $k \times k$ identity matrix; the corresponding component of $\tilde{\mathbf{c}}$ equals $\mathbf{c}^j\mathbf{v}$. Each basic feasible direction $\mathbf{w}$ of $\mathbf{S}^j\mathbf{x}^j = \mathbf{b}^j$, $\mathbf{l}^j \leq \mathbf{x}^j \leq \mathbf{u}^j$ gives rise to a column

$$\begin{bmatrix} \mathbf{L}^j\mathbf{w} \\ \mathbf{0} \end{bmatrix}$$

of $\tilde{\mathbf{A}}$, with $\mathbf{0}$ standing for the zero column vector of length k; the corresponding component of $\tilde{\mathbf{c}}$ equals $\mathbf{c}^j\mathbf{w}$. The column vector $\tilde{\mathbf{b}}$ has the form

$$\begin{bmatrix} \mathbf{b}^0 \\ \mathbf{1} \end{bmatrix}$$

with $\mathbf{1}$ standing for the column vector of length k, all of whose components are ones. Arguments motivating this definition are nearly the same as those given at the beginning of this chapter, except that Theorem 18.4 is now applied to each of the k systems $\mathbf{S}^j\mathbf{x}^j = \mathbf{b}^j, \mathbf{l}^j \le \mathbf{x}^j \le \mathbf{u}^j$ separately. The attendant changes in the decomposition algorithm are straightforward: delayed column generation is now implemented by solving any or all of the k subproblems

$$\text{maximize} \quad (\mathbf{c}^j - \mathbf{y}'\mathbf{L}^j)\mathbf{x}^j \qquad \text{subject to} \quad \mathbf{S}^j\mathbf{x}^j = \mathbf{b}^j, \quad \mathbf{l}^j \le \mathbf{x}^j \le \mathbf{u}^j.$$

In the economic interpretation, the resource manager now regards each of the k branches as a separate resource and considers the replies from their managers individually instead of lumping all of them into a single proposal.

Although this second kind of a master problem differs from the first kind introduced earlier in this chapter, the two definitions are related: as pointed out by A. F. Perold (in conversation), the second kind may be obtained by iterating the first kind with a suitable choice of $\mathbf{A}'$ and $\mathbf{A}''$ in each iteration. We leave the details for problem 26.8.

Practical Uses

Since the typical number of simplex iterations is believed to grow logarithmically with the number of variables, and since even fairly small problems may give rise to master problems with astronomically large numbers of variables, it is hardly surprising that the decomposition algorithm often takes a considerable number of iterations to reach an optimal solution. It seems fair to say that the simplex method is faster. Nevertheless, this statement requires qualification: the simplex method is faster on the problems it can solve at all. More precisely, even though the simplex method can handle sparse problems with several thousand rows quite comfortably, its use on truly huge problems may be out of question because of excessive memory requirements. Such problems, if suitably structured, may still be amenable to the decomposition algorithm. In this sense, the decomposition algorithm is not meant to compete with the simplex method, but rather to complement it by extending the range of applicability of linear programming. For a report on an advanced implementation of the algorithm, see J. K. Ho and E. Loute (1981).

PROBLEMS

26.1 Maximize $3x_1 + 2x_2 + 4x_3 + 2x_4 + x_5 + 2x_6$

subject to

$$\begin{aligned}
2x_1 + x_2 + 4x_3 + 3x_4 + 2x_5 + 2x_6 &= 23 \\
- x_2 + x_4 + x_5 - x_6 &= 1 \\
-x_1 + x_3 - x_5 + x_6 &= 2 \\
x_1 + x_2 - x_3 - x_4 &= -3 \\
x_1, x_2, \ldots, x_6 &\ge 0.
\end{aligned}$$

26.2 Maximize $6x_1 + 4x_2 + 3x_3 + 3x_4$

subject to

$$
\begin{aligned}
x_1 + x_2 + x_5 &\le 3\\
- x_2 + x_3 + x_4 - x_5 &\le 5\\
2x_1 + x_2 &\le 4\\
x_1 + x_2 + x_3 &\le 6\\
x_4 + x_5 &\le 3\\
x_4 + 2x_5 &\le 4\\
x_1, x_2, \ldots, x_5 &\ge 0.
\end{aligned}
$$

26.3 Maximize $x_1 + 4x_2 + x_3 + 4x_4 + 2x_5 + 3x_6$

subject to

$$
\begin{aligned}
x_1 + 2x_2 &\le 3\\
-x_1 - x_2 + x_3 + x_4 &\le 0\\
x_3 + 2x_4 &\le 3\\
- x_3 - x_4 + x_5 + x_6 &\le 0\\
x_5 + 2x_6 &\le 3\\
x_1, x_2, \ldots, x_6 &\ge 0.
\end{aligned}
$$

26.4 Prove: If in some iteration of the decomposition algorithm, the subproblem has optimal value s^* then every feasible solution $\mathbf{x}$ of the original problem (26.3) has $\mathbf{cx} \le s^* + \mathbf{y}'\mathbf{b}'$. Is this fact of any practical use?

26.5 Our example illustrating the economic interpretation of the decomposition algorithm terminates with the resource manager telling the production managers to implement $0.2P_1 + 0.3P_2 + 0.5P_3$. Instead, he could allocate 54 units of wood and 50 units of metal to the first branch, and 171 units of wood and 64 units of metal to the second branch, leaving the task of finding the actual programs to the production managers. This alternative is particularly attractive when the master problem of the second kind is used in solving the block angular problem (26.9). Describe how to implement this strategy.

26.6 Solve the furniture example by the decomposition algorithm of the second kind and interpret each iteration in economic terms.

26.7 Solve problem 1.7 by the decomposition algorithm interpreted in economic terms.

26.8 Construct the master problem of the first kind arising from (26.8) when $\mathbf{A}'$ consists of rows 1, 2, 5, 6 of $\mathbf{A}$, and $\mathbf{A}''$ consists of rows 3, 4. Then construct the master problem of the first kind arising from the master problem just constructed when $\mathbf{A}'$ consists of rows 1, 2, 5 of $\tilde{\mathbf{A}}$, and $\mathbf{A}''$ consists of rows 3, 4. Compare the result with the master problem of the second kind arising from (26.8). Generalize to arbitrary problems (26.9) with $\boldsymbol{l}^j = 0$ and $\boldsymbol{u}^j = +\infty$ for all j.

Appendix: The Ellipsoid Method

In the remainder of this book, we shall outline the ellipsoid method for finding optimal solutions of linear programming problems. Even though the method is eminently important from the theoretical point of view, its impact on the practice of linear programming seems to be nil. For this reason, our exposition will be quite sketchy.

The notion of a theoretically satisfactory algorithm requires the running time to be bounded from above by a fixed polynomial in the size of the data. The virtues and failings of this definition have been discussed in Chapter 4; now we shall comment on the complications arising when all computations are carried out exactly and each rational number is represented as the ratio of two integers. Is there a theoretically satisfactory algorithm for solving systems of linear equations? The affirmative answer is not justified by merely pointing out that Gaussian elimination solves

$$\sum_{j=1}^{n} a_{ij}x_j = b_i \qquad (i = 1, 2, \ldots, n) \tag{A.1}$$

with only some $n^3/3$ multiplications and the same number of additions: the time required to carry out each of these operations has to be taken into account as well. Eliminating x_1 from the first equation in (A.1), we replace the remaining $n - 1$ equations by

$$\sum_{j=2}^{n} a'_{ij}x_j = b'_i \qquad (i = 2, 3, \ldots, n) \tag{A.2}$$

with

$$a'_{ij} = a_{11}a_{ij} - a_{i1}a_{1j} = \det\begin{bmatrix} a_{11} & a_{1j} \\ a_{i1} & a_{ij} \end{bmatrix}$$

and

$$b'_i = a_{11}b_i - a_{i1}b_1 = \det\begin{bmatrix} a_{11} & b_1 \\ a_{i1} & b_i \end{bmatrix}.$$

Note that the integer coefficients a'_{ij} in (A.2) may get bigger than the integer coefficients a_{ij} in (A.1); if M stands for the largest $|a_{ij}|$, then we may have $|a'_{ij}| = 2M^2$ for some i and j. To put it differently, the number of digits in the largest coefficient may just about double with the elimination of a variable. Should this phenomenon persist through the subsequent iterations, the number of digits in the largest coefficient would grow exponentially fast, and the subsequent arithmetical operations could become prohibitively laborious. Fortunately, as J. Edmonds (1967) pointed out, large common divisors can be factored out of each system obtained after the elimination of x_k with $k = 2, 3, \ldots, n - 1$. The elimination of x_2 from the first equation in (A.2) yields the system

$$\sum_{j=3}^{n} a''_{ij}x_j = b''_i \qquad (i = 3, 4, \ldots, n) \tag{A.3}$$

with integers

$$a''_{ij} = \frac{1}{a_{11}} \det\begin{bmatrix} a'_{22} & a'_{2j} \\ a'_{i2} & a'_{ij} \end{bmatrix} \quad \text{and} \quad b''_i = \frac{1}{a_{11}} \det\begin{bmatrix} a'_{22} & b'_2 \\ a'_{i2} & b'_i \end{bmatrix},$$

the elimination of x_3 from the first equation in (A.3) yields the system

$$\sum_{j=4}^{n} a'''_{ij}x_j = b'''_i \qquad (i = 4, 5, \ldots, n)$$

with integers

$$a'''_{ij} = \frac{1}{a'_{22}} \det\begin{bmatrix} a''_{33} & a''_{3j} \\ a''_{i3} & a''_{ij} \end{bmatrix} \quad \text{and} \quad b'''_i = \frac{1}{a'_{22}} \det\begin{bmatrix} a''_{33} & b''_3 \\ a''_{i3} & b''_i \end{bmatrix}$$

and so on. When these cancellations (by $a_{11}, a'_{22}, a''_{33}$, and so on) are performed in each iteration, the integers in each of the resulting systems remain tolerably small: the number of digits in each is at most the total number of digits in the $n(n + 1)$ integers a_{ij} and b_i. These findings are intimately related to the well-known Cramer's rule expressing the solution of (A.1) as

$$x_1 = p_1/q, \quad x_2 = p_2/q, \ldots, x_n = p_n/q$$

such that $p_1, p_2, \ldots, p_n$ and q are certain integers. [In fact, Cramer's rule generalizes into a set of formulas expressing the coefficients in the system obtained after k iterations of Gaussian elimination. These formulas may be found, for instance, in Gantmacher (1959).] Again, it can be proved that the number of digits in each of these $n + 1$ integers is at most the total number of digits in the $n(n + 1)$ integers a_{ij} and b_i.

Next, we shall mention two facts on linear programming, suggested in problems 16.1–16.3 and known long before the publication of Khachian's paper. First, solving linear programming problems is no more difficult (from the theoretical point of view) than solving systems of linear inequalities: to find an optimal solution of the LP problem

$$\text{maximize} \quad \mathbf{cx} \qquad \text{subject to} \quad \mathbf{Ax} \leq \mathbf{b}, \quad \mathbf{x} \geq \mathbf{0} \tag{A.4}$$

we need only find a solution of the system

$$\mathbf{Ax} \leq \mathbf{b}, \quad \mathbf{x} \geq \mathbf{0}, \quad \mathbf{yA} \geq \mathbf{c}, \quad \mathbf{y} \geq \mathbf{0}, \quad \mathbf{cx} \geq \mathbf{yb}. \tag{A.5}$$

(If $\mathbf{x}$, $\mathbf{y}$ is a solution of (A.5), then $\mathbf{x}$ is a feasible solution of (A.4), $\mathbf{y}$ is a feasible solution of the dual problem, and the inequality $\mathbf{cx} \geq \mathbf{yb}$ guarantees the optimality of both.) Secondly, finding solutions to systems of linear inequalities is no more difficult (from the theoretical point of view again) than merely recognizing solvable systems. To clarify this point, let us consider the procedure in Box A.1, which, given an arbitrary system

$$\sum_{j=1}^{n} a_{ij}x_j \leq b_i \qquad (i = 1, 2, \ldots, m), \tag{A.6}$$

delivers a certain set I of the inequality subscripts i and a certain set J of the variable subscripts j.

BOX A.1 Finding Two Sets of Subscripts

Step 1. [Initialize the construction of J.] Set $j = 1$ and $J = \{1, 2, \ldots, n\}$.

Step 2. [Can j be deleted from J?] Let J^* denote J with the subscript j deleted. If the system

$$\sum_{j \in J^*} a_{ij}x_j \leq b_i \qquad (i = 1, 2, \ldots, m)$$

is solvable, then replace J by J^*.

Step 3. [All subscripts j looked at?] If $j = n$, then go to step 4; otherwise, replace j by $j + 1$ and return to step 2.

Step 4. [Initialize the construction of I.] Set $i = 1$ and $I = \varnothing$.

Step 5. [Can i be added to I?] Let I^* denote I with the subscript i added. If the system

$$\sum_{j \in J} a_{ij}x_j \leq b_i \qquad (i = 1, 2, \ldots, m)$$

$$\sum_{j \in J} a_{ij}x_j \geq b_i \qquad (i \in I^*)$$

is solvable, then replace I by I^*.

Step 6. [All subscripts i looked at?] If $i = m$, then stop; otherwise, replace i by $i + 1$ and return to step 5.

Clearly, the sets I and J delivered by this procedure are such that the system

$$\begin{aligned} \sum_{j \in J} a_{ij}x_j &= b_i \qquad (i \in I) \\ \sum_{j \in J} a_{ij}x_j &\leq b_i \qquad (i \notin I) \end{aligned} \tag{A.7}$$

is solvable. Furthermore, it is not difficult to see that the system

$$\sum_{j \in J} a_{ij}x_j = b_i \qquad (i \in I) \tag{A.8}$$

has a unique solution; otherwise (A.7) would have a solution satisfying either $x_j = 0$ for some $j \in J$ (contradicting our construction of J) or $\sum a_{ij}x_j = b_i$ for some $i \notin I$ (contradicting our construction of I). Hence, as soon as a subroutine for recognizing solvable systems of linear inequalities is available, (A.6) can be solved by first finding I and J and then solving (A.8).

The ellipsoid method provides a theoretically satisfactory algorithm for recognizing solvable systems of linear inequalities. (As our preliminary observations show, any such algorithm yields a theoretically satisfactory algorithm for finding optimal solutions of LP problems.) The method is best described in geometrical terms. The *unit sphere* in the n-dimensional Euclidean space R^n is the set of all the points $\mathbf{x} = [x_1, x_2, \ldots, x_n]^T$ such that $x_1^2 + x_2^2 + \cdots + x_n^2 \leq 1$ or, in matrix notation, simply $\mathbf{x}^T\mathbf{x} \leq 1$. An *affine transformation* of R^n, defined by a nonsingular $n \times n$ matrix $\mathbf{A}$ and by a point $\mathbf{c}$ of R^n, is the mapping that assigns the point $\mathbf{A}(\mathbf{x} - \mathbf{c})$ to every point $\mathbf{x}$. An *ellipsoid* is the image of the unit sphere under some affine transformation; thus, an ellipsoid is the set of all the points $\mathbf{x}$ satisfying

$$(\mathbf{x} - \mathbf{c})^T\mathbf{A}^T\mathbf{A}(\mathbf{x} - \mathbf{c}) \leq 1. \tag{A.9}$$

The point $\mathbf{c}$ is the *center* of this ellipsoid. By a *half-ellipsoid* $\frac{1}{2}E$, we mean the intersection of an ellipsoid E with any half-space whose bounding hyperplane passes through the center of E. In algebraic terms, the half-ellipsoid $\frac{1}{2}E$ is the set of all the points $\mathbf{x}$ satisfying (A.9) as well as

$$\mathbf{a}^T\mathbf{x} \geq \mathbf{a}^T\mathbf{c}$$

for some nonzero row vector $\mathbf{a}^T$. The crucial point of the ellipsoid method is contained in the following statement.

THE KEY LEMMA. Every half-ellipsoid $\frac{1}{2}E$ is contained in an ellipsoid $\tilde{E}$ whose volume is less than $e^{-1/2(n+1)}$ times the volume of E. ■

To take a rather transparent example, consider the intersection of the unit sphere with the half-space $x_1 \geq 0$; this half-sphere is contained in the ellipsoid

$$\left(\frac{n+1}{n}\right)^2\left(x_1 - \frac{1}{n+1}\right)^2 + \frac{n^2-1}{n^2}\sum_{i=2}^{n} x_i^2 \leq 1 \tag{A.10}$$

whose volume equals

$$\frac{n}{n+1}\left(\frac{n^2}{n^2-1}\right)^{(n-1)/2} \tag{A.11}$$

times the volume of the unit sphere, and (A.11) is easily shown to be less than $e^{-1/2(n+1)}$. This example is illustrated in Figure A.1 with $n = 2$.

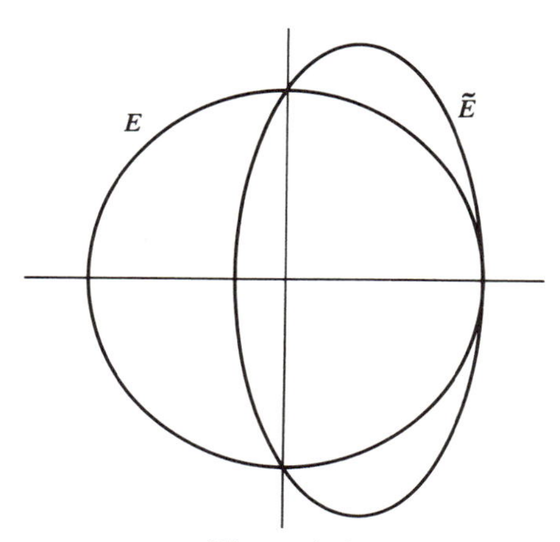

Figure A.1

The general version can be derived quite effortlessly from this illustrative special case: each half-ellipsoid $\frac{1}{2}E$ is the image of the half-sphere under some affine transformation and, since affine transformations preserve ratios of volumes, the volume of the image of (A.10) equals (A.11) times the volume of E.

The Key Lemma has an interesting corollary: the smallest ellipsoid E containing a polyhedron P has its center in P. (If the center of E lay outside P, then P would be contained in some $\frac{1}{2}E$ and, therefore, in a smaller ellipsoid $\tilde{E}$.) This corollary suggests an iterative search for a solution of a system of linear inequalities; in the kth iteration, we begin with an ellipsoid E_k containing the polyhedron P defined by our system and, unless the center of E_k belongs to P, we replace E_k by a smaller ellipsoid $E_{k+1} = \tilde{E}_k$. To make this idea work, we have to overcome a few technical obstacles. First, there is the problem of initialization: an unbounded P won't fit into any E_0. Fortunately, if

$$\sum_{j=1}^{n} a_{ij}x_j \leq b_i \qquad (i = 1, 2, \ldots, m) \tag{A.12}$$

has any solution at all, then it has a solution such that

$$-2^D \leq x_j \leq 2^D \qquad (j = 1, 2, \ldots, n) \tag{A.13}$$

with D standing for the total number of binary digits in the $m(n+1)$ integers a_{ij} and b_i. In other words, the system (A.12) is solvable if and only if the system consisting of (A.12) and (A.13) is solvable. Clearly, the polyhedron P defined by (A.12) and (A.13) fits into the sphere $\mathbf{x}^T\mathbf{x} \leq n4^D$. Secondly, there is the problem of termination. For instance, it is conceivable that P consists of a single point never quite coinciding with the center of any of our ellipsoids $E_0, E_1, E_2, \ldots$. In that case, the steady progress made by the rapidly shrinking ellipsoids is only illusory: essentially the trouble is that P may be nonempty and yet its volume may be zero. From this point of view, systems of strict linear inequalities behave much better: as soon as they are solvable, the sets of their solutions have positive volumes. In fact, a far stronger statement holds true.

THE LOCALIZATION LEMMA. If the system

$$\sum_{j=1}^{n} a_{ij}x_j < b_i \qquad (i = 1, 2, \ldots, m) \tag{A.14}$$

has any solution at all, then the volume of the set defined by (A.14) and (A.13) is at least $2^{-(n+1)D}$. ■

This lemma suggests a temporary shift of our attention from the systems (A.12) to the systems (A.14). An iterative procedure for solving (A.14) is outlined in Box A.2.

BOX A.2 Solving Systems of Strict Linear Inequalities (Geometric Description)

Step 0. [Initialize.] Let E stand for the sphere $\mathbf{x}^T\mathbf{x} \leq n4^D$.

Step 1. [Solution found?] Now the set S defined by (A.13) and (A.14) is enclosed in the ellipsoid E. If the center of E satisfies (A.14) then stop; otherwise, enclose S in a smaller ellipsoid $\tilde{E}$.

Step 2. [Ellipsoid too small?] If the volume of $\tilde{E}$ is less than $2^{-(n+1)D}$ then stop: (A.14) has no solution. Otherwise replace E by $\tilde{E}$ and return to step 1.

Let us describe the construction of $\tilde{E}$ in algebraic terms. First, consider the simple case where E is the unit sphere, and one of the inequalities in (A.14) reads $x_1 > t$

for some nonnegative t. As before, the set S can be enclosed in the ellipsoid (A.10). However, if $t > 0$, then S can be enclosed in an even smaller ellipsoid,

$$\left(\frac{n+1}{n(1-t)}\right)^2 \left(x_1 - \frac{1+nt}{n+1}\right)^2 + \frac{n^2-1}{n^2(1-t^2)} \sum_{i=2}^{n} x_i^2 \leq 1.$$

This construction is illustrated in Figure A.2 with $n = 2$.

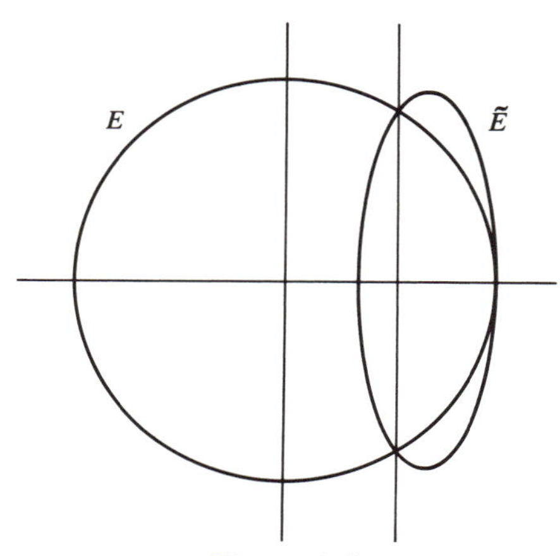

Figure A.2

More generally, write (A.14) as

$$\mathbf{a}_i^T\mathbf{x} < b_i \qquad (i = 1, 2, \ldots, m)$$

with $\mathbf{a}_i^T = [a_{i1}, a_{i2}, \ldots, a_{in}]$ and let E be defined by $(\mathbf{x} - \mathbf{c})^T\mathbf{B}^{-1}(\mathbf{x} - \mathbf{c}) \leq 1$ with $\mathbf{B}^{-1} = \mathbf{A}^T\mathbf{A}$. If the center of E violates (A.14), then $\mathbf{a}_i^T\mathbf{c} \geq b_i$ for some i. Define

$$t = (\mathbf{a}_i^T\mathbf{c} - b_i)\sqrt{\mathbf{a}_i^T\mathbf{B}\mathbf{a}_i}$$

$$\tilde{\mathbf{B}} = \frac{n^2(1-t^2)}{n^2-1}\left(\mathbf{B} - \frac{2(1+nt)}{(n+1)(1+t)} \frac{(\mathbf{B}\mathbf{a}_i)(\mathbf{B}\mathbf{a}_i)^T}{\mathbf{a}_i^T\mathbf{B}\mathbf{a}_i}\right) \tag{A.15}$$

$$\tilde{\mathbf{c}} = \mathbf{c} - \frac{1+nt}{n+1} \frac{\mathbf{B}\mathbf{a}_i}{\sqrt{\mathbf{a}_i^T\mathbf{B}\mathbf{a}_i}}.$$

It can be shown that the inequality $(\mathbf{x} - \tilde{\mathbf{c}})^T\tilde{\mathbf{B}}^{-1}(\mathbf{x} - \tilde{\mathbf{c}}) \leq 1$ defines an ellipsoid that contains the set

$$(\mathbf{x} - \mathbf{c})^T\mathbf{B}^{-1}(\mathbf{x} - \mathbf{c}) \leq 1, \quad \mathbf{a}_i^T\mathbf{x} < b_i$$

and whose volume $\tilde{v}$ is related to the volume v of E by the formula

$$\tilde{v} = v(1-t)(1-t^2)^{(n-1)/2} \frac{n}{n+1}\left(\frac{n^2}{n^2-1}\right)^{(n-1)/2}. \tag{A.16}$$

Thus the iterative procedure for solving (A.14) may be described in algebraic terms as given in Box A.3.

BOX A.3 Solving Systems of Strict Linear Inequalities (Algebraic Description)

Step 0. Set $\mathbf{B} = n4^D\mathbf{I}$, $\mathbf{c} = [0, 0, \ldots, 0]^T$ and $v = (2^{D+1}\sqrt{n})^n$.

Step 1. If $\mathbf{c}$ satisfies (A.14), then stop; otherwise, choose a subscript i with $\mathbf{a}_i^T\mathbf{c} \geq b_i$ and compute $\tilde{\mathbf{B}}$, $\tilde{\mathbf{c}}$, $\tilde{v}$ from (A.15), (A.16).

Step 2. If $\tilde{v} < 2^{-(n+1)D}$ then stop: (A.14) has no solution. Otherwise, replace $\mathbf{B}$, $\mathbf{c}$, v by $\tilde{\mathbf{B}}$, $\tilde{\mathbf{c}}$, $\tilde{v}$ and return to step 1.

Note that in each iteration, v is reduced by a factor bounded from above by $e^{-1/2(n+1)}$ and so the total number of iterations is at most a constant times n^2D.

Unfortunately, this procedure cannot be implemented as it stands, for the square roots in (A.15) cannot be computed exactly. Nevertheless, it can be modified so that all intermediate results are rounded to αnD digits following the decimal point for some constant α. Since the rounded version of ellipsoid (A.15) may fail to contain the set S, it is then necessary to multiply the right-hand side of the formula for $\tilde{\mathbf{B}}$ by a factor of $1 + \beta/n^2$ for some positive constant β. This multiplication amounts to expanding $\tilde{E}$ by a factor of $\sqrt{1 + \beta/n^2} \doteq 1 + \beta/2n$ about the fixed center $\mathbf{c}$. For a suitable choice of β, the rounded version of the new ellipsoid is large enough to contain S and yet small enough to guarantee a satisfactory progress towards termination.

THE ROUNDING LEMMA. There are positive constants α, β, γ such that the modified procedure delivers a correct answer and terminates within $\gamma n^2 D$ iterations. ■

Now we need only establish a link between the problem we can solve, recognizing solvable systems (A.14) of strict linear inequalities, and the problem we wish to solve, recognizing solvable systems (A.12) of weak linear inequalities. Trivially, if

$$\sum_{j=1}^{n} a_{ij}x_j \leq b_i \qquad (i = 1, 2, \ldots, m) \tag{A.17}$$

is solvable then

$$\sum_{j=1}^{n} a_{ij}x_j < b_i + \varepsilon \qquad (i = 1, 2, \ldots, m) \tag{A.18}$$

is solvable for every positive "perturbation" ε. The converse holds true whenever the perturbation is sufficiently small.

THE PERTURBATION LEMMA. The system (A.17) is solvable if and only if the system (A.18) with $\varepsilon = 1/mn2^D$ is solvable. ■

Thus any procedure for recognizing solvable systems of strict linear inequalities may be used to recognize solvable systems of weak linear inequalities: given a system (A.17) with integer data a_{ij}, b_i we may feed the procedure the system

$$\sum_{j=1}^{n} (mn2^D a_{ij})x_j < mn2^D b_i + 1 \qquad (i = 1, 2, \ldots, m) \tag{A.19}$$

whose data are integers. The increase in the data size incurred in passing from (A.17) to (A.19) is large but tolerable from the theoretical point of view.

Khachian's paper was preceded by a series of articles by A. Ju. Levin (1965), and by D. B. Judin and A. S. Nemirovskii (1976a, 1976b, 1976c), in which the ellipsoid method was gradually developed; this work is related to that of N. Z. Shor (1970a, 1970b, 1971a, 1971b, 1972a, 1972b, 1975, 1977a, 1977b). The method was originally intended to solve problems in a rather general area of convex programming (of which linear programming is a special case); Judin and Nemirovskii (1976b) proved that it approximates the exact solution within any positive ε in a time which is polynomial in the size of the data and in $\log(1/\varepsilon)$. Khachian showed that, when the method is restricted to LP problems with integer coefficients, even the exact solution can be found in a polynomial time. As his paper appeared in a journal limiting the lengths of publications to four pages, the details of proofs were not reproduced there. They were reconstructed by P. Gács and L. Lovász (1981) in a paper circulated at the 10th Mathematical Programming Symposium in Montreal in 1979; see also B. Aspvall and R. E. Stone (1980). The proof of the Rounding Lemma, missing from both of these expository papers, appears in M. Grötschel, L. Lovász, and A. Schrijver (1981). A wealth of information on the ellipsoid method can be found in R. G. Bland, D. Goldfarb, and M. J. Todd (1981).

It need not be emphasized that the ellipsoid method is eminently important in the context of pure mathematics. Nevertheless, it is hardly likely to become a practical tool for solving linear programming problems: the typical number of its iterations seems to be very large even on reasonably small problems, and each individual iteration may get prohibitively laborious. To clarify the second point, consider a 2000×1000 matrix $\mathbf{A}$ with only half a dozen nonzero entries in each column. The simplex method may be expected to solve $\mathbf{Ax} \leq \mathbf{b}$ quite routinely, with storage requirements in the neighbourhood of 20,000 numbers or so. However, the 1000×1000 matrix $\mathbf{B}$ used in the ellipsoid method will get dense almost instantly, calling for a million numbers to be stored and updated in each iteration. Even though more sophisticated implementations of the ellipsoid method require somewhat less storage, none of them takes significant advantage of sparsity.

• • •

Just as this book was going to print, a new exciting result was reported by Boris Yamnitsky and Leonid A. Levin of Boston University. They have shown that the role of ellipsoids in the ellipsoid method can be played by certain polyhedra, called *simplices.* The idea is not new, quite the contrary: A. Ju. Levin (1965) used simplices rather than ellipsoids in the article which started the gradual development of the ellipsoid method. The novelty of the Yamnitsky–Levin approach lies in implementing this idea so that the running time of the resulting algorithm remains bounded by a fixed polynomial in the size of the data.

To explain the details, we only need replace the key lemma of the ellipsoid method by its polyhedral version. A few definitions are required. Points $\mathbf{v}_0, \mathbf{v}_1, \ldots, \mathbf{v}_n$ in R^n are said to be in a *general position* if no hyperplane passes through all of them; the convex hull of a set of such points is called a *simplex.* (Thus, a simplex in R^2 is a triangle and a simplex in R^3 is a tetrahedron.) The *center* of this simplex is the point

$$\frac{1}{n+1}\sum_{i=0}^{n} \mathbf{v}_i.$$

By a *half-simplex* $\frac{1}{2}S$, we shall mean the intersection of a simplex S with any half-space whose bounding hyperplane passes through the center of S.

THE NEW KEY LEMMA. Every half-simplex $\frac{1}{2}S$ is contained in a simplex $\tilde{S}$ whose volume is less than $e^{-1/2(n+1)^2}$ times the volume of S.

To prove this lemma, let us say that S is the convex hull of points $\mathbf{v}_0, \mathbf{v}_1, \ldots, \mathbf{v}_n$ and that $\frac{1}{2}S$ is the intersection of S with the half-space $\mathbf{a}^T\mathbf{x} \leq b$. It will be convenient to write

$$e(\mathbf{x}) = b - \mathbf{a}^T\mathbf{x}$$

for all points $\mathbf{x}$. Let k be the subscript that maximizes $e(\mathbf{v}_k)$. (Since $\frac{1}{2}S$ is nonempty, we have $e(\mathbf{v}_k) > 0$.) Let us define points $\tilde{\mathbf{v}}_0, \tilde{\mathbf{v}}_1, \ldots, \tilde{\mathbf{v}}_n$ by

$$\tilde{\mathbf{v}}_i = \mathbf{v}_k + \frac{1}{d_i}(\mathbf{v}_i - \mathbf{v}_k)$$

with

$$d_i = 1 - \frac{e(\mathbf{v}_i)}{n^2 e(\mathbf{v}_k)}.$$

We claim that the convex hull $\tilde{S}$ of $\tilde{\mathbf{v}}_0, \tilde{\mathbf{v}}_1, \ldots, \tilde{\mathbf{v}}_n$ has the properties specified by the lemma.

To justify this claim, consider first an arbitrary point $\mathbf{x}$ in $\frac{1}{2}S$. Since $\mathbf{x} \in S$, we have

$$\mathbf{x} = \sum_{i=0}^{n} t_i \mathbf{v}_i$$

for some nonnegative numbers $t_0, t_1, \ldots, t_n$ whose sum equals 1. Let us set

$$\tilde{t}_i = \begin{cases} d_i t_i & \text{if } i \neq k \\ d_i t_i + \dfrac{e(\mathbf{x})}{n^2 e(\mathbf{v}_k)} & \text{if } i = k. \end{cases}$$

Since $e(\mathbf{x}) = \sum t_i e(\mathbf{v}_i)$, it is easy to verify that

$$\sum_{i=0}^{n} \tilde{t}_i \tilde{\mathbf{v}}_i = \mathbf{x} \quad \text{and} \quad \sum_{i=0}^{n} \tilde{t}_i = 1.$$

Since $e(\mathbf{v}_i) \leq e(\mathbf{v}_k)$ for all i, each $\tilde{t}_i$ with $i \neq k$ is nonnegative; since $\mathbf{a}^T\mathbf{x} < b$, we have $e(\mathbf{x}) > 0$, and so $\tilde{t}_k$ is nonnegative. Hence $\mathbf{x}$ belongs to $\tilde{S}$.

Now we need prove only that the ratio r between the volume of $\tilde{S}$ and the volume of S is less than $e^{-1/2(n+1)^2}$. For this purpose, note that $\tilde{\mathbf{v}}_k = \mathbf{v}_k$, that each $\tilde{\mathbf{v}}_i$ with $i \neq k$ lies on the line passing through $\mathbf{v}_k$ and $\mathbf{v}_i$, and that the distance from $\mathbf{v}_k$ to $\tilde{\mathbf{v}}_i$ equals the distance from $\mathbf{v}_k$ to $\mathbf{v}_i$ divided by d_i. Hence

$$r = \prod_{i \neq k} \frac{1}{d_i}.$$

Since $e(\mathbf{v}_i) \leq e(\mathbf{v}_k)$ whenever $i \neq k$, we have

$$d_i \geq 1 - \frac{1}{n^2} \quad \text{whenever} \quad i \neq k. \tag{A.20}$$

Since $e(\mathbf{c}) = 0$ for the center $\mathbf{c}$ of S, we have

$$\sum_{i \neq k} d_i = n + \frac{1}{n^2}. \tag{A.21}$$

From (A.20) and (A.21), it follows that

$$\prod_{i \neq k} d_i \geq \left(1 - \frac{1}{n^2}\right)^{n-1} \left(1 + \frac{1}{n}\right).$$

(More generally, suppose that we want to minimize a product of variables subject to (i) a positive lower bound on each variable, and (ii) the requirement that the sum of the variables equal a prescribed number. An easy argument shows that, in every optimal solution, at most one variable exceeds its lower bound.) The proof is completed by verifying that

$$\left(1 - \frac{1}{n^2}\right)^{n-1}\left(1 + \frac{1}{n}\right) = \left(1 - \frac{1}{n^2}\right)^{n}\left(1 - \frac{1}{n}\right)^{-1} > e^{1/2(n+1)^2}$$

which is a routine exercise in calculus.

Bibliography

AHO, A. V., J. E. HOPCROFT, and J. D. ULLMAN (1983). *Data Structures and Algorithms*. Reading, Massachusetts: Addison-Wesley.

ALI, A. I., R. V. HELGASON, J. L. KENNINGTON, and H. S. LALL (1978). "Primal simplex network codes: State-of-the-art implementation technology," *Networks 8*: 315–339.

ANDERSON, D., and W. L. STEIGER (1983). "A comparison of methods for discrete L_1 curve fitting," in press.

ANTOSIEWICZ, H. A., and A. J. HOFFMAN (1954). "A remark on the smoothing problem," *Management Science 1*: 92–95.

ASPVALL, B., and R. E. STONE (1980). "Khachiyan's linear programming algorithm," *Journal of Algorithms 1*: 1–13.

AVIS, D., and V. CHVÁTAL (1978). "Notes on Bland's pivoting rule," *Mathematical Programming Study 8*: 24–34.

BALINSKI, M. L. (1970). "On a selection problem," *Management Science 17*: 230–231.

BARTELS, R. H. (1971). "A stabilization of the simplex method." *Numerische Mathematik 16*: 414–434.

BARTELS, R. H., and G. H. GOLUB (1969). "The simplex method of linear programming using LU decomposition," *Communications of the Association for Computing Machinery 12*: 266–268.

BEALE, E. M. L. (1955). "Cycling in the dual simplex algorithm," *Naval Research Logistics Quarterly 2*: 269–275.

BIRKHOFF, G. (1946). "Tres observaciones sobre el algebra lineal," *Universidad Nacionale Tucumán Revista 5*: 147–151.

BLAND, R. G. (1977). "New finite pivoting rules for the simplex method," *Mathematics of Operations Research 2*: 103–107.

BLAND, R. G., D. GOLDFARB, and M. J. TODD (1981). "The ellipsoid method: A survey," *Operations Research 29*: 1039–1091.

BLASCHKE, W. (1916). *Kreis und Kugel*. Leipzig: Veit. [Reprint (1948): New York: Chelsea.]

BLATTBERG, R., and T. SARGENT (1971). "Regression with non-Gaussian stable disturbances: Some sampling results," *Econometrica 39*: 501–510.

BLOOMFIELD, P., and W. L. STEIGER (1980). "Least absolute deviations curve fitting," *SIAM Journal on Scientific and Statistical Computing 1*: 290–301.

BOGERT, L. J., G. M. BRIGGS, and D. H. CALLOWAY (1973). *Nutrition and Physical Fitness*. Philadelphia: W. B. Saunders.

BOREL, E. (1921). "La théorie du jeu et les équations intégrales à noyau symétrique," *Comptes Rendus de l'Académie des Sciences 173*: 1304–1308. [English translation by L. J. Savage: *Econometrica 21* (1953), 97–100.]

BOREL, E. (1924). "Sur les jeux où interviennent l'hasard et l'habileté des joueurs," in *Théorie des Probabilités*. Paris: Librarie scientifique, Hermann. [English translation by L. J. Savage: *Econometrica 21* (1953), 101–115.]

BOREL, E. (1927). "Sur les systèmes des formes linéaires à determinant symétrique gauche et la théorie générale du jeu," *Comptes Rendus de l' Acadẽmie des Sciences 184*: 52-53. [English translation by L. J. Savage: *Econometrica 21* (1953), 116–117.]

BORGWARDT, K.-H. (1982). "The average number of steps required by the simplex method is polynomial," to be published.

BOTTS, T. (1942). "Convex sets," *American Mathematical Monthly 49*: 527–535.

BOWMAN, E. H. (1956). "Production scheduling by the transportation method of linear programming," *Operations Research 4*: 100–103.

BRADLEY, G. H., G. G. BROWN, and G. W. GRAVES (1977). "Design and implementation of large scale primal transshipment algorithms," *Management Science 24*: 1–34.

CARATHÉODORY, C. (1907). "Über den Variabilitätsbereich der Koeffizienten von Potenzreihen, die gegebene Werte nicht annehmen," *Mathematische Annalen 64*: 95–115.

CHARNES, A. (1952). "Optimality and degeneracy in linear programming," *Econometrica 20*: 160–170.

CHARNES, A., W. W. COOPER, and D. FARR (1953). "Linear programming and profit preference scheduling for a manufacturing firm," *Journal of the ORSA 1*: 114–129.

CHARNES, A., W. W. COOPER, and B. MELLON (1952). "Blending aviation gasolines —A study in programming interdependent activities in an integrated oil company," *Econometrica 20*: 135–159.

CHVÁTAL, V. (1981). "Cheap, Middling, or Dear," in *The Mathematical Gardner*, D. A. Klarner, ed. Boston: Prindle, Weber and Schmidt, 44–50.

CHVÁTAL, V. (1983). "On the bicycle problem," *Discrete Applied Mathematics 5*: 165–173.

CLAERBOUT, J. F., and F. MUIR (1973). "Robust modelling with erratic data," *Geophysics 38*: 826–844.

Consumer Guide Magazine Editors (1974). *The Brand Name Food Game*. New York: New American Library Signet Book.

COOK, S. A. (1971). "The complexity of theorem-proving procedures," *Proceedings of the Third Annual ACM Symposium on Theory of Computing*, Association for Computing Machinery, New York, 151–158.

CUNNINGHAM, W. H. (1976). "A network simplex method," *Mathematical Programming 11*: 105–116.

CUNNINGHAM, W. H. (1979) "Theoretical properties of the network simplex method," *Mathematics of Operations Research 4*: 196–208.

CUNNINGHAM, W. H., and J. G. KLINCEWICZ (1983). "On cycling in the network simplex method." to appear in *Mathematical Programming*.

DANTZIG, G. B. (1951a). "A proof of the equivalence of the programming problem and the game problem," in *Activity Analysis of Production and Allocation*, T. C. Koopmans, ed. New York: John Wiley and Sons, 330–335.

DANTZIG, G. B. (1951b). "Application of the simplex method to a transportation problem," in *Activity Analysis of Production and Allocation*, T. C. Koopmans, ed. New York: John Wiley and Sons, 359–373.

DANTZIG, G. B. (1955). "Upper bounds, secondary constraints, and block triangularity in linear programming," *Econometrica 23*: 174–183.

DANTZIG, G. B. (1960). "Inductive proof of the simplex method," *IBM Journal of Research and Development 4*: 505–506.

DANTZIG, G. B. (1963). *Linear Programming and Extensions*. Princeton, New Jersey: Princeton University Press.

DANTZIG, G. B. (1980). "Expected number of steps of the simplex method for a linear program with a convexity constraint," Technical Report SOL 80-3, Stanford University.

DANTZIG, G. B., and D. R. FULKERSON (1954). "Minimizing the number of tankers to meet a fixed schedule," *Naval Research Logistics Quarterly 1*: 217–222.

DANTZIG, G. B., and W. ORCHARD-HAYS (1954). "The product form for the inverse in the simplex method," *Mathematical Tables and Other Aids to Computation 8*: 64–67.

DANTZIG, G. B., A. ORDEN, and P. WOLFE (1955). "The generalized simplex method for minimizing a linear form under linear inequality restraints," *Pacific Journal of Mathematics 5*: 183–195.

DANTZIG, G. B., and R. M. VAN SLYKE (1967). "Generalized upper bounding techniques," *Journal of Computer and System Sciences 1*: 213—226.

DANTZIG, G. B., and P. WOLFE (1960). "Decomposition principle for linear programs," *Operations Research 8*: 101–111.

DANZER, L., B. GRÜNBAUM, and V. KLEE (1963). "Helly's theorem and its relatives," in *Convexity*, V. Klee, ed. *Proceedings of Symposia in Pure Mathematics 7*. Providence, Rhode Island: American Mathematical Society, 101–118.

DE BUCHET, J. (1971). "How to take into account the low density of matrices to design a mathematical programming package," in *Large Sparse Systems of Linear Equations*, J. K. Reid, ed. New York: Academic Press, 211–217.

DIJKSTRA, E. (1959). "A note on two problems in connexion with graphs," *Numerische Mathematik 1*: 269–271.

DILWORTH, R. P. (1950). "A decomposition theorem for partially ordered sets," *Annals of Mathematics 51*: 161–166.

DINIC, E. A. (1970). "Algorithm for solution of a problem of maximum flow in a network with power estimation," *Soviet Mathematics Doklady 11*: 1277–1280.

DORFMAN, R., P. A. SAMUELSON, and R. M. SOLOW (1958). *Linear Programming and Economic Analysis*. New York: McGraw-Hill.

DUFFIN, R. J. (1974). "On Fourier's analysis of linear inequality systems," *Mathematical Programming Study 1*: 71–95.

DYER, M. E., and L. G. PROLL (1977). "An algorithm for determining all extreme points of a convex polytope," *Mathematical Programming 12*: 81–96.

EDMONDS, J. (1965). "Paths, trees and flowers," *Canadian Journal of Mathematics 17*: 449–467.

EDMONDS, J. (1967). "Systems of distinct representatives and linear algebra," *Journal of Research of the National Bureau of Standards 71 B*: 241–245.

EDMONDS, J., and R. M. KARP (1972). "Theoretical improvements in algorithmic efficiency for network flow problems," *Journal of the Association for Computing Machinery 19*: 248–264.

EGERVÁRY, J. (1931). "Matrixok kombinatorikus tulajdonságairól," *Mathematikai és Fizikai Lápok 38*: 16–28.

EISEMANN, K. (1957). "The trim problem," *Management Science 3*: 279–284.

EISENHART, C. (1961). "Boscovich and the Combination of Observations," in *Roger Joseph Boscovich*, L. L. Whyte, ed. New York: Fordham University Press.

ELIAS, P., A. FEINSTEIN, and C. E. SHANNON (1956). "Note on maximum flow through a network," *IRE Transactions on Information Theory IT-2*, 117–119.

EVES, H. (1964). *An Introduction to the History of Mathematics*. New York: Holt, Rinehart and Winston.

FAMA, E. P. (1965). "The behavior of stock market prices," *Journal of Business 38*: 34–105.

FARKAS, J. (1902). "Theorie der einfachen Ungleichungen," *Journal für die reine und angewandte Mathematik 124*: 1–27.

FORD, L. R., JR., and D. R. FULKERSON (1956). "Maximal flow through a network," *Canadian Journal of Mathematics 8*: 399–404.

FORD, L. R., JR., and D. R. FULKERSON (1957). "A simple algorithm for finding maximal network flows and an application to the Hitchcock problem," *Canadian Journal of Mathematics 9*: 210–218.

FORD, L. R., JR., and D. R. FULKERSON (1958a). "Constructing maximal dynamic flows from static flows," *Operations Research 6*: 419–433.

FORD, L. R., JR., and D. R. FULKERSON (1958b). "A suggested computation for maximal multi-commodity network flows," *Management Science 5*: 97–101.

FORD, L. R., JR., and D. R. FULKERSON (1962). *Flows in Networks*. Princeton, New Jersey: Princeton University Press.

FORREST, J. J. H., and J. A. TOMLIN (1972). "Updating triangular factors of the basis to maintain sparsity in the product form simplex method," *Mathematical Programming 2*: 263–278.

FORSYTHE, G., and C. B. MOLER (1967). *Computer Solution of Linear Algebraic Systems*. Englewood Cliffs, New Jersey: Prentice-Hall.

FOURIER, J. B. J. (1890). "Second extrait," in *Oeuvres*, G. Darboux, ed. Paris: Gauthiers-Villars, 325–328. [English translation by D. A. Kohler (1973).]

FULKERSON, D. R. (1956). "Note on Dilworth's decomposition theorem for partially ordered sets," *Proceedings of the American Mathematical Society 7*: 701–702.

GÁCS, P., and L. LOVÁSZ (1981). "Khachiyan's algorithm for linear programming," *Mathematical Programming Study 14*: 61–68.

GADDUM, J. W., A. J. HOFFMAN, and D. SOKOLOWSKY (1954). "On the solution of the caterer problem," *Naval Research Logistics Quarterly 1*: 223–229.

GALE, D. (1957). "A theorem on flows in networks," *Pacific Journal of Mathematics 7*: 1073–1082.

GALE, D. (1960). *The Theory of Linear Economic Models*. New York: McGraw-Hill.

GALE, D. (1963). "Neighborly and cyclic polytopes," in *Convexity*, V. Klee, ed. *Proceedings of Symposia in Pure Mathematics 7*. Providence, Rhode Island: American Mathematical Society, 225–232.

GALE, D., H. W. KUHN, and A. W. TUCKER (1951). "Linear programming and the theory of games," in *Activity Analysis of Production and Allocation*, T. C. Koopmans, ed. New York: John Wiley and Sons, 317–329.

GALIL, Z. (1978). "A new algorithm for the maximal flow problem," *Proceedings of the Nineteenth Annual Symposium on Foundations of Computer Science*, IEEE, 231–245.

GANTMACHER, F. R. (1959). *The Theory of Matrices*, Vol. 1. New York: Chelsea.

GAREY, M. R., and D. S. JOHNSON (1979). *Computers and Intractability*. San Francisco: W. H. Freeman and Company.

GARFINKEL, R. S., and G. L. NEMHAUSER (1972). *Integer Programming*. New York: John Wiley and Sons.

GASS, S. I. (1975). *Linear Programming*. New York: McGraw-Hill.

GASS, S. I., and T. SAATY (1955). "The computational algorithm for the parametric objective function," *Naval Research Logistics Quarterly 2*: 39–45.

GASSNER, B. J. (1964). "Cycling in the transportation problem," *Naval Research Logistics Quarterly 11*: 43–58.

GAUSS, C. F. (1906). "Theoria Motus Corporum Caelestium in Sectionibus Conicis Solem Ambientium," in *Werke*, Vol. 7. Leipzig: Teubner.

GAY, D. M. (1978). "On combining the schemes of Reid and Saunders for sparse LP bases," in

Sparse Matrix Proceedings 1978, I. S. Duff and G. W. Stewart, eds. Philadelphia: The Society for Industrial and Applied Mathematics, 313–334.

GILMORE, P. C., and R. E. GOMORY (1961). "A linear programming approach to the cutting-stock problem," *Operations Research 9*: 849–859.

GILMORE, P. C., and R. E. GOMORY (1963). "A linear programming approach to the cutting-stock problem—Part II," *Operations Research 11*: 863–888.

GIVENS, W. (1954). "Numerical computations of the characteristic values of a real symmetric matrix," Oak Ridge National Laboratory, ORNL-1574.

GLASSEY, C. R., and V. K. GUPTA (1974). "A linear programming analysis of paper recycling," *Management Science 21*: 392–408.

GLOVER, F., D. KARNEY, and D. KLINGMAN (1974). "Implementation and computational comparisons of primal, dual and primal-dual computer codes for minimum cost network flow problems," *Networks 4*: 191–212.

GOLDFARB, D., and J. K. REID (1977). "A practicable steepest-edge simplex algorithm," *Mathematical Programming 12*: 361–371.

GORDAN, P. (1873). "Über die Auflösung linearer Gleichungen mit reelen Coefficienten," *Mathematische Annalen 6*: 23–28.

GRATTAN-GUINESS, I. (1970). "Joseph Fourier's anticipation of linear programming," *Operational Research Quarterly 21*: 361–364.

GREENE, J. H., K. CHATTO, C. R. HICKS, and C. B. COX (1959). "Linear programming in the packing industry," *Journal of Industrial Engineering 10*: 364–372.

GRÖTSCHEL, M., L. LOVÁSZ, and A. SCHRIJVER (1981). "The ellipsoid method and its consequences in combinatorial optimization," *Combinatorica 1*: 169–197.

HALL, P. (1935). "On representatives of subsets," *Journal of the London Mathematical Society 10*: 26–30.

HAMMERSLEY, J. M., and J. G. MAULDON (1956). "General principles of antithetic variates," *Proceedings of the Cambridge Philosophical Society 52*: 476–481.

HANSSMANN, F., and S. W. HESS (1960). "A linear programming approach to production and employment scheduling," *Management Technology 1*: 46–52.

HARRIS, P. M. J. (1973). "Pivot selection methods of the Devex LP code," *Mathematical Programming 5*: 1–28.

HAWKINS, T. (1974). "The theory of matrices in the 19th century," in *Proceedings of the International Congress of Mathematicians*, Vancouver, 561–570.

HELLERMAN, E., and D. RARICK (1971). "Reinversion with the preassigned pivot procedure," *Mathematical Programming 1*: 195–216.

HELLERMAN, E., and D. RARICK (1972). "The partitioned preassigned pivot procedure," in *Sparse Matrices and Their Applications*, D. J. Rose and R. A. Willoughby, eds. New York: Plenum Press, 67–76.

HELLY, E. (1923). "Über Mengen konvexer Körper mit gemeinschaftlichen Punkten," *Jahresbericht Deutsche Mathematische Vereinungen 32*: 175–176.

HITCHCOCK, F. L. (1941). "The distribution of a produce from several sources to numerous localities," *Journal of Mathematical Physics 20*: 224–230.

HO, J. K., and E. LOUTE (1981). "An advanced implementation of the Dantzig–Wolfe decomposition algorithm for linear programming," *Mathematical Programming 20*: 303–326.

HO, J. K., and A. S. MANNE (1974). "Nested decomposition for dynamic models," *Mathematical Programming 6*: 121–140.

HOFFMAN, A. J. (1953). "Cycling in the simplex algorithm," National Bureau of Standards, Report 2974. See also Dantzig (1963), pp. 228–229.

HOFFMAN, A. J., and W. JACOBS (1954). "Smooth patterns of production," *Management Science 1*: 86–91.

HOFFMAN, A. J., and H. W. WIELANDT (1953). "The variation of the spectrum of a normal matrix," *Duke Mathematical Journal 20*: 37–39.

HOTELLING, H. (1943). "Some new methods in matrix calculation," *Annals of Mathematical Statistics 14*: 1–34.

JACOBS, W. (1954). "The caterer problem," *Naval Research Logistics Quarterly 1*: 154–165.

JEROSLOW, R. G. (1973). "The simplex algorithm with the pivot rule of maximizing criterion improvement," *Discrete Mathematics 4*: 367–377.

JEWELL, W. S. (1958). "Optimal flow through networks," Interim Technical Report No. 8, Massachusetts Institute of Technology.

JOHNSON, D. B. (1977). "Efficient algorithms for shortest paths in sparse networks," *Journal of the Association for Computing Machinery 24*: 1–13.

JOHNSON, D. S. (1973). "Near-optimal bin packing algorithms," Ph.D. Thesis, Electrical Engineering Dept., Massachusetts Institute of Technology. [See also D. S. Johnson, A. Demers, J. D. Ullman, M. R. Garey, and R. L. Graham, "Worst-case performance bounds for simple one-dimensional packing algorithms," *SIAM Journal on Computing 3* (1974): 299–325.]

JUDIN, D. B., and A. S. NEMIROVSKII (1976a). "Estimation of the informational complexity of mathematical programming problems" (in Russian), *Ekonomika i Matematicheskie Metody 12*: 128–142. [English translation: *Matekon: Translations of Russian and East European Mathematical Economics 13*: 3–25, Winter 1976–1977.]

JUDIN, D. B., and A. S. NEMIROVSKII (1976b). "Informational complexity and effective methods for the solution of convex extremal problems" (in Russian), *Ekonomika i Matematicheskie Metody 12*: 357–369. [English translation: *Matekon: Translations of Russian and East European Mathematical Economics 13*: 25–45, Spring 1977.]

JUDIN, D. B., and A. S. NEMIROVSKII (1976c). "Informational complexity of strict convex programming" (in Russian), *Ekonomika i Matematicheskie Metody 12*: 550–559.

KANTOROVICH, L. V. (1939). "Mathematical methods in the organization and planning of production." [English translation: *Management Science 6* (1960): 366–422.]

KHACHIAN, L. G. (1979). "A polynomial algorithm in linear programming" (in Russian), *Doklady Adademiia Nauk SSSR 244*: 1093–1096. [English translation: *Soviet Mathematics Doklady 20*: 191–194.]

KIRCHBERGER, P. (1903). "Über Tschebyschefsche Annährungsmethoden," *Mathematische Annalen 57*: 509–540.

KLEE, V., and G. J. MINTY (1972). "How good is the simplex algorithm?" in *Inequalities–III*, O. Shisha, ed. New York: Academic Press, 159–175.

KNUTH, D. E. (1973). *The Art of Computer Programming*. Vol. 3: *Sorting and Searching*. Reading, Massachusetts: Addison-Wesley.

KOHLER, D. A. (1973). "Translation of a report by Fourier on his work on linear inequalities," *Opsearch 10*: 38–42.

KÖNIG, D. (1916). "Über Graphen und ihre Anwendung auf Determinantentheorie und Mengenlehre," *Mathematische Annalen 77*: 453–465.

KÖNIG, D. (1931). "Graphen und Matrizen," *Matematikai és Fizikai Lápok 38*: 116–119.

KÖNIG, D. (1936). *Theorie der Endlichen und Unendlichen Graphen*. Leipzig: Akademische Verlagsgesellschaft. [Reprint (1950): New York: Chelsea.]

KOTIAH, T. C. T., and D. I. STEINBERG (1978). "On the possibility of cycling with the simplex method," *Operations Research 26*: 374–376.

KOTZIG, A. (1956). *Súvislosť a Pravidelná Súvislosť Konečných Grafov*. Bratislava: Vysoká Škola Ekonomická.

KUHN, H. W. (1950). "A simplified two-person poker," in *Contributions to the theory of games I*, H. W. Kuhn and A. W. Tucker, eds. *Annals of Mathematics Studies 24*: 97–103.

KUHN, H. W. (1955). "The Hungarian method for the assignment problem," *Naval Research Logistics Quarterly 2*: 83–97.

KUHN, H. W. (1956). "Solvability and consistency for linear equations and inequalities," *American Mathematical Monthly 63*: 217–232.

KUHN, H. W., and R. E. QUANDT (1963). "An experimental study of the simplex method," *Proceeding of Symposia in Applied Mathematics 15*: 107–124.

LANDAU, H. G. (1953). "On dominance relations and the structure of animal societies. III. The condition for a score structure," *Bulletin of Mathematical Biophysics 15*: 143–148.

LAPLACE, P. S. (1895). "Sur les degrés mesurés des méridiens et sur les longuers observées du pendule," in *Oeuvres*, Vol. 11, 493–516. Paris: Gauthiers-Villars.

LAWLER, E. L. (1976). *Combinatorial Optimization: Networks and Matroids*. New York: Holt, Rinehart and Winston.

LEMKE, C. E. (1954). "The dual method of solving the linear programming problem," *Naval Research Logistics Quarterly 1*: 36–47.

LEVIN, A. JU. (1965). "On an algorithm for the minimization of convex functions," *Soviet Mathematics Doklady 6*: 286–290.

MACLAGAN, R. C. (1901). *The Games and Diversions of Argyleshire*. London: David Nutt.

MCMULLEN, P. (1970). "The maximum number of faces of a convex polytope," *Mathematika 17*: 179–184.

MAGEE, J. F. (1953). *Studies in Operations Research I: Application of Linear Programming to Production Scheduling*. Cambridge, Massachusetts: Arthur D. Little Inc.

MALHOTRA, V. M., M. P. KUMAR, and S. N. MAHESHWARI (1978). "An $O(|V|^3)$ algorithm for finding maximum flows in networks," *Information Processing Letters 7*: 277–278.

MANDELBROT, B. (1963). "The variation of certain speculative prices," *Journal of Business 36*: 394–419.

MANDELBROT, B. (1966). "Forecasts of future prices, unbiased markets, and 'martingale' models," *Journal of Business 39*: 242–255.

MANDELBROT, B. (1967). "The variation of some other speculative prices," *Journal of Business 40*: 393–413.

MARKOWITZ, H. M. (1957). "The elimination form of the inverse and its application to linear programming," *Management Science 3*: 255–269.

MARSHALL, K. T., and J. W. SUURBALLE (1969). "A note on cycling in the simplex method," *Naval Research Logistics Quarterly 16*: 121–137.

MASUDA, S. (1970). "The bicycle problem," University of California, Berkeley: Operations Research Center Technical Report ORC 70-35.

MATTHEISS, T. H., and D. S. RUBIN (1980). "A survey and comparison of methods for finding all vertices of convex polyhedral sets," *Mathematics of Operations Research 5*: 167–185.

MATTHEISS, T. H., and B. K. SCHMIDT (1980). "Computational results on an algorithm for finding all vertices of a polytope," *Mathematical Programming 18*: 308–329.

MINKOWSKI, H. (1911). *Gesammelte Abhandlungen*. Leipzig: Teubner.

MULVEY, J. M. (1978). "Pivot strategies for primal-simplex network codes," *Journal of the Association for Computing Machinery 25*: 266–270.

MUNKRES, J. (1957). "Algorithms for the assignment and transportation problems," *Journal of the Society for Industrial and Applied Mathematics 5*: 32–38.

MURTAGH, B. A. (1981). *Advanced Linear Programming: Computation and Practice*. New York: McGraw-Hill.

MURTY, K. G. (1980). "Computational complexity of parametric linear programming," *Mathematical Programming 19*: 213–219.

ORDEN, A. (1956). "The transhipment problem," *Management Science 2*: 276–285.

PEROLD, A. F., and G. B. DANTZIG (1978). "A basis factorization method for block triangular linear programs," in *Proceedings of the Symposium on Sparse Matrix Computations*, I. S. Duff and G. W. Stewart, eds. Knoxville, Tennessee, 283–312.

PICARD, J.-C. (1976). "Maximal closure of a graph and applications to combinatorial problems," *Management Science 22*: 1268–1272.

PRAGER, W. (1956). "On the caterer problem," *Management Science 3*: 15–23.

PRATT, W. A., J. N. TOAL, G. W. RUSHIZKY, and H. A. SOBER (1964). "Spectral characterization of oligonucleotides by computational methods," *Biochemistry 3*: 1831–1837.

PRÉKOPA, A. (1972). "On the number of vertices of random convex polyhedra," *Periodica Mathematica Hungarica 2*: 259–282.

Recommended Dietary Allowances (1974). Washington, D.C.: National Academy of Sciences.

REID, J. K. (1982). "A sparsity-exploiting variant of the Bartels-Golub decomposition for linear programming bases," *Mathematical Programming 24*: 55–69.

RHYS, J. (1970). "A selection problem of shared fixed costs and network flows," *Management Science 17*: 200–207.

RILEY, V., and S. I. GASS (1958). *Bibliography of Linear Programming*. Baltimore: Johns Hopkins Press.

ROCKAFELLAR, R. T. (1970). *Convex Analysis*. Princeton, New Jersey: Princeton University Press.

ROSE, D. J., and R. E. TARJAN (1978). "Algorithmic aspects of vertex elimination on directed graphs," *SIAM Journal on Applied Mathematics 34*: 176–197.

RYSER, H. J. (1957). "Combinatorial properties of matrices of zeros and ones," *Canadian Journal of Mathematics 9*: 371–377.

SAUNDERS, M. A. (1976a). "The complexity of LU updating in the simplex method," in *The Complexity of Computational Problem Solving*, R. S. Anderssen and R. P. Brent, eds. Queensland: Queensland University Press, 214–230.

SAUNDERS, M. A. (1976b). "A fast, stable implementation of the simplex method using Bartels-Golub updating," in *Sparse Matrix Computations*, J. Bunch and D. Rose, eds. New York: Academic Press, 213–226.

SCRIMSHAW, N. S., and V. R. YOUNG (1976). "The requirements of human nutrition," in *Food and Agriculture*, A Scientific American Book. San Francisco: W. H. Freeman and Company.

SENETA, E., and W. L. STEIGER (1983). "A new LAD curve fitting algorithm: Slightly overdetermined equation systems in L_1," to appear in *Discrete Applied Mathematics*.

SHARPE, W. F. (1971). "Mean-absolute-deviation characteristic lines for securities and portfolios," *Management Science 18*: B1–B13.

SHOR, N. Z. (1970a). "Utilization of the operation of space dilatation in the minimization of convex functions" (in Russian), *Kibernetika 6*: 6–12. [English translation: *Cybernetics 6*: 7–15.]

SHOR, N. Z. (1970b). "Convergence rate of the gradient descent method with dilatation of the space" (in Russian), *Kibernetika 6*: 80–85. [English translation: *Cybernetics 6*: 102–108.]

SHOR, N. Z. (1971a). "A minimization method using the operation of extension of the space in the direction of the difference of two successive gradients" (in Russian), *Kibernetika 7*: 51–59. [English translation: *Cybernetics 7*: 450–459.]

SHOR, N. Z. (1971b). "Certain questions of convergence of generalized gradient descent" (in Russian), *Kibernetika 7*: 82–84. [English translation: *Cybernetics 7*: 1033–1036.]

SHOR, N. Z. (1972a). "Solution of minimax problems by the method of generalized gradient decent with dilatation of the space" (in Russian), *Kibernetika 8*: 82–88. [English translation: *Cybernetics 8*: 88–94.]

SHOR, N. Z. (1972b). "A class of almost-differentiable functions and a minimization method for functions in this class" (in Russian), *Kibernetika 8*: 65–70. [English translation: *Cybernetics 8*: 599–606.]

SHOR, N. Z. (1975). "Convergence of a gradient method with space dilatation in the direction of the difference between two successive gradients" (in Russian), *Kibernetika 11*: 48–53. [English translation: *Cybernetics 11*: 564–570.]

SHOR, N. Z. (1977a). "Cut-off method with space extension in convex programming problems," *Kibernetika 13*: 94–95. [English translation: *Cybernetics 13*: 94–96.]

SHOR, N. Z. (1977b). "New development trends in nondifferentiable optimization," *Kibernetika 13*: 87–91. [English translation: *Cybernetics 13*: 881–886.]

SLEATOR, D. D. K. (1980). "An O(*mn* log *n*) algorithm for maximum network flow," Ph.D. Thesis, Computer Science Department, Stanford University.

SMALE, S. (1982). "On the average speed of the simplex method of linear programming," to be published.

SMITH, D. E. (1953). *History of Mathematics. Volume II: Special Topics of Elementary Mathematics.* Boston: Ginn and Company.

STIEMKE, E. (1915). "Über positive Lösungen homogener linearer Gleichungen," *Mathematische Annalen 76*: 340–342.

STRUIK, D. J. (1967). *A Concise History of Mathematics.* New York: Dover.

STRUM, J. E. (1972). *Introduction to Linear Programming.* San Francisco: Holden-Day.

SWANSON, E. R., and K. FOX (1954). "The selection of livestock enterprises by activity analysis," *Journal of Farm Economics 36*: 78–86.

TEWARSON, R. P. (1973). *Sparse Matrices.* New York: Academic Press.

TOMLIN, J. A. (1972). "Modifying triangular factors of the basis in the simplex method," in *Sparse Matrices and Their Applications*, D. J. Rose and R. A. Willoughby, eds. New York: Plenum Press, 77–85.

TUCKER, A. W. (1956). "Dual systems of homogeneous linear equations," in *Linear Inequalities and Related Systems*, H. W. Kuhn and A. W. Tucker, eds. *Annals of Mathematics Studies 38*: 3–18.

TURING, A. M. (1948). "Rounding-off errors in matrix processes," *Quarterly Journal of Mechanics 1*: 287–308.

VILLE, J. A. (1938). "Sur la théorie générale des jeux ou intervient l'habileté des jouers," in E. Borel, *Traité du Calcul des Probabilités et des ses Applications.* Paris: Gauthiers-Villars.

VON NEUMANN, J. (1928). "Zur Theorie der Gesselchaftschpiele," *Mathematische Annalen 100*: 295–320.

VON NEUMANN, J. (1953). "A certain zero-sum game equivalent to the optimal assignment problem," in *Contributions to the Theory of Games I*, H. W. Kuhn and A. W. Tucker, eds. *Annals of Mathematics Studies 28*: 5–12.

VON NEUMANN, J., and H. H. GOLDSTINE (1947). "Numerical inverting of matrices of high order," *Bulletin of the American Mathematical Society 53*: 1021–1099.

WAGNER, H. M. (1959). "Linear programming techniques for regression analysis," *Journal of the American Statistical Association 54*: 206–212.

WARDLE, P. A. (1965). "Forest management and operational research: A linear programming study," *Management Science 11*: B260–B270.

WATT, B. K., and A. L. MERRILL (1963). *Composition of Foods.* Washington, D.C.: U.S. Dept. of Agriculture.

WHITE, W. C., M. B. SHAPIRO, and A. W. PRATT (1963). "Linear programming applied to ultraviolet absorption spectroscopy," *Communications of the ACM 6*: 66–67.

WILKINSON, J. H. (1963). *Rounding Errors in Algebraic Processes.* London: Her Majesty's Stationery Office.

WOLFE, P. (1963). "A technique for resolving degeneracy in linear programming," *Journal of the Society for Industrial and Applied Mathematics 11*: 205–211.

WOOLSEY, R. E. D. (1972a). "A candle to St. Jude, or, Four real world applications of integer programming," *Interfaces 2* (no. 2): 20–27.

WOOLSEY, R. E. D. (1972b). "Operations research and management science today, or, Does an education in checkers really prepare one for a life of chess?" *Operations Research 20*: 729–737.

WOOLSEY, R. E. D. (1973). "A novena to St. Jude, or, Four edifying case studies in mathematical programming," *Interfaces 4* (no. 1): 32–39.

WOOLSEY, R. E. D., and H. S. SWANSON (1975). *Operations Research for Immediate Application: A Quick and Dirty Manual.* New York: Harper and Row.

ZADEH, N. (1973). "A bad network problem for the simplex method and other minimum cost flow algorithms," *Mathematical Programming 5*: 255–266.

Solutions to Selected Problems

1.4 a. $s > 0$ and $t > 0$, b. never, c. $s \leq 0$ or $t \leq 0$ or both.

1.5 True.

1.6

$$\begin{array}{lrcrcrcrcrcrl}
\text{Maximize} & 6x_1 & + & 3x_2 & + & 8x_3 & + & 3x_4 & + & 9x_5 & + & 5x_6 & \\
\text{subject to} & x_1 & + & x_2 & & & & & & & & & \leq 480 \\
 & & & & & x_3 & + & x_4 & & & & & \leq 400 \\
 & & & & & & & & & x_5 & + & x_6 & \leq 230 \\
 & x_1 & & & + & x_3 & & & + & x_5 & & & \leq 420 \\
 & & & x_2 & & & + & x_4 & & & + & x_6 & \leq 250 \\
 & & & & & & & & & & x_1, x_2, \ldots, & x_6 & \geq 0.
\end{array}$$

1.9 t = arrival time of the last person.
x_j = time spent by person j walking towards the destination.
x'_j = time spent by person j walking away from the destination.
y_j = time spent by person j bicycling towards the destination.
y'_j = time spent by person j bicycling away from the destination.

2.1 Optimal dictionaries:

a.
$$\begin{aligned}
x_1 &= 2.5 - 0.5x_5 - 1.5x_3 \\
x_2 &= 1.5 + 0.5x_5 - 0.5x_3 - x_4 \\
x_6 &= 0.5 + 0.5x_5 + 0.5x_3 + x_4 \\
\hline
z &= 10.5 - 0.5x_5 - 1.5x_3 - 2x_4
\end{aligned}$$

b.
$$\begin{aligned}
x_1 &= 1 - 2x_6 + x_5 - x_3 - 5x_4 \\
x_2 &= 2 + x_6 - x_5 - x_3 + 2x_4 \\
\hline
z &= 17 - 4x_6 - x_5 - 2x_3 - 5x_4
\end{aligned}$$

c.
$$\begin{aligned}
x_3 &= 3 - 2x_1 - 3x_2 \\
x_4 &= 1 - x_1 - 5x_2 \\
x_5 &= 4 - 2x_1 - x_2 \\
x_6 &= 5 - 4x_1 + x_2 \\
\hline
z &= 2x_1 + x_2
\end{aligned}$$

3.1 Unbounded: try $x_1 = 0$, $x_2 = 5 + 0.5t$, $x_3 = t$.

3.9 Final dictionaries:

a.
$$\begin{array}{rcl} x_1 &=& 1 - x_4 - x_5 \\ x_2 &=& 2 + 2x_4 + x_5 \\ x_3 &=& \quad\; 3x_4 + 2x_5 \\ \hline z &=& 1 - x_4 - x_5 \end{array}$$

b.
$$\begin{array}{rcl} x_2 &=& 2.5 - 1.5x_1 + 0.5x_4 - 0.5x_5 \\ x_0 &=& 0.5 + 0.5x_1 + 0.5x_4 + 0.5x_5 \\ x_3 &=& 2 - 2x_1 + x_4 \\ \hline w &=& -0.5 - 0.5x_1 - 0.5x_4 - 0.5x_5 \end{array} \quad \text{(infeasible)}$$

c.
$$\begin{array}{rcl} x_4 &=& 4 + 3x_3 - 2x_5 \\ x_1 &=& 3 + x_3 - x_5 \\ x_2 &=& 4 + 2x_3 - x_5 \\ \hline z &=& 3 + x_3 - x_5 \end{array} \quad \text{(unbounded)}$$

4.1
a. largest-coefficient wins (2 iterations vs. 3).
b. largest-increase wins (1 iteration vs. 2).
c. smallest-subscript wins (2 iterations vs. 3).

5.2 $x_1^* = 0.6, \quad x_2^* = 0.$

5.3 a. not optimal, b. optimal.

5.4 Optimal.

5.5 Optimal.

6.2 $x_1 = 1, \quad x_2 = 0, \quad x_3 = -1, \quad x_4 = 1, \quad x_5 = 1, \quad x_6 = 0.$

6.3 $x_1 = 1, \quad x_2 = x_3 = x_4 = x_5 = x_6 = 2.$

6.4 $\mathbf{L}_4\mathbf{L}_3\mathbf{L}_2\mathbf{L}_1\mathbf{A} = \mathbf{U}$ with

$$\mathbf{L}_1 = \begin{bmatrix} 1 & & & \\ -1 & 1 & & \\ -1 & & 1 & \\ 1 & & & 1 \end{bmatrix}, \quad \mathbf{L}_2 = \begin{bmatrix} 1 & & & \\ & 0.5 & & \\ & & 1 & \\ & & & 1 \end{bmatrix}, \quad \mathbf{L}_3 = \begin{bmatrix} 1 & & & \\ & 1 & & \\ & & -1 & \\ & & 1 & 1 \end{bmatrix},$$

$$\mathbf{L}_4 = \begin{bmatrix} 1 & & & \\ & 1 & & \\ & & 1 & \\ & & & -0.5 \end{bmatrix}, \quad \text{and } \mathbf{U} = \begin{bmatrix} 1 & 3 & -2 & 4 \\ & 1 & 0.5 & 0.5 \\ & & 1 & -2 \\ & & & 1 \end{bmatrix}.$$

Now $\mathbf{x} = [-17, 7, -7, -3]^T, [3, 1, 1, 1]^T, [0, 1, 1, 0]^T$ and $\mathbf{y} = [3, 1, 4, 1], [1, 1, 0, 0]$.

7.5 $[1, 2, 3] \rightarrow [16, 2, 3] \rightarrow [16, 2, -19] \rightarrow [45, 2, -19] \rightarrow [45, -114, -19]$.
$[1, 2, 3]^T \rightarrow [-11, 4, -13]^T \rightarrow [-5.5, 9.5, 9]^T \rightarrow [-14.5, -17.5, 9]^T \rightarrow$
$[29, -104.5, -20]^T$.

8.1
a. $x_1 = 0, x_2 = 6, x_3 = 0, x_4 = 15, x_5 = 2, x_6 = 1, x_7 = 1, x_8 = 0.$
b. Unbounded: try $x_1 = 15 + 2t, x_2 = -6 - t, x_3 = x_4 = 0.$
c. Unbounded: try $x_1 = 9t, x_2 = 2, x_3 = 0, x_4 = -3, x_5 = 1 - 5t, x_6 = 2 - 2t.$
d. Infeasible: try $\mathbf{y} = [1, -2.5, -1, -2, 0]$.
e. $x_1 = \frac{25}{14}, x_2 = 0, x_3 = 0, x_4 = 0, x_5 = 0, x_6 = 0, x_7 = \frac{9}{14}.$

9.4 $x_1 = 0.5, x_2 = 2, x_3 = 4.5$.

10.2 (i) 39 desks, 9 chairs, 69 bedframes.
(ii) 51 bookcases, 72 chairs.
(iii) 13 coffee tables, 22 chairs, 82 bedframes.
(iv) 31 bookcases, 8 desks, 40 chairs, 38 bedframes.
(v) 15 bookcases, 90 bedframes.

11.1 The January and March constraints are also redundant.

12.1 Workforce: 265, 255, 220, 207, 208, 248, 288, 328, 368, 368, 360, 320.

13.2 a. 164 raws suffice, b. 23 raws suffice.

14.1 L_1: guanine 0.0642, uracil 0.0700.
L_2: guanine 0.0648, uracil 0.0702.
L_∞: guanine 0.0660, uracil 0.0715.

15.1 A should hide x and y with probabilities $y/(x + y)$ and $x/(x + y)$, respectively; B should hide x and y with probabilities 1/2.

15.3 Choose 1, 2, or 3 with probabilities 1/3.

16.6 The system is unsolvable.

16.8 Numbers x_1, x_2 satisfy the system if and only if

$$\begin{aligned} x_1 &= -2p_1 - 3p_2 + q_1 - 3q_2 \\ x_2 &= 2p_1 + 3p_2 - 2q_1 + q_2 \end{aligned}$$

for some nonnegative p_1, p_2, q_1, q_2 such that $p_1 + p_2 = 1$.

18.2 A point $\mathbf{x}$ belongs to the convex hull of $\mathbf{v}^1, \mathbf{v}^2, \ldots, \mathbf{v}^6$ if and only if

$$\begin{aligned} x_2 - x_4 &\le 0 \\ -2x_1 + 8x_2 + x_3 - 4x_4 &\le 0 \\ 10x_1 - x_2 - 8x_3 - x_4 &\le 0 \\ -4x_1 - 5x_2 + 2x_3 + x_4 &\le 0 \\ 4x_1 + 5x_2 - 5x_3 + 5x_4 &\le 0 \\ -10x_1 - 8x_2 + 5x_3 + 4x_4 &\le 0 \\ x_1 - x_2 + x_3 - x_4 &= 9. \end{aligned}$$

19.1 (i) $\mathbf{x} = [1, 1, 2, 0, 0, 1, 6, 0, 0, 0, 0, 3]^T, \mathbf{y} = [0, 3, 3, 1, 3, 4, 5, 5]$,
(ii) unbounded, (iii) infeasible.

20.1 $\mathbf{x} = [1, 0, 0, 4, 0, 0, 0, 0, 4]^T, \mathbf{y} = [-2, 0, 0, -1, 0, 1]$.

21.1 (i) $\mathbf{x} = [2, 0, 2, 0, 2, 5, 2, 1, 0]^T, \mathbf{y} = [1, 4, 4, 0, 1, 3]$,
(ii) unbounded, (iii) infeasible.

23.14 Introduce an artificial node w and artificial arcs wk, kw with $u_{wk} = u_{kw} = +\infty$ and $c_{wk} = c_{kw} = c^*$ for each node k; the symbol c^* represents some extremely large positive number (whose actual value need not be specified).

23.15 *Step 1.* Choose an arc ij such that $x_{ij} > 0$, set $v_1 = i, v_2 = j, k = 2$.
Step 2. As long as v_k is different from $v_1, v_2, \ldots, v_{k-1}$, repeat the following operations: set $i = v_k$, find an arc ij with $x_{ij} > 0$, set $v_{k+1} = j$, replace k by $k + 1$.
Step 3. Now $v_k = v_t$ for some t with $1 \le t < k$. If $C = v_t v_{t+1} \cdots v_k v_t$ is a negative cycle then stop. Otherwise, let d be the smallest x_{ij} on C, subtract d from each x_{ij} on C, and return to step 1.

Index

Alternative terms or abbreviations are indicated as follows:

Term (alternative)